John Gould
THE BIRD MAN
A Chronology and Bibliography

John Gould
THE BIRD MAN
A Chronology and Bibliography

by

Gordon C. Sauer

University Press of Kansas

© Gordon C. Sauer 1982

Designed by David Deakin, Wandiligong
Edited by Tim Bass
Typeset by Savage & Co., Brisbane.
Printed in Australia by
The Craftsman Press Pty Ltd, Victoria
Bound in Australia by
Apollo Library Binding, Victoria

ISBN 0 7006 0230 5

Exclusive distribution in the U.S. and Canada
by the University Press of Kansas is
made possible by the Kansas University
Endowment Association

NATURAL HISTORY MUSEUM
library
OF LOS ANGELES COUNTY

Frontispiece

Plaster plaque of John Gould, probable age early forties.
On the right is the inscription 'John Gould FRS.' This plaque
measures 190mm (7½ inches) in diameter.
Courtesy of Mr and Mrs Aubrey Twyman of Salisbury.

Contents

List of figures and plates	vi
Lists of abbreviations	x
Gould's major published works	x
Institutions	xi
Preface	xiii
Introduction	xv
Acknowledgements	xix

Part One — 1
Genealogy of John Gould and Elizabeth Coxen Gould — 3

Part Two — 11
John Gould's major published works — 13

Part Three — 87
Chronology of the life and works of John Gould — 89

Part Four — 157
Bibliography of John Gould, his family and associates — 159
Postscript — 394
Index — 395

List of Figures and Plates

The source of each illustration is acknowledged in the text. I am grateful to the many individuals and institutions who granted me permission to reproduce them.

Mr Larry Brown, photographer at Research Hospital of Kansas City, Missouri took the majority of the photographs of the books, maps, lithographs and drawings in the author's collection. Mr E. Richard Marolf, chief of the photography department at the University of Kansas, with the able assistance of L. E. James Helyar, librarian, took the colour and black-and-white photographs of the Spencer Library–Ellis Collection. My sincere appreciation again.

Figures

FRONTISPIECE Plaster plaque of John Gould
1   Dr and Mrs Geoffrey Edelsten.
2   Lyme Regis, Dorsetshire.
3   Cover folder for Gould's *Century of Birds*.
4   Prospectus for *Birds of Europe*.
5   Cover for the first part of Gould's *Trogonidae*.
6   Cover for the first part of Gould's *Synopsis*.
7   Prospectus for Gould's *Synopsis*.
8   Cover for the first part of Gould's *Birds of Australia and adjacent islands* (Cancelled Parts).
9   Cover for the first part of Gould's *Icones avium*.
10 Cover for the second part of *Icones avium*.
11 Pencil sketch attributed to John Gould of *Ardea leucophaea* for *Birds of Australia*.
12 Cover for the seventh part of *Birds of Australia*.
13 Prospectus for the Sturm edition of *Monographie der Ramphastiden*.
14 Cover for the first part of the Sturm edition of *Monographie der Ramphastiden*.
15 Pencil sketch attributed to John Gould of *Lagorchestes conspicillata* for *Macropodidae*.
16 Cover of the first part of Gould's *Macropodidae*.
17 Prospectus for Hinds's *The zoology of the voyage of H.M.S. Sulphur*.
18 Cover for No. III Hinds's *Voyage of the Sulphur*.
19 Cover for the 'Atlas of plates' for Hinds's *Voyage of the Sulphur*.
20 Prospectus for Gould's *Odontophorinae*.
21 Pencil sketch attributed to John Gould of *Petaurista taguanoides*.
22 Pencil sketch attributed to John Gould of *Petrogale xanthopus*.
23 Cover for the first part of Gould's *Mammals of Australia*.
24 Prospectus for Gould's *Birds of Great Britain* and other works.
25 Cover of book *An introduction to Mammals*.

LIST OF FIGURES AND PLATES

26 Prospectus for the *Handbook*.
27 Cover label for the *Introduction to Gould's Birds of Asia*.
28 Gardens of the Zoological Society in Regent's Park, 1829.
29 A section of London in 1832.
30 Inset from 1832 map of London.
31 Hobart Town, Van Diemen's Land (Tasmania) in 1834.
32 Map of Van Diemen's Land in 1833.
33 Map of Van Diemen's Land, showing locations visited or mentioned by Gould.
34 Flinders Island grass trees.
35 Map of New South Wales showing principal localities visited by Gould.
36 The Coxen homestead at Yarrundi.
37 Map of Western Australia in 1833.
38 Map of South Australia showing principal localities visited by Gould.
39 Drawing of a plant by Mrs Gould.
40 An 1846 map of the Thames River from Oxford to Kew.
41 Elizabeth Coxen Gould, wife of John Gould.
42 Map of Journey of the Leichhardt Expedition.
43 Mural tablet in memory of John Gilbert.
44 John Gould at the age of 45.
45 Bronze plaque of Gould.
46 Plan of the Zoological Gardens in 1851.
47 Interior of the Humming-bird house.
48 Pencil sketch of interior of Humming-bird house.
49 Exterior of Gould's Humming-bird house in the Gardens.
50 A balloon view of London in 1851.
51 Drawing in pencil, signed 'J. Gould Jan. 7, '52'.
52 Diploma of the German Ornithological Society granted to Gould in 1853.
53 Dinner given by Waterhouse Hawkins set in the belly of an Iguanodon model.
54 Dr Snow's map showing cholera deaths in the Broad Street pump area in 1854.
55 Broad Street buildings.
56 Watercolour signed 'J. Wolf 1856' of Ptarmigan in summer plumage.
57 Watercolour signed 'J. Wolf 1856' of King Duck [Eider].
58 Photograph of John Gould, probably taken in Philadelphia.
59 Preliminary watercolour portrait of John Gould by Jas. E. McClees of
60 Philadelphia.
61 A gallery of six portraits of John Gould.
    Dr Franklin Gould.
62 Charles Gould.
63 Two portraits of John Gould by Maull and Fox in 1875.
64 Edwin C. Prince.
65 Eliza Gould Muskett family.
66 Portrait in oils of John Gould by Robertson.
67 Gould's personal cabinet built to house his imperial folio works and other volumes.
68 Portrait of the Gould's three daughters.
69 First certificate of The Gould League of Bird Lovers of Victoria.
70 Pages concerning Gould's works from Jean Anker. 1938. Bird books and bird art.
71 Lithographic stone used for printing of *Pitta coccinea* for *Birds of Asia*.
72 Cover and first page of 'Gould Commemorative Issue' of *The Emu*, October 1938.

73 First page of Puttick and Simpson's 1881 sale catalogue of Gould's library.
74 Pages concerning Gould's works from Ripley and Scribner 1961 Ornithological books in the Yale University Library.
75 Advertisement pages from R. B. Sharpe's 1893 *Analytical Index*.
76 Cover for Sotheran's 1934 'Piccadilly Notes' No. 9.
    Cover for Sotheran's 1937 'Piccadilly Notes' No. 22.
77 Pages concerned with Gould's works from Casey Wood 1931 An introduction
78 to the literature of vertebrate zoology.
79 John Gould's pencil sketch of 'A new form among the Trochilidae'.
80 Pages concerned with Gould's works from John Todd Zimmer 1926 Catalogue of the Edward E. Ayer ornithological library.

Plates

1 Drawing in ink and watercolour, signed E. Gould, of *Noctua cuculoides*, for *Century of Birds*.
2 Drawing in watercolour, signed E. Gould, of *Garrulus lanceolatus*, for *Century of Birds*.
3 Drawing in watercolour, attributed to E. Lear, of White-tailed or Sea Eagle, for *Birds of Europe*.
4 Drawing in pencil and watercolour, attributed to John Gould, of *Pteroglossus castanotis*, for *Ramphastidae* 2nd edn.
5 Drawing in pencil and watercolour, attributed to John Gould, of *Aulacorhamophus prasinus*, for *Ramphastidae* 2nd edn.
6 Drawing in pencil and watercolour, attributed to John Gould, of *Harpactes duvauceli*, for *Trogonidae*, 2nd edn.
7 Drawing in pencil and watercolour, attributed to Mrs Gould, of *Falco frontatus*, for *Birds of Australia*, I.
8 Drawing in pencil, ink and watercolour, attributed to Mrs E. Gould, of Australian Goshawk, for *Birds of Australia*, I.
9 Drawing in pencil and chalk, attributed to John Gould, of *Melithreptus melanocephalus*, for *Birds of Australia*, IV.
10 Drawing in pencil, ink and watercolour, attributed to Mrs E. Gould, of unidentified *Turnix*, unpublished [Australian].
11 Drawing in pencil and chalk, attributed to John Gould of *Eudyptes chrysocome*, for *Birds of Australia*, VII.
12 Drawing in watercolour, attributed to H. C. Richter, of Rose-breasted Robin, for *Birds of Australia*, III.
13 Drawing in pencil, ink and watercolour, attributed to Mrs Gould, of Short-tailed Petrel, for *Birds of Australia*, VII.
14 Drawing in pencil, ink and crayon, attributed to John Gould, of *Platycercus semitorquatus*, for *Birds of Australia*, V.
15 Drawing in pencil, ink and watercolour, attributed to W. Hart, of *Androdon aequatorialis*, for *Supplement to Trochilidae*.
16 Drawing in pencil and watercolour, attributed to W. Hart, of *Helianthea dichroura*, for *Supplement to Trochilidae*.
17 Drawing in pencil and watercolour, attributed to John Gould, of *Pteroglossus humboldtii*, similar to *Ramphastidae*, 2nd edn, plate 22.
18 Drawing in pencil and watercolour, attributed to John Gould, of *Ramphastos ambiguus*, for *Ramphastidae*, 2nd edn.
19 Drawing in pencil and watercolour, attributed to John Gould, of *Pharomacrus pavoninus*, for *Trogonidae*, 2nd edn.

LIST OF FIGURES AND PLATES

20 Drawing in pencil and watercolour, attributed to John Gould, of *Trogan elegans*, for *Trogonidae*, 2nd edn.
21 Drawing in pencil and watercolour, signed J. Gould, F.R.S., of *Cyanecula suecica*, for *Birds of Great Britain*, II.
22 Drawing in pencil and watercolour, attributed to W. Hart, of *Ptilotis frenata*, for *Birds of New Guinea,* III.
23 Drawing in pencil and watercolour, signed by H. C. Richter, of *Rhinolophus aurantius*, for *Mammals of Australia*, III.
24 Drawing in watercolour, attributed to H. C. Richter, of Rose-breasted Cockatoo, for *Birds of Australia*, V.
25 Drawing in pencil and watercolour, artist unknown, of *Pteroglossus langsdorffi*, for *Ramphastidae*.
26 Drawing in pencil and watercolour, attributed to John Gould, of *Pteroglossus mariae*, for *Ramphastidae*, 2nd edn.
27 Drawing in pencil and watercolour, artist unknown, of *Garrulus taivanus*, for *Birds of Asia*, V.
28 Drawing in pencil and watercolour, signed by Wolf, of *Galloperdix lunulosa*, for *Birds of Asia*, VI.
29 Three steps in the development of a 'Gould' lithograph of a bower-bird.
30 Drawing in pencil, watercolour and chalk, artist unknown, of *Uria Troile*, for *Birds of Great Britain*, V.
31 Drawing in watercolour, attributed to H. C. Richter, of Gouldian Finch, for *Birds of Australia*, III.
32 Drawing in watercolour, signed E. Lear 1832 June 28, of White Stork, for *Birds of Europe*, IV.
33 Drawing in watercolour, attributed to Mrs E. Gould, species not labelled [Bohemian Waxwing].
34 Drawing in ink and watercolour, attributed to Mrs E. Gould, of *Coccothraustes icterioides*, for *Century of Birds*.
35 Drawing in watercolour, signed W. Hart, of *Lophorhina minor*, for *Birds of New Guinea*, I.
36 Drawing in watercolour, signed W. Hart del, of *Parotia sexpennis*, for *Birds of New Guinea*, I.

List of Abbreviations

Gould's major published works

Birds of Asia	The Birds of Asia (1849–83)
Birds of Australia	The Birds of Australia (1840–48)
Birds of Australia and adjacent islands	The Birds of Australia and the adjacent islands (1837–38)
Birds of Australia, Supplement	The Birds of Australia, Supplement (1851–69)
Birds of Europe	The Birds of Europe (1832–37)
Birds of Great Britain	The Birds of Great Britain (1862–73)
Birds of New Guinea	The Birds of New Guinea (1875–88)
Century of Birds	A century of birds hitherto unfigured from the Himalaya Mountains (1830–33)
Handbook	Handbook to the birds of Australia (1865)
Icones avium	Icones avium, or figures and descriptions of new and interesting species of birds from various parts of the globe (1837–38)
Introduction to Birds of Australia	An introduction to the birds of Australia (1848)
Introduction to Birds of Great Britain	An introduction to the birds of Great Britain (1873)
Introduction to Gould's Birds of Asia	Introduction/to/Gould's Birds of Asia./by/R. Bowdler Sharpe (1883)
Introduction to Mammals	An introduction to the mammals of Australia (1863)
Introduction to Trochilidae	An introduction to the Trochilidae, or family of humming-birds (1861)
Macropodidae	A monograph of the Macropodidae, or family of kangaroos (1841–42)
Mammals of Australia	Mammals of Australia (1845–63)
Monographie der Ramphastiden	J. Gould's Monographie der Ramphastiden oder Tukanartigen Voegel (1841–47)

LIST OF ABBREVIATIONS

Odontophorinae	*A monograph of the Odontophorinae, or partridges of America* (1844–50)
Pittidae	*Monograph of the Pittidae* (1880–81)
Ramphastidae	*A monograph of the Ramphastidae, or family of toucans* (1833–35)
Ramphastidae, 2nd edn	*A monograph of the Ramphastidae, or family of toucans. Second Edition* (1852–54)
Supplement to Ramphastidae	*Supplement to the first edition of a monograph on the Ramphastidae, or family of toucans* (1855)
Supplement to Trochilidae	*A monograph of the Trochilidae, or family of humming-birds. Supplement* (1880–87)
Synopsis	*A synopsis of the birds of Australia, and the adjacent islands* (1837–38)
Trochilidae	*A monograph of the Trochilidae, or family of humming-birds* (1849–61)
Trogonidae	*A monograph of the Trogonidae, or family of trogons* (1835–38)
Trogonidae, 2nd edn	*A monograph of the Trogonidae, or family of trogons. Second edition* (1858–75)
Voyage of the Sulphur	Richard B. Hinds. *The zoology of the voyage of H.M.S. Sulphur. Birds,* by John Gould. (1843–44)
Zoology of the Beagle	Charles Darwin. *The zoology of the voyage of H.M.S. Beagle. Birds.* (1838–41)

Institutions

Acad. Nat. Sciences Phila.	Academy of Natural Sciences of Philadelphia, Pennsylvania
Amer. Mus. Nat. Hist.	American Museum of Natural History, New York, New York
Amer. Philosophical Soc.	American Philosophical Society, Philadelphia, Pennsylvania
Brit. Assoc.	British Association for the Advancement of Science
Brit. Mus. (N.H.)	British Museum (Natural History), Cromwell Road, London
Cambridge Univ. Mus. Zoology	Cambridge University Museum of Zoology, Cambridge, England
Mus. Nat'l. d'Hist. Nat.	Muséum National d'Histoire Naturelle, Paris
Nat'l Lib. Canberra	National Library of Australia, Canberra, A.C.T.

Nat'l Mus. Victoria	National Museum of Victoria, Melbourne, Victoria
Newton Library	Newton Library, Cambridge University, Cambridge
Roy. Scot. Mus.	Royal Scottish Museum, Edinburgh, Scotland
Zool. Soc. London	Zoological Society of London, Regent's Park, London

Preface

'A satisfactory biography of Gould remains to be written' — so wrote Whittell in 1954. This is still true in 1982. But in order for one to write a good biography, some facts are necessary. Amazingly, there has never been any major study of this very important naturalist. As an initial effort in this direction, here is a compendium on John Gould that will include a bibliography in combination with a genealogy and chronology of Gould's life and works.

My interest in John Gould began almost forty years ago. It began insidiously with an interest in bird study, but more specifically following the acquisition of a few of Gould's bird prints from an Illinois antique dealer. They were impressive. Who was this man? Where could I find more prints?

In the mid-1940s my wife and I began to explore antique shops, and our entrée was 'Do you have any John Gould bird prints?' A few did. In Charleston, West Virginia, where I was stationed after the war was over, this question asked of a dealer produced, to our great surprise, volume one of Gould's *Birds of Great Britain*. The dealer was reluctant to sell it intact. She planned on tearing out the prints of these great owls and hawks to sell them separately, and thereby realize more money on her investment. But she sold the volume to us, and we paid the $300.00 over a few months. Volume one has many introductory pages, so this provided some insight into this man Gould and his team.

The die was cast. The hunt was on. The quest for Gouldiana was to become a consuming vocation, but with many ebbs and flows.

During some continuing education years in New York City I had access to the New York Public Library, the American Museum of Natural History, and print dealers such as Kennedy Gallery and Newman's Old Print Shop. In New York I discovered the Sharpe *Analytical Index* of Gould's works, and the Henry Sotheran Ltd publications of 'Piccadilly Notes'. My information on and interest in John Gould was expanding.

Then came a move back to the Midwest, to St. Joseph, Missouri. The quest for information on Gould could certainly continue from anywhere, but the natural history centres and great library collections were not in this area, or so I thought. Can anyone imagine, then, my shock and disbelief, but very pleasant surprise, to learn that the largest collection of Gouldiana in the world was but 125 kilometres away at the University of Kansas? The Ralph Ellis Ornithological Collection contained over 1200 original sketches and drawings for Gould's works, plus a complete set of all of Gould's folio and other volumes. Several of the sets are in their original parts as they were issued. Many supporting volumes and journals were also in the collection. That did it! Even the fates seemed to will it! They were on my side, or Gould's. My life's vocation seemed to be laid out for me.

Now, some thirty years later, the accumulated Gouldiana files and books (no imperial folio volumes) occupy over seven metres of shelf space in my office and home. There was a move to Kansas City, Missouri, with many trips to LAWRENCE, Kansas and the new Kenneth Spencer Research Library which houses the fabulous Ellis Collection. There was also much correspondence with American, British and Australian libraries and museums, plus personal contacts with those interested in John Gould. Most rewarding were the visits with Gould's great-grandson and his wife, Dr and Mrs Geoffrey Edelsten in Winchester, England, and when they in turn visited us in Kansas City. They have been a joy to know. Geoffrey has done everything he could to help me in my quest for Gouldiana.

The past eight years have been the period of greatest activity. As my interest and collection expanded, the idea developed in my mind that I was accumulating this material for a major biography of John Gould. Over the years I had published a few articles on Gould, and one small book. Also, as the time passed, I became aware that a Pandora's box had been opened. For a man who did not even write a cursory autobiography, I uncovered literally hundreds of his letters, plus many articles and books written about him and about his collector, John Gilbert. This wealth of information — coupled with the extensive imperial folio-sized volumes, and articles, written and published by Gould himself — began to overwhelm me. A decision had to be made. The Gould material would have to be published in several volumes. The first volume would contain the material accumulated in the course of my research on this man, namely, extensive bibliographical material and related Gouldiana that had been amassed.

Some have said that with the publication of this bibliography another or others would hurry and write the definitive biography. More power to them. Any help would be appreciated. But there is much more that needs to be studied and researched before this could be done in an acceptable manner. Further work must include an annotated study of Gould's correspondence; this I have almost completed. In addition we need to study carefully Gould's artistic and technical methods as revealed (but not too clearly) in almost two thousand extant original drawings by Gould himself, by his wife and by the other artists of the Gould team. Finally, these original drawings must be catalogued more completely than has been done so far, not only those in the University of Kansas Spencer Library–Ellis Collection, but also those scattered throughout the world in various collections.

Gordon C. Sauer
6400 Prospect Avenue
Kansas City
Missouri 64132
1982

Introduction

All ornithologists are not artists. Many artists are not successful at business. But the Englishman John Gould was an ornithologist, artist and successful businessman. In the field of natural history the accomplishments of this man in his 76 years of life from 1804 to 1881 are truly monumental. No other ornithologist has ever exceeded (or will ever exceed) the number of Gould's bird discoveries and the magnitude and splendour of his folio publications.

What about John Gould the ornithologist? Some think of him as an armchair ornithologist, but not the Australians. At the age of 34, already established as one of Britain's prominent ornithologists, he embarked for the relatively unknown and forbidding continent of Australia to collect and study known and unknown species of birds and, incidentally, mammals, insects and plants. His wife, his eldest child, and his assistant, John Gilbert, accompanied him on this bold venture. At the end of Gould's first eight months in Australia his collection consisted of '800 specimens of birds, 70 of quadrupeds (several of which are new), more than 100 specimens preserved whole in spirits, and the nests and eggs of above 70 species of birds, together with the skeletons of all the principal forms'. By this time, John Gilbert had gone to Western Australia. At the conclusion of Gould's Australian venture of nineteen months in the frontier lands of Tasmania, South Australia and New South Wales, he had collected and made notes on most of Australia's *known* bird species and in addition had discovered over 300 *new* species. A hundred of these 'new' species, however, were in later years to be reduced to the category of subspecies. He also made a major contribution to the study of the mammals of Australia.

Gould and his family returned to England in 1840. Within four months he published the first of thirty-six parts of *Birds of Australia*. For the next eight years Gould carried out research for the work, wrote the text, with extensive synonymy, and roughly drew for his artists almost 600 species of Australian birds that were to be included in the thirty-six-part work. For an edition of 250 copies, his artists and staff would, having drawn the birds and foliage on lithographic stone, print 1500 plates a month, and then hand-colour them. An additional 1500 text-pages would also be printed every month. This exacting schedule was to be religiously adhered to so that four parts a year could be issued. At the conclusion of the publication, in 1848, the complete set of 600 imperial folio-sized plates, with text-pages, was to be bound in seven volumes. During this same eight-year period, Gould also published a volume on Kangaroos and began work on his three-volume *Mammals of Australia*.

The work on Australian birds was in itself an impressive achievement, but Gould eventually was to publish fifteen such folio sets in forty-nine volumes, containing over 3000 plates depicting birds from all over the world and mammals from Australia. In addition he wrote over 300 scientific articles and several smaller books.

Obviously, John Gould's contribution to nineteenth-century ornithology was monumental. In his lifetime he was the recognized authority on the birds (and mammals) of Australia, the very complex humming-birds, the birds of Asia, and the birds of New Guinea. He also had major publications on the trogons, toucans and partridges of America, the birds of Europe and those of Great Britain.

Gould was a recognized and careful systematist. He was an upholder of the Law of Priority, and if he found that an earlier naturalist, such as Lathan, Gmelin or others, had described a specimen Gould had thought to be new, he unhesitatingly restored the earlier name. Gould did establish many new genera, particularly for the birds of Australia. According to Mathews and Iredale (1938) Gould was very 'qualified to judge differences . . . thus his work as a systematist claims praise from every serious investigator after a century's trial'.

Gould's views on the future decline or extinction of certain species of birds and mammals were profoundly prophetic. The animal *homo sapiens* was to successfully compete for and take over the natural habitats of many of the other species of living beings. In his *Mammals of Australia*, volume 1, Gould wrote about the thylacine, or Tasmanian tiger, predicting that

when the comparatively small island of Tasmania becomes more densely populated, and its primitive forests are intersected with roads from the eastern to the western coast, the numbers of this singular animal will speedily diminish, extermination will have its full sway, and it will then, like the wolf of England and Scotland, be recorded as an animal of the past.

The last Tasmanian tiger died in 1933 in a zoo.

There are many similar comments by Gould about the unnecessary destruction of birds and mammals. On the Bullfinch in *Birds of Great Britain* Gould wrote, 'throughout the work I have been a champion for our poor persecuted birds and defended them as well as I could in words on account of the great amount of good they effect.'

John Gould the artist is less impressive. Since many people know of Gould only through the beautiful hand-coloured lithographs framed on their walls, this statement will be surprising. However, it is well established that Gould himself never executed a finely finished drawing for any of his works. He drew rough pencil or watercolour sketches that were to serve as guides for his artists. There are numerous extant examples of Gould's preliminary sketches and, while rough, many are more exciting and true to life than the finely-finished developments. Sharpe, a renowned ornithologist and personal friend of Gould, said that 'He was always able to sketch, somewhat roughly perhaps, the positions in which the birds were to be drawn on the plates, and no one could have a better "eye" for specific differences.' Allan McEvey, the Australian ornithologist and art historian, after the most careful evaluation of Gould's 'native talent [which] though limited was real', concluded (in 1973) that 'the plates, in their grand conception, their genesis, their spirit and their achievement, cannot in these senses fairly be attributed to anyone but Gould'.

John Gould the successful businessman is well documented. It is probably this known financial success, combined with the fact that the fine artistry on his plates was done by others employed by him, that has made a few uninformed ornithologists not fully appreciative of Gould's scientific achievements. The struggles of John James Audubon for financial security are well known, and we tend to applaud struggle. Ponder for a moment, however, the position in which Gould found himself early in life. He was born the son of a working gardener, had no so-called formal education, and was to live at a time when social class was to many an insurmountable barrier

INTRODUCTION

to achievement and acceptance. As was written in one of his obituaries by a socially-conscious author and scientist (and friend), 'In the character of the man we must look for the secret of his success because it is well known that he possessed neither the advantages of wealth nor education at the commencement of his career, and yet he has left behind him a series of works the like of which will probably never be seen again' (Obituary, *Nature*, 1881).

Notwithstanding these artificial social and educational barriers, Gould by his talent, ability, and motivation, had achieved enough recognition by the age of 21 to be stuffing birds for King George IV of England. At the age of 22 he was competitively chosen as the first 'Curator Preserver' of the newly-formed museum of the Zoological Society of London. At the age of 26, in conjunction with N. A. Vigors and his wife as artist, he published his first folio-sized volume *A century of birds of the Himalaya Mountains*. This first work was subscribed by over 300 of his contemporary naturalists, many museums in England and Europe, and good share of the nobility of England. When he left on his Australian expedition at the age of 34, it is said that he had made seven thousand pounds from his publications. He was to continue to do well financially; he was a successful businessman. But it was the result of hard work, unique ability, and great motivation.

To complete this short introduction, let me consider for a moment John Gould, the man. For almost forty years it has been my pleasant avocation to collect, discover, read, sort, catalogue and write on John Gould, his family, his works, his associates, and his times. The man was a genius. He controlled his mind, and obviously several minds around him, so that he was able to initiate, organize, direct and bring to fruition with overwhelming success the compilation and publication of a massive amount of natural-history material. He was goal oriented, to say the least.

For an avocation he had fishing. Apparently he was very successful at that also, earning an enviable reputation among his fishing friends. He would sit for long periods on the bank of a stream smoking a cigar, studying the feeding habits of a trout he hoped to 'kill' the next day.

As a person he was praised by all his contemporaries and respected by most. R. Bowdler Sharpe said that he was 'somewhat brusque, but those who were intimate with him were aware that under a rough exterior he concealed a very kind heart'. Professor Richard Owen, P. L. Sclater, and Sir William Jardine wrote of Gould as a close friend. But two of the artists in his employ, famous personages in themselves, left critical comments about him. Edward Lear, an accomplished artist, but most famous for his 'Nonsense Rhymes', thought Gould 'a harsh and violent man', a 'persevering hard working toiler...but ever as unfeeling for those about him'. If Lear really thought so poorly of Gould as quoted, then Lear was a hypocrite, because his letters to Gould show none of this feeling. Joseph Wolf, said to be the greatest of all animal painters, thought Gould to be gruff and calculating. Palmer, writing about Wolf in 1895, relates several uncomplimentary incidents concerning Gould to support these opinions.

Gould, on his part, as seen in his writings and correspondence, hardly ever evinced pettiness or unkindness toward any of his associates or fellow scientists. If he disagreed with someone about the identification of a bird, for instance, he was firm and knowledgeable in defence of his theory, and on many occasions admitted his errors. Thus it could be said that as a man, he was a good man.

But I cannot refrain from some further objective remarks. Aside from his birds and his occasional comments on fishing and shooting, Gould very rarely made any mention in his letters or writings about anything pertaining to politics, religion, society,

art or history. He never commented on Darwin's theory of evolution, but he supplied the bird drawings for one of Darwin's books. He never made any observations on people or on places. He evinced very little humour. From what I know, if conversation with him got off the subject of birds, or possibly fishing, the intercourse would dull. But this judgement may be unfair. As evidenced in his correspondence, Gould was a frequently invited guest to many of his friends' homes or country places for visiting, fishing and shooting. Possibly his conversation was not dull, only his letters.

Something must be said of his family. His wife of eleven years was also his first artist and was thought by all to be very accomplished. She died of puerperal fever after delivery of her eighth child. Two children died in infancy. Gould's letter to Sir William Jardine after her death is very sad and moving, but the remaining two-thirds of this same black-bordered letter is about birds. The recollection of one of the daughters was of a father who was constantly busy on his science, with a house-keeper and a cook sharing the responsibility of raising the six children. That he loved his family there is no question; but he lived his birds.

This is a preliminary account of a great ornithologist and naturalist. To do justice to his accomplishments would require several volumes. Before he died, he asked that his epitaph be 'Here lies John Gould, the Bird Man'. Gould was a favourable product of his times. His times were Victorian.

Acknowledgements

How can one adequately express proper appreciation to the many persons who have rendered assistance to this amateur natural-history bibliophile? Many have freely given of their time and knowledge, have researched books, letters and other references, made photocopies if necessary, and have been of great assistance in so many ways. The important thanks were said at the moment, but here is another thank-you in the form of a completed work.

The earliest years of earnest search for Gouldiana were in the 1940s. Some of those who helped me then may now be deceased.

To begin, this book could not be a reality without someone to type it in its varying stages of maturity, with the many, many preceding drafts and other crazy literary efforts that, through the years, have erupted from fits of supposed brilliance. One person was responsible for the great majority of this typing. She is my office manager, Mrs Ruby (Wilkie) Steele. She knows of many personal thank-yous, but I acknowledge here again her invaluable role.

No man ('person' is more appropriate) is an Island. This is especially true for a man with a wife and family. My love and gratitude to them have been expressed personally, daily, I hope.

Some other individuals need to be named specifically. The three great-grandchildren of John Gould have been of great help, most encouraging and extremely cooperative. This category includes the late Dr Alan Edelsten and, more recently, his wife Grace, also Mrs Doris Edelsten Twyman and, most of all, Dr Geoffrey Edelsten with his charming wife Peggy. After years of correspondence with Geoffrey, my

Figure 1 Dr and Mrs Geoffrey Edelsten, taken in October 1977 at Mr J. L. Wade's Purple Martin Junction, Griggsville, Illinois (Courtesy Mr Wade).

wife and I had the pleasure of spending a few days with them in September 1974. Then in October 1977 they graced us with a visit to Kansas City when we spent a week together (Figure 1). Interestingly, and fortunately, Geoffrey shares with me the excitement of new discoveries about his famous forebear. I have profited much from this association.

Other members of the family on the Gould side to whom I owe thanks are Mr and Mrs Lionel Lambourne and Mrs Olive Pagett.

On the Coxen side of the family I am greatly indebted for genealogical information from Miss Marion Coxen, with whom I have corresponded quite a bit, and also Mrs Nancé Morton, and Mr and Mrs Harold Crawford, all of Australia.

Especial thanks go to members of the University of Kansas Kenneth Spencer Research Library staff. Specifically I would name and thank Miss Alexandra Mason, Mrs Carol Chittenden, L.E. James Helyar, Dr Robert Mengel, cataloguer of the Ellis Collection, William Mitchell, Ann Hyde, Sally Hocker and the late Joseph Rubinstein. It is pleasant to reminisce about the days in the 1950s when Joe Rubinstein and I explored the recently opened safes, file boxes, and piles of books that made up the Ralph Ellis Ornithological Collection. Month after month, I would transport sets of separate humming-bird and birds-of-paradise plates to my home in St. Joseph to catalogue them.

Then I began to catalogue a few of the original drawings, but soon realized the magnitude of the task and had to stop. This irreplaceable material is now housed in a new library, the Kenneth Spencer Research Library, climatically controlled, and under the watchful eye of Sandy Mason and her staff. Sandy has been most encouraging and always patient in bearing with an amateur bibliophile. Carol Chittenden has worked most intimately with the Ellis Collection of original drawings. Her catalogue of them was a pioneer job and we shared great enthusiasm over this task. A grant from the Dible Foundation through William Hickok enabled Carol to complete her task more quickly. James Helyar has been most helpful, especially with the illustrations obtained from the Spencer Library. Other staff at the University of Kansas have been most helpful. Dr Raymond Hall, more than any other person, was responsible for getting the eccentric Ralph Ellis to deposit his ornithological book and manuscript collection at Kansas. Also Robert Vosper, head of the libraries at Kansas for years, and more recently, Jim Ranz, in that same position, have been most supportive. John Langley, head of the Regents Press of Kansas has been very encouraging and is a close personal friend. Of the ornithology staff at Kansas, Dr Harrison B. Tordoff in the earlier years, and more recently Dr Philip S. Humphrey, have also been of assistance.

The library staff at the Watson Library at the University of Kansas have always been ready to assist. The small but important ornithological library at the Dyche Museum at Kansas has a good journal collection, some of it from the Ellis Collection. My thanks to Dr and Mrs Mengel who worked with me there.

Another great library with unexpected treasures, especially in old natural history journals, is the Linda Hall Library in my present home town, Kansas City, Mo. Here I have also spent many hours, and have been greatly assisted by Tom Gillies and his very competent staff.

Three individuals who have researched and written much on Gould have been correspondents of mine for many years. Correspondence with the late Alec Chisholm of Sydney, Australia, began in 1954 and continued until the year of his death in 1977. More than any other individual he succeeded in publicizing John Gould and John Gilbert. I know the thrill he experienced when, after an advertisement in the

ACKNOWLEDGEMENTS

London *Times*, he was invited to visit Dr Alan Edelsten, Gould's great-grandson, near Winchester, in England. Here he found letters and diaries of John Gould, Elizabeth Gould and John Gilbert which, with other memorabilia, were graciously given to him to take back to Australia. That was in 1938. Based on this valuable material Chisholm, in the ensuing years, wrote many articles and several books on Gould, his wife, Elizabeth, and John Gilbert.

The current Australian authority on Gould is Allan McEvey of the National Museum of Victoria in Melbourne. We began corresponding in 1969. In 1971 he visited at the University of Kansas and also spent a day with my collection in Kansas City. Allan has been a very questioning and stimulating researcher, especially since, with his background of art-history and ornithology, he has helped to establish John Gould's direct and indirect role in the artistry of the 'Gould prints'.

More recently, in England, Mrs Christine Jackson has written a fact-filled volume on early bird-lithographers (1975). Our correspondence began in 1975 and she has provided me with quite a few excerpts from letters, by or about Gould, not then known to me.

Another individual deserves especial mention and thanks. He is Adrey Borell of Golden, Colorado. He was a personal tutor and companion for Ralph Ellis in the field of natural history, when Ralph lived in Berkeley, California. Through several interviews and a written biographical account of Ralph, Adrey has provided a personal story of the man who had the foresight and money to purchase in 1937-38 the more than 1200 original Gould drawings and many other related items soon after they were re-discovered in the basement of Henry Sotheran, Ltd, in London. The life of Ralph Ellis was a wild and fascinating one. Information from Borell's accounts, from Ralph's personal extensive no-holds-barred files at the Spencer Library, and from the printed Supreme Court of Kansas legal proceedings relative to the ornithological collection left to the University of Kansas after Ralph's early death, would provide ample facts for a good writer to pen a best-selling biography.

As another source of information on Ralph Ellis, and for a very pleasant evening dinner and visit, I wish to thank Dr Portia Bell Hume. She was one of Ralph's psychiatrists in Berkeley.

In the 1950s Mr J. L. Baughman of Rockport, Texas had accumulated quite a bit of Gouldiana which I was able to acquire from him. This has been very helpful to me.

Most sincere appreciation goes out to many institutions, and to the individuals who make them run, for freely answering my many questions, searching for Gouldiana letters, and, if necessary, photocopying them for me.

Personal contacts have been most rewarding and stimulating. In England there was a visit in 1974 to the Newton Library, Cambridge University, where I spent a pleasant but busy day with Mr Ron Hughes; several hours were spent at the British Museum (Natural History) Department of Zoology where Mrs Ann Datta was and has been most helpful. I also visited the General Library of the British Museum (Natural History), the print room of the Victoria and Albert Museum, and the Zoological Society of London with Mr Fish, Librarian.

Through the kindness of Lord Derby, three busy but exciting days were spent by us at the Old Knowsley Hall Library near Liverpool. We were pleasantly assisted by J. R. Hickish and Agnes Rose. It was exciting to have dinner in the 13th Lord Derby's retreat 'the Nest'. The original drawings in the present Lord Derby's library are second only to the University of Kansas Ellis Collection. I have catalogued them.

A visit with Mr Peter Morgan of the Merseyside County Museum at Liverpool, where are deposited some Gould and Lord Derby bird skins, was rewarding, after I was able to find a place to park my car.

In Scotland my wife and I spent a stimulating afternoon with Ian H. J. Lyster of the Royal Scottish Museum in Edinburgh where we uncovered several boxes of Gould and Jardine correspondence. A few days later a pleasant hour at tea was spent with the present Sir William and Lady Jardine and their son Alec at their home near Dumfriesshire.

Through correspondence I have gratefully received assistance from the British Museum; the British Library; Linnean Society of London; the Athenaeum (Gould's Club); the Windsor Castle Royal Archives and Library; the Bath Library; St. Bartholomew's Hospital Medical Library and Mr John Thornton, relating to the cholera epidemic near Gould's residence on Broad Street, Golden Square; the Royal College of Surgeons, where there are some of Gould's bird and mammal skeletons; the Library of the University of Birmingham; the University Museum of Zoology, Cambridge University; the Cambridge University Library; the Department of Zoology at Cambridge University; the Scott Polar Research Institute, Cambridge; and the Ipswich Museum and H. Mendel.

My Australian contacts have been numerous also but, unfortunately, except for Allan McEvey, not personal. Allan McEvey and Alec Chisholm have been mentioned earlier as my most important sources of information and assistance. My sincere thanks go also to the Royal Australasian Ornithologists Union where in the 1950s the late Mr D. Dickison was very helpful, and more recently Judy Gilmore and Tess Kloot have provided many answers to my queries. I am grateful also for considerable assistance from the National Library of Australia in Canberra and Phyllis Mander Jones, Suzanne Mourot, and Barbara Perry. Particularly helpful has been Jean Dyce of the Mitchell Library of the State Library of New South Wales in Sydney. The Gouldiana in these two libraries is extensive. The Mitchell Library has many manuscript items of John Gould, Mrs Gould, and Edwin Prince, Gould's faithful secretary. In the National Library of Canberra there are over 200 pattern plates for Gould's Australian works with some original drawings.

Assistance has also come from Gilbert P. Whitley of the Australian Museum of Sydney; from the Museum of Applied Arts and Sciences of Sydney; from Marjorie Walker and C. T. Sheehan of the John Oxley Library of Brisbane, who provided me with photocopies of several Gouldiana articles; from J. H. Love of the State Library of South Australia, Adelaide; from the La Trobe Collection, State Library of Victoria in Melbourne; from the Library of Flinders University, of Bedford Park, South Australia; from the National Museum of Victoria in Melbourne; from R. H. Green of the Queen Victoria Museum and Art Gallery, Launceston; from Miss P. S. King, University of Tasmania Library, Hobart; and especial help from Mrs D. A. Lewis of the Royal Society of Tasmania, Hobart.

My thanks go also to L. B. Holthuis of the Rijksmuseum van Natuurlijke Historie, Leiden, and to Yves Laissus of the Museum National d'Histoire Naturelle, Paris.

In the United States, in addition to the fabulous University of Kansas Spencer Library and the Linda Hall Library, I have visited in and corresponded with the Library of the Academy of Sciences in Philadelphia (where Venia T. Phillips, Susan Klimley, who also visited with me in Kansas City, Carol Spawn and Anita Loscalzo were most kind and helpful); the John Crerar Library in Chicago; the Bancroft Library at the University of California at Berkeley; the Beinecke Library at Yale University with S. Dillon Ripley and Suzanne L. Rutter; the American Museum of Natural

ACKNOWLEDGEMENTS

History in New York City with John T. Zimmer; the New York Public Library; the Library of Congress in Washington, where in the Reading Room one is overwhelmed by the beauty and majesty of the architecture; the Library of the Smithsonian Institution; the National Museum of Natural History, with Dr George Watson and J. Phillip Angle; the library at the Louisiana State University at Baton Rouge; the Nature Society, Griggsville, Illinois with Mary Hallock and J. L. Wade; and finally the Research Hospital, Kansas City, Missouri, with the librarian Gerald Kruse and the photographer Larry Brown. I am grateful to all for their interest, assistance and subsequent correspondence.

Through correspondence only I have received assistance in my Gouldiana searching from Mary Baum at the library of the Cleveland Museum of Natural History; from G. W. Cottrell, Jr and Rodney G. Dennis at the Houghton Library, Harvard University; from the Missouri Botanical Garden Library; from the Historical Society of Pennsylvania; from the American Philosophical Society, Philadelphia; and from Elizabeth Baer of the John Work Garrett Library of Johns Hopkins University, Baltimore.

In Canada, it has been my pleasure to correspond with and visit with Eleanor MacLean, at the famous Blacker-(Casey) Wood Library of McGill University in Montreal, Quebec; and to correspond with Mrs Rachel Grover of the Thomas Fisher Rare Book Library at the University of Toronto, Ontario. Mr David Spalding of the Provincial Museum of Alberta, and H. F. Chritchley of the Riveredge Foundation, Calgary, have been helpful correspondents.

Finally, no bibliographer or biographer could exist without the valuable cooperation and knowledge of book dealers, and also in this instance, print dealers. The catalogues published by these dealers are, in themselves, mines of information. To all of the following I am most grateful.

In Kansas City, Missouri, the late Frank Glenn and Mrs Ardis Glenn, of the Glenn Book store, have always been interested and helpful. Other dealers include John Howell Books in San Francisco, where Warren Howell provided personal facts about Ralph Ellis; Taylor Clark of Baton Rouge; Pierce Books of Iowa; John Johnson of Vermont; and in New York City, the Kennedy Gallery, H. P. Kraus, and the Old Print Shop; my thanks to them.

In England the bookseller firm of Henry Sotheran Ltd, London, has been supplying me with catalogues, books, prints and original 'Gould' drawings for over thirty years. I have corresponded with several officers in the past, and then, in 1974, had two personal visits with Mr L. G. Simpson. This firm had purchased Gould's publishing rights and remaining unpublished stock after his death in 1881. The Gould collection story is most interestingly related in their 'Piccadilly Notes' catalogues. Ralph Ellis bought over 1200 original 'Gould' drawings from Sotheran which, on his death, were left to the University of Kansas.

More recently, David Evans of Winchester, England, has been a friend and interested dealer, for which I am grateful. Mr C. Kirke Swann, of Wheldon & Wesley, kindly answered some questions about Ralph Ellis and his company's association with Ralph.

In Australia I have dealt with N. H. Seward and Gaston Renard of Melbourne, with Berkelouw of Sydney, and with the late Doris Eddey of Glen Iris.

Finally, this book owes its completion to the competent staff of Lansdowne Editions, especially Terry Greenwood, Publisher, and Tim Bass, Editor. Gratefully, because this is such a fact-filled compendium, they deleted almost none of the manuscript. Thus, any superfluity or inadequacy is the responsibility of the author. A very sincere thank you with personal appreciation.

I apologize for the length of this acknowledgement section but, over a forty-year period of study of an individual and his works, one accumulates an indebtedness to many, named and unnamed. Appreciation is extended to the institutions or individuals who have allowed publication rights for any quotations from letters, books or journals under their ownership. They are also appropriately acknowledged in the text of this volume. If any acknowledgement has been omitted, I would appreciate this being brought to my attention.

Part One

Genealogy of John Gould and Elizabeth Coxen Gould

Ornithology is a science that developed as an outgrowth of an interest in and a concern for the natural history of birds. Gould's family background of farming and gardening provided an obvious environment and heritage for this vocation.

John Gould's grandfather, Robert Gould, was born at Whitestaunton, Somerset, in the mid-eighteenth century. His wife, Ann, was born at Stocklinch, near Ilminster, in 1757. According to a genealogical chart prepared by Agnes Olive Griffiths Pagett, there were ten children of this marriage, all born between 1781 and 1805 in Ilminster, Somerset. Robert was a farmer. He died at the age of 82, and presumably was buried at Ilminster. His wife Ann died at Ilminster on 11 February 1847 at the age of 90. Both of these grandparents of John Gould lived long enough to be quite aware of their grandson's early and already burgeoning accomplishments in ornithology. There is no allusion to them, however, in any known correspondence.

Robert and Ann's second child and first son was named John. He was born in 1783 and died at Windsor. There is almost no other information on him. Apparently the senior John Gould had died by the time his son, John, was in Australia, since the letters from Australia back home to England in 1839 mention only his mother and sisters (Chisholm 1944, *Elizabeth Gould*, p. 40). John Gould, senior, was a gardener and obviously capable enough in this vocation to receive a 'good appointment' at the Royal Gardens at Windsor Castle in 1818. Possibly he was one of the foremen. He was buried at Windsor, but the date is unknown.

His wife was Elizabeth. Likewise, very little is known of her. She was living during her son's Australian venture from 1838 to 1840, and received letters from him, as her daughter-in-law Elizabeth so stated in her correspondence. The whereabouts of these letters and what happened to Mrs Gould, or when she died, is not known.

John Gould, the 'Bird Man', was apparently the only son of this family. Sharpe, in his biographical memoir of John Gould (1893), states that when Gould was born on 14 September 1804, the family was living in Lyme Regis, a small fishing village in Dorsetshire on the southern coast of England (Figure 2). Soon afterwards the family moved to Stoke Hill near Guildford and, according to Sharpe, 'it was in this beautiful neighbourhood that the child first imbibed his notions of the beauty of natural life'.

Besides the son, John, there were some daughters in this family. Chisholm (1944, *Elizabeth Gould*, p. 15) mentions John's younger sister, Sarah, who married John Wilson. She died in December 1841 at the age of 25, shortly after the death of John Gould's wife. It is evident from Elizabeth Gould's letters to her mother from Australia, that John had more than one sister.

There is a considerable amount of information in my files on John and Elizabeth Coxen Gould's family of six children who attained adulthood. However, this will

Figure 2 Lyme Regis in Dorsetshire, England, birthplace of John Gould. From a 1796 engraving, eight years before Gould's birth.

have to wait for a future volume. But as a thank-you to several of the Gould and Coxen descendants, for sharing their facts with me, a genealogical table will follow.

As mentioned before, I am most grateful to Dr Geoffrey Edelsten, Gould's great-grandson. As part of a long and close association with him, through correspondence and in conversation, Geoffrey has provided most of the data on the Gould family. I am also much indebted to a great-grand-daughter, Mrs Doris Twyman, and especially to Dr Alan Edelsten's widow, Grace Edelsten, for her cooperation and assistance. I also corresponded with Dr Alan Edelsten and his letters are most interesting. Mrs Olive Pagett of Bristol, a descendant of John Gould's uncle, Robert, prepared a genealogical chart of the Gould family, especially as it related to his antecedents, which she shared with me. Her assistance is greatly appreciated.

On the Coxen family side the genealogical data come mainly from Australia. Dr Geoffrey Edelsten gave me a copy of the Coxen family genealogy prepared in 1906 by Henry William Coxen of Brisbane, Australia, and also one prepared in 1939 by Walter Adams Coxen of Melbourne. Using these as a base, additional information on the Coxen family was kindly supplied by Miss Marion Coxen, who has also sent me several other interesting items on the Gould and Coxen family, by Mr and Mrs Harold Crawford and by Mrs Nancé Morton. I am highly indebted to all of them.

PART ONE GENEALOGY OF JOHN GOULD AND ELIZABETH COXEN GOULD

Gould Family

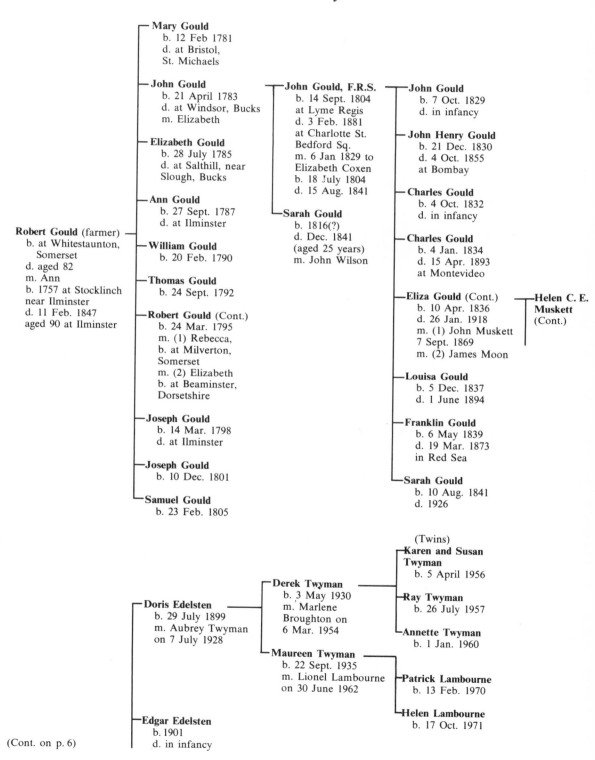

(Cont. on p. 6)

JOHN GOULD — THE BIRD MAN

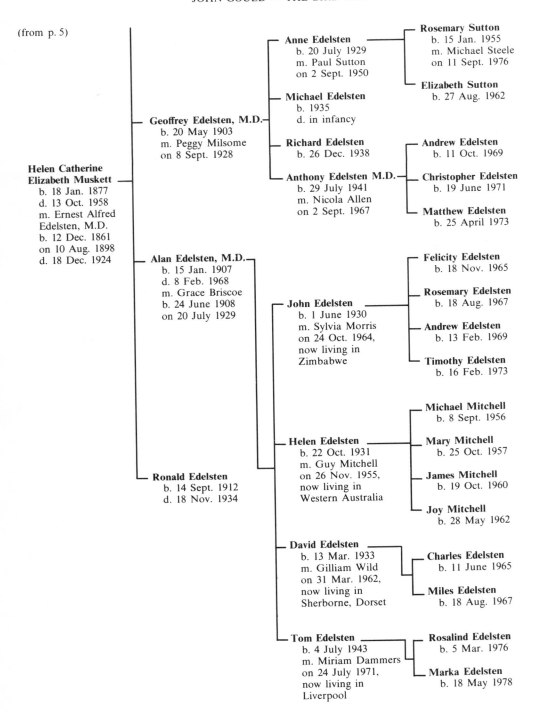

PART ONE GENEALOGY OF JOHN GOULD AND ELIZABETH COXEN GOULD

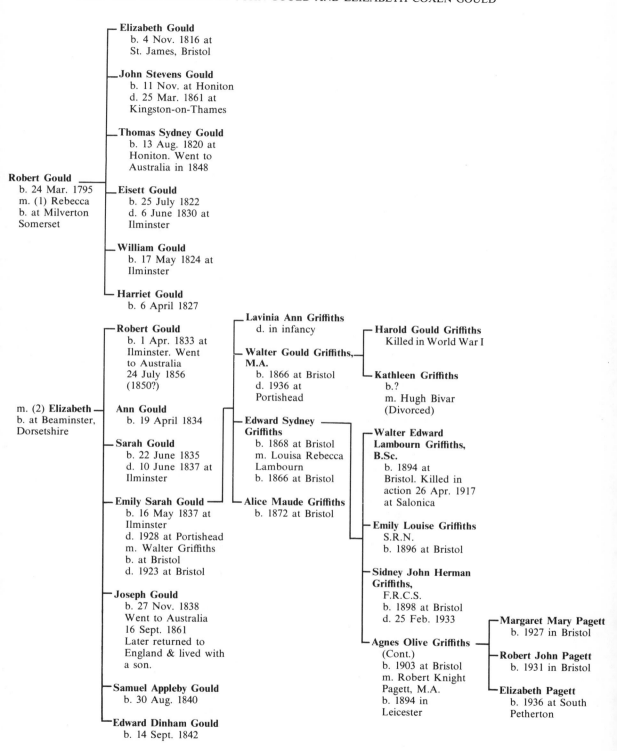

JOHN GOULD — THE BIRD MAN

Coxen Family

Capt. Ebenezer Coxen
b. c. 1736.
Lived in Ramsgate
Isle of Thanet, Kent.
d. 28 Mar. 1821,
aged 85.
Buried in one grave
with wife in the
Parish of
St. Lawrence,
Isle of Thanet,
Kent.
m. Ann
b. c. 1741
d. 14 May 1819,
aged 78.

— Son
b. 12 Nov. 1763
(died without name)

— **Nicholas Coxen**
b. 30 Nov. 1765
d. 26 Apr. 1833
in London
m. Elizabeth Tomkins
of Lickhamstead,
Bucks
11 May 1796

— **Francis Coxen**
b. 25 Apr. 1769
d. in West Indies.
Captain of a
merchant ship

— **Henry Coxen**
b. 10 Mar. 1771
d. c. 1837 in Ireland.
Not married. Captain
in Navy.

— **Ann Coxen**
b. 8 Dec. 1772
d. 1787
at boarding school,
Cheshunt, Hurts

— **Stephen Coxen**
b. 24 Mar. 1776
d. 1815
Slain at Waterloo,
Lieutenant in 23rd
Dragoons

— **Ebenezer Coxen**
(called Edward)
b. 11 Dec. 1779
d. c. 1863
Captain in the Army,
60th Rifles

— **Charles Coxen**
b. 4 Mar. 1782
d. May 1801
in Jamaica.
Midshipman in Navy,
H.M. Frigate
Magecien

— **Henry Holman**
Conygham Coxen
(cont.)
b. 18 Mar. 1797
at Ramsgate
d. 15 Sept. 1825
at Eton, Bucks.
Buried Lambeth New
Church, Surrey.
Lieut. in 14th Infantry
m. Eliza Ann Adams
at St. George's
Hanover Square, London
15 May 1822.
(afterwards she
m. John Bevil.
Went to S.Africa)

— **John Coxen** (Twin)
b. 18 Nov. 1798
at Ramsgate
d. Infancy

— **Stephen Coxen** (Twin)
(cont.)
d. 1845
m. Sarah Freeman
of Englefield Green,
Surrey

— **Ann Coxen**
b. 6 June, 1800
at Ramsgate
d. 21 Jan. 1801
at Ramsgate

— **Mary Coxen**
b. 9 July 1802
at Ramsgate
d. 18 July 1802
at Ramsgate

— **Elizabeth Coxen**
(see Gould genealogy)
b. 18 July 1804
at Ramsgate
d. 15 Aug. 1841
at Egham, Surrey
m. John Gould
6 Jan. 1829
at St. James' Piccadilly

— **Mary Ann Coxen**
b. 19 Mar. 1806 at
Farnborough, Kent
d. 8 June 1806 at
Shoreham, Kent

— **Charles Coxen**
b. 22 Apr. 1808 at
Shoreham, Kent
d. 17 May 1875 at
Bulimba, Queensland
m. Elizabeth Frances
Isaac 1851

— **Mary Coxen**
b. 26 Dec. 1811 at
Shoreham, Kent
d. 16 Sep. 1812 at
Shoreham

— **Henry William Coxen**
b. 3 Mar. 1823
at Croydon, Surrey
d. 25 Aug. 1915,
buried in Sherwood
Cemetery, Oxley,
Brisbane.
m. Margaret
Moorhead
1 Feb. 1866 at
St. Andrew's Church
Brighton, Vic., Aust.
b. 12 Nov. 1839 at
Meerut, India E.
d. 3 Apr. 1902

— **Charles Coxen**
b. 16 Nov. 1824 at
Croydon, Surrey.
To South Africa in
Commissariat
Department, married
and had a large
family

— **Stephen Coxen**
b. 1826 at Lambeth

— **Charles Coxen**
b. ? in
New South Wales,
Aust. Family left
England for Aust.
Feb. 1827

PART ONE GENEALOGY OF JOHN GOULD AND ELIZABETH COXEN GOULD

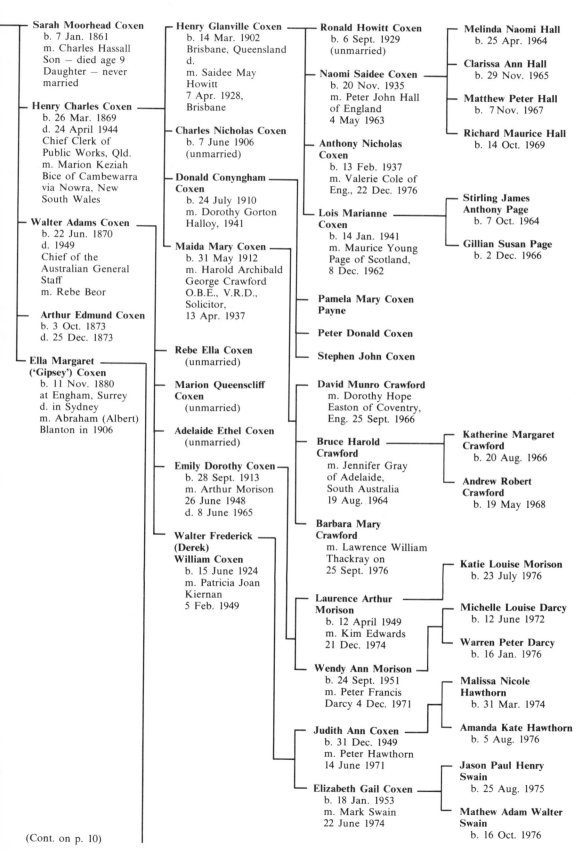

(Cont. on p. 10)

JOHN GOULD — THE BIRD MAN

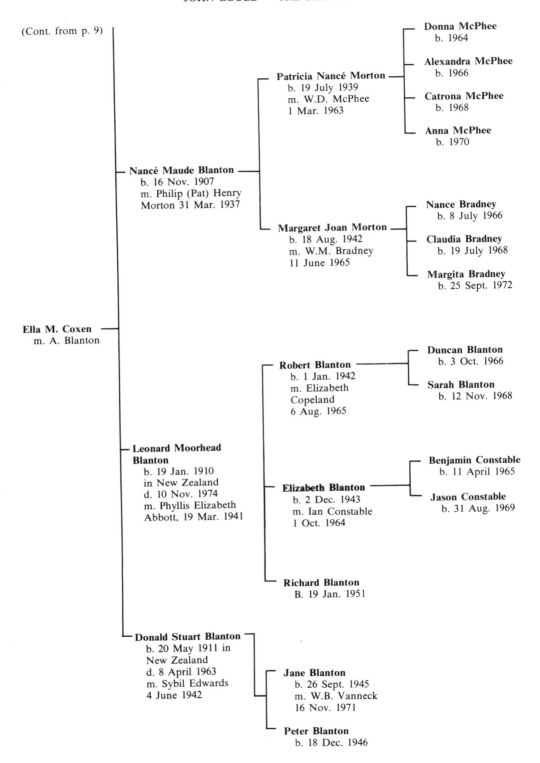

Part Two

John Gould's
Major Published Works

A considerable amount of new information has been found concerning the dates of publication, the contents of the original parts, and other pertinent data relating to Gould's works. This has been gleaned from Gould's unpublished correspondence, from reviews of his works, and especially from unbound copies of Gould's publications as they originally appeared in separate parts.

Earlier printed sources of data on Gould's works have been invaluable. They include the frequently quoted volumes by John T. Zimmer (1926), Casey A. Wood (1931), Jean Anker (1938), and S. Dillon Ripley and Lynette Scribner (1961). To enable one to refer to these four important sources more easily, all the pages devoted to Gould's works in these catalogues have been reprinted in their entirety in Part Four.

Other authors have also provided information on certain of Gould's works; these authors include Carus, Coues, Englemann, Mathews, Nissen, Sharpe, Taschenburg, and Waterhouse. The 1938 Gould commemorative issue of *Emu* was freely consulted.

The tabulation of this expanded information from Gould's correspondence and other sources has been quite a time-consuming and exacting task. But Gould's contribution to the nineteenth-century ornithology was monumental.

Every major work by Gould, even his quarto publications, was published in parts, to be bound later by the subscriber. For his imperial-folio works, a 'List of Plates' with a sheet of directions 'To the Binder' was supplied with the final part. This 'List of Plates' gave the proper order for the plates to be bound in each volume. This meant that the binder was to go through every separate part, take out the appropriate plate and text page, assemble them in the order as indicated on the 'List of Plates', and then bind these sheets in the correct volume. For *A monograph of the Trochilidae, or family of humming-birds*, for instance, there were twenty-five parts, with a total of 360 plates with text pages to be bound in five volumes. Not only did Gould publish monumental works, he also provided a considerable task for the binder.

When you consider that all of these imperial-folio works came out in parts, that the parts were published at varying intervals scheduled for each work, that some works such as the *Birds of Asia* were published in this manner over a thirty-three-year period, and that Gould stressed in his Prospectus that 'no person will commence the Subscription without continuing it to its close', it is amazing that any of his works were eventually bound or, at least, bound by the original subscriber. Some obviously were not. Some subscribers died, and many individuals, and even institutions, had a change of fortune or a lapse in interest over the period of the issue.

When I read Zimmer or Wood and see the great difficulty bibliographers have had in ascertaining correct publication dates for many, if not all of the early natural-history works, the problem with Gould's dates of publication would seem relatively simple. Only two of his works, those on the toucans and trogons, were published in second editions or revisions. Nevertheless, there are problems. There is no complete list of the dates on which the separate parts appeared, especially for the earlier works. Also, copies of Gould's works in the original parts are rare and, even when found, some of these were not dated. As seen from Gould's correspondence, especially with Sir William Jardine, some parts came out earlier, some later, than Gould planned in his timetables — and he varied his timetables. To Gould's great credit, however, many of the parts came out earlier than scheduled. As a publisher he ran a tight shop.

Frederick Herschel Waterhouse in 1885 published *The dates of publication of some of the zoological works of the late John Gould*. Included were Gould's two works on mammals, the works on the birds of Great Britain, Asia, Australia, his monographs on the *Trochilidae, Odontophorianae* and the revised edition of his *Trogonidae*. Waterhouse gave the year, but not the month or day, of publication for the various parts. This year-date is valuable, however, since the final arrangement of the plates in Gould's volumes was without regard for the sequence of issue. Unfortunately, his compilation covered only eight of Gould's works, but even worse, it is not accurate.

With this introductory information, the dates of publication and other pertinent information concerning Gould's major works, including his small 'Introduction' books, will be given as completely as possible. Zimmer (1926) has been the best published source for this information in the past. But, as mentioned earlier, having uncovered a considerable amount of Gould's unpublished correspondence and having access to the Gouldiana in the University of Kansas Spencer Library–Ellis Collection (where some of the editions are still in the original parts as published), a considerable amount of new information has been found concerning the plates, text, dates of publication of the parts, numbers of parts, reviews, etc. Nevertheless, some gaps remain. It is to be hoped that more of Gould's correspondence and editions in the original parts will be discovered and that these will yield additional specific data.

Gould's scientific papers will not be listed here. There were over 300. The list by Salvadori (1881), reprinted in R. B. Sharpe's *An analytical index to the works of the late John Gould* of 1893, is valuable but not complete. That area of Gouldiana will have to wait for future research.

All works, unless otherwise specified, were imperial-folio size, that is, 22 × 16 inches (559 × 406 mm), and were published in London by the author.

All the plates are hand-coloured lithographs. In the lower-left corner of almost every plate is printed the reference to the artist (del.) and the lithographer (lith.) usually in these phrases: 'Drawn from nature and on stone by J. & E. Gould'; 'J. Gould and H. C. Richter, del. et lith.'; 'E. Lear, del. et lith.'; 'J. Wolf and H. C. Richter, del. et lith.'; 'J. Gould and W. Hart, del. et lith.'; or 'W. Hart, del. et lith.'.

The printer's name appeared in the right lower corner of most plates as follows: C. Hullmandel, Hullmandel and Walton, Walter and Cohn, Walter, or Mintern Bros.

1 1830–1833. *A century of birds hitherto unfigured from the Himalaya Mountains*, 1 volume, 80 plates.

A copy in the University of Kansas Spencer Library–Ellis Collection (Ellis F-163) exists in twelve separate brown-paper folders with twenty numbers marked in ink on the covers (Figure 3).

PART TWO JOHN GOULD'S MAJOR PUBLISHED WORKS

Prospectus: One dated 'Nov. 1st. 1830' is listed in Wheldon & Wesley's catalogue No. 28, 1932.

Date on title-page: The date of 1831 is from Wheldon & Wesley's catalogue No. 28, of 1932. The date 1832 is the date on most copies according to Zimmer. An 1834 date is mentioned in the 1961 British Museum general catalogue of printed books.

Number of plates: Eighty. Information on the plates in the numbers or parts as issued will be elaborated on in a following section.

A note on the colouring of the plates is found in a letter of 5 February, 1844 written by Gould to Lord Derby (Wagstaffe & Rutherford 1954-55, letter number 17):

You will probably recollect that in my first work on the Birds of the Himalayan Mountains neither the plants or [sic] Backgrounds were coloured; In order to render the Series of my Publications complete ... I have had those parts coloured in the few copies I have left, and ... I have this day sent off a fine copy for your Inspection.

Dates of publication of the parts: Preliminary information. Apparently this work was published both in twenty numbers and in four parts.

William Wood, in an 1832 work entitled 'Catalogue of an extensive and valuable collection of the best works on natural history', lists Gould's *Century of Birds, hitherto unfigured, from the Himalaya Mountains*. He states that 'This elegant work is publishing in Nos., each containing 4 coloured plates, 12.s.' Evidence for the work also being issued in four parts will be presented later.

Since the Ellis Collection copy is unique, to my knowledge, I will collate it. The Ellis F-163 copy untrimmed measures 21⅞ × 15 inches wide (555 × 380 mm).

The first folder has 'No 1 2 & 3' written in ink in the upper-left corner of the cover (Figure 3). Then follows:

A/Century of Birds/hitherto unfigured/from the/Himalaya Mountains./Dedicated by Permifsion, to/Their Majesties./by/John Gould,/A.L.S./____ London./Published by the Author, 20, Broad Street, Golden square./1831/____ Printed at C. Hullmandel's Lithographic Establishment, No. 49, Gt Marlborough Street.

The folder contains twelve plates, with the background and foliage uncoloured. There is no text. The plates are sewn loosely in this brown-paper folder.

'No 4' is written in ink in the upper-left corner of the next cover. The printing on this cover, and on 'No 5', is different from all of the other covers in that these two do not contain the lines 'Dedicated by Permifsion, to/Their Majesties'.

Numbers 4 to 12, inclusive, contain four plates each. 'No 13 & 14' are together in one folder and contain eight plates. The last folder has 'Nos 15 to 20 which completes the work' written on the cover. The writing is not by John Gould, in my opinion. There are twenty-four plates in this last set.

There is no text in any of the folders. This absence of text lends support to the new information elaborated on later, that the text by Vigors was published more than eighteen months after the final numbers of the plates. There are no dates on these twenty numbers, unfortunately.

An early mention of this work by Gould is in a letter from Prideaux John Selby to Sir William Jardine of 30 November 1830 in the Newton Library, Cambridge (pers. corres. Christine Jackson):

Gould is about to publish Illustrations of ornithology from the Himalayan Mountains, he has received an importation of very valuable skins from that interesting part of Asia, the majority of them quite new, the figures are of natural size drawn and lithographed by Mrs. E. Gould, and the letter press is to be furnished by Vigors. The specimen he sent me is very tolerable.

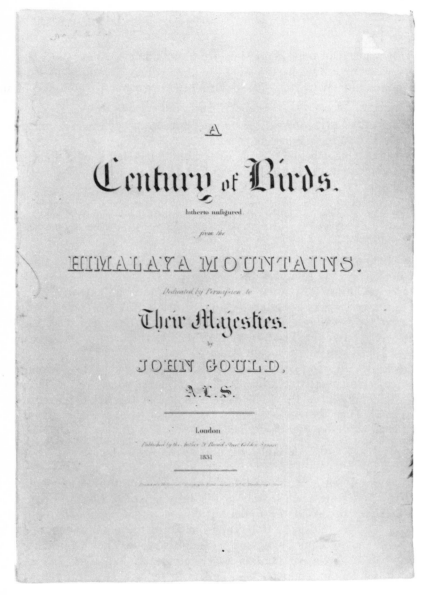

Figure 3 Cover folder for 'No. 1, 2 & 3' of Gould's *Century of Birds* (University of Kansas Spencer Library-Ellis Collection).

Some plates of *Century of Birds* were exhibited at a meeting of the Zoological Society on 23 November 1830.

Number 1 was published on or before 20 December 1830, according to a letter from Gould to Sir William Jardine in the Royal Scottish Museum, Edinburgh.

I have this day forwarded according to your directions a Box ... with the first no. of my new work on the birds of the Himalaya Mountains which I hope will meet with your approbation and support ... I am very well supported by the scientific gentlemen of London, etc. having allready [sic] 50 subscribers ... I shall limit my number of coppys [sic] to about 200 and then destroy the stones, therefore every coppy sold will make the work more scarse [sic] and consequently more sought after. From the support I have allready received I do not fear selling the whole before the work is completed, which will handsomely repay me ... It is Mrs. Goulds very first attempt at stone drawing which I hope you will take into consideration.

Plate 1 Drawing in ink and watercolour, signed E. Gould, of *Noctua cuculoides*, for *Century of Birds*, plate 4, Kansas-Ellis-Gould MS 994.

Plate 2 Drawing in watercolour, signed E. Gould, of *Garrulus lanceolatus*, for *Century of Birds*, plate 40, Kansas-Ellis-Gould MS 1067.

PART TWO JOHN GOULD'S MAJOR PUBLISHED WORKS

In the letter dated 10 January 1831 from P. J. Selby to Jardine (Newton Library, pers. corres., Christine Jackson) Selby writes 'I have... from Gould... the No. of his own Century of Himalayan Birds, very good indeed, Mrs. G. is improving every day in her drawing and attitudes, it is the same size as Lears... '

The final Number or Part appeared on or before 17 April 1832. This is suggested by a letter of this date from Gould to Jardine (Roy. Scot. Mus.) 'I have herewith forwarded to you the remaining portion of my work on the Himalayan Birds and the letter press shall be forwarded very soon.' That the 'work was completed at the beginning of the present year' of 1832 is also stated in the *Magazine of Natural History* (5:535).

The frequency of the publication of the numbers of the *Century of Birds* can be inferred to have been monthly. Vigors indicated this at a meeting of the Zoological Society on 27 December, 1831 (*Proceedings of the Committee of Science and Correspondence of the Zoological Society of London*, 1:170-6).

One could calculate, then, that number 1 appeared in December 1830 but was scheduled for January 1831, that number 2 was scheduled for February 1831, number 3 for March, and so on down to number 12 scheduled for December 1831; that number 13 was scheduled for January 1832 but published with number 14 (as in the Ellis set) either in January or February, and the final numbers 15 to 20 published in March 1832. This scheme is, of course, only a supposition, but the beginning and ending dates are supported.

Further evidence that this calculated schedule is an accurate supposition is the statement in the 1868 Prospectus and List (Acad. Nat. Sciences Phila.) that the *Century of Birds* was 'commenced in January 1831, and completed in August 1832'. This would fit the monthly publication of each of the twenty numbers perfectly, except that as noted above, the first number appeared earlier, in November or December 1830, and the last six numbers came out at the 'beginning of the present year [1832]'. It would be natural for Gould, thirty-six years later, to forget or regard as unimportant the fact that he had published the last numbers of his *Century of Birds* earlier than called for on his well-planned schedule.

A confusing fact, however, is that Gould billed Jardine for 'Parts 3 & 4 of the Birds of the Himalaya Mountains' at a price equalling essentially one-half of the cost of the whole set. So while the plates were published in 'Numbers', Gould obviously thought of them as being issued in 'Parts' also; each 'Part' undoubtedly was to contain twenty plates. However, there is no other evidence whatsoever of their being issued in this 'Part' form.

Letterpress: This was published on or before 5 November 1833 according to the letter of this date from Gould to Jardine (Roy. Scot. Mus.):

I have forwarded in a parcel to Mr. [James] Wilson of Canaan the Letterpress descriptions to my work on the "Birds of the Himalayan Mountains", which will enable you to have your copy bound. I regret that it has been so long in hand which must be attributed to Mr. Vigor's political career which, I am sorry to say has taken him almost entirely from the pursuit of Natural History.

[Bill appended for:]
Part 3 & 4 of the Birds of the Himalaya Mountains   6.5.0
Letterpress to ditto   1.5.0

The only letterpress for Gould's works written by anyone other than Gould was this one done by Nicholas A. Vigors (1787-1840).

In Wheldon & Wesley's catalogue No. 28 of 1832, there was a copy for sale for £12 without the text. In a prospectus dated 1 November 1830 bound in this copy there is the following statement: 'The descriptions of the birds will be supplied by N. A. Vigors, Esq., F.R.S., Secretary of the Zoological Society, in a *separate publication*

to correspond with the plates.' It was published a year and a half after the final part 4 was completed, as seen from the letter of 5 November 1833, but not as a 'separate publication' in the true sense. Vigors presented six papers on the subject of Himalayan birds to the Zoological Society of London on the following dates: 23 November 1830, 11 January 1831, 8 February 1831, 22 February 1831, 22 March 1831 and 27 December 1831. These dates are at variance with those given by Zimmer (1926). (*See* Journal. *Proceedings of the committee of science* ... 1830-32.)

Contents of the separate numbers: The University of Kansas Spencer Library copy (Ellis F-163) has the plates loosely bound in folders as issued in twenty numbers. Since this is the only copy of the *Century of Birds* noted to date in this form, I will tabulate the numerical 'List of Plates' given for binding in the single volume (which appeared with the letterpress in 1833), along with the numbers of the plates in order of publication in the twenty separate numbers. The date of publication appended is probably accurate, as discussed earlier.

List of Plates in 'Century of Birds'

Bound plate number	Species name	Number as issued	Plate in Number	Date of issue
1	*Haematornis undulatus*	15–20	2	March 1832
2	*Falco Chicquera*	13 & 14	5	Jan.–Feb. 1832
3	*Otus Bengalensis*	13 & 14	2	Jan.–Feb. 1832
4	*Noctua cuculoīdes*	1, 2 & 3	4	Dec. 1830–March 1831
5	*Alcedo guttatus*	4	4	April 1831
6	*Muscicapa melanops* (male and female)	15–20	18	March 1832
7	*Muscipeta princeps*	13 & 14	1	Jan.–Feb. 1832
8	*brevirostris* (male and female)	5	3	May 1831
9	*peregrina* (male and female)	15–20	7	March 1832
10	*Hypsipetes psaroīdes*	15–20	5	March 1832
11	*Lanius erythropterus* (male and female)	1, 2 & 3	7	Dec. 1830–March 1831
12	*Collurio Hardwickii*	15–20	13	March 1832
	erythronotus	15–20	13	March 1832
13	*Turdus erythrogaster* (male and female)	10	4	October 1831
14	*poecilopterus* (male and female)	15–20	23	March 1832
15	*Cinclosoma ocellatum*	6	1	June 1831
16	*variegatum*	12	4	December 1831
17	*erythrocephalum**	15–20	12	March 1832

PART TWO JOHN GOULD'S MAJOR PUBLISHED WORKS

Bound plate number	Species name	Number as issued	Plate in Number	Date of issue
18	leucolophum†	8	1	August 1831
19	Petrocincla cinclorhyncha‡	7	4	July 1831
20	Myophonus Horsfieldii	4	1	April 1831
21	Temminckii	8	4	August 1831
22	Zoothera monticola	15–20	6	March 1832
23	Pitta brachyura	15–20	10	March 1832
24	Cinclus Pallasii	13 & 14	4	Jan.–Feb. 1832
25	Phoenicura rubeculoïdes	10	2	October 1831
	caeruleocephala	10	2	October 1831
26	frontalis	6	3	June 1831
	leucocephala	6	3	June 1831
27	Enicurus maculatus	1, 2 & 3	8	Dec. 1830–March 1831
28	Scouleri	13 & 14	8	Jan.–Feb. 1832
29	Parus xanthogenys	4	3	April 1831
	monticolus	4	3	April 1831
30	erythrocephalus	11	3	November 1831
	melanolophus	11	3	November 1831
31	Fringilla rodopepla	8	3	August 1831
	rodochroa	8	3	August 1831
32	Pyrrhula erythrocephala	15–20	15	March 1832
33	Carduelis caniceps	13 & 14	7	Jan.–Feb. 1832
	spinoides	13 & 14	7	Jan.–Feb. 1832
34	Lamprotornis spilopterus (male and female)	7	3	July 1831
35	Pastor Traillii (male and female)	13 & 14	3	Jan.–Feb. 1832
36	Nucifraga hemispila	6	4	June 1831
37	Garrulus striatus	1, 2 & 3	9	Dec. 1830–
38	Garrulus bispecularis	1, 2 & 3	12	Dec. 1830–March 1831
39	lanceolatus (male)	9	1	September 1831
40	(female)	1, 2 & 3	1	Dec. 1830–March 1831
41	Pica erythrorhyncha	5	2	May 1831
42	vagabunda	7	1	July 1831
43	Sinensis	15–20	11	March 1832
44	Buceros cavatus	15–20	1	March 1832
45	Coccothraustes icterioides (male and female)	1, 2 & 3	3	Dec. 1830–March 1831

* Spelled *Cinclostoma erythrocephala* in the plates.
† Erroneously named *Garrulus leucolophus* in the plates.
‡ Erroneously named *Phoenicura cinclorhyncha* in the plates.

Bound plate number	Species name	Number as issued	Plate in Number	Date of issue
46	*Bucco grandis*	1, 2 & 3	5	Dec. 1830–March 1831
47	*Picus occipitalis*	11	1	November 1831
48	*squamatus*	5	4	May 1831
49	*Shorii*	10	1	October 1831
50	*hyperythrus* (male and female)	1, 2 & 3	11	Dec. 1830–March 1831
51	*Mahrattensis* (male and female)	15–20	21	March 1832
52	*brunnifrons* (male and female)	12	3	December 1831
53	*Cuculus sparverioïdes*	15–20	9	March 1832
54	*Himalayanus*	11	4	November 1831
55	*Pomatorhinus erythrogenys*	9	4	September 1831
56	*Cinnyris Gouldiae*	9	3	September 1831
57	*Vinago sphenura*	15–20	4	March 1832
58	*militaris* (male and female)	15–20	16	March 1832
59	*Columba leuconota*	5	1	May 1831
60	*Lophophorus Impeyanus* (male)	13 & 14	6	Jan.–Feb. 1832
61	(female)	15–20	22	March 1832
62	*Tragopan Satyrus*	15–20	17	March 1832
63	*Hastingsii* (male)	12	1	December 1831
64	(young male)	1, 2 & 3	2	Dec. 1830–Mar. 1831
65	(female)	12	2	December 1831
66	*Phasianus albo-cristatus* (male)	1, 2 & 3	6	Dec. 1830–March 1831
67	(female)	10	3	October 1831
68	*Staceii*	4	2	April 1831
69	*Pucrasia* (male)	7	2	July 1831
70	(female)	8	2	August 1831
71	*Perdix Chukar*	6	2	June 1831
72	*Otis nigriceps*	9	2	September 1831
73	*Himalayanus* (male)	1, 2 & 3	10	Dec. 1830–March 1831
74	(young male)	15–20	20	March 1832
75	(female)	11	2	November 1831
76	*Totanus glottoïdes*	15–20	19	March 1832
77	*Parra Sinensis*	15–20	3	March 1832
78	*Vanellus Goensis*	15–20	8	March 1832
79	*Ibidorhyncha Struthersii*	15–20	14	March 1832
80	*Anser Indicus*	15–20	24	March 1832

PART TWO JOHN GOULD'S MAJOR PUBLISHED WORKS

In a preliminary cataloguing of the 1523 original drawings and other items for Gould's works in the Ellis Collection by Carol Chittenden (1976) there were noted original drawings or studies for the *Century of Birds* plates numbers 4, 38, 40, 45 and 78. The drawings for plates 4, 38 and 40 are signed 'E. Gould'.

Review: In D. W. Mitchell's expansive review of 1841, the *Century of Birds* and the Himalaya area are both discussed at length.

2 1832–1837. *The birds of Europe,* 5 volumes, 22 parts, 448 plates.

Prospectus: Gould, in a letter of 17 April 1832 to Jardine (Roy. Scot. Mus.), writes

You will perceive by the accompanying Prospectus that I have commenced another work of much greater magnitude [than the Himalayan birds]; for my own part I should have been more anxious to have gone on with unfigured foreign birds and by that means have added so much the more interest to the science of ornithology, but the greater number of the subscribers to my other work not paying [word heavily crossed out] attention to birds generally but limiting themselves to those of our own country, they have frequently reiterated their request that I should commence a similar work on the Birds of ['this country' crossed out] Europe and this has been the only motive for my undertaking so laborious a task. I had not the most remote idea of doing it when on my journey to Scotland. I have now, however, commenced it with considerable spirit, and I trust I shall produce a work that will merit the patronage of the public. I have much new information to communicate, even on the object by which we are so nearly surrounded altho' so much has already been said on them, and should not the health of myself and Mrs. G. fail I hope to complete it in the short space of 3 or 4 years. I intend to make an alteration from the statement in my Prospectus, viz to bring out the letter press with the plates and the same size.

I shall take an early opportunity of sending you a copy of the First part when I shall be glad to have your opinion of it and any suggestions you may please to favor me with.

A prospectus for *Birds of Europe* dated 24 October 1831 was bound in the University of Kansas Spencer Library copy (Ellis F-163) of Gould's *Century of Birds*. Since this is the earliest prospectus that I have seen, it appears as Figure 4. There are alterations in the wording in ink as shown, reducing the number of plates per part from twenty-five to twenty, and a reduction in the price per part.

In the letter to Jardine of 17 April 1832 mentioned earlier, Gould also writes about another change, namely that the letterpress will appear with the plates and be of the same size. This was the plan of publication that Gould was to follow for all later works.

An abstract of this prospectus appeared as an announcement in the March 1832 issue of the *Magazine of Natural History* (5:190-1).

It is also interesting that Gould stated in the same prospectus that 'Subscribers desirous of possessing the British Birds only, are requested to signify their wishes'. There is no evidence suggesting that the plates for the British birds were ever published separately from the birds of Europe. Obviously, as noted from the correction in Gould's letter to Jardine of 17 April 1832, and noted later in a Havell letter to Audubon, Gould was apparently torn between a work on British birds and European birds. Of course, many European birds are also indigenous to Great Britain. A partial solution to this dilemma was his statement in the prospectus, that fifteen (of the twenty plates in a Part) 'will be devoted to the representation of British Birds, and five to those of the European continent'.

Havell, Jr. on 30 November 1831 wrote to John Audubon (Ford 1964 p. 292) that Gould 'has just issued a prospectus and as soon as I have one I will send it to you, announcing a work on English birds ... next June ... first part, 25 plates, price 3/16/0 [sic] ... his conceit leads him beyond common sense'. Here again is a reference to Gould's work on 'English' birds.

A short single-sheet prospectus dated April 1836 is bound in the Kansas-Ellis copy (Ellis Aves E62) of part 1 of *A synopsis of the birds of Australia and the adjacent islands.* 'The publication, commenced on the 1st of June 1832, is now fast drawing to a close: fifteen Parts are already published and it will be completed about the end of the present year.'

Number of plates: Gould states in his Prospectus and List of 1 January 1868 that there were 449 plates in this work, but plates 447 and 448 are on one plate, so the correct total is 448 plates. Twenty plates were to be published in each part, as stated in the Prospectus of 24 October 1831.

The Kansas copy of *Birds of Europe* is Ellis F-100.

Dates of publication of the parts: Gould states in his Prospectus of 24 October 1831 that the first part was to appear on 1 June 1832 and the remainder on the first of each third month thereafter. Thus, parts would be issued on September 1, December 1, March 1 and June 1 of each year, being completed by 1 September 1837. From Gould's letters to Jardine it is evident he kept quite well to that schedule.

In the Library of the Zoological Society of London is an unpublished manuscript by F. H. Waterhouse listing alphabetically the species in *Birds of Europe* with volume

Figure 4 a & b Prospectus for *Birds of Europe* dated 24 October 1831 (University of Kansas Spencer Library-Ellis Collection).

On the 1st of June, 1832, will be published,

PART I.

(CONTAINING TWENTY-FOUR PLATES,)

OF

THE BIRDS OF EUROPE;

By JOHN GOULD, A.L.S.

TO BE CONTINUED EVERY THREE MONTHS.

THE condescending patronage bestowed by Their Most Gracious Majesties upon the "Illustrations in Ornithology from the Himalaya Mountains"; the extraordinary degree of favour with which that Work has been honoured by the Nobility and the Public; the very flattering approbation conferred upon it by all classes of the Subscribers, and the request of many most valuable friends, have induced the Author to propose the publication of an entirely new Work on the Birds of Europe.

It has frequently been matter of remark, and even of censure, that the productions of distant countries have received a much larger share of attention than those objects by which we are more immediately surrounded; and it is certainly true, that while numerous and costly illustrations have made us acquainted with the ornithology of all the other parts of the world, the birds of Europe, in which we are, or ought to be, most interested, have remained, by comparison at least, neglected, unfigured, and in proportion unknown.

Eight years almost exclusively devoted to Ornithology, more particularly that of Great Britain; extensive acquaintance and constant communication with the most celebrated cultivators of this branch of Natural History; and resources in art beyond those which have already given such universal satisfaction, ensure to the Author advantages of no common extent. Assisted by experienced collectors at all the most favourable localities, it is intended that the artists employed on this Work shall have, as far as

PART TWO JOHN GOULD'S MAJOR PUBLISHED WORKS

number, plate number, part number and year of publication. Thus Waterhouse performed this indexing also for *Birds of Europe* as he had for some other volumes as published in his 1885 'Dates of Publication...'. In going over this manuscript listing, however, I find that there is a considerable variation in the number of plates issued for each part. For some parts there are nineteen plates and many have twenty-one or more plates listed per part. These seeming discrepancies need to be checked with a copy still in the original parts, but I am not aware of any such copy. The final part 22 is recorded by Waterhouse to have twenty-eight plates. This is correct, as shown on a cover for this part in the Blacker-Wood Library at McGill University (pers. corres. MacLean 19 May 1978).

A listing follows of each part with the number of plates and the scheduled date of publication. It is assumed that each part had twenty plates except for the final part which had twenty-eight plates. Additional information gleaned from Gould's correspondence and reviews will be appended.

PART 1 Twenty plates, 1 June 1832. Part 1 was reviewed in the July 1832 issue of the *Magazine of Natural History*, (5:535-6). Thus this part could have appeared in June 1832. It is stated that twenty plates were in this part. Interestingly, the price is listed as '£2.10s plain; £3.3s coloured', but I have never seen a listing of an uncoloured copy. The review is very favourable, 'admirably executed . . . equal, if not superior, to any other ornithological production in Europe'.

PART 2   Twenty plates, 1 September 1832.

PART 3   Twenty plates, 1 December 1832. Parts 2 and 3 were reviewed together in the March 1833 issue of the *Magazine of Natural History*, (6:135-6), and 'bear evident proof of increasing excellence as the numbers proceed'. Both parts contained twenty plates and were also offered uncoloured. As a criticism, the reviewer felt that there was 'a slight appearance of stiffness in the drawing of some of the figures, and the tone of colouring in one or two instances, is too uniform'.

PART 4   Twenty plates, 1 March 1833.

PART 5   Twenty plates, 1 June 1833.

PART 6   Twenty plates, 1 September 1833. In the letter of 5 November 1833 from Gould to Jardine (Roy. Scot. Mus.) he writes '6th part published already'.

PART 7   Twenty plates, 1 December 1833. A letter of 19 December 1833 from Gould to Jardine (Roy. Scot. Mus.) states that 'the next Part (7) of my "Birds of Europe" will be published in a few days'. Gould was three weeks behind schedule.

PART 8   Twenty plates, 1 March 1834.

PART 9   Twenty plates, 1 June 1834.

PART 10   Twenty plates, 1 September 1834.

PART 11   Twenty plates, 1 December 1834. Parts 8, 9, 10 and 11 were recently sent off to Jardine by Gould according to the letter of 13 January 1835 (Newton Library).

PART 12   Twenty plates, 1 March 1835.

PART 13   Twenty plates, 1 June 1835.

PST 14   Twenty plates, 1 September 1835.

PART 15   Twenty plates, 1 December 1835.

PART 16   Twenty plates, 1 March 1836.

PART 17   Twenty plates, 1 June 1836. In a letter of 29 February 1836 to Jardine (Newton Library) Gould wrote that 'parts 16 & 17 will be published in a month or so'. If this statement by Gould was adhered to, part 16 would have appeared later than scheduled, but part 17 would have been published earlier.

In this letter Gould also billed Jardine for parts 1 to 15 for £51.0.0, so Jardine was also behind schedule.

Lear wrote on 11 March 1836 to Jardine (Roy. Scot. Mus.): 'Gould has been so clamorous lately at my not having done any Birds of Europe for him, that I must do a batch for him without further delay.'

In a letter to Jardine of 15 March 1836 (Newton Library) Gould states: 'the Birds of Europe will extend to 20 Pts. at £3.8.0 each; but the whole will be finished during the present year.' It was extended to twenty-two parts and completed in 1837.

PART 18   Twenty plates, 1 September 1836. A letter to Jardine of 2 November 1836 (Newton Library) contains the information that Part 18 was being sent to him.

PART 19   Twenty plates, 1 December 1836. Gould writes on 16 January 1837 to Jardine (Newton Library) 'I have this day forwarded to you direct by Coach a small box containing Part 19.'

PART 20   Twenty plates, 1 March 1837.

PART 21   Twenty plates, not scheduled, but probably published in March.

In a letter of 27 February 1837 to Jardine (Newton Library) Gould writes 'Part 20 and 21 . . . will appear together very shortly . . . [may] extend my work to 23 [22?]

Plate 3 Drawing in watercolour, attributed to E. Lear, of White-tailed or Sea Eagle, for *Birds of Europe*, I: 10, Blacker-Wood Library, McGill. Many pencilled notes by John Gould.

Plate 4 Drawing in pencil and watercolour, attributed to John Gould, of *Pteroglossus castanotis*, for *Ramphastidae*, 2nd edn, plate 19, Kansas-Ellis-Gould MS 960.

Plate 5 Drawing in pencil and watercolour, attributed to John Gould, of *Aulacorhamophus prasinus*, for *Ramphastidae*, 2nd edn, plate 47, Kansas-Ellis-Gould MS 962.

Plate 6 Drawing in pencil and watercolour, attributed to John Gould, of *Harpactes duvauceli*, for *Trogonidae*, 2nd edn, plate 40, Kansas-Ellis-Gould MS 936.

PART TWO JOHN GOULD'S MAJOR PUBLISHED WORKS

Parts'. This statement by Gould was undoubtedly the basis for the following announcement by editor Jardine. Under Bibliographical Notices in the 1837 *Magazine of Zoology and Botany* (1:572-3) is the report that parts 20 and 21 of Gould's *Birds of Europe* 'are nearly completed, and will appear together ... another (Part 22) will still be necessary'.

PART 22 Twenty-eight plates, not scheduled, but probably published in July 1837. In August [?] 1837 Gould wrote to Jardine (Newton Library):

Dear Sir William

I am happy to say that my work on the "Birds of Europe" is at length brought to a close; its extension to 22 Parts will I trust be looked upon with leniency by the subscribers as it includes all the recently discovered species and no pains have been spared on my part to render it complete up to the present time and I am led to hope or rather to believe that it has given very general satisfaction. The extent of its sale and the fact of entire sets having been ordered very lately is perhaps some evidence of its merits.

That part 22 had twenty-eight plates is evident from a photocopy of the cover for this part received from Eleanor MacLean of the Blacker-Wood Library at McGill University, Montreal. On this cover the number '22' is written in ink after 'Part' and then follows on the next line a handwritten sentence 'Containing 8 Extra Plates etc.' The date on the cover is 1837.

Contents of the separate parts: The MS of F. H. Waterhouse in the Library of the Zoological Society of London contains the years of issue of the parts, but as stated earlier, it appears there are some errors. The Blacker-Wood Library at McGill University apparently had the work as issued in the twenty-two parts (Wood 1931), but their copy is now bound to conform to the binder's directions (pers. corres., MacLean 19 May 1978). However, they do have the covers of the twenty-two parts. No species are listed on these covers, so the order of publication of each plate is unavailable.

Reviews: The reviews are mentioned under the appropriate part. This work was also included in D. W. Mitchell's elaborate review of 1841.

Seven manuscript volumes, apparently used by Gould as a 'filing cabinet' for notes and sketches for the *Birds of Europe* and the *Birds of Great Britain*, are in the libraries of the British Museum (N.H.) and the Academy of Natural Sciences of Philadelphia. (*See also* Gould, John. Manuscript. [No date.] '... for *Birds of Europe*', and 'Europe/insessores ...', and Sauer (1976 and 1978).) It would be of considerable interest to compare these notebooks with the published works.

3 1833-1835. *A monograph of the Ramphastidae, or family of toucans*, 1 volume, 3 parts, 33 coloured plates, plus one uncoloured plate on the anatomy of the toucan accompanying an article by Richard Owen.

The University of Kansas Spencer Library copy is Ellis Aves H17, item 1.

Prospectus: Not seen.

Number of plates: Thirty-three coloured, one uncoloured. Part 1, twelve plates; part 2, eleven plates; part 3, ten plates (Zimmer 1926).

Dates of publication of the parts:

PART 1 In a letter of 5 November 1833 Gould wrote to Jardine (Roy. Scot. Mus.) 'I have lately published the first part of a "Monograph of the Ramphastidae"'. Thus, part 1 was published before 5 November 1833.

It may have been scheduled for December 1833 because D. W. Mitchell in his review (1841) states that this work was 'commenced in December 1833 and finished in December 1835'.

PART 2 This work was originally to be in two parts only, as seen from a corrected 'To the binder' sheet listed in Zimmer. The original plan for two parts only is corroborated in a letter to Wm. Swainson dated 30 April 1834 (Linnean Society) where Gould writes:

> From the description of the bird in your possession I am led to believe it will differ from any I have seen and if so will add another species to the interesting family; the addition of which, together with an expected collection of Birds from Guatemala wherein I am assured will be some species of Toucans and several Trogons new to the collections of this country will prevent my closing, with any degree of satisfaction to myself, my Monograph with the next part, the more so as I have in my own collection the heads of two if not three species which I have never seen in a perfect state.

So Gould altered his plans and was to publish three parts because of additional specimens being acquired.

For the date of publication of part 2, Gould then continues in his letter to Swainson and makes reference to the second part of the Toucans coming out in a fortnight: 'As the one [a new species of Toucan] I brought from Paris (which is in all probability the *Cuvieri* of Wagler) completes the 12 pls. I intend bringing them out immediately and hope to have them ready for delivery in a fortnight as I am now far behind the promised time of appearance... I will send you the second Part as soon as published.'

Zimmer mentions that in a paper read on 8 July 1834, there is a statement to the effect that Gould's monograph *Ramphastidae* 'is just completed' (*Proc. Zool. Soc. London*, 1824, p. 79).

Thus, from the letter of 30 April 1834 and this paper read on 8 July, we can fix the approximate date for the publication of part 2 as between 15 May and 8 July 1834.

PART 3 As seen in a letter of 28 September 1835 to Jardine (Newton Library) Gould again wrote that he had revised his thinking and decided to add a third part.

> I have also nine new species [from my trip to the Continent] to add to my Monograph of the Ramphastidae, which will appear in the form of a supplement in the course of a few weeks. This vast addition to the tribe cannot but be acceptable to all those who take an interest in Ornithology particularly as many of them are additional examples of a peculiar form and style of colouring of which we had not before received one or two species.

D. W. Mitchell in his 1841 review, as related earlier, stated that this work 'commenced in December 1833 and finished in December 1835'.

In a letter of 29 February 1836 from Gould to Jardine (Newton Library), Jardine is billed for 'Pt. 3 Toucans 2.0.0.'. Thus Part 3 had been published by that February date. Based on this information, and especially that from Mitchell's review, I would place the date of publication of the third and final part as in December 1835.

Contents of separate parts: The species figured in each part are listed completely in Zimmer as taken from a copy in the library of the Munich Museum.

Review: The entire work of three parts is reviewed at length [by Jardine?] in the 1837 issue of *Magazine of Zoology and Botany* (1:187-92), and by D. W. Mitchell in 1841.

4 1835–1838. *A monograph of the Trogonidae, or family of trogons*, 1 volume, 3 parts, 36 plates.

In the University of Kansas Spencer Library copy Ellis F-159 is the monograph in the three parts as issued (Figure 5). The only other reference to a copy in this form

PART TWO JOHN GOULD'S MAJOR PUBLISHED WORKS

is an incomplete set containing parts 2 and 3, belonging to Len Audubon of Australia. Iredale (1951) writes interestingly of this association of Gould with John James Audubon, and then proceeds to collate the two parts in hand.

Prospectus: Not seen.

Number of plates: Thirty-six. Part 1, eleven plates; part 2, twelve plates; part 3, thirteen plates.

Dates of three parts:

PART 1   Letter of 16 January 1834, from Gould to Jardine (Roy. Scot. Mus.) 'I am now paying some attention to the Trogons and will ere long write you a few lines

Figure 5 Cover for the first part of Gould's *Trogonidae* (University of Kansas Spencer Library-Ellis Collection).

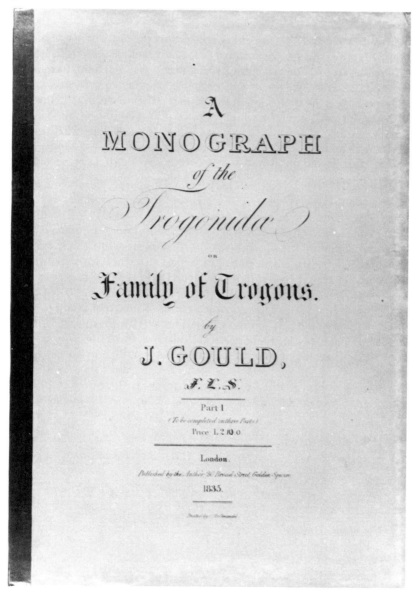

on the subject'. In the letter of 27 January 1835 from Jardine to Selby (Newton Library, pers. corres. from Christine Jackson) Jardine writes that Gould 'had with him here [St. Boswell, Roxburgh, Scotland], plates of his monograph of the *Trogons* by far the best plates he has yet done'.

The letter of 17 March 1835 from Gould to Jardine (Newton Library) contains the statement that 'My work on the *Trogonidae* will appear in about a fortnight'. Mitchell (1841) wrote in his review also that this work was 'begun in April 1835'. Thus, part 1 was published around 1 April 1835. In the letter of 29 February 1836 to Jardine (Newton Library), Gould lists a bill for 'Pt 1 Trogons ⸺ 2.10.0'.

PART 2 Gould wrote in a letter of 29 February 1836 to Jardine (Newton Library) 'the second part of the Trogons will be published in a few days'.

In a letter of 15 [?] March 1836 to Jardine (Newton Library) Gould said he had sent 'a small parcel for you containing the 2nd Part of my Monograph of the Trogonidae which has appeared this day... The trogons will be completed in 3 Pts. at £2.10 each.'

Iredale (1951) reports that on the part 2 copy given to or exchanged with J. J. Audubon, the cover bears at the top 'J. J. Audubon, March 21st, 1836, London' in three lines of Gould's handwriting. Thus, Part 2 was published around 15 March 1836.

PART 3 Gould writes in the letter of 14 March 1838 to Jardine (Newton Library) 'I have this day sent a parcel to Highley's for you containing the third concluding part of the Trogons and the second part of the Birds of Australia.'

From D. W. Mitchell (1841) we have the information that it was 'finished March 1838'. Thus, part 3 was published on or before 14 March 1838.

Contents of separate parts: Iredale (1951) has collated the plates and text from J. J. Audubon's personal copy of parts 2 and 3, and extrapolated for the missing part 1. The numerical order of the plates in part 1 was not known to Iredale.

From the Kansas-Ellis Collection copy of the complete three parts as issued I am able to give the order of the species for the first part. Unfortunately, the order of the species as listed by Iredale for the third part does not correspond to the order in the Ellis copy. The only adequate way to handle this difference is to list all of the plates, with the now known order for part 1 and then for part 3 list the numerical order for both the Iredale-Audubon copy and for the Kansas-Ellis copy. Perhaps someone will find a third copy that agrees with either one of these and thus put an end to the confusion. Each of the three separate parts of the Ellis copy is firmly bound in printed boards. Iredale makes no comment on whether or not Audubon's copy was bound.

In the following table it will be seen that for part 3, the Ellis plates follow in numerical order. This would add a modicum of support to the Ellis order being correct.

Bound plate number	Species name	Part as issued	Plate in part Ellis	Plate in part Audubon	Date of issue
1	*Trogon mexicanus* (adult male)	1	1		1 April 1835
2	*T. mexicanus* (young adult and female)	1	2		1 April 1835

PART TWO JOHN GOULD'S MAJOR PUBLISHED WORKS

Bound plate number	Species name	Part as issued	Plate in part Ellis	Plate in part Audubon	Date of issue
3	*T. elegans*	1	4		1 April 1835
4	*T. ambiguus*	1	8		1 April 1835
5	*T. collaris*	2	11	11	15 March 1836
6	*T. variegatus*	2	4	4	15 March 1836
7	*T. caligatus*	3	1	2	March 1838
8	*T. atricollis*	1	3		1 April 1835
9	*T. meridionalis*	2	12	12	15 March 1836
10	*T. melanopterus* (male and female)	2	10	10	15 March 1836
11	*T. melanopterus* (young male)	2	8	8	15 March 1836
12	*T. melanocephala*	2	6	6	15 March 1836
13	*T. citreolus*	3	2	7	March 1838
14	*T. aurantius*	3	3	3	March 1838
15	*T. surucura*	2	5	5	15 March 1836
16	*T. massena*	3	4	1	March 1838
17	*T. macroura*	3	5	11	March 1838
18	*T. melanura* (erroneously on plate as *T. nigricaudata*)	3	6	12	March 1838
19	*T. (Temnurus) albicollis* (on plate as *Trogon Temnurus*)	2	1	1	15 March 1836
20	*T. (Tem.?) roseigastor*	3	7	5	March 1838
21	*T. (Calurus) resplendens*	1	6		1 April 1835
22	*T. (Cal.) pulchellus* (on plate as -elus)	3	8	4	March 1838
23	*T. (Cal.) pavoninus*	2	7	7	15 March 1836
24	*T. (Cal.) fulgidus*	3	9	6	March 1838
25	*T. (Cal.) neoxenus*	3	10	10	March 1838
26	*T. (Apaloderma) narina*	2	2	2	15 March 1836
27	*T. (Apalo.) reinwardtii*	1	9		1 April 1835
28	*T. (Apalo.) gigas*	3	11	13	March 1838
29	*T. (Harpactes) temminckii*	1	11		1 April 1835
30	*T. (Har.) diardii*	2	9	9	15 March 1836
31	*T. (Har.) malabaricus*	1	5		1 April 1835
32	*T. (Har.) duvaucelii*	2	3	3	15 March 1836
33	*T. (Har.) erythrocephalus*	1	7		1 April 1835
34	*T. (Har.) hodgsonii*	3	12	8	March 1838
35	*T. (Har.) ardens*	3	13	9	March 1838
36	*T. (Har.) oreskios*	1	10		1 April 1835

   The prices on the printed board covers of the Kansas–Ellis parts are corrected in ink on part 1 from £2.12.0 to £2.10.0. On part 2 the price is written in ink '2.10.0', and on part 3 it is also written in ink '2.18.0'.

Reviews: This work on the *Trogonidae* was reviewed by D. W. Mitchell in 1841.

5 1837–1838. *A synopsis of the birds of Australia, and the adjacent islands*, 1 volume, imperial octavo, 4 parts, 73 plates.

The 73 plates contain figures mainly of the heads of birds. The University of Kansas Spencer Library–Ellis Collection has item Ellis Aves E62 which is the *Synopsis* in its original four parts as issued (Figure 6). This is a rare item but other copies must exist. Whittell (1954) apparently had access to a copy in the four parts.

Another copy (Ellis Aves D185) in the Ellis Collection is bound. On the 'Contents' page of this bound volume the four separate parts are listed, with a date of publication and a list of all the species included in each part. The confusing aspect is that 'Part II' on this 'Contents' page is dated as having been published 'January, 1837'. This is in error, but it accounts for Zimmer quoting this date for part 2.

Collation of the title page of the Kansas-Ellis Aves E62 unbound copy will be unnecessary since the first part is illustrated here. In this copy, part I is dated January 1837, part II is dated April 1837, and parts III and IV both are dated April 1838. Parts I and II have advertising of Gould's works on the back cover but the back covers of parts III and IV are blank. At the end of part IV are eight pages of a 'Description of New Species of Australian Birds'.

Prospectus: Bound in part I of the *Synopsis* of the Ellis Aves E62 copy is the Prospectus, undated (Figure 7).

Number of plates: Seventy-three plates representing the heads of 168 birds. Part 1, nineteen plates; part 2, eighteen plates; part 3, eighteen plates; part 4, eighteen plates. Gould stated in the Prospectus that each part would contain eighteen plates. All the parts have tan-coloured covers.

Dates of publication of the parts: Information on these publication dates is found in the Ellis copies and in Gould's correspondence.

PART 1 In the Prospectus is this statement:
> The First Part will be published on the 2nd of January, 1837, and will comprise Illustrations and Descriptions of Forty-four Species... the price of each Part, £1.5s. coloured, 15s. uncoloured; to appear at intervals of Three Months... it is impossible to state the precise number of parts to which the work will extend... Should the present publication meet the degree of support to which the Author trusts its merit will entitle it, he... contemplates visiting Australia, New Zealand, etc. for the space of two years, in order to investigate and study the natural history of those countries.

The date on the cover of part 1 is January 1837.

In a letter of 16 January 1837 from Gould to Jardine (Newton Library) he writes:
> I have this day forwarded to you direct by coach a small box containing... the first Part of my Synopsis ... my Synopsis will contain as many more new species as have yet been characterized... What do you think of the Synopsis... I think it might be extended to other countries and ultimately form the general history I had in contemplation.

(For an additional comment on this contemplated 'general history' see the letter of 29 February 1836 under *Icones avium*.)

Gould also wrote to William Swainson on 21 January 1837 (Linnean Society):
> Will you do me the favour to accept the first part of my Synopsis of the Birds of Australia, which appeared on the first of the present month. I hope to render this little book useful both to the collector and the scientific ornithologists. I think it will also be peculiarly adapted for the use of Curators of Museums, etc.
> You are perhaps aware that I have two of Mrs. Gould's brothers in Australia engaged in collecting the natural product of that fine country, nearly the wholeeof which are consigned to myself... I think I shall add at least as many more species as are at present known... I have a larger publication on the same subject in hand and I would hope to visit the southern hemisphere.

PART TWO JOHN GOULD'S MAJOR PUBLISHED WORKS

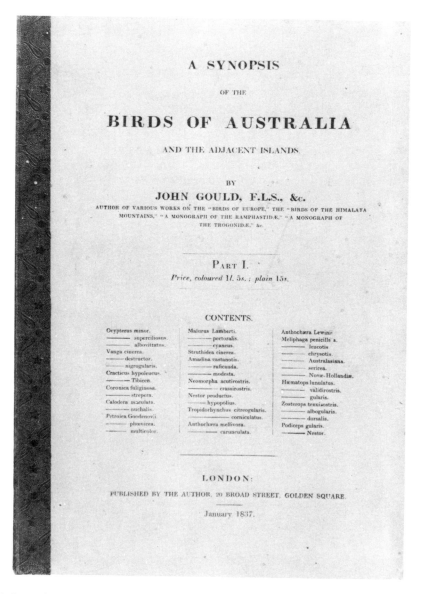

Figure 6 Cover for the first part of Gould's *Synopsis* (University of Kansas Spencer Library-Ellis Collection).

Further comment by Gould on his *Synopsis* is in the letter of 27 February 1837 to Jardine (Newton Library):

I am obliged to you for your last letter and also for your remarks on the Synopsis which I am quite certain were communicated in the true spirit of kindness. The work might perhaps have had a more appropriate Title but I was really at a loss what to call it, being anxious to avoid giving it a too voluminous one, lest it might interfere too much with my larger publication on the same subject. It may also appear dear but the coloring is really so expensive that I could not in justice to myself publish it at a lower rate. Are you aware that it may be had uncolored for 15s. and the Letterpress alone for 5? Your suggestion of placing more than 3 heads on a Plate would have diminished the price very little as I should have been obliged to charge a higher price in order to cover the extra expense of colouring, and letterpress. I am happy to add that it is very much approved of and am led to believe it will have an extensive circulation.

> PROSPECTUS.
>
> A SYNOPSIS
> OF THE
> **BIRDS OF AUSTRALIA,**
> AND THE ADJACENT ISLANDS:
> By JOHN GOULD, F.L.S., &c.
> AUTHOR OF VARIOUS WORKS ON THE BIRDS OF EUROPE, THE BIRDS OF THE
> HIMALAYA MOUNTAINS, A MONOGRAPH OF THE RHAMPHASTIDÆ OR
> FAMILY OF TOUCANS, A MONOGRAPH OF THE TROGONIDÆ
> OR FAMILY OF TROGONS, &c.
>
> THE science of Ornithology has now become so popular that it has attracted within its influence the talents of many scientific men in all parts of the world, and we find several of these highly gifted individuals engaged in illustrating particular portions of this interesting branch of study. Thus, independently of his works on general science, Mr. Swainson is engaged on the Birds of Brazil and Western Africa; Mr. Audubon on the Birds of the United States of America; Dr. Rüppell of Frankfort on those of Abyssinia; while the deservedly celebrated M. Temminck is occupied upon the Birds of Japan, besides adding yearly to the stores of science by recording new species in his "Planches Coloriées"; and M. Natterer, after a residence of sixteen years in the Brazils, has just returned to Vienna with an exceedingly rich collection of the zoological productions of that vast country, the novelties comprised in which will in all probability be immediately made known. Thus while we find the Ornithology of almost every other portion of the globe occupying the attention of various talented individuals, that of Australia and its islands, although not forgotten, remains almost unheeded. When we reflect upon the almost continental extent of this distant country, one of our most important and rapidly growing colonies, and upon the little that has yet been done towards elucidating its natural history *, the Author trusts that the present work will not be deemed uncalled for; more especially as not only are we less acquainted with the natural forms inhabiting this portion of the globe; differing as they do so widely from all others, whether belonging to the Old or to the New World, but less interest seems to have been excited by a knowledge of the few extraordinary species with which we are acquainted than their strange and anomalous nature would seem to warrant; among the Mammalia we may instance the Ornithorhynchus,
>
> * The only works of any importance on the subject being Lewin's Birds of New Holland, comprising only a few Species; the masterly essay on the same subject by Messrs. Vigors and Horsfield in the 15th vol. of the Linn. Trans., and occasional Memoirs dispersed in various scientific publications.

Figure 7 a & b Prospectus for Gould's *Synopsis* (University of Kansas Spencer Library-Ellis Collection).

The whole of the Birds procured by the Gents. proceeding on the new expedition to the interior of Australia are promised me which will of course render it so much the more complete. The Beagle is commissioned and will sail in a month or so. I am preparing a few copies of the 2nd Part of the Synopsis purposely that they may be taken out with them: it will contain many new and interesting species. Full generic characters will be given at or near the close of the work (this is named in the Prospectus); my reason for not giving them now is that I intend to devote a page to that purpose and to add some little history of the group, the number of species, their affinities, etc. and perhaps, a few words on their habits.

Part 1 of the *Synopsis* was reviewed in the 1837 *Magazine of Zoology and Botany* (1:571-2) and in the January 1837 issue of the *Magazine of Natural History* (51-2).

Plate 7 Drawing in pencil and watercolour, attributed to Mrs Gould, of *Falco frontatus*, for *Birds of Australia*, I: 10, Kansas-Ellis-Gould MS 496.

Plate 8 Drawing in pencil, ink and watercolour, attributed to Mrs E. Gould, of Australian Goshawk, for *Birds of Australia*, I: 17, Author's Collection MS 16GS. This fine drawing is listed in both the Henry Sotheran, Ltd No. 9 and No. 32 'Piccadilly Notes', as item 780 and 2608 respectively, and was attributed to Mrs Gould. This was purchased in 1954 for £8.8s.

PART TWO   JOHN GOULD'S MAJOR PUBLISHED WORKS

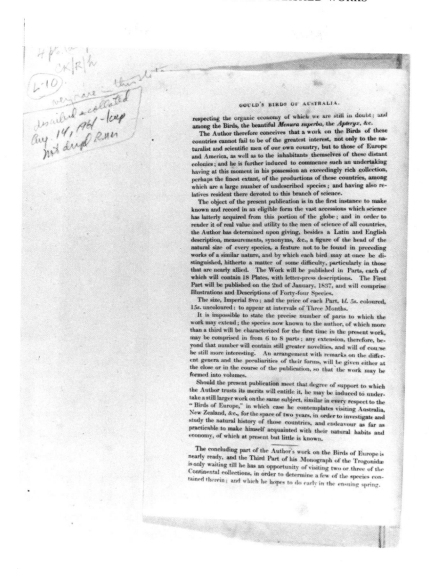

PART 2   From the above letter and the reviews it is obvious that part 2 was not published with part 1 as stated by Zimmer, but separately around the end of March 1837 when the *Beagle* was to sail. Ellis Aves E62 is dated April 1837.

Part 2 is reviewed in the 1838 *Magazine of Zoology and Botany* (2: 266-7): 'We have illustrations of the characters of forty species [in] this peculiarly managed work.' It is also reviewed in the May 1837 *Magazine of Natural History*, page 270.

PARTS 3 AND 4   Parts 3 and 4 in the Kansas copy (Ellis Aves E62) are dated April 1838 on the covers. In a letter of 10 August 1838 from Edwin Prince, Gould's secretary, to Jardine (Roy. Scot. Mus.) we find: 'Mr. Yarrell will also bring you Parts 3 & 4 of the Synopsis of the Birds of Australia which I hope will merit your approbation.'

Contents of separate parts: When this work was finally bound as one volume, the four parts were bound consecutively. Therefore, no separate listing of the order of the plates is necessary.

In the University of Kansas Spencer Library copy (Ellis Aves D185), which is a bound copy, the species for each part are listed in a 'Contents' page. For part 1, forty-five species are listed, for part 2, forty species, for part 3, thirty-six species, and for part 4, forty-three species, or a total of 164 species. On counting the illustrations of the heads, however, there are 165 figured. One plate has the male and female of *Polytelis barrabandi* figured separately.

Whittell (1954) and Mathews (1920–25) list the *new* species in each part.

At the conclusion of part 4 is a 'Description of New Species of Australian Birds' on eight pages. Gould's paper on these birds was presented to the Zoological Society of London on 5 December 1837 but was not published until a year later, on 7 December 1838. Both Zimmer and Whittell comment on this and state that the several new genera and thirty-six new species described thus date from Gould's part 4 published in April 1838.

Reviews: These are mentioned earlier under parts 1 and 2. I have not found any reviews for parts 3 and 4.

6 1837–1838. *The birds of Australia, and the adjacent islands*, 1 volume, 2 parts, 20 plates.

The University of Kansas Spencer Library copy (Ellis F-107) is in the original two parts as issued (Figure 8).

Prospectus: Not seen.

Number of plates: Twenty. Part 1, ten plates; part 2, ten plates. Both parts have tan-coloured covers.

Dates of the [two] parts:
PART 1 In Gould's letter of 2 November 1836 he wrote to Jardine 'would not a work on the Birds of Australia be interesting? . . . [I] have an idea of making it my next illustrative work. I have even some serious intention of visiting the colony for two years for the purpose of making observations, etc. I have several plates of my new species already drawn and coloured' (Newton Library).

In Gould's next letter to Jardine, dated 16 January 1837 (Newton Library) he mentions Mrs Gould having two brothers in New South Wales and that 'I am slowly preparing a large work on the Birds of Australia'.

By the time of the writing of Gould's letter of 3 November 1837, to Jardine (Newton Library) the first part of his *Birds of Australia and the adjacent islands* had been published.

August 1837 is the date on the cover of the first part of this work in the Kansas–Ellis Collection (Ellis F-107).

Review of Part 1. This appeared in the 1838 *Magazine of Zoology and Botany* (2:357-8) and the ten plates are listed with minor comments.

PART 2 In a letter of 3 November 1837 Gould wrote 'Mrs. Gould is just placing a drawing of the *Apteryx* on stone which will appear in the next part of the Birds of Australia'.

PART TWO JOHN GOULD'S MAJOR PUBLISHED WORKS

On 14 March 1838 Gould wrote to Jardine (Newton Library) 'I have this day sent a parcel to Highley's for you containing the...second part of the Birds of Australia.'

The second part was supposedly published in February 1838, according to the date on the title-page of the Kansas–Ellis Collection copy, but March may be more correct.

Review of Part 2. This appeared in the 1838 issue of *Annals of Natural History* (1: 223-4). The ten birds figured are commented upon and listed. The author [Jardine?] also alludes with enthusiasm to Gould's forthcoming Australian trip.

Contents of the [two] parts: The species are listed on the covers of both parts in the Kansas copy (Ellis F-107), and they are listed in the reviews. Zimmer records that plates 6, 7, 9 and 10 of part one, and plates 3, 4, 8, 9 and 10 of part two were

Figure 8 Cover for the first part of Gould's *Birds of Australia, and the adjacent islands* (Cancelled Parts), (University of Kansas Spencer Library-Ellis Collection).

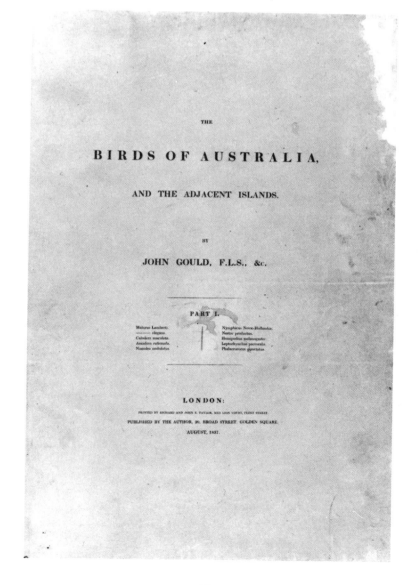

reprinted with occasional minor alterations in colours in Gould's later work on the *Birds of Australia*, but the remainder of the eleven plates were redrawn.

This 1837-38 issue of two parts was truly a 'suppressed' or 'cancelled' work. In Gould's letter of 2 July 1841 to Jardine (Roy. Scot. Mus.) he writes:

I am rather surprized that you were not aware that the two former parts of the work are received back from the subscribers and the price allowed off the new series, such plates only then again worked into the publication as are accurately drawn. By this you will perceive that I am a looser [sic] and that to a considerable amount, and not the former subscribers. When an opportunity occurs for you to return yours I shall be glad to receive them.

Further on this subject, in a letter of 5 February 1844 to Lord Derby, (Wagstaffe & Rutherford, letter number 17) Gould wrote:

In some instances I received those cancelled parts back from my Subscribers, and gave them the first part of the new Series in lieu of them. In others they were retained... the cancelled parts, may, if you please, be bound together, and entitled "Illustrations of Birds from Australia["], and from the few copies in the hands of the Public, they will some day be of value, though, it is true, more to the Book-Collector, than the Naturalist.

This statement was profoundly prophetic.

Reviews: As mentioned under parts 1 and 2.

7 1837–1838. *Icones avium, or figures and descriptions of new and interesting species of birds from various parts of the globe* ['*world*' in part 2], 1 volume, 2 parts, 18 plates.

A copy in the University of Kansas Spencer Library–Ellis Collection is in the two parts as issued (copy Ellis F-161) (Figures 9 & 10).

Prospectus: Not seen.

Number of plates: Eighteen. Part 1, ten plates; part 2, eight plates.

Dates of the [two] parts: Preliminary information. This work was in Gould's mind for at least two years before it was published. In a letter of 17 March 1835 Gould wrote to Jardine (Newton Library) 'This work will appear under the title of Supplement to my other works'.

In his letter of 29 February 1836 to Jardine (Newton Library) he writes:

After the Birds of Europe are completed, I propose (in accordance with the requests of several scientific friends) to commence a general history of Birds, a most laborius [sic] undertaking you will say. If I prosecute this intention I shall illustrate one species of every genus (dividing much farther than has hitherto been done) and that species to be far as possible, either entirely new or hitherto unfigured: the details of the remaining species to be merely descriptive: the size Imperial Folio and Imp. 4to. I have for a long time had such a work in contemplation but have hitherto refrained from commencing it in consequence of the arduous nature of the undertaking: particularly as I wish to continue a series of Monographs. I have several drawings already engraved which I can bring out at a short notice and as we have acknowledged that there is much to be done that we may both be employed I shall be most happy to render you any assistance in my power: if agreeable to you perhaps it would be advisable for each of us to advertize the other of what birds we intend to publish in the ensuing nos. of our publications and by this means obviate the chance of our each figuring the same species. I shall never publish more than one of a form. The plates I have ready and which will probably appear in the first Pt. comprise the genera *Yunx, Sialia, Paradoxornis, Eurylaimus (Crossodera) Dalhousiae, Nanodes, Todus, Numida, Accentor,* & *Pitta*.

In the letter of 15 [?] March 1836 to Jardine (Newton Library) Gould writes 'I also enclose for your inspection some drawings which will form a part of my new works. Perhaps you will oblige me with your opinion respecting them'. In the 1837 issue of the *Magazine of Zoology and Botany* (1: 108-9) a condensation of these last two letters appears as an Announcement for a 'New general History of Birds' and

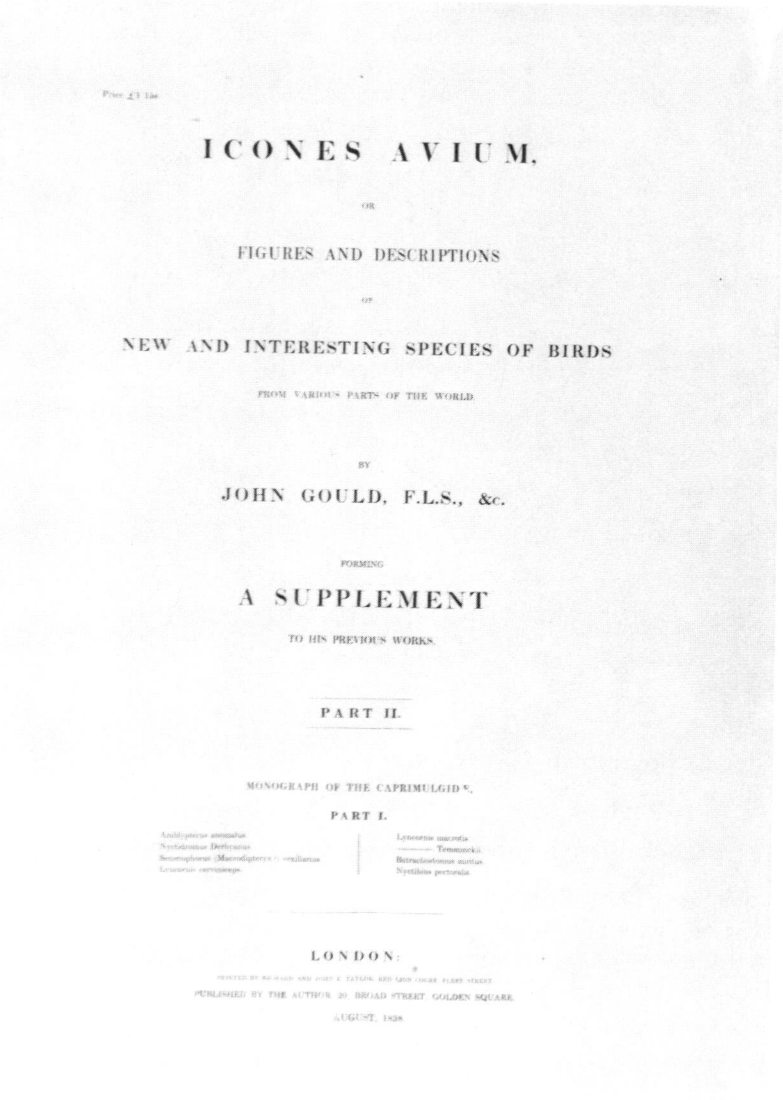
ICONES AVIUM,

OR

FIGURES AND DESCRIPTIONS

OF

NEW AND INTERESTING SPECIES OF BIRDS

FROM VARIOUS PARTS OF THE WORLD.

BY

JOHN GOULD, F.L.S., &c.

FORMING

A SUPPLEMENT

TO HIS PREVIOUS WORKS.

PART II.

MONOGRAPH OF THE CAPRIMULGIDÆ,

PART I.

Amblypterus anomalus.
Nyctidromus Derbyanus.
Semeiophorus (Macrodipteryx?) vexillarius.
Lyncornis cervinieps.

Lyncornis macrotis.
——— Temminckii.
Batrachostomus auritus.
Nyctibius pectoralis.

LONDON:

PRINTED BY RICHARD AND JOHN E. TAYLOR, RED LION COURT, FLEET STREET.

PUBLISHED BY THE AUTHOR, 20, BROAD STREET, GOLDEN SQUARE.

AUGUST, 1838.

Figure 9 Cover for the first part of Gould's *Icones avium* (University of Kansas Spencer Library-Ellis Collection).

the author, obviously Jardine, states that 'The plates we have seen are lithographic, drawn by Mrs. Gould, and beautifully executed.'

PART 1 In a letter of [?] August 1837 from Gould to Jardine (Newton Library) there is evidence that part 1 of *Icones Avium* had been published: 'I send for your inspection one of my Box Books in which is a portion of my Gen. History.'

The date on the cover of part 1 of the Kansas copy (Ellis F-161) is August 1837.

A review of part 1 was in the 1838 issue of *Magazine of Zoology and Botany* (2: 357-8) and the ten species were listed and commented upon.

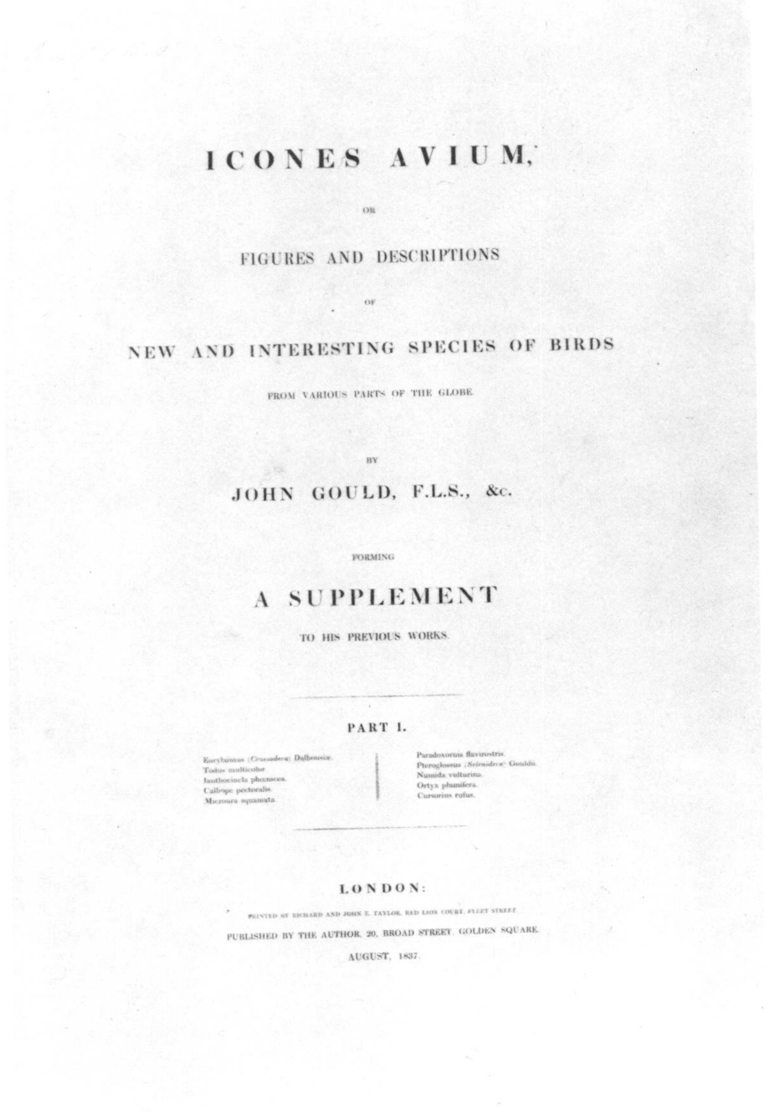

ICONES AVIUM,

OR

FIGURES AND DESCRIPTIONS

OF

NEW AND INTERESTING SPECIES OF BIRDS

FROM VARIOUS PARTS OF THE GLOBE.

BY

JOHN GOULD, F.L.S., &c.

FORMING

A SUPPLEMENT

TO HIS PREVIOUS WORKS.

PART I.

Eurylaimus (*Crossodera*) Dalhousiæ.
Todus multicolor
Ianthocincla phœnicea.
Callrope pectoralis.
Microura squamata.

Paradoxornis flavirostris.
Pteroglossus (*Selenidera*) Gouldii.
Numida vulturina.
Ortyx plumifera.
Cursorius rufus.

LONDON:

PRINTED BY RICHARD AND JOHN E. TAYLOR, RED LION COURT, FLEET STREET.

PUBLISHED BY THE AUTHOR, 20, BROAD STREET, GOLDEN SQUARE.

AUGUST, 1837.

Figure 10 Cover for the second part of Gould's *Icones avium* on the *Caprimulgidae* (University of Kansas Spencer Library-Ellis Collection).

PART 2 The earliest mention of this work on the *Caprimulgidae* or Goatsuckers was in a letter of 3 November 1837 from Gould to Jardine (Newton Library): 'I am deep among the *Caprimulgidae* and have collected together a great many species among which are some of extreme interest, in fact I do not know that I have ever been so much interested in any group before . . . I have already drawn some of them . . . Will you be so good as to send me all the *Caprimulgidae* you possess to compare with mine . . . '

In the letter dated 20 November 1837 (Newton Library) Gould acknowledges receipt of some specimens.

PART TWO JOHN GOULD'S MAJOR PUBLISHED WORKS

In a letter of 14 March 1838 (Newton Library) Gould states 'The specimens of *Caprimulgus* which you have lent me shall be returned before I go [to Australia] and I hope also to have a part of the work ready which I may probably get you to exhibit at the New Castle meeting [of the British Association for the Advancement of Science] and plead an excuse for my not being there as promised.'

According to the Report of the 1837 meeting of the British Association, 'The following reports and monographs were requested... on the Caprimulgidae, by Mr. Gould.' Thus, this request by Gould that Jardine exhibit part of his work at the Newcastle meeting in 1838.

In the next letter of 10 August 1838 to Jardine (Roy. Scot. Mus.), Edwin Prince, Gould's secretary, writes that Gould wished to be excused for his apparent neglect in writing Jardine before he left for Australia, but would

express his hope that you would... oblige him by having the goodness to lay the first Part of his Monograph of the Caprimulgidae before the meeting of the British Association... make the best apology you can for its not being farther advanced in consequence of the subject proving far more voluminous than was originally supposed [even Gould had this problem] and his leaving England... I have had the Plates carefully printed and colored by the original drawings and the Part will be ready on Monday [August 13] next.

Thus part 2 of *Icones Avium* appeared around 13 August 1838. The date on the Kansas copy (Ellis F161) is August 1838.

As a follow-up on this second part in a letter dated 20 July 1839 (Roy. Scot. Mus.) Prince wrote 'I presume you were kind enough to lay the 1st Part of the *Caprimulgidae* before the British Association but as I have found no notice of it in the Reports of the Meeting and have not had the pleasure of hearing from you since I could not allude to this subject in my Letters to Mr. Gould.' True, there is no evidence that Jardine carried out Gould's request.

A review of part 2 appeared in the 1839 *Annals and Magazine of Natural History* (2: 222-3). There were eight species figured and described, with no less than five new generic names.

That Gould wished to continue publication of parts of *Icones Avium* is evident from a letter of 19 January 1841 to Jardine (Roy. Scot. Mus.) where Gould wrote 'The Birds I spoke of at India House are very curious. Several of them new, two additional species of *paradoxornis*. Dr. Horsfield and I are talking of publishing them in my "Icones Avium"'.

Further on the subject of a projection of *Icones Avium* is the letter of 5 February 1844 from Gould to Lord Derby (Wagstaffe & Rutherford, letter number 17): 'The *Icones Avium* is a Work which will always be in progress. A Title Page for a Volume will be given when 80 or 100 plates have been published... The third part is nearly ready, and will contain nine new species of Toucans'.

Also in a letter to Lord Derby on 3 January 1851 (Wagstaffe and Rutherford, letter 26) Gould still had *Icones Avium* in his plans: '*Balaeniceps Rex*... I shall figure it of the Natural Size in my *Icones Avium*'.

Contents of the two parts: The species are listed on the covers of the two parts in the Kansas copy (Ellis F161), and in the two reviews, and also in Zimmer (1926). D. W. Mitchell (1841) also included this work in his long review.

8 1838-1841. *Charles Darwin. The zoology of the voyage of H.M.S. Beagle. Birds, Part III, John Gould, Esq. F.L.S.*, 1 volume, 5 numbers, 50 plates, medium quarto.

The University of Kansas Spencer Library-Ellis Collection copy (Ellis Aves E10) is bound in five volumes. The five numbers on birds are in volume three.

Prospectus: Not seen.

Number of medium quarto size plates: Fifty. Each of the five numbers contained ten plates and text (Sherborn 1897). There is no reference to the artist, lithographer or printer on any of the coloured plates. The measurement of the Ellis Aves E10 copy trimmed is 12 3/16 × 9 5/8 inches wide (309 × 244 mm).

Dates of publication of the numbers: Preliminary information. Darwin had returned on H.M.S. *Beagle* on 2 October 1836. Gould's letter of 16 January 1837 to Jardine (Newton Library) contains the first extant mention of the fact that Darwin asked Gould to describe his bird collection:

I must not omit to tell you that Mr. Darwin's Collection of Birds (made during the late survey under Capn Fitzroy) are exceedingly fine: they are placed in my hands to describe: some of the forms are very singular particularly those from the Gallipagos [sic]. I have one family of ground Finches in which there are 12 or 14 species all new: many of the Magellanic Birds however are now being described by D'Orbigny* whose work you have of course seen.

Darwin's account of Gould's role in this volume on the birds appears in the Advertisement on pages i and ii:

When I presented my collection of Birds to the Zoological Society, Mr. Gould kindly undertook to furnish me with descriptions of the new species and names of those already known. This he has performed, but owing to the hurry, consequent on his departure for Australia, — an expedition from which the science of Ornithology will derive such great advantages, — he was compelled to leave some part of his manuscript so far incomplete, that without the possibility of personal communication with him, I was left in doubt on some essential points. Mr. George Robert Gray, the ornithological assistant in the Zoological department of the British Museum, has in the most obliging manner undertaken to obviate this difficulty, by furnishing me with information with respect to some parts of the general arrangement, and likewise on that most intricate subject, — the knowledge of what species have already been described, and the use of proper generic terms... As some of Mr. Gould's descriptions appeared to me brief, I have enlarged them, but have always endeavoured to retain his specific character; so that, by this means, I trust I shall not throw any obscurity on what he considers the essential character in each case; but at the same time, I hope, that these additional remarks may render the work more complete.

The accompanying illustrations, which are fifty in number, were taken from sketches made by Mr. Gould himself, and executed on stone by Mrs. Gould, with that admirable success, which has attended all her works.

Number I of Part III Birds, being number 3 of the total of nineteen parts as issued, contained pages 1 to 16, ten plates, and appeared in July 1838 (Freeman 1965).

Number II, being number 6 as issued, contained pages 16 to 32, ten plates and appeared in January 1839.

Number III, being number 9 as issued, contained pages 33 to 56, ten plates and appeared in July 1839.

Number IV, being number 11 as issued, contained pages 57 to 96, ten plates, and appeared in November 1839.

Number V, was number 15 as issued, contained pages 97 to 164, ten plates, and appeared in March 1841. (The information on these numbers is taken from Freeman, (1965) and Sherborn (1897).)

Contents of the numbers: The plates conform to the text pages as issued from July 1838 to March 1841. Zimmer (1926) comments at length on the proper accreditation for the generic and species names, i.e. between Gould, Darwin or G. R. Gray.

The species names for the entire fifty plates are listed by Coues in the 1879 second instalment of his *Ornithological Bibliography*, pages 258-9.

---

*1835–47, *Voyage dans l'Amérique médidionale*... Alcide D. D'Orbigny wrote the section on birds, with sixty-seven coloured plates.

Plate 9 Drawing in pencil and chalk, attributed to John Gould, of *Melithreptus melanocephalus*, for *Birds of Australia*, IV: 75, Kansas-Jardine-Gould MS 1393.

Plate 10 Drawing in pencil, ink and watercolour, attributed to Mrs E. Gould, of unidentified *Turnix*, unpublished [Australian], Kansas-Sauer-Gould MS 2GS. This drawing is listed in the Henry Sotheran Ltd 1937 Catalogue No. 22 'Piccadilly Notes' as item 2627, by Mrs Gould, and was for sale for £6.6s.

Plate 11 Drawing in pencil and chalk, attributed to John Gould, of *Eudyptes chrysocome*, for *Birds of Australia*, VII: 83, Kansas-Jardine-Gould MS 1521.

Plate 12 Drawing in watercolour, attributed to H. C. Richter, of Rose-breasted Robin, for *Birds of Australia*, III: 2, Blacker-Wood, McGill.

Plate 13 Drawing in pencil, ink and watercolour, attributed to Mrs Gould, of Short-tailed Petrel, for *Birds of Australia*, VII: 56, Kansas-Sauer-Gould MS 18GS. This drawing is listed in both the Henry Sotheran, Ltd No. 9 and No. 22 'Piccadilly Notes', as item 791 and 2616 respectively, and was attributed to Mrs Gould. This was offered for sale and purchased for £7.7s.

In the right lower corner is the pencilled inscription 'Latt 39. 22S, Long. 57. 55W, J. & E. Gould'.

Plate 14 Drawing in pencil, ink and crayon, attributed to John Gould, of *Platycercus semitorquatus*, for *Birds of Australia*, V: 19, Kansas-Jardine-Gould MS 1420.

PART TWO   JOHN GOULD'S MAJOR PUBLISHED WORKS

9 1840–1848. *The birds of Australia*, 7 volumes, 36 parts, 600 plates (Figure 11).

Prospectus: Chisholm (*Emu* 42: 74-84) has written quite a detailed article with illustrations on a Prospectus dated at Sydney, 6 March 1840, and also mentions one published two years earlier in England.

Number of plates: 600. Gould wrote in his 1868 Prospectus that each part contained seventeen plates. This obviously is not correct because 612 plates would then have been issued. Whittell (1954, p. 288) and Waterhouse (1885) both state that part 36 had only nine plates. If the other thirty-five were to have seventeen each, this still leaves three in excess. An examination of an unbound set of the parts would solve this minor dilemma, but I have not seen such a set.

Dates of publication of the parts: The first part was scheduled to appear on 1 December 1840. Future parts were to appear every three months, namely, on 1 March, 1 June, 1 September, and 1 December. The four final parts, according to both Zimmer and Whittell, appeared on 1 December 1848. These dates were obviously those printed on the separate covers of the parts. Mathews (1925) conjectured that since the octavo 'Introduction' was published 1 August 1848, 'apparently pt. 36 was completed at that date, and apparently the last five parts were sent out soon after Sept. 1st, 1848, being dated Dec. 1st, 1848, for purposes of payment.'

Here follows a listing of the separate parts and the dates of publication, combined with any other information gleaned for Gould's correspondence.

PART 1   1 December 1840. Reviewed in 1841 *Annals and Magazine of Natural History* 6:337-9. It is mentioned that there were seventeen plates. It was also reviewed by D. W. Mitchell in 1841.

In the letter of 9 September 1840, a month after Gould's return from Australia, he wrote to Jardine (Newton Library): 'I am now busily engaged in preparing the first part of my work which I hope will be ready before Christmas, in fact I mean to publish it if I can by the 1st of December'.

In a letter to Sir John Franklin (Scott Polar Res. Inst.) of 4 January 1841, Gould wrote 'It is with great pleasure that I have this day shipped for Hobarton... a package containing the first part of my works on the Ornithology of Australia being the first portion of the results of my labours in that interesting country'.

In a letter to Prince Charles Lucien Bonaparte dated 2 February 1841 (Mus. Nat'l. d'Hist. Nat., Paris) Gould wrote 'Since my return I have been heavily employed in *recommencing* my work on the Birds of Australia... the first part of the Birds is before the public and was forwarded to you some time [ago].'

PART 2   1 March 1841. In the letter of 19 January 1841 Gould wrote to Jardine (Roy. Scot. Mus.): 'You will see in the next part that I have dedicated it by permission to the Queen and Her Ct [Consort] Prince Albert.'

Concerning part 2, in his letter of 13 February 1841 to Jardine (Roy. Scot. Mus.) Gould wrote: 'I am glad to say that the second part of the Birds [of Australia] will be ready by the first of March.'

'The second part of the work was left at Highley's before I left [London] I hope you have received it' — so wrote Gould to Jardine from Leiden on 4 March 1841 where Gould went to visit some collections of Kangaroos (Roy. Scot. Mus.). Jardine had not received part 2 when Gould wrote him on 29 March 1841 (Roy. Scot. Mus.): 'It was delivered to Highleys a month ago last Saturday.'

PART 3   1 June 1841.

PART 4   1 September 1841. Mrs. Gould died on 15 August 1841 after giving birth to their sixth living child. In a black-bordered letter of 1 October 1841 to Jardine (Roy. Scot. Mus.), the majority of which is concerned with his bereavement, Gould does write that he had sent part 4 of his *Birds of Australia* by parcel.

PART 5   1 December 1841. In a letter of 3 November 1841 to Hugh Strickland, (Cambridge Univ. Mus. of Zoology) Gould wrote 'I have delayed forwarding your parcel until this day in order that I might send Part 5 at the same time, having a few copies ready although the time of its appearing is not until the first of the next month.' Thus, this part was published almost a month earlier than scheduled. In the same letter Gould wrote further 'I shall be glad of a line saying how you like the present part; almost the last of the work of my Dear and never to be forgotten partner.'

PART 6   1 March 1842. A review of parts 1 to 6 appeared in the June 1842 *Ann. Mag. Nat. Hist.*, 9:337-9. It was lengthy and favourable.

Figure 11 Drawing in pencil, attributed to John Gould, of *Ardea leucophaea*, for *Birds of Australia*, VI: 55, Kansas-Jardine-Gould MS 1471.

PART TWO   JOHN GOULD'S MAJOR PUBLISHED WORKS

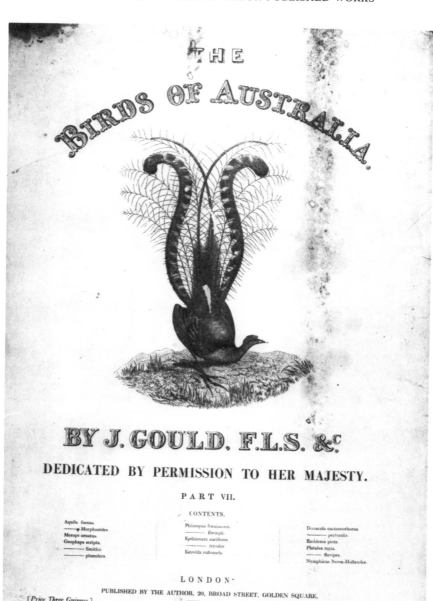

Figure 12 Cover for the seventh part of Gould's *Birds of Australia* (Author's Collection).

PART 7   1 June 1842 (Figure 12).
PART 8   1 September 1842.
PART 9   1 December 1842. Reviews of parts 8 and 9 appeared in the 1843 *Ann. Mag. Nat. Hist.*, 11:58.

In Gould's letter of 2 December 1842 to Jardine (Newton Library) he mentions 'a parcel for you containing parts 8 and 9 of the Birds of Australia'. The schedule called for part 9 to be published on 1 December 1842, so Gould's work was on schedule.

PART 10   1 March 1843.
PART 11   1 June 1843.
PART 12   1 September 1843.
PART 13   1 December 1843. Part 13 was published before a letter of 6 December 1843 to Strickland (Cambridge Univ. Mus. of Zoology) since Gould wrote 'before this [letter] reaches you you will have received the thirteenth part'.

As conjectured earlier, it must have been difficult for the subscribers to wait for years to get a work bound. In a letter of 5 February 1844 (Wagstaffe and Rutherford, letter number 17) Gould answered Lord Derby's concern with 'I must ask the great favour of your Lordships yet waiting a year or two longer before you attempt to bind it, as by that time I hope to have it so far advanced that I may be quickly able to give your Lordship and a few other Subscribers, who wish to get it bound, a scientific arrangement, Title, etc.'

PART 14   1 March 1844. A letter dated 14 March 1844 from Gould to Jardine (Newton Library) states 'Part 14 of the Birds of Aust. is out'.
PART 15   1 June 1844. Part 15 was sent to Lord Derby on 3 July 1844 (Wagstaffe, etc., letter number 22).
PART 16   1 September 1844.
PART 17   1 December 1844. Regarding Parts 16 and 17, in a letter to Bonaparte dated 26 November 1844 (Mus. Nat'l. d'Hist. Nat., Paris) Gould wrote 'I trust that my opus magnum, the "Birds of Australia," continues to please you. It is now far advanced ... About a month ago a package was sent to you ... containing among other things, parts 16 and 17 of the Birds of Australia ... I have now scarcely 150 good subscribers and 250 copies are all that ever will be printed.' Part 17 was scheduled for 1 December 1844 so Gould was quite early in publishing it around 26 October.

Parts 16 and 17 were sent around 1 December 1844 to C. J. Temminck according to Gould's letter to Temminck of 30 January 1845 (Rijksmuseum, Leiden).

PART 18   1 March 1845.
PART 19   1 June 1845.
PART 20   1 September 1845.
PART 21   1 December 1845. Gould was apparently referring to parts 20 and 21 due in September and December when he wrote to Jardine on 30 July [1845] (Newton Library): 'I am however just running the descriptions of the two autumnal parts of my work through the press and consequently shall be kept in town a short time longer.'
PART 22   1 March 1846.
PART 23   1 June 1846.
PART 24   1 September 1846.
PART 25   1 December 1846.
PART 26   1 March 1847.
PART 27   1 June 1847.
PART 28   1 September 1847.
PART 29   1 December, 1847.
PART 30   1 March 1848.
PART 31   1 June 1848.
PART 32   1 September 1848.
PARTS 33, 34, 35 AND 36   are all dated 1 December 1848. As stated before, Mathews conjectured that these four final parts may have been sent out around 1 September 1848. But it would have been almost a physical impossibility to have completed

the work by this time. As it was, by 1 December 1848, Gould was to complete the thirty-six parts nine months ahead of schedule.

Part 36 had only nine plates.

In the letter of 28 June [1848] Gould wrote to Jardine (Roy. Scot. Mus.) that he was 'much occupied with the completion of the birds of Australia'.

Contents of the parts: The date of issue by year for each plate is in F. H. Waterhouse's 1885 tabulation.

Reviews: Part 1 was reviewed, as related earlier, in the 1842 *Ann. Mag. Nat. Hist.* (6: 471-2), undoubtedly by Jardine.

Parts 1 to 6 were reviewed together in the June 1842 *Ann. Mag. Nat. Hist.* (9: 337-9). Some pertinent quotations from the reviews will be found in the bibliographic listing under 'Gould, John. Reviews'.

Figure 13 a & b Prospectus for the Sturm edition of J. Gould's *Monographie der Ramphastiden*, dated September 1840 (University of Kansas Spencer Library-Ellis Collection).

10 1841–1847. *J. Gould's Monographie der Ramphastiden oder Tukanartigen Voegel*, 1 volume, crown folio, 4 parts, 38 plates (36 coloured and 2 plain), published by the Sturm brothers in Nuremberg.

A copy of this work as issued in four parts is in the Ellis Collection at the Kenneth Spencer Research Library, University of Kansas, item Ellis E-15. This set was owned by Gould, and has his signature on each of the four parts. Included are two letters from the Sturm brothers to Gould. This item, with the letters, was listed in the 1932 Wheldon & Wesley Catalogue No. 28 as for sale for £8.

Prospectus: A prospectus dated September 1840 (Figure 13) is included in the set of four parts in the Ellis collection.

PART TWO JOHN GOULD'S MAJOR PUBLISHED WORKS

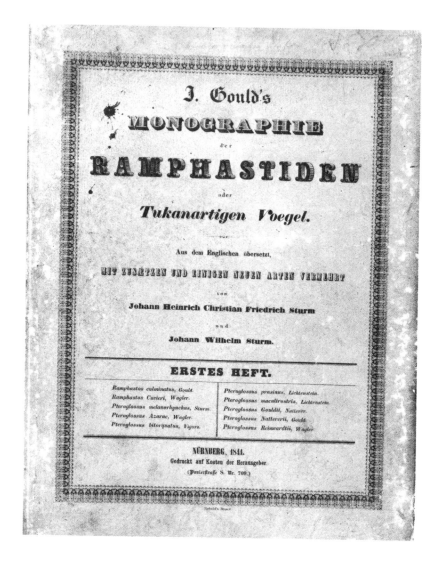

Figure 14 Cover for the first part of the Sturm edition of *Monographie der Ramphastiden* (University of Kansas Spencer Library-Ellis Collection).

Number of plates: Thirty-six, coloured. Parts 1, 2 and 3 contained ten plates each, and part 4 contained six plates.

Dates of the publication of the parts: In the Kansas–Ellis set the dates of 1841, 1841, 1842 and 1847 appear on the covers (Figure 14). But two letters from Frederic and John Sturm with the Ellis set contain more explicit information on publication dates.

In a three-page letter dated 31 July 1841 from Nuremberg the Sturms wrote that they would have acquainted Gould with their design to publish this work on the *Ramphastidae*, but he was in Australia. Now, on his return 'we take the liberty to send you the first and second volume of our German edition... which have just now been published.' Thus, parts 1 and 2 were published before 31 July 1841.

Gould replied that he was 'pleased with the execution of the first two volumes', but that their German edition 'found its way to Turin', and thus deprived him of a sale. In their reply to Gould of 12 November 1842, the Sturm brothers said they regretted that this had happened. Further in the letter they wrote 'we have the honour of sending you subjoined the third volume of our Ramphastidae just now published, it is the first copy we send away, and we should be happy if it would likewise receive your approbation. We believe it superior to the first number and are convinced that the addition[s] of Mr. Natterer are very valuable.' Thus, part 3 was published on or before 12 November 1842.

Part 4, according to the date on the cover of the part, was published in 1847.

Contents of the parts: Zimmer (1926) lists the species figured and described in each part. They are also on the covers of the four parts in the Kansas–Ellis Collection.

Review: In the *Ann. Mag. of Nat. Hist.*, A.W. [Alfred Waterhouse?] reviews the 1849 *Narrative of an expedition into Central Australia ... with ornithological notices by John Gould* (3 (second series): 56-61). In a final paragraph of the review, A.W. comments on the German publication of Gould's *Ramphastidae* and adds that 'it is a pity that so spirited and talented a man should not have all the results of the profit of such books'. A.W. was obviously an early advocate of more strict copyright laws.

According to Coues's *Ornithological bibliography* (third instalment, pp. 722-3), Heft II was reviewed in the 1842 *Isis* (35: 235-6); Heft III in the 1843 *Isis* (36: 558-9), and Heft IV in the 1848 *Isis* (41: 695).

Lying in this Kansas library German set was a third letter. It is dated 8 October 1851 and was written by W. Laurence to Gould. Apparently Laurence had been asked by Gould to translate the German description in these four parts. Laurence wrote 'my ignorance of the technical terms of ornithology has no doubt often led me into very awkward periphrases [sic], so that should you wish to incorporate any of the German addenda into your own work it would be highly necessary that I should with your assistance just look over the translation again before publication'.

11 1841–1842. *A monograph of the Macropodidae, or family of kangaroos*, 1 volume, 2 parts, 30 plates (Figure 15).

The University of Kansas Spencer Library–Ellis Collection has two copies. One copy (Ellis F-155) is in the original two parts as issued (Figure 16) and the second copy is bound with the two covers. Sharpe (1893) gives 1841–44 as the dates of publication, but 1844 is in error.

Prospectus: Not seen.

Number of plates: Thirty. The Kansas–Ellis Collection copy of this work in the two parts has fifteen plates in each part. Waterhouse (1885) also tabulates fifteen plates for each part. Sharpe (1893) erroneously states that there was a total of forty-five plates.

Dates of publication of the parts: Preliminary information. This work was not completed, so there was no title page, table of contents, introductory text or index. When the *Mammals of Australia* was published, Gould reprinted thirteen of these thirty plates without changes for this larger work, and on two more plates had only the positions of the animals altered. The remaining fifteen Kangaroo plates were totally revised for the larger work (Dickison *Emu* 38: 126).

PART TWO JOHN GOULD'S MAJOR PUBLISHED WORKS

On the cover of part 1 of the Kansas-Ellis Collection copy is the date 1 August 1841. On the cover of part 2 the date is 1 May 1842. From Gould's correspondence there is some additional information on the publication dates.

In a letter of 9 September 1840 from Gould to Jardine (Newton Library) he writes:

I had fine sport with these interesting animals [Kangaroos]; they are a most remarkable tribe of quadrupeds many possessing extreme beauty. I made it a point to collecting all the information I could respecting them and having been very successful in so doing, considering the meager accounts that are before the public, I have determined to publish a Monograph of the family with figures of every species as large as the size (folio) will admit... My collection alone contains 22 or 23 species of course including the small *hypsiprymni*. I have also a fine general collection of quadrupeds.

Gould went on to say that he had read a paper before the Zoological Society of London on 25 August where he 'exhibited and named 6 new species of Kangaroos'. (In the *Proceedings* of 1840, pages 92 to 94, only five new species are listed, however, not six.)

In Gould's letter of 4 March 1841 from 'Leyden' (Roy. Scot. Mus.), he writes: 'The fact is I could not with satisfaction to myself publish the first part of my Monograph of the Kangaroos until I had visited the Collections of Leyden and Paris.'

Figure 15 Drawing in pencil, attributed to John Gould, of *Lagorchestes conspicillata*, for *Macropodidae*, plate 2, and *Mammals of Australia*, II: 59, Kansas-Ellis-Gould MS 636.

PART 1 The date on the cover of the Kansas-Ellis Collection copy is 1 August 1841. Gould wrote on 1 October 1841 to Jardine (Roy. Scot. Mus.): 'As I am anxious that you should receive part 4 of my Birds of Australia and the first part of my Monograph of the Kangaroos I intend in <u>this instance</u> putting you to the expense of a parcel by way of Carlisle'.

Thus, part 1 was scheduled to be published on 1 August 1841, but possibly it did not appear until just before 1 October 1841.

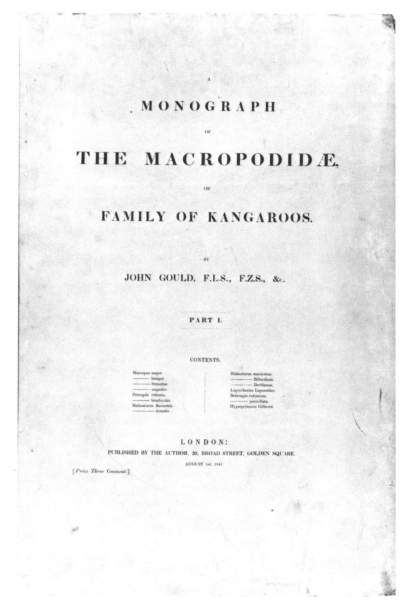

Figure 16 Cover of the first part of Gould's *Macropodidae* (University of Kansas Spencer Library-Ellis Collection).

PART 2 The date on the cover is 1 May 1842. In Gould's letter to Richard Owen of 9 May 1842 (Nat'l Library, Canberra) he writes 'I send to you herewith the second part of a "Monograph of the Macropodidae" which you will oblige me by accepting... At the close of the work I hope to render the generic distinctions very clear by an illustration of the cranium, feet, etc. of each genus... I have at least a dozen good species still to figure and expect more.'

Regarding a proposed, but not completed, third part of the *Macropodidae*, Gould wrote to Lord Derby on 9 January 1843, (Wagstaffe & Rutherford 1954-55) that the 'third and last part of the Kangaroos [will be published] about the end of the year'.

PART TWO JOHN GOULD'S MAJOR PUBLISHED WORKS

Gould wrote again on 5 February 1844 to Lord Derby (Wagstaffe, etc.): 'As regards the Kangaroos I must state that until Gilbert gets round to the East and North Coasts, (which I hope he will accomplish in another year,) this work cannot be brought to a close. I shall certainly complete it the moment I can; therefore I must beg you Lordship to keep the two parts you have unbound, until you receive the next with Titles, etc.' A third part was not published.

Contents of the two parts: The thirty species on the thirty plates are listed on the front of the appropriate covers in the Ellis set in parts. The plates are also listed in Waterhouse (1885), under the two years of 1841 and 1842. Waterhouse noted that eleven of the Kangaroo plates were republished in the *Mammals of Australia*. These eleven were republished as numbers 2, 5, 9, 23, 25, 27, 30, 32, 36, 42 and 46 in the *Mammals of Australia*. Dickison had said, as noted earlier, that thirteen plates were reprinted essentially unchanged.

12 1843–1844. *Richard B. Hinds. The zoology of the voyage of H.M.S. Sulphur.* Numbers III and IV. *Birds*, by John Gould, 1 volume, two parts, 16 plates, royal quarto.

The University of Kansas Spencer Library–Ellis Collection contains copy Ellis Aves E90 in the original ten parts as issued. There is also a second copy of an 'atlas of plates' only, where the plates are uncoloured.

Prospectus: In the University of Kansas Spencer Library copy Ellis Aves E90 there are two identical copies (Figures 17a & b). 'The work will extend to about Twelve Parts, one of which will appear on the 1st of every Third Month...price of Ten Shillings [each]...Part 1 will be published on the 1st of April [no year]'. Only ten parts were published.

Number of [royal quarto] plates: Sixteen. Eight plates (numbered 19 to 26) in No. III, *Birds*, Part 1, and eight plates (numbered 27 to 34) in No. IV, *Birds*, Part 2. Of the sixteen plates, plates 19, 20, 24, 26, and 27 to 34 have printed in the lower-left corner 'Drawn by J. Gould, on Stone by B. Waterhouse Hawkins'. Plates 21, 22, 23 and 25 have printed 'Drawn from Nature on Stone by B. Waterhouse Hawkins'. In the lower-right corner of all the plates is 'Printed by C. Hullmandel'. The measurement of Ellis Aves E90 untrimmed is $12\frac{7}{16} \times 9\frac{15}{16}$ inches (315 × 250 mm).

Dates of publication of the numbers: The dates are shown on the covers. No. III, *Birds*, Part 1 was published in October 1843, and No. IV, *Birds*, Part 2 was published in January 1844.

The complete zoological report consisted of ten numbers published from April 1843 to October 1845. Further information on the entire publication will follow.

The only letter from Gould on this bird collection was on 5 July 1845 to Jardine (Newton Library) where he sent a skin of *Fratercula placientius* 'for comparison after which you will do well to burn it. It is from the Sulphur collection and I am sorry to say many of the specimens are in a similar state'.

Contents of the two numbers: There is no published list of the species figured on each plate. Having access to a copy in the original parts, a list follows. The names are taken from the plates.

No. III, *Birds* by John Gould, Part 1, October 1843.
  Plate 19   *Halcyon saurophaga*: Gould
  Plate 20   *Pipra linearis*: Bonap

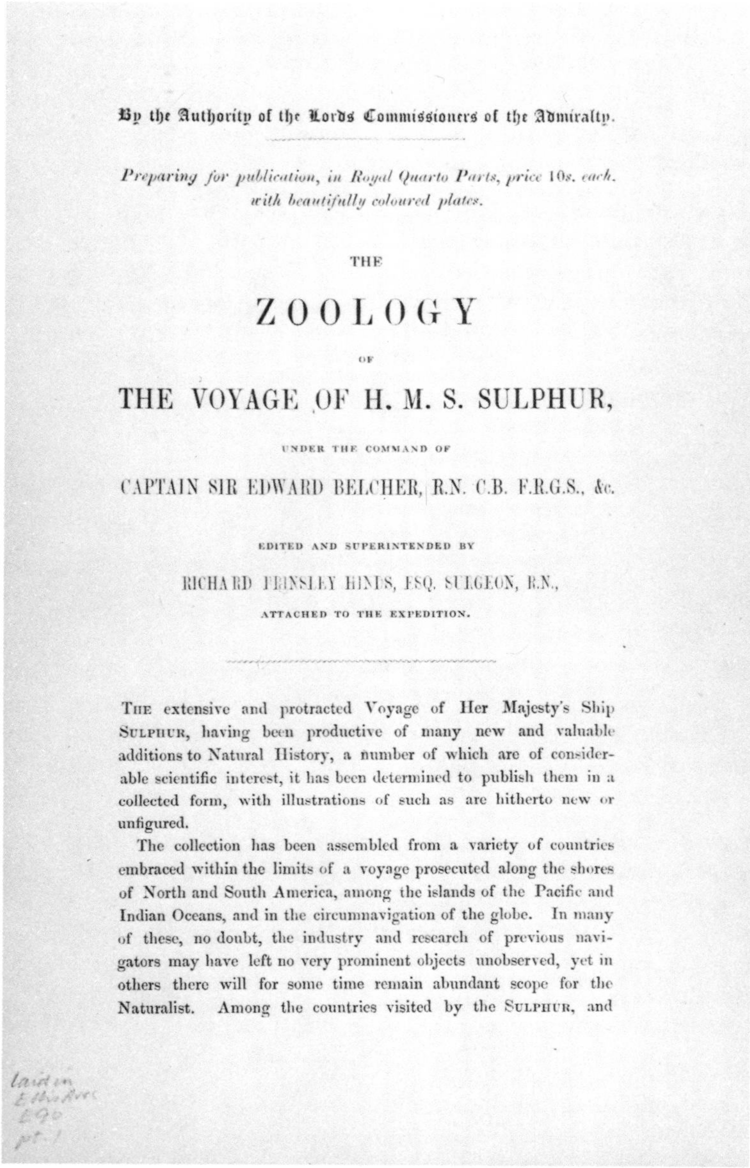

> By the Authority of the Lords Commissioners of the Admiralty.
>
> *Preparing for publication, in Royal Quarto Parts, price 10s. each, with beautifully coloured plates.*
>
> THE
>
> # ZOOLOGY
>
> OF
>
> THE VOYAGE OF H. M. S. SULPHUR,
>
> UNDER THE COMMAND OF
>
> CAPTAIN SIR EDWARD BELCHER, R.N. C.B. F.R.G.S., &c.
>
> EDITED AND SUPERINTENDED BY
>
> RICHARD BRINSLEY HINDS, ESQ. SURGEON, R.N.,
>
> ATTACHED TO THE EXPEDITION.
>
> THE extensive and protracted Voyage of Her Majesty's Ship SULPHUR, having been productive of many new and valuable additions to Natural History, a number of which are of considerable scientific interest, it has been determined to publish them in a collected form, with illustrations of such as are hitherto new or unfigured.
>
> The collection has been assembled from a variety of countries embraced within the limits of a voyage prosecuted along the shores of North and South America, among the islands of the Pacific and Indian Oceans, and in the circumnavigation of the globe. In many of these, no doubt, the industry and research of previous navigators may have left no very prominent objects unobserved, yet in others there will for some time remain abundant scope for the Naturalist. Among the countries visited by the SULPHUR, and

Figure 17 a & b Prospectus for R. B. Hinds's *Voyage of the Sulphur* (University of Kansas Spencer Library-Ellis Collection).

Plate 21   *Pipra vitelluia*: Gould
Plate 22   *Linaria* (?) *coccinea*
Plate 23   *Leucosticte griseogenys*: Gould
Plate 24   *Nectarinia flavigastra*: Gould
Plate 25   *Cactornis inornatus*: Gould
Plate 26   *Coryphilus dryas*: Gould

No. IV, *Birds* by John Gould, Part II, January 1844.
Plate 27   *Psittacus flavinuchus*: Gould
Plate 28   *Pteroglossus erythropygius*: Gould

PART TWO JOHN GOULD'S MAJOR PUBLISHED WORKS

> 2
>
> which in the present state of science are invested with more particular interest, may be mentioned the CALIFORNIAS, COLUMBIA RIVER, the NORTHWEST coast of AMERICA, the FEEJEE GROUP (a portion of the Friendly Islands), NEW ZEALAND, NEW IRELAND, NEW GUINEA, CHINA, and MADAGASCAR.
>
> Animated by a devotion to science, the following gentlemen have liberally engaged to undertake those departments with which each respectively is best acquainted. The MAMMALIA will thus be described by Mr. J. E. GRAY; BIRDS, by Mr. GOULD; FISH, by Dr. RICHARDSON; CRUSTACEA, by Mr. BELL; SHELLS, by Mr. HINDS; RADIATA, by Mr. J. E. GRAY.
>
> **PLAN OF PUBLICATION.**
>
> I. The work will extend to about *Twelve Parts*, one of which will appear on the 1st of every Third Month.
>
> II. The Parts will be published at the uniform price of *Ten Shillings*, and it is intended that each department shall, as far as possible, be complete in itself.
>
> PART I.
>
> Will be published on the 1st of APRIL, containing
>
> # MAMMALIA.
>
> By JOHN EDWARD GRAY, ESQ., F.R.S. &c.
>
> KEEPER OF THE ZOOLOGICAL COLLECTION IN THE BRITISH MUSEUM.
>
> *⁎⁎* *In order to secure to science the full advantage of these Discoveries, the Lords Commissioners of Her Majesty's Treasury have been pleased to sanction a liberal grant of money towards defraying part of the expenses of their publication. It has in consequence, been undertaken on a scale worthy of the high patronage thus received, and is offered to the public at a much lower price than would otherwise have been possible.*
>
> LONDON:
> SMITH, ELDER AND CO., 65, CORNHILL.

Plate 29   *Coccyzus ferrugineus*: Gould
Plate 30   *Pterocles personatus*: Gould
Plate 31   *Penelope albiventer*: Lefs
Plate 32   *Phalocrocorac perspicilatus*: Pall*   *In text as *perspicillatus*.
Plate 33   *Thalassidroma furcata*
Plate 34   *Larus brachyrhynchus*

Zimmer comments on the priority of the dates of the two *Birds* numbers, and how these dates relate to a paper by Gould given at the Zoological Society of London, published in the Proceedings for 1843, issued in December 1843.

Concerning the entire publication: The Kansas–Ellis copy in ten parts appears to be unique. Zimmer (1926) and Anker (1938) both mention that the *Birds* numbers on which they report are in their original covers. But they do not give any information on the entire set in the original wrappers except that both state, in error, that the work was published in twelve parts. According to the Prospectus, twelve parts were planned but only ten parts were published. The Kansas–Ellis Collection copy is in ten parts only. The National Union Catalog of Pre-1956 Imprints, entry number NH 0382107 also lists the Hinds's *Voyage of the Sulphur* as in ten parts. Also, on page 118 of Bernard Quaritch's General Catalogue of November 1881, item 441 is a list of a set of Hinds's 'Sulphur' work for sale. There is some confusion in the listing, but ten parts can be counted. This record also agrees with the Kansas– Ellis parts.

The copies mentioned in Wood (1931) and Ripley and Scribner (1961) are bound.

A separate publication appeared in six parts titled, *The botany of the voyage of H.M.S. Sulphur* by G. Bentham, from January 1844 to May 1846.

For interest and completeness, here are some facts from the ten parts of Hinds's *Zoology of the voyage of H.M.S. Sulphur* in the University of Kansas Spencer Library.

No. I   *Mammalia* by John Edward Gray, Part I April 1843, artist B. Waterhouse Hawkins, plates nos. 1-8 coloured and no. 17 uncoloured of skulls.

No. II   *Mammalia* by J. E. Gray, Part II July 1843, artist B. Waterhouse Hawkins, plates nos. 9-16 coloured and no. 18 uncoloured of skulls.

No. III   *Birds* by John Gould, Part I October 1843, artists Gould and Hawkins as stated earlier, plates nos. 19-26 coloured (Figure 18).

No. IV   *Birds* by John Gould, Part II January 1844, artists Gould and Hawkins, plates nos. 27-34 coloured.

No. V   *Ichthyology* by Dr John Richardson, Part 1 April 1844, artist W. Mitchell, plates nos. 35-44. All uncoloured.

No. VI   *Mollusca* by R. B. Hinds, Part I July 1844, artist G. B. Sowerby, Jun., plates nos. 1-7 coloured. A title-page is bound in with 'Vol. II' on it.

No. VII   *Mollusca* by R. B. Hinds, Part II October 1844, artist Sowerby, plates nos. 8-14 coloured.

No. VIII   *Mollusca* by R. B. Hinds, Part III January 1845, artist Sowerby, plates nos. 15-21 coloured. Bound in Number VIII are some extra pages. There are four pages of an 'Explanation of the plates' with a list of the Mollusca species on the twenty-one plates. Then follows a page consisting of an 'Index of Genera'. After this the text continues, and then the seven plates for this number.

No. IX   *Ichthyology* by Dr John Richardson, Part II April 1845, artist W. Mitchell, plates nos. 45-54 uncoloured.

No. X   *Ichthyology* by Dr John Richardson, Part III October 1845, artist W. Mitchell, plates nos. 55-64 uncoloured.

All of the ten numbers contain text pages.

The final aspect of Hinds's *Voyage of the Sulphur* to be mentioned is the set of uncoloured plates forming an 'Atlas' in the Kansas–Ellis Collection. Ripley and Scribner (1961) list a similar copy in the Coe Collection at Yale. A collation of the title as printed on the pale-yellow hard-cover is as follows:

The /Zoology/ of/ the voyage of H.M.S. Sulphur,/ during the years 1836-42./_____/Atlas of plates /Mammalia, Birds, Ichthyology/_____/ [18 lines of text]/ London:/Published by Smith, Elder and Co., 65, Cornhill/1843-1846/(Figure 19).

Ellis writes on the inside of the cover that this was purchased from 'Edwards 30/10/0 net Sept. 25.37 Pls. 1-64... Case Wheeler $4.00 May 1942'. The question is: was the atlas sold with this printed title cover, or did Ellis or someone at a later date

PART TWO   JOHN GOULD'S MAJOR PUBLISHED WORKS

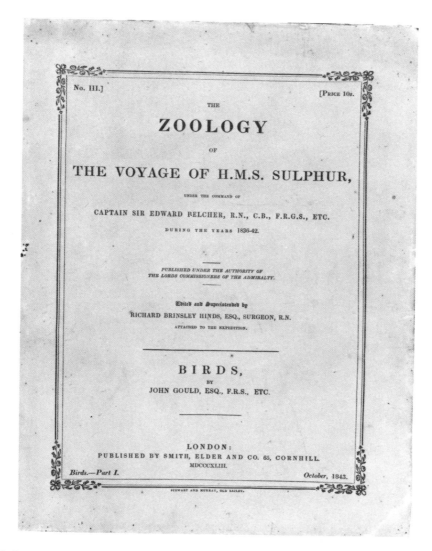

Figure 18 Cover for Hinds's *Voyage of the Sulphur* No. III Birds by John Gould, Part I (University of Kansas Spencer Library-Ellis Collection).

than 1843–46 have it printed? The fact that there is another copy at Yale would tend to support the first premise, that the atlas was sold with this printed cover.

Loose in the folder are sixty-four uncoloured plates. Note that the Mollusca plates are not included in the 'Atlas'.

Review: None found.

13 1844–1850. *A monograph of the Odontophorinae, or partridges of America*, 1 volume, 3 parts, 32 plates.

Prospectus: A copy of a prospectus for this monograph is in the University of Kansas-Ellis Collection (MS.P463 B:1) (Figure 20). It is a small sheet measuring 220 × 143 mm with text as follows: 'Preparing for Publication/ A Monograph/of/the

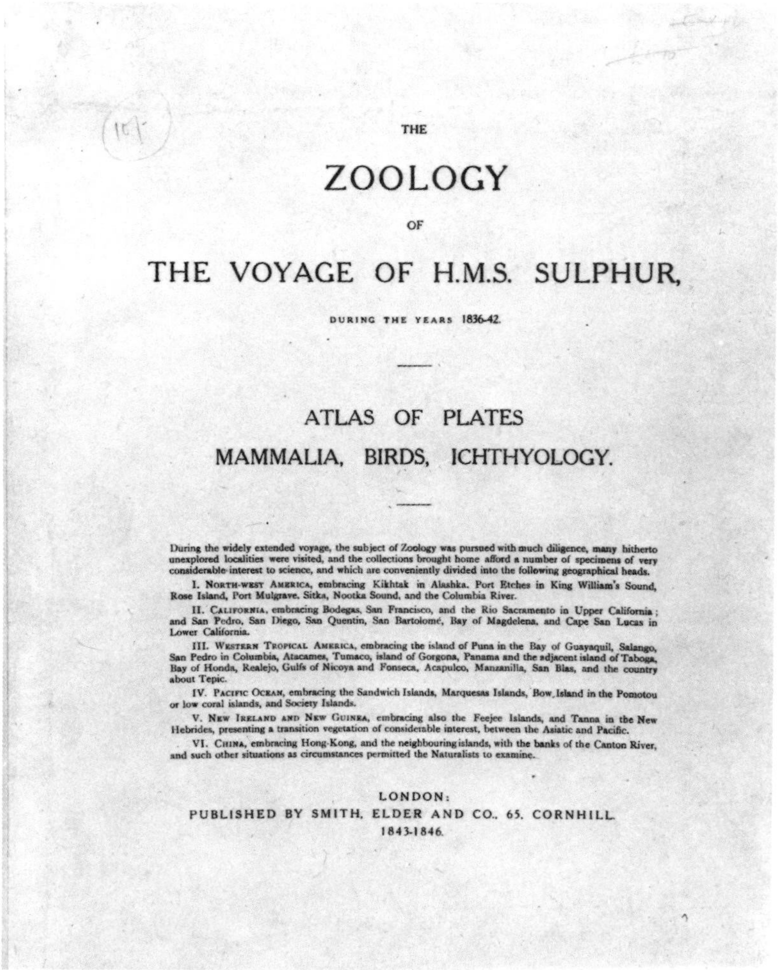

THE

ZOOLOGY

OF

THE VOYAGE OF H.M.S. SULPHUR,

DURING THE YEARS 1836-42.

ATLAS OF PLATES
MAMMALIA, BIRDS, ICHTHYOLOGY.

During the widely extended voyage, the subject of Zoology was pursued with much diligence, many hitherto unexplored localities were visited, and the collections brought home afford a number of specimens of very considerable interest to science, and which are conveniently divided into the following geographical heads.

I. North-west America, embracing Kikhtak in Alashka. Port Etches in King William's Sound, Rose Island, Port Mulgrave. Sitka, Nootka Sound, and the Columbia River.

II. California, embracing Bodegas, San Francisco, and the Rio Sacramento in Upper California; and San Pedro, San Diego, San Quentin, San Bartolomé, Bay of Magdelena, and Cape San Lucas in Lower California.

III. Western Tropical America, embracing the island of Puna in the Bay of Guayaquil, Salango, San Pedro in Columbia, Atacames, Tumaco, island of Gorgona, Panama and the adjacent island of Taboga, Bay of Honda, Realejo, Gulfs of Nicoya and Fonseca, Acapulco, Manzanilla, San Blas, and the country about Tepic.

IV. Pacific Ocean, embracing the Sandwich Islands, Marquesas Islands, Bow Island in the Pomotou or low coral islands, and Society Islands.

V. New Ireland and New Guinea, embracing also the Feejee Islands, and Tanna in the New Hebrides, presenting a transition vegetation of considerable interest, between the Asiatic and Pacific.

VI. China, embracing Hong-Kong, and the neighbouring islands, with the banks of the Canton River, and such other situations as circumstances permitted the Naturalists to examine.

LONDON:
PUBLISHED BY SMITH, ELDER AND CO., 65, CORNHILL.
1843-1846.

Figure 19 Cover for the 'Atlas of plates' for Hinds's *Voyage of the Sulphur* (University of Kansas Spencer Library-Ellis Collection).

Ortyginae/or/Partridges of America./ . . . It is proposed to publish this Monograph in three Parts, each Part to contain Ten Coloured Plates . . . the price of each Part to subscribers £2:10s . . . The first part will appear on the 1st of September, and the succeeding Parts at intervals of three months . . . April 4, 1844'.

A larger-sized prospectus dated 1 November 1844 measures 443 × 280 mm (Ellis F-122). Essentially the same information is contained in it as in the smaller one.

Number of plates: Thirty-two. The first two parts had ten plates each, and the third part had twelve plates.

Dates of publication of the parts: Preliminary information. In Gould's letter of 2 December 1842 to Jardine (Newton Library) he wrote:

Having many years paid considerable attention to and in fact always been much interested in the South American partridges "*Ortyx*" I am preparing a Monograph of this group with figures with the intention of publishing it. If you could give me the loan of *Macroura* or of any ['other' crossed out] species from the West Indies you think I do not possess I should feel obliged . . . I may start for the Continent in a week or so to examine the *Ortyxes* in the collections there.

PART TWO JOHN GOULD'S MAJOR PUBLISHED WORKS

PREPARING FOR PUBLICATION,

A MONOGRAPH

OF

THE ORTYGINÆ,

OR

PARTRIDGES OF AMERICA.

DEDICATED TO HIS HIGHNESS
CHARLES LUCIEN BONAPARTE, PRINCE OF CANINO AND MUSIGNANO.

BY

JOHN GOULD, F.R.S. &c.,

Author of "A Century of Birds from the Himalaya Mountains," "The Birds of Europe,"
"The Birds of Australia," "Monographs of the Toucans, Trogons, Kangaroos," &c.

PROSPECTUS.

In taking up the present tribe of Birds as the subject of his fourth Monograph, the Author is impressed with the belief that it will be found fraught with the highest interest, not only to the scientific Ornithologist, but to those also who are laudably endeavouring to add to the number of our Gallinaceous Birds, either as ornaments of the aviary or as naturalized introductions to Europe; and he feels assured that when the numerous elegant species in his possession are illustrated and before the Public, the tribe in question will become no less interesting to the sportsman than to the lover of Ornithology. The Author may here observe, that in the study of this group of Birds, and which he has sedulously investigated, he has had access to every public institution of importance in Europe, as well as to every extensive private collection, and has in every instance received the most liberal co-operation and assistance.

It is proposed to publish this Monograph in three Parts, each Part to contain Ten Coloured Plates, representing as many species of the natural size, accompanied with appropriate scenery, descriptive letter-press, &c. The size will be imperial folio, and the price of each Part to subscribers £2 : 10s. The present publication will not be found inferior in any one respect to those of the Author's previous works, with which it will correspond. The first Part will appear on the 1st of September, and the succeeding Parts at intervals of three months.

London:—Published by the Author, 20 Broad Street, Golden Square, where Subscribers' names are requested to be forwarded, in order that they may be included in the list, which will be given at the conclusion of the work.

April 4, 1844.

Figure 20 Prospectus for Gould's *Odontophorinae* (University of Kansas Spencer Library-Ellis Collection).

A short letter of 15 December 1842 to Jardine (Newton Library) follows up on the 2 December letter: 'The time of my departure however is not yet determined. I should like to receive the *Ortyxes* as soon after your next visit to Jardine Hall as you can ... Did Dr. Parnell bring any *Ortyxes* or have you any from the West Indies?'

Gould's secretary, Edwin Prince, wrote to Jardine on 8 August 1843 (Roy. Scot. Mus.) saying 'Mr. Gould is on the Continent and writes me word he has gained much information respecting the *Orticidae* [sic].'

Gould's letter of 10 October 1843 to Jardine (Newton Library) relates that:

During my last Continental trip I visited Paris, Brussels, Frankfurt, Nuremburg, Leipsig, Berlin, Hanover, Amsterdam and Leyden ... I found many new *Ortyx's* nearly the whole of which I was allowed to bring with me to London and I consequently have a surprizingly fine collection. I shall therefore at once set about my proposed monograph.

In Lord Derby's letter from Gould of 5 February 1844 (Wagstaffe and Rutherford, letter number 17) we find a rare reference to a specific cost of publication:

> The Ortyzida will be published the first moment I have to spare. Surely, my Lord, I shall have to avoid the error of getting too many Irons in the fire.[!] The preparation of the last-mentioned Work will involve an expence [sic] of from 5 to 7 hundred Pounds before it appears in public. It will, however, appear at no lengthened period, I hope, before the Year is out.

In Gould's letter of 12 September [1844] to Jardine (Newton Library) he writes: 'I am at this moment very busy preparing my monograph of the *Ortyxes* which I will read at York if the committee wishes me to do so. I have taken an immense deal of pains to work out all the species, their synonyms, etc.'

On pages 61 and 62 of the 1844 Transactions of the Sections of the British Association for the Advancement of Science is a summary of Gould's paper entitled 'A Monograph of the Sub-family Odontophorinae, or Partridges of America'. The paragraph concludes 'The subject was fully illustrated with drawings of most species'. The paper was read around 26 September 1844.

McGill University Libraries, according to Casey Wood (1931), has the 'original manuscript in his own handwriting ... of a paper which he read at the British Association meeting in 1844'.

PART 1 Ten plates. Scheduled for publication 1 September 1844. In a letter to Prince Bonaparte dated 26 November 1844 (Mus. Nat'l. d'Hist. Nat. Paris) Gould wrote that

> about a month ago a package was sent to you ... containing among other things ... the first part of the *Odontophorinae*, the ... work I have dedicated to you as a slight token of the value I entertain for your friendship and as a just tribute of respect for your talents as a Scientific Naturalist.

It would appear from this letter that the first part appeared around 26 October 1844, instead of September first as proposed in the Prospectus.

Gould mailed Part 1 to C. J. Temminck about 1 December 1844 according to letter of 30 January 1845 in the Rijksmuseum, Leiden.

PART 2 Ten plates. Zimmer and Waterhouse give the date as 1846. From a letter of 1 April 1846 to Jardine (Newton Library) 'part 2 of the *Odontophorinae* [has been published some time]'; apparently it was published before 1 April 1846.

PART 3 Twelve plates. The Preface in the Linda Hall Library copy of this volume is dated 15 October 1850.

In a letter of 30 July 1846 to Jardine (Roy. Scot. Mus.) Edwin Prince wrote that Gould had just visited France, Germany and Switzerland and 'brought with him a fine and varied collection among which are some new Odontophorinae'. This statement may account for the fact that the final third part contained twelve plates instead of the originally planned ten plates.

On 30 January 1850 Gould wrote to Bonaparte 'I have written a line to M. Temminck soliciting the loan of two or three *Ortyges*. When I receive these I shall be able to complete my Monograph which you will probably recollect is dedicated to yourself.'

Writing to Bonaparte on 18 November 1850 (Mus. Nat'l. d'Hist. Nat., Paris) Gould informed him that 'at length my Monograph of the *Odontophorinae* which I had the pleasure of dedicating to you has been brought to a close and I have this day forwarded to the call of Mr. E. Verreaux a very fine copy handsomely bound which I beg you will do me the favour to accept for the reasons assigned

PART TWO JOHN GOULD'S MAJOR PUBLISHED WORKS

Figure 21 Drawing in pencil and pastel, attributed to John Gould, (or H. C. Richter) of *Petaurista taguanoides*, for *Mammals of Australia*, I: 22, Kansas-Ellis-Gould MS 688.

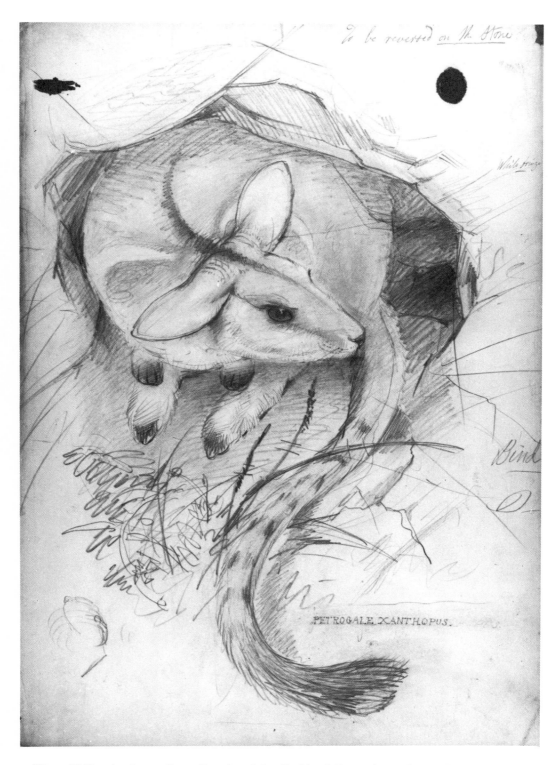

Figure 22 Drawing in pencil, attributed to John Gould, of *Petrogale xanthopus*, for *Mammals of Australia*, II: 43, Kansas-Ellis-Gould MS 679.

in the dedication'. Thus, part 3 was published before 18 November 1850, possibly even before 15 October 1850, as stated in the Preface.

Contents of Parts: Waterhouse (1885) allocates the plates to each part.

Reviews: None seen.

14   1845–1863. *Mammals of Australia*, 3 volumes, 13 parts, 182 plates (Figures 21 & 22).

The University of Kansas Spencer Library–Ellis Collection has a copy (Ellis F-156) in the separate parts but lacks part thirteen (Figure 23).

Prospectus: Not seen.

Number of plates: 182. According to the 1868 Prospectus and List, at the Academy of Natural Sciences in Philadelphia *Mammals of Australia* was 'completed in Thirteen Parts, each containing Fifteen Plates, price £3 3s. or in Three Volumes price £41.' The last part had only two plates according to Waterhouse (1885).

Dates of the publication of the parts: From the Kansas copy (Ellis F-156) in twelve parts (the thirteenth is missing) the dates of issue and the contents of each part are printed on the cover board. The plates and text are bound in each part. This date of publication information is consistent with any additional data obtained from Gould's correspondence, prospectus or reviews.

Preliminary information. In a letter to Prince Bonaparte dated 26 November 1844 (Mus. Nat'l. d'Hist. Nat., Paris) Gould wrote

My present intention is to follow it [the *Birds of Australia*] up with a general history of the Mammals of Australia of which I have lately received a surprizing number of new species... Gilbert, who is still collecting for me, is on the Darling Downs... I have no doubt he will reap a rich harvest in this new field for his exertions.

PART 1   Fifteen plates, 1 May 1845. As further corroboration for this date there is in the 'Introduction to the Birds of Australia' (Kansas copy (Ellis K-16)) a six-page prospectus of Gould's works. On page 6 is a paragraph on *Mammals of Australia* stating that 'the First Part of the work appeared on the 1st of May 1845'.

PART 2   Fifteen plates, 1 September 1849. In a letter from Edwin Prince to Jardine of 24 July 1846 and from Gould to Jardine on 30 July 1846 (both Roy. Scot. Mus.) the question of Jardine returning part 1 is answered: 'With respect to Pt 1 of the Mammals, you are at perfect liberty to retain or to return them as you may yourself wish.' Thus part 2 had not been published by July 1846.

PART 3   Fifteen plates, 1 December 1851. In the Prospectus of 1 May 1852 (Acad. Nat. Sciences, Phila.) Gould writes that 'of this work three Parts have been published ... the work to be completed in Ten or Twelve Parts'.

PART 4   Fifteen plates, 1 December 1852.

PART 5   Fifteen plates, 1 November 1853.

PART 6   Fifteen plates, 1 December 1854. By 1 December 1854 Gould had sent Part 6 of *Mammals of Australia* to C. J. Temminck, according to Gould's letter of 26 July 1855 (Rijksmuseum, Leiden).

PART 7   Fifteen plates, 1 July 1855.

PART 8   Fifteen plates, 1 October 1856.

Figure 23 Cover for the first part of Gould's *Mammals of Australia* (University of Kansas Spencer Library-Ellis Collection).

PART 9   Fifteen plates, 1 October 1857. In a letter from Gould to Professor McCoy of 7 December 1857 (Nat'l. Mus. Victoria) he writes '9 parts are published at £3-3 each and 2 or 3 more pts will complete it'.

PART 10   Fifteen plates, 1 October 1858.

PART 11   Fifteen plates, 1 September 1859.

PART 12   Fifteen plates, 1 November 1860. In a letter of 22 February 1861 from Gould to the Librarian of the University of Melbourne (Nat'l. Mus. Victoria), Gould wrote that 'In December 1859 I had the pleasure of forwarding...pts 2 to 11 of my work on the Mammals of Australia for the University Library.' Part 1 had already been sent. The letter continues by saying that now Part 12 was being sent 'and by this time next year I hope to send you the conclusion with titles and every requisite to form the work into volumes'.

PART 13   Two plates, 1 May 1863. This information is from Waterhouse (1885) and from a photocopy of the cover of part 13 sent to me by Eleanor MacLean of the Blacker-Wood Library at McGill University. This part in addition contained the title pages, for the three volumes, the dedication, list of subscribers, preface, and introduction, etc. Also included were '28 sheets of Letterpress descriptive of the life-sized heads of animals published in the preceding Parts'.

To Professor Frederick McCoy on 13 May 1863 (Nat'l. Mus. Victoria) Gould said 'I must now tell you that I have completed my work on the Mammals of Australia and that the concluding part with the Hairy-nosed Wombat figured therein will shortly be sent out to you.'

Contents of the parts: Waterhouse (1885) lists the plates as they appeared in each part over the nineteen years of publication, but there are errors in this tabulation.

The University of Kansas copy (Ellis F-156) in the separate parts has the species in each part printed on the board covers.

Carus and Engelmann (1861) list all the species for the first nine parts.

Iredale (1951) wrote about the pattern plates for *Mammals of Australia* that are now deposited in the National Library at Canberra.

The Gould Collection of mammals in the British Museum (Natural History) is elaborated upon in the Museum's 1906 history of the collection, page 35.

15   1848. *An introduction to the birds of Australia*, 1 volume, no plates, post octavo.

The material in this octavo volume is identical, with minor exceptions, with the introductory material appearing in the larger imperial-folio edition of Gould's *Birds of Australia*. A copy in the Kansas-Ellis Collection (Ellis Aves C694) has a six-page addition at the end of the book consisting of one page of reviews from newspapers and journals concerning Gould's Australian trip, and five pages of prospectus of Gould's other previously published works. Another copy (Ellis Aves C446) lacks these six pages.

Date of publication: Zimmer (1926) quotes Mathews as affixing the date of publication as 1 August 1848, but says that Mathews gives no citation of source of his information. The source is available from the Kansas copy (Ellis Aves C694). The date of 1 August 1848 appears on the final prospectus page of the volume. However, at the conclusion of the Preface is the date of 12 June 1848.

Some information in the correspondence is interesting and relevant. In a letter to Strickland of 5 May 1848 (Cambridge Univ. Mus. Zoology) Gould wrote he had begun working on 'the preface and introduction — a rather long job'.

From a letter to Strickland of 23 June 1848 (Cambridge Univ. Mus. Zoology) it is apparent the 'Introduction' had not been published by this date. Gould wrote that he was enclosing 'proof of the preface...it has been set up in the present form for the sake of convenience...I am very busy with the introduction a long affair... Please to return the proof'. Thus, positive proof for a definite date of publication is lacking. As seen with the other 'Introductions', a date on a Prospectus sheet is not necessarily the date of publication. One can only assume this 'Introduction' was published between 23 June 1848 and 1 December 1848, when the final part 36 of the *Birds of Australia* was apparently published.

Comments on contents: Zimmer comments at length on the question of priority of some new names listed in the volume. It was published four months before the similar introductory material for the imperial-folio edition. He feels that various new names of species in this octavo volume have priority over the later-appearing folio volume. Sherborn (1922) disagrees and calls this octavo edition simply 'proof sheets'. Mathews (1920-25) agrees with Sherborn, but gives the five new names with both references.

16 1849–1861. *A monograph of the Trochilidae, or family of humming-birds*, 5 volumes, 25 parts, 360 plates.

Prospectus: A separate prospectus has not been seen. The work is referred to, however, in all of the copies of the 'Prospectus and List of Subscribers' from 1852 on.

Number of plates: 360. According to the 1870 Prospectus and List of Subscribers in the National Library of Australia at Canberra 'This Monograph was published in 25 Parts, containing Fifteen Plates, with descriptive letterpress, price £3 3s. each Part, ... or complete in 5 volumes, price £78 15s'. The final part had no plates.

Dates of publication of the parts: Preliminary information. Gould first wrote about his interest in humming-birds in a letter to Jardine (Newton Library) of 19 February [1846] when he said 'I have now been confined to the house more than three months time ... all the time engaged with the *Trochilidae* and have now a collection worth seeing ... [George] Loddige's collection will be kept in the family.'

In a letter from Edwin Prince to Jardine (Roy. Scot. Mus.) of 30 July 1846, Prince wrote that Gould 'has visited France, Germany and the finest parts of Switzerland has enjoyed himself much ... [has just returned] and brought with him ... large accessions to the group of Trochilidae of great beauty and interest'.

A very complete compilation of Gould's five-volume work on the humming-birds, is printed in Coues's *Ornithological bibliography*, third instalment (pp. 670-4). First, Coues listed the species as published in each of the twenty-five parts with the dates taken from the covers of each part. Then he listed every species in the order in which the plates were to be bound in each of the five volumes.

The Beinecke Rare Book and Manuscript Library at Yale University has a set in the original parts with the dates on the covers.

For completeness and for chronological purposes, I will list each part. The number of plates, and the dates of publication of each part are taken from Coues's list. In addition, if any other information is available from Gould's correspondence or any prospectus, it will be appended.

PART 1 Fifteen plates, 1 June 1849. In an undated letter (but written before 12 May 1849), Gould wrote to Prince Bonaparte (Mus. Nat'l. d'Hist. Nat., Paris) that 'The Humming Birds new series are also out and look well'. Thus this first part was apparently published before 12 May, earlier than scheduled.

A further note on this new publication is in a letter of 30 January 1850 to Bonaparte (Mus. Nat'l. d'Hist. Nat., Paris) where Gould wrote 'Humbolt has promised to get me the Queen of Prussia as a subscriber to the Humming Birds and

Plate 15 Drawing in pencil, ink and watercolour, attributed to W. Hart, of *Androdon aequatorialis*, for *Supplement to Trochilidae*, Plate 1, Author's Collection 33 GS. The pencil alterations of two of the figures are undoubtedly by John Gould.

Plate 16 Drawing in pencil and watercolour, attributed to W. Hart, of *Helianthea dichroura*, for *Supplement to Trochilidae*, plate 19, Kansas-Ellis-Gould MS 969. This fine drawing was offered for sale as item 1844 by Henry Sotheran, Ltd in No. 9 'Piccadilly Notes', for £4.4s.

as I have already obtained the Queen of Saxony, the Princess of Wied and several other noble ladies for whom the work is especially adapted'.

PART 2   Fifteen plates, 1 November 1851. In the Prospectus and List of Subscribers of 1 May 1852 (Acad. Nat. Sciences, Phila.) it is stated that 'the First and Second Parts are ready, and the Third is in preparation'.

PART 3   Fifteen plates, 1 May 1852.
PART 4   Fifteen plates, 1 October 1852.
PART 5   Fifteen plates, 1 May 1853.
PART 6   Fifteen plates, 1 October 1853.
PART 7   Fifteen plates, 1 May 1854.
PART 8   Fifteen plates, 1 October 1854. As of 1 December 1854 Gould had sent parts 7 and 8 to Temminck, according to a letter of 26 July 1855 in the Rijksmuseum, Leiden.
PART 9   Fifteen plates, 1 May 1855.
PART 10   Fifteen plates, 1 September 1855. From the Prospectus and List of 1 January 1856 (Yale University) is the statement 'Ten Parts having appeared, and the eleventh and twelfth being in press'.
PART 11   Fifteen plates, 1 May 1856.
PART 12   Fifteen plates, 1 September 1856.
PART 13   Fifteen plates, 1 May 1857.
PART 14   Fifteen plates, 1 September 1857. Gould wrote to Professor McCoy on 7 December 1857 (Nat'l. Mus. Victoria) that 'the Monograph of the Humming Birds of which 14 pts. are ready is the most popular and perhaps the most beautiful of my works... There is not one in your part of Australia'.
PART 15   Fifteen plates, 1 May 1858. Reviewed in the January 1859 *Ibis* 1(1):99-101.
PART 16   Fifteen plates, 1 September 1858. Reviewed in the January 1859 *Ibis* 1(1):99-101.
PART 17   Fifteen plates, 1 May 1859. Reviewed in the October 1859 *Ibis* 1(1):454-5.
PART 18   Fifteen plates with sixteen species, 1 September 1859. Reviewed in the October 1859 *Ibis* 1(1):454-5.
PART 19   Fifteen plates, 1 May 1860.
PART 20   Fifteen plates, 1 September 1860.
PART 21   Fifteen plates, 1 May 1861.
PART 22   Fifteen plates, 1 July 1861.
PART 23   Fifteen plates, 1 September 1861.
PART 24   Fifteen plates, 1 September 1861.
PART 25   No plates, 1 September 1861. A review of *Trochilidae* on its completion is in the January 1862 *Ibis* 4(1):72-4.

Waterhouse (1885) also has an allocation of the plates to the various parts.

Reviews: Though not a review in the true sense, Coues, in an 'Addendum to Trochilidae' on pages 690 to 692 of the third instalment of his *Ornithological bibliography*, wrote a brilliant commentary concerning the entire 'hummer' field from the ancients to Elliot. He spoke of 'the malignant epidemic which we may call the genus-itch, which broke out simultaneously in 1849, from two foci of contagion, in France and in Germany... some authors had simply amused themselves in "playing chess" with the names of Hummers... the fog of synonymy had completely enveloped the subject.' Our hero, Gould, however, with Swainson and Lesson, established 'many

genera of Hummers [which] had been found acceptable, and, indeed, necessary'. The needed reform in nomenclature, however, really got its impetus from the studies of D. G. Elliot and Osbert Salvin. Coues concluded these comments with a lengthy index of the genera of the trochilidae.

The reviews occurring in *Ibis*, though short, are listed under the part reviewed.

17 1849–1883. *The birds of Asia*, 7 volumes, 35 parts, 530 plates.

The University of Kansas Spencer Library copy is Ellis F-101. This copy is not in the separate parts as issued, but in the back of volume one are bound some valuable 'temporary' pages, elaborated on later.

Prospectus: No separate prospectus has been seen. The Prospectus and List of Subscribers of 1852, 1856, 1866, 1868, and 1870 all contain mention of this work.

Number of plates: 530. Gould began by issuing seventeen plates with each part but eventually reduced the number to as low as twelve plates, with the final part 35 containing only seven plates.

Dates of publication of the parts: Preliminary information. Gould wrote to Hugh Strickland 5 May [1848] (Cambridge Univ. Mus. Zoology) 'I am making a complete list of all the birds of Asia preparatory to my great work on that subject. I shall therefore be glad to see all Capt. Boy's [collection] named.'

In a letter to Lord Derby of 8 February 1850 (Wagstaffe & Rutherford, p. 362), Gould wrote:

I have but now returned from a long Journey on the Continent, in the course of which I visited Holland, Northern Prussia, Saxony, and France to inspect their museums, obtain subscribers to my works, & establish Correspondents for the acquisition of Specimens from Siberia and the northern parts of Asia for my Work on the birds of that Continent, on all which points I have been very successful and moreover obtained many fine things.

The thirty-five parts were issued over a space of thirty-four years from 1849 to 1883. R. B. Sharpe prepared the last three parts after Gould's death in 1881.

In Zimmer (1926) there is a typographical error regarding the dates of publication. He states that parts 3 to 28 appeared 'at the rate of one each year from 1851 to 1866', which last date should read '1876'.

Since there was no definite schedule I will tabulate the separate parts, with the numbers of plates issued, the year and month of issuance, and any separate data from correspondence, prospectus, and reviews.

When R. Bowdler Sharpe published the final part 35 of this work with the title pages and other introductory material, he included under the 'List of Plates' for each volume the month and year and the part wherein the plate was published.

Also, quite specific information on this work is available from the Kansas copy (Ellis F-101). At the end of volume 1, three important items are bound in. First, there are seven sheets of temporary titles, collated as follows: '(*Temporary Title.*) /The/Birds of Asia/by/John Gould, F.R.S./Parts I.-V.[VI.-X.; XI.-XV.; XVI.-XX.; XXI.-XXV.; XXVI.-XXX.; XXXI.-XXXV] /London:/Printed by Taylor and Francis, Red Lion Court, Fleet Street,/Published by the author, 26, Charlotte Street, Bedford Square./1850-1853./[1854-1858; 1859-1863; 1864-1868; 1869-1873; 1873-1877; 1873-1877 [sic]].'

Secondly, there is a single sheet with the heading 'List of Plates/in the first four *temporarily* arranged volumes'. This one sheet contains a tabulation for four volumes

PART TWO JOHN GOULD'S MAJOR PUBLISHED WORKS

with eighty species in each, with the listing of each species name, the part in which it appeared, and temporary numerical arrangement in the first four volumes. At the bottom of this sheet is a tabulation of the month, day and year of publication of each of the first twenty parts. This sheet is dated 1 May 1868.

The third item also bound in at end of Volume 1 of the Kansas copy Ellis F-101 is six sheets listing the 'contents' for parts 1 to 5, 6 to 10, 11 to 15, 16 to 20, 21 to 25, and 26 to 30. The sheet with the contents for parts 31 to 35 is missing or may never have been published. Here then is a complete list of every species as published in each separate part, except for parts 31 to 35. The plates for these last five parts are ascertainable from the index Sharpe supplied with part 35.

An explanation of these 'temporary' sheets is a statement in the Prospectus and List (Yale University) dated 1 January 1856

it will be published in Parts, containing Sixteen or Seventeen Plates, with descriptive letter-press of the species thereon represented; price Three Guineas; every six parts will comprise One-Hundred Species; and it is proposed to deliver to each Subscriber who may wish it without charge, Titles and Indices so that they may be temporarily formed into a volume for the purpose of security.

PART 1   Seventeen plates, before 5 December 1849. In a letter from Gould to Bonaparte, not dated, but received by the Prince on 5 December 1849 (Mus. Nat'l. d'Hist. Nat., Paris) Gould wrote 'The first part of the Birds of Asia is ready for delivery certainly the best number of Birds I have ever issued ... It shall be sent ... almost immediately.' This information moves the date of beginning publication of Gould's *Birds of Asia* back a year from the usually accepted date of 1850. It is conceivable that Gould scheduled the first part for January 1850, but it was ready in December 1849 and apparently delivered then. Gould was ahead of schedule.

PART 2   Seventeen plates, 1 July 1850. In a letter of 6 May 1850 to Lord Derby (Wagstaffe & Rutherford, p. 363) Gould wrote:

In the next part of my Birds of Asia which will not be long before it appears I intend to publish, if possible, the whole of the longtailed *Cinnyri* or *Nectariniae* of Continental India, and it will very much aid my illustration of the Group if your Lordship would kindly allow me a sight of any you may possess, especially the females.

Therefore, part 2 was published soon after this date of 6 May 1850.

PART 3   Seventeen plates, 2 June 1851. In the Prospectus and List of 1 May 1852 (Acad. Nat. Sciences Phila.) there is also the statement that 'three Parts are published'.

PART 4   Seventeen plates, 1 November 1852.

PART 5   Seventeen plates, 1 October 1853.

PART 6   Seventeen plates, 1 July 1854. In a letter to Temminck of 26 July 1855 (Rijksmuseum, Leiden) Gould wrote that as of 1 December 1854 he had sent part 6 to Temminck.

PART 7   Seventeen plates, 2 April 1855. In the 1856 Prospectus and List dated 1 January 1856 (Yale University) is the statement 'of this work seven Parts are published, and for the present it will appear at the rate of not more than one or two parts a year'.

PART 8   Sixteen plates, 1 May 1856.

PART 9   Sixteen plates, 1 May 1857.

PART 10   Sixteen plates, 1 June 1858. Reviewed in the January 1859 *Ibis* 1(1):99-101.

PART 11   Sixteen plates, 1 May 1859. Reviewed in the July 1859 *Ibis* 1(1):319-20.

PART 12   Sixteen plates, 1 June 1860.

PART 13   Sixteen plates, 1 May 1861.
PART 14   Sixteen plates, 1 May 1862.
PART 15   Sixteen plates, 1 June 1863.
PART 16   Sixteen plates, 1 April 1864. Reviewed in the April 1865 *Ibis* 1(2):220.
PART 17   Sixteen plates, 1 April 1865. Reviewed in the October 1866 *Ibis* 2(2):409-10. In the Prospectus and List dated 1 January 1866 (Library of Congress) is the statement that 'of this work seventeen Parts are published'.
PART 18   Sixteen plates, 1 April 1866. Reviewed in the July 1867 *Ibis* 3(2):371-2.
PART 19   Sixteen plates, 1 May 1867. Reviewed in the April 1868 *Ibis* 4(2):216-18. In the Prospectus and List of 1 January 1868 (Acad. Nat. Sciences of Phila.) there is the statement that nineteen parts had been published at 'one or 2 Parts a year'.
PART 20   Sixteen plates, 1 April 1868. Reviewed in the October 1868 *Ibis* 4(2):472.
PART 21   Fifteen plates, 1 April 1869. Reviewed in the January 1870 *Ibis* 6(2):118-21. In the 1 January 1870 Prospectus and List (Nat. Library of Australia, Canberra) is the statement that 'Twenty-one Parts have been published'.
PART 22   Sixteen plates, 1 March 1870. Reviewed in the October 1871 *Ibis* 1(3):437-8. In a letter to George N. Lawrence of 18 February 1870 (Amer. Mus. Nat. Hist.) Gould wrote 'Pt. 22 will be on the first of next month'.
PART 23   Fifteen plates, 1 March 1871. Reviewed in the October 1872 *Ibis* 2(3):428-9.
PART 24   Fifteen plates, 1 March 1872. Reviewed in the supplement to the 1873 *Ibis* 3(3):453-4.
PART 25   Fifteen plates, 1 March 1873.
PART 26   Fifteen plates, 1 August 1874. Reviewed in the October 1874 *Ibis* 4(3):452-3. There is no list of species in this review.
PART 27   Fifteen plates, 1 March 1875. Gould in a letter of 15 April 1876 (Nat. Mus. Victoria) wrote to McCoy that 'on the 16th of February a parcel was sent to you through Mr. Bain, Maymarket, containing... Pt. 27 of the "Birds of Asia" which I imagine is for one of the other Institutions, which you say are amalgamated'. Here is further evidence that Gould's parts were published ahead of schedule.
PART 28   Thirteen plates, 1 July 1876. With this part ended the regular annual issuing of the Birds of Asia parts, maintained since part 3.
PART 29   Thirteen plates, 1 April 1877. A somewhat critical review appeared in the July 1877 *Ibis* 1(4):377-8.
PART 30   Thirteen plates, 1 October 1877. Reviewed in the January 1878 *Ibis* 2(4):109-10.
PART 31   Thirteen plates, 1 July 1879. Reviewed in the January 1880 *Ibis* 4(4):134-5.
PART 32   Twelve plates, 1 July 1880. Reviewed in the January 1881 *Ibis* 5(4):160.
PART 33   Thirteen plates, February 1882. R. Bowdler Sharpe issued this and the remaining two parts. Reviewed in the July 1882 *Ibis* 6(4):461.
PART 34   Thirteen plates, January 1883. Reviewed in the July 1883 *Ibis* 1(5):379.
PART 35   Seven plates, August 1883. Reviewed in the January 1884 *Ibis* 2(5):104. In this same issue pages 49 to 60 are a reprint of R. B. Sharpe's 'Introduction to Gould's "Birds of Asia"' that appeared in this concluding part of Gould's work. Waterhouse (1885) also states that this part contained only seven plates.

Contents of the separate parts: As stated earlier the Kansas copy (Ellis F-101) has in volume 1 some 'temporary title' pages that contain quite a bit of information on

PART TWO JOHN GOULD'S MAJOR PUBLISHED WORKS

the dates of publication and contents of the separate parts. Also Sharpe gave the month and year of appearance of each plate in the index to each volume which he published with part 35 containing the title pages and other introductory material. The years of issue of each plate are listed in Waterhouse (1885) but I have not checked them carefully.

Reviews: The reviews from *Ibis*, though short, are mentioned under each appropriate part. In every review, except for the review of part 26, there is a list of the species of birds that were illustrated in that particular part.

18 1851–1869. *The birds of Australia, supplement*, 1 volume, 5 parts, 81 plates.

Anker (1938) lists the University Library at Copenhagen as having this Supplement in the original five parts with covers.

Prospectus: None seen.

Number of plates: Eighty-one plates, one of which is folded. There were sixteen plates in parts 1, 3, 4 and 5, and seventeen plates in part 2, according to Anker.

The 'pattern plates' for the Supplement are in the National Library of Australia at Canberra. Iredale (1951) wrote on this subject.

Dates of the five parts: Mathews (Mathews 1925 *Birds of Australia*, Suppl. 4, p.48) lists the month and day of issue of each part. Apparently these dates were taken from the covers of the separate parts.

Having gained additional data on the dates of publication of these parts, I will tabulate the five parts.

PART 1   Sixteen plates, 15 March 1851.
PART 2   Seventeen plates, 1 September 1855.
PART 3   Sixteen plates, 1 September 1859. Reviewed in the October 1859 *Ibis* 1(1):454-5 where fifteen species are listed as illustrated. One species *Casuarius Bennetti*, is illustrated on two plates, one with heads only.
PART 4   Sixteen plates, 1 December 1867. Reviewed in the April 1868 *Ibis* 4(2):216-18 with a list of sixteen species illustrated.

In the *Zoological Record* of 1865, page 65, it is stated in error that seventeen species were included in this part.

In Gould's letter to Professor McCoy of 20 November 1867 (Nat'l. Mus. Victoria) he wrote

You will see by the enclosed strip that a new part of the Suppt. to the Birds of Australia will appear on the 1st of December... I have a sufficient number of novelties to form another part and would issue it at once with Titles etc. so that the 5 pts. might be bound in a volume did not the colouring of the plates interfere[sic] with the other publications I have in hand. The Birds of Great Britain, B of Asia etc.

In the next letter to McCoy, dated 18 February 1868, Gould wrote that part 4 was shipped 'about 6 weeks ago', or around early January 1868.

Gould also wrote to F. G. Waterhouse of Adelaide, Curator of the Museum of the South Australian Institute, on 19 November 1867 (Sutton 1929) 'On the 1st of the next month I shall publish a 4th part of the supplement to the Birds of Australia.'

Further support for the December 1867 publication is in the 1 January 1868 Prospectus and List (Acad. Nat. Sciences Phila.) with the statement that 'Parts I,

II, III & IV, price £3 3s. each have been published; and when Part V is ready, Titles and every requisite to form the whole into a Volume will be furnished'.

PART 5 Sixteen plates, 1 August 1869. Reviewed in the January 1870 *Ibis* 6(2):118-21 where fourteen species are listed as illustrated. Two species, *Casuarius australis* and *C. uniappendiculatus*, are illustrated on four plates.

In the *Zoological Record* of 1869, page 45, it is also stated that fourteen species were included in this part.

In a letter of 28 January 1870 to F. G. Waterhouse (Sutton 1929), Gould stated that 'On the nineteenth of August I sent to your Institute [South Australian Institute] through Mr. Pitman of Paternoster Row the 5th part of the Supplement'.

The Prospectus and List of 1 January 1870 (Nat'l. Lib. Canberra) contains the statement that part 5 had now been completed with title page, etc.

Evidence that Gould planned another part as continuation of the Supplement to his *Birds of Australia* is in his letter to Professor McCoy of 18 February 1870 (Nat'l. Mus. Victoria). Gould had written that part 5 had been shipped and then added, 'I have nearly sufficient for another part. If it should appear it will be under the title of part 1 of Vol. 2 of the Suppt. to the Birds of Australia.' It never appeared.

Contents of the separate parts: The plates as issued are listed in Waterhouse (1885). The species illustrated in parts 3, 4, and 5 are also listed in the *Ibis* reviews.

Zimmer comments on the priority for the new names that appeared in this *Supplement*. Gould had published a paper entitled 'A brief account of the Researches in natural history of John M'Gillivray, Esq.' which appeared in Jardine's *Contributions to Ornithology* for 1850, pages 92 to 105. According to Zimmer, several species in part 1 (numbered 5, 6, 12, 16 and 45 in the List of Plates) along with number 36 in part 3, and 67 in part 2 were included in this paper. If Jardine's *Contributions to Ornithology* were published before 15 March 1851 then those species named in part 1 would definitely have priority from the *Contributions to Ornithology*. For the two names appearing in parts 2 and 3, there is no question of the priority of the *Contributions to Ornithology* over the 'Supplement'.

Editor Jardine states that the paper by Gould on MacGillivray's researches was read in 1850 at the British Association meeting. On page 86 of the 1850 *Contributions to Ornithology* the date of the meeting is given as 30 July 1850. The only complication is that there is no record in the 1850 British Association Report of Gould having read this paper. However, phrases in the contents of the paper lend support to Jardine's statement that the Gould paper was read at the 1850 British Association meeting. Further support for Zimmer's contention on priority would be the examination of the original parts as issued of Jardine's *Contributions to Ornithology* for a more definite date.

Reviews: The reviews in *Ibis* and in the *Zoological Record* are mentioned under the appropriate part.

19 1852-1854. *A monograph of the Ramphastidae, or family of toucans*, 1 volume, 4 parts, 51 coloured plates, 1 uncoloured.

Gould writes in the Preface that 'after an interval of twenty years [following the first edition] a Second Edition [was] a matter of necessity ... with new drawings of the old species, and figures and descriptions of no less than eighteen others'. The Preface is dated 1 May 1854 in the Kansas copy (Ellis F-123).

PART TWO JOHN GOULD'S MAJOR PUBLISHED WORKS

Prospectus: No separate sheet seen. The Prospectus and List of 1 May 1852 (Acad. Nat. Sciences Phila.) contains the information that 'a new edition' was to be published in 'Four Parts, each containing 14 Plates and descriptive letter-press, price £3 3s., at intervals of [sheet torn] months commencing with the 1st of May, 1852'. In the 1856 Prospectus and List (Yale University) is the statement that this work 'was published in 1854 at the price of £12 12s'.

Number of plates: Fifty-one coloured plates and one uncoloured anatomical plate to accompany a text by Richard Owen. Part 1 contained fourteen plates, but there is no definite information on the number of plates in the other parts.

Dates of publication of the parts: Preliminary information. Gould wrote to Lord Derby in a letter of 5 February 1844 (Wagstaffe & Rutherford, letter number 17) that 'the "Icones Avium" is a Work which will always be in progress... the third part is nearly ready, and will contain nine new species of Toucans'. A third part of *Icones Avium* was not published, and Gould waited eight years to issue a new monograph on the toucans. He did publish several articles on new species in the *Proc. Zool. Soc. of London*. Coues lists these papers in his *Ornithological bibliography*, instalment three, pages 722-3.

PART 1   Fourteen plates, 1 May 1852. In the Prospectus and List of 1 May 1852. as mentioned earlier, this 'new edition' of *Ramphastidae* was announced to appear in four parts, each with fourteen plates, beginning on 1 May 1852. Zimmer lists these fourteen species and states that Part 1 was reviewed under the date of 1852 in *Naumannia*, 1853, pages 237-40.

In a letter of 16 April 1852 to Jardine (Brit. Mus. (Nat. Hist.)) Gould wrote that he had sent part 1 of his new edition but that 'all the new species will be published separately for the original subscriber and I suppose you would wish to have them to complete your copy'. Apparently then Gould was just sending this first part for Jardine's enlightenment. It was to be returned, since Jardine would receive the 1855 Supplement as an original subscriber to the 1833-35 first edition.

PART 2   Number of plates and date unknown.
PART 3   Number of plates and date unknown.
PART 4   Number of plates unknown. 1 May 1854. The date that appears on the Preface is 1 May 1854, so part four was scheduled to be completed on this date.

Gould, in a letter of 2 May 1854 to Bonaparte (Mus. Nat'l. d'Hist. Nat., Paris), spoke 'of the new Edition of the Toucans' as soon to be published.

In Gould's letter of 18 May 1854 to Bonaparte he wrote: 'Knowing that you are anxious to see the introductory matter to my work on Toucans, I have sent off an uncorrected proof this day... a fine copy shall be sent to the binders for you immediately.' These two letters support a later May date, for publication of the final part.

In a letter in the Rijksmuseum, Leiden, dated 26 July 1855, to Temminck (to whom Gould dedicated this second edition of toucans), Gould wrote that as of 1 December 1854 he had sent part 4 to Temminck.

Further research is necessary on the question of the dates of publication and the number of plates in each part.

Contents of the parts: As stated previously, Zimmer cites a review in *Naumannia* listing the fourteen species in part 1. Zimmer also prints this list of species.

20   1855. *Supplement to the first edition of a monograph on the Ramphastidae, or family of toucans*, 1 volume, 2 parts, 20 plates.

As related earlier, Gould wrote to Jardine on 16 April 1852 that all the new species of toucans that Gould had published in the second edition would be published separately for the original subscribers to the first 1833-35 edition. This supplement was the separate publication to which Gould referred. But, as Zimmer (1926) states, and as is obvious from a studied comparison, the combination of the first edition of 1833-35 with the supplement of 1855 does not equal the second edition of 1852-54. In the second edition there are various alterations of the species that appeared originally in the first edition. The supplement does not note these changes.

At Kansas this supplement is bound with the 1833-35 Ramphastidae which is Ellis Aves H17.

Prospectus: Not seen. There is a note about the *Supplement to Ramphastidae* in the 1852 Prospectus and List (Acad. Nat. Sciences Phila.) but part of the page has been torn away, making it impossible to quote completely. The 1856 Prospectus and List (Yale University) states that this *Supplement* 'will be supplied separately [from the 1852-54 edition], if required, to the subscribers to the original Monograph [of 1833-35], at the price of £5 5s'.

Number of plates: Twenty. According to Zimmer, from a copy in the Bavarian State Museum, part 1 had fourteen plates and part 2 had six plates.

Dates of publication of the parts: Zimmer gives the date of 1 January 1855 as appearing on the title-page. It usually was Gould's custom to give the date on the title-page as the date of publication of the last part. If this is true for the *Supplement to Ramphastidae*, part 1 might have been published in 1854.

Contents of the separate parts: The fourteen species in part 1 and the six species in part 2 are listed by Zimmer from the Bavarian State Museum copy.

Reviews: None found.

21   1858-1875. *A monograph of the Trogonidae, or family of trogans*, 1 volume, 4 parts, 47 plates.

This is the 'new' or second edition of a monograph that first appeared in 1836-38 with thirty-six plates. According to Wood (1931) the Blacker Wood Library at McGill University had this copy in the four original parts, but it now is bound in the conventional manner (pers. corr. Eleanor MacLean).

Prospectus: No separate sheet seen, but it is listed in both the Prospectus and List of 1 January 1866 and that of 1 January 1868 with the comment that the 'new edition will be completed in four Parts, at £3 3s, the first of which is now ready for delivery'.

Number of plates: Forty-seven. In the 'List of Plates' Gould lists the numbers for forty-six plates, but he inserts an additional plate as 4a, an immature male of *Pharomacrus auriceps*. Waterhouse (1885) allocates each plate to the individual part as issued.

Date of publication of the four parts: Waterhouse, Wood and Zimmer all give the dates in years for the first two parts, and Wood and Zimmer give the month and year for parts 3 and 4.

Plate 17 Drawing in pencil and watercolour, attributed to John Gould, of *Pteroglossus humboldtii*, similar to *Ramphastidae*, 2nd edn, plate 22, Kansas-Ellis-Gould MS 298.

Plate 18 Drawing in pencil and watercolour, attributed to John Gould, of *Ramphastos ambiguus*, for *Ramphastidae*, 2nd edn, plate 5, Kansas-Ellis-Gould MS 957.

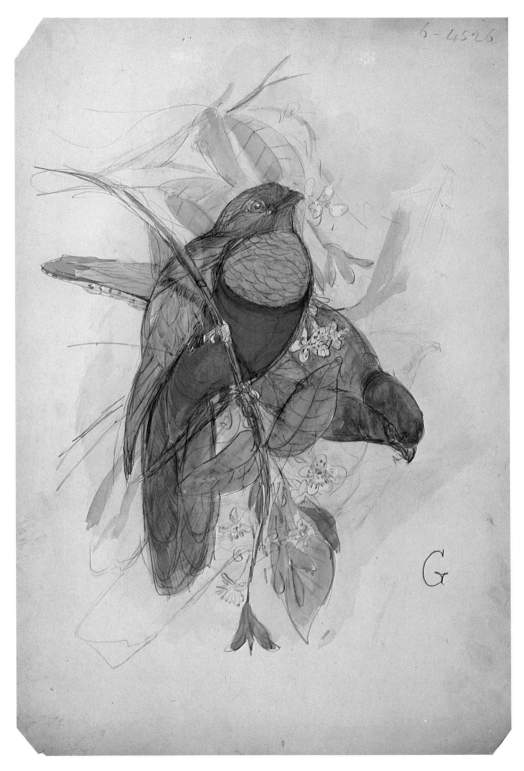

Plate 19 Drawing in pencil and watercolour, attributed to John Gould, of *Pharomacrus pavoninus*, for *Trogonidae*, 2nd edn, plate 5, Kansas-Ellis-Gould MS 938.

Plate 20 Drawing in pencil and watercolour, attributed to John Gould, of *Trogon elegans*, for *Trogonidae*, 2nd edn, plate 9, Kansas-Ellis-Gould MS 929.

PART TWO JOHN GOULD'S MAJOR PUBLISHED WORKS

PART 1   Thirteen plates, 1858. In the January 1859 *Ibis* 1(1):99-101 there is mention of the issuance in 1858 of 'Trogons, first part of a new edition'. There is no list of the species or the number that was illustrated.

PART 2   Thirteen plates, 1 March 1869. Reviewed in the January 1870 *Ibis* 6(2):118-21 with the date of issue of 1 March 1869 and thirteen species listed as illustrated.

In a letter to Professor McCoy of 24 February 1869 (Nat'l Mus. Victoria) Gould wrote

> among the Books you will find the first part of a Monograph of that beautiful Family of Birds the *Trogonidae* which I trust you will subscribe for, for your Museum Library. The 2nd part will be out in a fortnight and the whole affair when finished will not cost more and perhaps less than £12.12 so it will not ruin you.

Here is an inference that this part might not appear before 10 March or so.

The Prospectus and List of 1 January 1870 (Nat'l. Library of Australia, Canberra) contains the statement the 'first and second [parts]... now ready for delivery'.

According to Gould's letter to McCoy of 18 February 1870 (Nat'l. Mus. Victoria) part 2 had been shipped.

PART 3   Eleven plates, June 1875. Reviewed with part 4 in the 1875 *Zoological Record*, as next.

PART 4   Ten plates, September 1875. Parts 3 and 4 were reviewed together in the 1875 *Zoological Record* 12:34 and 66. On page 34 is a short review giving the dates of publication as June and September, 1875 respectively. Mention is made that 'with these two parts, the author completes the second edition of his Monograph of the Trogons, the first part of which was issued in 1858, the second in 1869.' A Preface is included. The total is '47 plates [to] illustrate the 46 species recognized'. On page 66 of this issue of *Zoological Record*, under *Trogonidae* is a combined list of the twenty-one species figured in these two parts.

Parts 3 and 4 were shipped to McCoy on 16 February 1876, according to Gould's letter of 15 April 1876 (Nat'l. Mus. Victoria).

Contents of the separate parts: The review of part 2 in *Ibis* contains a list of the thirteen species illustrated. The *Zoological Record* review lists the twenty-one species illustrated in parts 3 and 4. Waterhouse (1885) also gives the year of issue for each plate, but since the final two parts appeared in the same year, 1875, the breakdown for parts 3 and 4 is not accomplished in his volume either.

Review: The reviews of the parts in *Ibis* and *Zoological Record* are referred to under the appropriate parts.

22   1861. *An introduction to the Trochilidae, or family of humming-birds*, 1 volume, post octavo.

There are two copies of the work in hand, one bound in red cloth and the other in dark green, otherwise the volumes are identical. A Kansas copy (Ellis Aves C447a) bound in black cloth is 'Gould's own proof copy'. There is another copy (Ellis Aves C447b) in red cloth.

In my copies there is a title-page dated 1861, a page of dedication, a 'Notice' of two pages, dated 1 September 1861, a 'Preface' of four pages, dated 1 September 1861, and then an 'Introduction' of 182 pages.

Following the long introduction is an 'Explanation of the abbreviations, and list of the authors and works referred to' taking up five pages, a four-page 'List of generic and specific names adopted' with the page numbers, and then an eighteen-page 'Index of specific names of humming-birds' with page numbers. The book closes with a 'Prospectus of the works on ornithology, and on the Mammalia of Australia by John Gould, F.R.S. etc.' This prospectus covers ten works on four pages and is dated 1 September 1861.

The University of Kansas Spencer Library copy (Ellis Aves C447a) is unique. Ellis writes on the inside front-cover that this is 'Gould's own proof copy'. Mengel (MS) agrees and adds 'compared with Gould MS on hand and evidently true'. There are a few notations throughout the book, some in Gould's hand (pp. iii, 14, 62) but the majority are by E. Prince (pp. 1, 63, 99). Some sections of the book have been removed by scissors. This volume is in an original black binding. According to a note by Ellis it was purchased from 'H. Sotheran £0/18/9 Mar. 11, 37'.

A second copy in the Kansas collection (Ellis Aves C447b) has many annotations and comments written in a shaky hand, that is not Gould's hand. The author of the annotations especially comments on grammar (pp. 3 & 4), on genera types, and on species names (pp. 40-1). Some comments are clever, namely 'Type specimens, Type species, Type forms, Type genera, type fiddles and fiddlesticks'. This copy is in an original red binding. Ellis noted that it was purchased from 'H. Sotheran £0/18/9 Mar. 11, 37'.

The Yale-Coe Collection (Ripley and Scribner 1961) has two copies, one containing MS notes by George Lawrence.

Prospectus: There would have been no prospectus for this small book because it was not for sale.

Date of publication: The date of 1 September 1861 appears in three places in this work, so this should be accepted as the date of publication.

There are only two notes in the available Gould correspondence with reference to this small book. According to a letter of 16 October 1861 to Dr Schlegel (Rijksmuseum, Leiden), Gould wrote that he had 'sent off a few days ago...a little work of mine which please to accept...I have paid attention to the subject both as regards the general history of the species, the synonymy [sic] etc. This latter has been a work of great labour I assure you.' In the box of birds sent to the Rijksmuseum on 15 October 1861 this 'Introduction' was also listed.

In a letter of 18 February 1870 to George Lawrence (Amer. Mus. Nat. Hist.), Gould wrote 'I have written to Prof. Orton and sent for his acceptance a copy of the 8vo Introduction to the Trochilidae.' In Coues's *Ornithological bibliography* (third instalment, p. 681), Coues lists several 'species given without a reference' in Gould's 'Introduction'.

Review: None found.

23   1862–1873. *The birds of Great Britain*, 5 volumes, 25 parts, 367 plates.

Gould was especially proud of this work on the birds of his native land. As a 'novelty' he included quite a few drawings with figures of young birds.

As would be expected, there were more subscribers for this set than any other; 468 were listed in *An Introduction to the Birds of Great Britain*. This was the only time Gould made a reference to a 'second edition'. In a letter dated 13 May 1863 (Nat'l. Mus. Victoria) Gould wrote to Professor McCoy that he had sent in a box 'two copies of parts 1 and 2 of my new work on the Birds of Great Britain, a publication that has so pleased everyone, and had so large a sale here that a second edition was called for in 4 months'.

PART TWO JOHN GOULD'S MAJOR PUBLISHED WORKS

Prospectus: A full-page prospectus appears at the end of Gould's *Introduction to the Mammals of Australia* (personal copy), with other works listed on a second page, and is dated 1 August 1862 (Figures 24a & b). The work was to be published 'at the rate of two Parts a year, price Three Guineas each,... the precise number of

PROSPECTUS

OF

"THE BIRDS OF GREAT BRITAIN."

BY

JOHN GOULD, F.R.S., ETC.

THE Author has been induced to commence a work under the above title at the urgent request of numerous scientific Friends and Subscribers to his former publications, and he trusts he shall not disappoint them in their desire for a standard work on our native birds.

"The Birds of Great Britain" will be published in Imperial Folio, at the rate of two Parts a year, price Three Guineas each; the Subscription will therefore be Six Guineas per annum. The precise number of Parts cannot be stated, but it is expected that the work will be completed in eight or nine years. With the last Part a full introduction to the subject will be published, together with Titles and every requisite to form the whole into five volumes; the first of which will comprise the Raptores or Birds of Prey; the second and third, the Insessores or Perching Birds; the fourth, the Rasores and Grallatores; and the fifth, the Natatores or Swimming Birds.

It must not be supposed that this work is a continuation of, or similar to, the "Birds of Europe;" the subjects are very differently treated as to the illustrations, the letterpress is much more voluminous, and the figures of the infantine states of most of the genera render it quite distinct.

The Author makes it a *sine quâ non* that the name of every Subscriber shall be given, and the Subscription paid on delivery, or at least annually; and he trusts that no person will commence the Subscription without continuing it to its close.

Any additional information may be obtained by letter or by personal application to the Author at his house, 26 Charlotte Street, Bedford Square, London, W.C.

[*Turn over.*

August 1, 1862.

Figure 24 a & b A two-page prospectus on *Birds of Great Britain* and other Gould works, appearing at the end of Gould's *Introduction to Mammals* (Author's Collection).

2

The Author's other Publications, all in Imperial Folio, are—

A CENTURY OF BIRDS FROM THE HIMALAYA MOUNTAINS. 1 Vol. 1832 . . . . £14 14	
All sold.	
THE BIRDS OF EUROPE. 5 Vols. 1837	76 8
All sold.	
THE BIRDS OF AUSTRALIA. In 36 Parts, forming 7 Vols. 1848	115 0
Very few Copies remain.	
SUPPLEMENT TO DITTO. 3 Parts, each £3 3s.	9 9
THE BIRDS OF ASIA. This important work is issued for the present at the rate of one Part per annum. 15 Parts are published, price £3 3s. each	47 5
The precise number of Parts cannot at present be determined.	
A MONOGRAPH OF THE TROCHILIDÆ OR HUMMING BIRDS. In 25 Parts, forming 5 Vols. 1861	78 15
A MONOGRAPH OF THE RAMPHASTIDÆ, OR FAMILY OF TOUCANS. 1 Vol. 1834	7 0
SECOND EDITION, all the Plates redrawn, and many new Species. 1854	12 12
A MONOGRAPH OF THE TROGONIDÆ, OR FAMILY OF TROGONS. 1 Vol. 1838	8 0
SECOND EDITION. Part I.	3 3
To be completed in 4 Parts; all the Plates redrawn, and many additional Species.	
A MONOGRAPH OF THE ODONTOPHORINÆ, OR PARTRIDGES OF AMERICA. 1 Vol. 1850	8 8
THE MAMMALS OF AUSTRALIA. In 13 Parts (forming 3 Vols.)	41 0
12 Parts published.	

The Author will be happy to supply any of the above he may have to those who may be desirous of completing their series of his works; or of supplying, so far as lies in his power, the Parts which may be required to perfect any one of them which may have been left incomplete. He would particularly call attention to the lately finished MONOGRAPH OF THE TROCHILIDÆ, OR HUMMING BIRDS, one of the most beautiful and interesting of his Publications.

As no just conception of the character of these Works can be formed from a Prospectus, the Author is willing to send a Part or Parts for the inspection of those who may be desirous of possessing them.

LONDON: PUBLISHED BY THE AUTHOR
AT 26 CHARLOTTE STREET, BEDFORD SQUARE, W.C.

Figure 24 b

Parts cannot be stated but it is expected that the work will be completed in eight or nine years'. A separate prospectus of this date is in the La Trobe Collection. (*See* Gould, John. Correspondence 1857-81 ... item 6.)

In the October 1862 *Ibis* 4(1):392 there is an announcement that 'Mr. Gould is engaged in preparing for publication the two first numbers of the ... *Birds of Great Britain*'.

PART TWO JOHN GOULD'S MAJOR PUBLISHED WORKS

Number of plates: 367. Waterhouse (1885) stated that 'Part XXV contains the title pages, etc. and eight plates of Young Birds, to be inserted in their proper places in vols. II & III'.

Unfortunately, from a study of Waterhouse, and the reviews in *Ibis* and *Zoological Record*, and in the absence of a copy of this work in the original parts as issued, there is no absolute information now available to give the correct number of plates for each part. The confusion arising from the Waterhouse, *Ibis* and *Zoological Record* data, I will try to make more evident as I list the individual parts. The most prudent assumption is that, aside from the last part, each of the other twenty-four parts except one contained fifteen plates, and the exception had only fourteen plates. However, other mathematical combinations are possible.

Dates of publication of the parts: Zimmer (1926) states that 'parts I & II appeared on October 1, 1862, thereafter two parts appeared annually (on August 1 and September 1 in 1866 to 1872 and probably also in the other years), and the last part... was issued in December, 1873'. This information is supported by the *Ibis* reviews.

The following list is a schedule of the dates of publication of the parts, plus any additional information I have acquired from Gould's correspondence, prospectus and lists of subscribers, reviews, etc. The assumption is made that fifteen plates were published in most parts.

PART 1   Fifteen plates, 1 October 1862.
PART 2   Fifteen plates, 1 October 1862. Parts 1 and 2 were reviewed together in the January 1863 *Ibis* 5(1):102-3 as being issued on 1 October 1862.

In a letter dated 13 August 1862 to Professor McCoy (Nat'l. Mus. Victoria) Gould wrote 'the first and second parts will soon be ready'.

In a letter to George Williams of 27 February 1863 (Nature Society, Griggsville, Illinois) Gould thanked him for subscribing to his *Birds of Great Britain* and stated that 'the first two parts have been forwarded to Trevina'.

PART 3   Fifteen plates, 1863.
PART 4   Fifteen plates, 1863. Parts 3 and 4 were reviewed together in the January 1864 *Ibis* 6(1):116 with no date of issue but with the fifteen species in each part listed.
PART 5   Fifteen plates, 1864. The Waterhouse listing adds up to only thirteen species.
PART 6   Fifteen plates, 1864. Parts 5 and 6 were reviewed together in the January 1865 *Ibis* 1(2):98-9 with no dates of issue, but with the fifteen species in each part listed.
PART 7   Fifteen plates, 1865.
PART 8   Fifteen plates, 1865. Parts 7 and 8 were reviewed together in the October 1865 *Ibis* 1(2):526 with no date but with a list of the fifteen species for each part.
PART 9   Fifteen plates, 1 August 1866. This date and the subsequent more specific dates are taken from reviews in *Ibis*.
PART 10   Fifteen plates, 1 September 1866. Parts 9 and 10 were reviewed together in the April 1867 *Ibis* 3(2):237-8. Here and in every subsequent *Ibis* review the list of species for each part is included in the review.
PART 11   Fifteen plates, 1 August 1867.
PART 12   Fifteen plates, 1 September 1867. Parts 11 and 12 were reviewed together in the April 1868 *Ibis* 4(2):216-18.

In the 1868 Prospectus and List of Subscribers (Acad. Nat. Sciences Phila.), dated 1 January 1868, is the statement 'Twelve Parts have now been published', at the rate of two per year.

PART 13 Fifteen plates, 1 August 1868.

PART 14 Fifteen plates, 1 September 1868. Parts 13 and 14 were reviewed together in the January 1869 *Ibis* 5(2):108-9.

PART 15 Fifteen plates, 1 August 1869.

PART 16 Fifteen plates, 1 September 1869. Parts 15 and 16 were reviewed together in the January 1870 *Ibis* 6(2):118-21.

Gould's letter to McCoy dated 18 February 1870 (Nat'l. Mus. Vict.) states that he had shipped parts 15 and 16.

PART 17 Fifteen plates, 1 August 1870.

PART 18 Fifteen plates, 1 September 1870. Parts 17 and 18 were reviewed together in the October 1871 *Ibis* 1(3):437-8 with only 1870 as the date. The day and month of issue are shown in the 1870 *Zoological Record* review (7:23).

PART 19 Fourteen plates, August 1871. In a letter to George Lawrence of New York dated 22 June 1871 (Amer. Mus. Nat. Hist.), Gould wrote that he was considering publishing more on the humming-birds but that 'it will be impossible for me to do so until after the completion of the Birds of Great Britain; for I have almost all the colourers in London engaged in colouring my works in hand and more I cannot do'.

PART 20 Fifteen plates, September 1871. Parts 19 and 20 were reviewed together in the October 1872 *Ibis* 2(3):428-9. The twenty-nine species illustrated are listed together, so the number for each part is indefinite. But according to the data I obtained from Waterhouse (1885), part 19 had fourteen plates and part 20 had fifteen plates.

PART 21 Fifteen plates, August 1872.

PART 22 Fifteen plates, September 1872. Parts 21 and 22 were reviewed together in the supplement to the 1873 *Ibis* 3(3):453-4 with twenty-six species listed together. In both of these parts, according to the 'List of Plates' in the bound volumes, there were multiple illustrations for three species. These extra illustrations number four, so if the *Ibis* number is correct, 26 plus 4 equals 30 for these two parts.

PART 23 Fifteen plates, 1 August 1873.

PART 24 Fifteen plates, 1 September 1873.

PART 25 Eight plates, before 3 December 1873. In a letter to T. C. Eyton dated 4 December 1873 (Library, Univ. Birmingham, England) Gould wrote 'I yesterday forwarded the concluding parts of my work on The Birds of Great Britain'.

In the April 1874 *Ibis* 4(3):172-3 is a short statement that Gould's 'great work on the Birds of Great Britain is now complete ... no bird-book, it is whispered, has ever had such a financial success'.

Contents of the parts: An allocation of the plates to their respective parts is made by Waterhouse (1885). The reviews in *Ibis* also list each plate except for those in the first two parts and in the last three parts. As stated earlier, the data from Waterhouse, *Ibis* and the *Zoological Record* show areas of disagreement.

Reviews: The *Ibis* reviews are referred to under each appropriate part.

The *Zoological Record* reviews begin in 1864 and cover parts 5 and 6. This first review is a very laudatory long paragraph appearing on pages 42 to 43 of volume 1. Subsequent reviews are of one sentence only. Under the section of Neossology is listed the species of the young birds illustrated in Gould's *Birds of Great Britain* for that year.

PART TWO   JOHN GOULD'S MAJOR PUBLISHED WORKS

As mentioned earlier under the section on the *Birds of Europe*, seven manuscript volumes (apparently used by Gould as a 'filing cabinet' for notes and sketches for the *Birds of Europe* and the *Birds of Great Britain*), are in the libraries of the British Museum (Natural History) and the Academy of Natural Sciences of Philadelphia. *See also* Gould, John. Manuscript. [no date.] '...for "The Birds of Europe,"' and "Europe/insessores...," and Sauer (1976 and 1978). It would be of considerable interest to compare these notebooks with the published works.

24   1863. *An introduction to the mammals of Australia*, 1 volume, post octavo.

Gould states in the 'Notice': 'The Preface and Introduction to my "Mammals of Australia" having been set up in small type for facility of correction, I have had a limited number of copies printed in an octavo form, for distribution among my scientific friends and others... they must however still regard it rather in the light of a proof-sheet, inasmuch as it contains many imperfections, which have been corrected in the folio edition' (Figure 25).

Figure 25 Cover of post octavo book *Introduction Mammals* (Author's Collection).

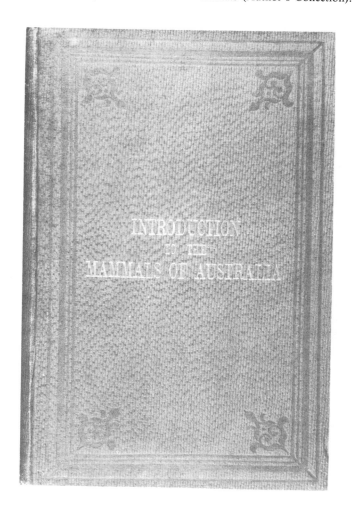

Date of publication: The date of publication was before 13 May 1863 as evidence this letter to Professor McCoy of this date (Nat'l Mus. Vict.) where Gould wrote 'I sent by the post an 8vo Edition of the Introduction for your own acceptance'. The final part 13 of the *Mammals of Australia* was published on 1 May 1863. The date of 1 August 1862 on the final page of the volume under a Prospectus of *Birds of Great Britain*, I would assume to date the Prospectus and not *Introduction to the Mammals*.

Comments on contents: A simple collation from my personal copy is as follows: title-page dated 1863; dedication-page to His Royal Highness, the Prince Consort, with a statement overleaf about the 'untimely loss' of the Prince; a notice-page; a preface (pp. vii–xii); an introduction (pp. 1–51) containing some general remarks on the mammals of Australia, and then a listing, with a few comments, of the species figured in the folio edition, and finally, a page of Prospectus of *Birds of Great Britain* dated 1 August 1862. Overleaf is a list with prices of the 'Author's other publications'.

25   1865. *Handbook to the birds of Australia*, 2 volumes, royal octavo.

The Preface contains Gould's statement for this work. Twenty years had elapsed since his folio work on the *Birds of Australia* was completed. 'During that period many new species have been discovered, and much additional information acquired respecting those comprised therein; consequently it appeared to me that a careful résumé of the entire subject would be acceptable.'

Prospectus: At the back of my copy of volume 1 is a sheet pasted in. Recto is a prospectus of this work (Figure 26). Verso is a prospectus of *Birds of Great Britain*, and a listing of ten of 'the Author's other Publications', with prices.

The first paragraph of the prospectus for the *Handbook* gives important information: 'This day is published, Vol. I of the above work, in royal octavo, containing 636 pages, price £1 5s.; and on the 2nd of December will be issued the second and concluding volume, at the same price.' The date at the bottom of this sheet is 'September 1st, 1865'.

Number of plates: None.

Dates of publication of the two volumes: As stated above, volume 1 was published 1 September 1865 and volume 2 scheduled for 2 December 1865. Here is some additional information from Gould's correspondence.

In a letter to Professor McCoy of Melbourne dated 16 September 1865 (Nat'l. Mus. Victoria), Gould wrote 'I am deep in a publication on the Birds of Australia of which the accompanying prospectus will give you the particulars. Now I do expect and hope this work will have a large sale in the Colonies. Robertson the Bookseller of Melbourne has subscribed for 25 Copies to begin with...a copy of the first volume has been sent overland.'

McCoy's letter of 20 December 1865 (Nat'l. Mus. Victoria) from Gould had the statement 'The second [volume] is now out...the second is by far the most important as it contains a table of the distribution of the species, Index. etc.'

Of interest regarding the cost of publication is a letter of 24 January 1866 to F. G. Waterhouse of Adelaide (Sutton 1929). Gould wrote 'I have just published a Handbook to the Birds of Australia – price £2.10:...I would send you a copy free of charge but the expense of printing it, over £600, precludes my so doing.'

Plate 21 Drawing in pencil and watercolour, signed J. Gould, F.R.S., of *Cyanecula suecica*, for *Birds of Great Britain*, II: 49, Kansas-Ellis-Gould MS 1075. This fine drawing was offered for sale as item 1833A by Henry Sotheran, Ltd in No. 9 'Piccadilly Notes' for £5.5s.

Plate 22 Drawing in pencil and watercolour, attributed to W. Hart, of *Ptilotis frenata*, for *Birds of New Guinea*, III: 49, Kansas-Ellis-Gould MS 978.

Plate 23 Drawing in pencil and watercolour, signed by H. C. Richter, of *Rhinolophus aurantius*, for *Mammals of Australia*, III: 35, Kansas-Ellis-Gould MS 730.

Plate 24 Drawing in watercolour, attributed to H. C. Richter, of Rose-breasted Cockatoo, for *Birds of Australia*, V: 4, Blacker-Wood, McGill.

PART TWO JOHN GOULD'S MAJOR PUBLISHED WORKS

HANDBOOK

TO THE

BIRDS OF AUSTRALIA.

BY

JOHN GOULD, F.R.S., &c.

This day is published, Vol. I. of the above work, in royal octavo, containing 636 pages, price £1 5s.; and on the 2nd of December will be issued the second and concluding volume, at the same price.

Extract from Preface, vol. i.

" Nearly twenty years have elapsed since my folio work on the Birds of Australia was completed. During that period many new species have been discovered, and much additional information acquired respecting those comprised therein; it therefore appeared to me that a careful *résumé* of the entire subject" [in an octavo form] " would be acceptable to the possessors of the former edition, as well as to the many persons in Australia who are now turning their attention to the ornithology of the country in which they are resident. Indeed I have been assured that such a work is greatly needed to enable the explorer during his journeyings, or the student in his quiet home, to identify the species that may come under his notice, and as a means by which the curators of the museums now established in the various colonies may arrange and name the collections entrusted to their charge."

PUBLISHED BY THE AUTHOR,

AT 26 CHARLOTTE STREET, BEDFORD SQUARE, W.C.

September 1st, 1865.

Figure 26 Prospectus for Gould's *Handbook* (Author's Collection).

Contents: Whittell (1954) lists the new names of genera and species that appear in these two volumes.

Review: The two volumes are reviewed in the January 1866 *Ibis* 2(2):111-13. There are some critical comments regarding excessive genus formation.

A review in the 1865 *Zoological Record* 2:74-5 is the longest review of any of Gould's work in this Record. 'This will most likely prove to be the most important work on

ornithology published during the past year, and, if we except Dr. Jerdon's "Birds of India," it might even be safely said the most important published for several years.' Gould's work on the synonymy of birds is greatly praised, but his 'diagnoses, whether generic or specific, are in almost every case wanting'.

Two long reviews with extensive quotes are by Edward Newman in the 1865 and 1866 *Zoologist* 23(series 1):9785-9 and 1(series 2):41-58.

26   1873. *An introduction to the birds of Great Britain*, 1 volume, post octavo.

On the page headed 'To the Reader' Gould writes 'I have had the *Introduction to the Birds of Great Britain* set up in small type for the convenience of correction before printing it for the folio work.' It is the only octavo 'Introduction' having a price on the title-page; 'Price Five Shillings and Sixpence'.

Four copies are in my personal collection. There is only one variation:
Copy 1 has the 'Preface' first, then a page 'To the Reader'.
Copy 2 has the 'To the Reader' page before the 'Preface', and has the bookplate of John Eliot Thayer.
Copy 3 also has the 'To the Reader' page first. It is inscribed 'Major Godwin Austen /with the Authors Kind Regards/ 19th Dec. 1873'.
Copy 4 has the 'Preface' first. It is inscribed 'Mrs. Amhurst/with the Authors Kind Regards/Oct. 21st 1874'.

Date of publication: The date of 1 November 1873 concludes the preface to the book.

Comments on contents: A simple collation of Copy 1 is as follows: the title-page dated 1873; dedication page to Rowland, Viscount Hill, of Hawkstone; preface of two pages dated 1 November 1873; 'To the Reader' page; introduction of 135 pages with a listing of all the species illustrated in the folio work, with paragraphs of text; a 'List of Subscribers' on fourteen pages (in Copy 3 the number 468 is pencilled correctly next to the last name on this list of subscribers), and finally, there is a four-page 'Prospectus/of the works/on ornithology, etc.,/by/John Gould, F.R.S.' which is dated August 1873.

Review: A short paragraph review is in the October 1874 *Ibis* 4(3):450.

Edward Newman wrote a long review in the 1874 *Zoologist* 9(series 2): 4085-97, but it is mainly made up of quotes from this 'Introduction'.

27   1875-1888. *The birds of New Guinea*, 5 volumes, 25 parts, 320 plates.

R. Bowdler Sharpe contributed more to this unfinished Gould work than to any of the other four in which he assisted. Sharpe assumed authorship of the work after the twelfth part had been issued in 1881.

Prospectus: No single sheet seen.

Number of plates: 320. Each part, according to Anker (1938), contained thirteen plates except for twelve each in parts 8 and 9, and ten plates only in the last part.

Dates of publication of the parts: In the 'Contents' for each of the five volumes, the species illustrated is listed with the year of issue of the original part in which

PART TWO JOHN GOULD'S MAJOR PUBLISHED WORKS

it appeared. Anker (1938) has also tabulated this information from a copy in the original parts.

For chronological interest I will list each part, the number of plates, date of issue, and any further information obtained from correspondence or reviews.

PART 1   Thirteen plates, 1 December 1875.
PART 2   Thirteen plates, 1 January 1876.
PART 3   Thirteen plates, 1 May 1876. Mentioned as 'preparing for issue' in the July 1876 *Ibis* 6(3):363, with thirteen species listed.

In a letter to Professor McCoy dated 15 April 1876 (Nat'l. Mus. Victoria) Gould wrote that parts 1 and 2 were sent to him on the 16th of February and that 'Pt 3 is ready'. This would indicate that this part was ready ahead of schedule.

PART 4   Thirteen plates, 1 January 1877. Reviewed in the July 1877 *Ibis* 1(4):377, with thirteen species listed.
PART 5   Thirteen plates, 1 June 1877. Reviewed in the January 1878 *Ibis* 2(4):109, with thirteen species listed.
PART 6   Thirteen plates, 1 February 1878.
PART 7   Thirteen plates, 1 June 1878.
PART 8   Twelve plates, 1 October 1878. Parts 6, 7 and 8 were reviewed in the January 1879 *Ibis* 3(4): 94-5, with thirteen, thirteen and twelve species respectively listed. 'In spite of severe health, Mr. Gould continues his great work on the birds of New Guinea.'

In a letter to J. Noble dated 17 October 1878 (Nature Society, Griggsville, Illinois) Gould wrote 'Yesterday I sent Part 8 of "New Guinea" '.

PART 9   Twelve plates, 1 March 1879. Reviewed in the July 1879 *Ibis* 3(4):356, with twelve species listed.
PART 10   Thirteen plates, 1 September 1879. Reviewed in the January 1880 *Ibis* 4(4):134-5, with twelve species listed. One species of Bird of Paradise was portrayed on two plates.
PART 11   Thirteen plates, 1 February 1880. Reviewed in the October 1880 *Ibis* 4(4):474-5, with thirteen species listed.
PART 12   Thirteen plates, 1881. Reviewed in the January 1882 *Ibis* 6(4):168, with a list of twelve species. One species of Bird of Paradise was portrayed on two plates.
PART 13   Thirteen plates, 1882. Reviewed in the October 1882 *Ibis* 6(4):601-2, with a list of thirteen species.
PART 14   Thirteen plates, 1883. Reviewed in the October 1883 *Ibis* 1(5):568, with thirteen species listed.
PART 15   Thirteen plates, 1883. Reviewed in the April 1884 *Ibis* 2(5):208-9, with thirteen species listed.
PART 16   Thirteen plates, 1884. Reviewed in the July 1884 *Ibis* 2(5):341-2, with thirteen species listed.
PART 17   Thirteen plates, 1884.
PART 18   Thirteen plates, 1884. Parts 17 and 18 were reviewed in the April 1885 *Ibis* 3(5):227-8, both having thirteen species.
PART 19   Thirteen plates, 1885. Reviewed in the July 1885 *Ibis* 3(5):316, with thirteen species listed.
PART 20   Thirteen plates, 1885. Reviewed in the April 1886 *Ibis* 4(5):194-5, with thirteen species listed.
PART 21   Thirteen plates, 1886.

PART 22   Thirteen plates, 1886. Both parts 21 and 22 were reviewed in the January 1887 *Ibis* 5(5):108, with thirteen species in each part.
PART 23   Thirteen plates, 1887.
PART 24   Thirteen plates, 1888. Parts 23 and 24 were both reviewed in the April 1888 *Ibis* 6(5):271-2, with thirteen species in each part.
PART 25   Ten plates, 1888. Reviewed in the April 1889 *Ibis* 1(6):247, with ten species listed. A comment is made that it is a pity to stop when 'only some 300 species have been figured out of a total of 1030' now known.

Contents of the parts: As stated before, the contents of each part are listed in an index to each volume, and in the *Ibis* reviews.

Reviews: The completed work was reviewed by J. A. Allen in the *Auk* 6:269 as a 'magnificent contribution to ornithology'. The separate parts, except for parts 1 and 2, were reviewed in *Ibis*, as mentioned under the appropriate part.

28   1880–1881. *Monograph of the Pittidae*, 1 volume, 2 parts, 10 plates.

The first part with ten plates and text was published in 1880 before Gould died. In my bound copy, there are an additional ten text pages in a second part written by Gould but edited by Sharpe. Both parts were enclosed in a pale-blue cover.

Prospectus: None seen. Under item 2323 of the 1881 Quaritch *General Catalogue* is the statement 'The Pittas will be finished in four parts, At Three Guineas a part. Professors of the Birds of Asia and New Guinea do not require this work, as it contains no new plates, but those already published there.'

Number of plates: Ten. On the original cover of the first part is the statement 'The Illustrations are principally taken from the Author's works on the "Birds of Asia", "Australia", and "New Guinea".' There are no illustrations in the second part. The measurement of the trimmed pages of the volume in the author's collection is $21\frac{7}{8} \times 15\frac{1}{8}$ inches wide (555 × 384 mm).

Dates of publication of the two parts: The date on the original wrapper of the first part is 1 October 1880. On the wrapper for the second part the date is 1881.

Contents of the parts: The ten species illustrated in the first part are printed on the original cover, and also in Zimmer (1926). Text pages accompany the illustrations.

In the second part, the ten species are also printed on the original cover, but this part contains the text pages only. All ten are initialled at the end with '[R.B.S.]', but from comments in the text it is obvious that they were written by Gould.

No title page (or other introductory material) is included.

Reviews: None found.

29   1880–1887. *A monograph of the Trochilidae, or family of humming-birds*. Supplement, 1 volume, 5 parts, 58 plates.

In 1881, after Gould died, Henry Sotheran & Co. purchased the publication rights to Gould's works, along with other items of his estate. Only part 1 appeared before Gould's death, so the remaining four parts were published by Sotheran & Co., with

PART TWO   JOHN GOULD'S MAJOR PUBLISHED WORKS

R. B. Sharpe responsible for the text, with W. Hart as the artist, and with Osbert Salvin who gave 'advice in planning the present work and ... supervised the proofs of each Part'.

Prospectus: Not seen. In Bernard Quaritch's *General Catalogue* of November 1881, item 2587 concerns this 'Supplement'. The first part is listed for sale at £3.3s. 'This Supplement will be completed in four parts; the second part is forward and will shortly be published.'

Plates: Fifty-eight. Sixty species were illustrated on fifty-eight plates.

Dates of publication of the parts: Specific dates and the number of species figured in each of the five parts are given in *Ibis* reviews. The years of publication of each of the first four parts are also given by Waterhouse (1885).

PART 1   Twelve plates, 1 August 1880. Reviewed in the October 1880 *Ibis* 4(4):474-5. There is the statement that Gould was starting this work 'undeterred by serious illness'.

PART 2   Ten plates, 1881. Reviewed in the January 1882 *Ibis* 6(4):167-8. A statement is made that, due to the death of Gould, Henry Sotheran & Co. would continue the work with the help of Salvin and Sharpe.

PART 3   Twelve plates, 1883. Reviewed in the April 1883 *Ibis* 1(5):214. Fourteen species are listed as being illustrated. These are on twelve plates (Waterhouse 1885). An additional sixteen species are mentioned in the review as being described only, not illustrated.

PART 4   Twelve plates, 1885. Reviewed in the July 1885 *Ibis* 3(5):316-17.

PART 5   Twelve plates, 1887. Reviewed in the July 1887 *Ibis* 5(5):352-3. Though only twelve species were illustrated, an additional twenty-nine other species were described in the text. The review further states that the title-page, preface, and contents are also included in this final part. 'The total number of Humming-birds treated of in the Supplement is 122, of which 60 are figured. In the original work 360 species were illustrated. This would make the total number of species about 482.'

Contents of the parts: The *Ibis* review for each of the five parts contains a listing of each species illustrated. This information is also available for the first four parts (and by deduction for the fifth part) from Waterhouse in his 1885 'Dates of Publication of some of the zoological works of the late John Gould'.

Reviews: The dates and pages of the reviews from *Ibis* have been listed under each appropriate part. J. A. Allen very favourably reviews the entire *Supplement to Trochilidae* in the 1889 *Auk* 6:334.

30   [1883]. *Introduction/to/Gould's Birds of Asia./by/R. Bowdler Sharpe*, 1 volume, imperial folio, paper wrapper.

This 'Introduction' was advertised as item 48 in the 1967 No. 945 'An illustrated catalogue of fine books' published by Henry Sotheran, Ltd of London. I have three copies of this separate. It is not listed in any printed collection of Gould's works.

The information on this item from Sotheran's 1967 catalogue is as follows: '48. Gould (John), and Richard Bowdler Sharpe. Introduction to Gould's Birds of Asia, with a specimen handcoloured lithograph plate, imperial folio (pp. 9), original paper wrapper, cloth back, £6. [1883]. Printed off separately; consisting of an interesting résumé of the progress of Oriental ornithology since Gould began his Birds of Asia in 1850.'

It would appear that Sharpe decided to provide an 'Introduction' for the *Birds of Asia* as Gould had done before for his works on Australian birds, Australian mammals, humming-birds and birds of Great Britain. The 'Introductions' by Gould, however, were post-octavo size, cloth bound, contained *all* the introductory material that was to appear in the first volume of the folio work, and had no coloured plate. This Asia item is imperial-folio size, contained in a tan paper wrapper with lavender cloth-backing, and consists only of the nine-page 'Introduction' identical with that appearing in the first volume of the *Birds of Asia*. The further difference is that this Asia separate has bound in a specimen hand-coloured lithographic plate. One of my three copies, however, is lacking the plate.

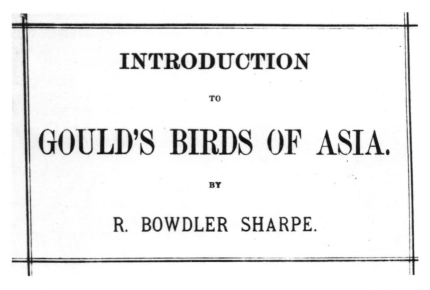

Figure 27 Cover label for the *Introduction to Gould's birds of Asia* (Author's Collection).

A similar small label for this 'Introduction' item appears on the paper wrapper cover of all of my three copies (Figure 27). The label measures 95 × 140 mm. The collation of this title appears at the beginning of this section. The work is not dated, but obviously it appeared in 1883 when the final 35th part of *Birds of Asia* appeared. R. Bowdler Sharpe's name as author of this introductory text appears on the final page, page nine.

Part Three

Chronology of the life and works of
John Gould

With neither the time nor the ability to write a flowing biography of John Gould, it was felt that a chronological listing would be of some value. Gould's correspondence, which I hope will soon form a second volume, will provide a more detailed chronology of his life and works. Until that time, however, this incomplete record will have to suffice.

The basis for this chronology is the listing of any specific date that refers to John Gould, his family, his associates, or any of their relevant activities. Thus, there may be some extended periods without any apparent activity. These gaps could have been filled in with the dates of many of his letters, but, as I said, that compilation is not yet complete.

The dates have been taken from many sources: from letters to and from the Gould relatives, his associates, and his subscribers, from his *Handbook to the Birds of Australia*, from the 1938 *Emu* Commemorative Issue, and from diaries, etc. All of these sources are acknowledged. Gould or his family left no extensive diary or other formal record encompassing any extended period. Mrs Gould did write a short diary over a five-week period from 21 August to 30 September 1839, while on the Australian venture.

Other writers have commented on Gould's lapses regarding a few dates, particularly as listed in his *Handbook*, and these lapses have been corrected.

To give a flavour of the Victorian times a few selected dates of noteworthy events or situations have been interspersed. Many more could have been added, but, as evidenced from Gould's correspondence, I do not feel he was keenly aware of the world about him, unless it was the world of ornithology, and even then his interest was esoteric and businesslike. For example, the content of Gould's correspondence was the antithesis of John Audubon's or Edward Lear's long, gossipy, personal and informative epistles.

Gould's life, as he wrote quite a few times, was his birds and his publications. Therefore it was deemed appropriate in this chronology also to concurrently tabulate the dates of issuance of his published volumes or parts of volumes. The present chronology will not include the dates of reading or publication of Gould's scientific papers.

Introduction

In 1804, the year of Gould's birth, King George III and Queen Charlotte were on the throne of England. England was at war with France where Napoleon Bonaparte was Emperor. Thomas Jefferson was President of the United States.

In New Holland (Australia), Hobart Town in Van Diemen's Land (Tasmania) was settled by the English. Just a year before this (1803) they had tried unsuccessfully

to establish a settlement on Port Phillip Bay, some distance from the present site of Melbourne. Sydney, in the Port Jackson-Botany Bay area, was slowly developing into a city from an English convict colony begun by 'transportation' some 16 years earlier.

England, at this time, was still an agricultural nation and only beginning to build up its industrial strength. The population of London in 1804 was almost one million while that of Britain was ten million. Transportation was on dirt or rock roads until a few macadam roads were laid after 1815. Travel on these roads was in horse-drawn carriages until the early 1900s. Train coaches on the early railroads were also drawn by horses, or on short routes were pulled by cable from a fixed point, until the steam-engine began to be used in the 1830s. By law these early trains could not exceed 4 miles per hour (6.4 km/h) and a man with a flag had to walk in front of the engine. Ocean travel was by wind-power until the 1830s when steam-power began to be used in combination with sails. Farming was by wooden plough until the 1820s. A few streets in London and the larger cities were lit by whale-oil lamps, but gas lamps had gradually replaced them by the 1830s.

Printing was done with cumbersome, slow hand-presses until after the 1850s. Gould's large imperial-folio sheets were undoubtedly always printed on these hand-presses.

The flint-lock muzzle-loaded shotgun was in use in 1804. One usually had two guns when shooting for sport, so that a 'loader' could reload one gun while you fired the other. The percussion gun using a detonating powder was not invented until 1807. The powder was not smokeless and you were enveloped in a cloud of smoke that had to clear before you could fire again.

Viewing of birds or other objects at a distance was done by telescope.

Medicine in 1804 was still mainly a barber-surgeon science. Anaesthesia was not known until after 1809. Diseases were thought to be spread by 'humours' in the air, and cells and 'organisms' were thought to reproduce by spontaneous generation. Smallpox vaccination had been found to be successful in 1796, but many physicians ridiculed those others who used it. Epidemic diseases killed millions of people in the 1800s. Yellow fever, typhus, cholera, tuberculosis, and influenza occurred in pandemics.

In the early 1800s, concomitant with a proliferation of weekly and daily newspapers, it became faddish for persons in the middle and upper classes to attend scientific lectures given by the renowned scientists of the time. Private museums displaying the latest scientific discoveries were popular. This burgeoning interest in science was no more evident than in the field of natural history. In the 1750s Linnaeus had devised a system of nomenclature for all living creatures, using Latin generic and species terms. Thus, the binomial system developed. Linnaeus described and named the birds he knew, but others such as Latham, Gmelin, Horsfield, John Edward Gray, and Vigors, were to add many more to the list before Gould wrote his first scientific notice in the 1829-30 *Zoological Journal*.

The Victorians especially loved birds. They wore their feathers in their hats, which Gould spoke out against very strongly. They built aviaries. They bought large folio-sized volumes by Buffon, Levaillant, Selby, Lear and Audubon to adorn their libraries and homes. The stage was set for an indefatigable worker with a systematic mind and a carefully chosen team of associates literally to flood the market with a never-to-be-repeated series of imperial-folio works on birds and mammals.

PART THREE CHRONOLOGY OF THE LIFE AND WORKS OF JOHN GOULD

Chronology

1804
September 14 — John Gould was born at Lyme Regis, Dorsetshire, a small fishing village on the south coast of England (Figure 2).
October 7 — Gould was baptised at the Parish church.

1806
After 1806, while John was still an infant, the Gould family moved to Stoke Hill near Guildford, Surrey.

It was on the wild commons and heaths of that district that at the early age of five to six years he ran about in search of flowers and insects. Soon he commenced to make collections of them, and at the age of bird-nesting to ramble in search of eggs, which he strung and hung about the cottage window. (Reeve 1863-67)

1811
(In this year the Prince of Wales was installed as Regent, due to the illness of King George III.)

1812
May 12 — Edward Lear, the artist and author, was born at Holloway, near London.

1816
John Gould's sister, Sarah, was born.

1818
When Gould was 14 his father was appointed as a gardener in the Royal Gardens at Windsor Castle. The Gould family apparently lived in neighbouring Eton. '[Gould's] early [bird] specimens obtained ...when he was little more than fourteen years of age, show great skill in preparation.' (Obituary. 1881. *Proc. Linn. Soc. London*)

1820
January 21 — Joseph Wolf (one of the artists employed by Gould) was born at Moerz, near Coblenz, Germany.
January 29 — (King George III died, and the Regent became King George IV.)

***c*.1822**
At this time young Gould was sent to Sir William Ingleby's seat at Ripley Castle, Yorkshire, under the care of the gardener, Mr Legge, to learn the art of forcing plants.

***c*.1824**
Around the age of 20, Gould moved to London.
The Zoological Journal began in 1824. It was the organ of the Zoological Club, a section of the Linnean Society.

1825
In the Windsor Castle Royal Archives there is a receipt from John Gould 'For preserving a Thick knee'd Bustard £1.5.0.' This was done for King George IV.
(In September George Stephenson's *Locomotion* hauled a 34-wagon train up to speeds of 15 mph (24 kilometres per hour) on the Stockton and Darlington railroad.)

1826

April 29 — 'The first General meeting [of the Zoological Society of London] was held at the House of the Horticultural Society, Regent Street' (Scherren 1905, p. 20).

May 20 & August 30 — There is a billing in the Windsor Castle Royal Archives for work on these two dates for His Majesty King George IV, for preserving 'two Mouse Deer, two Cotamundy's' and an ostrich. The charge was £41.12.5½. Gould's address on the bill was No. 11 Broad St., Golden Square, London.

1827

April 25 — The first scientific meeting of the Zoological Society of London was held at the headquarters at 33 Bruton Street.

July — (The first part of *The Birds of America* by J. J. Audubon was issued.)

Gould, at 23, after a competitive exhibition of the skill of various applicants, was appointed 'Curator and Preserver' of the newly-formed Zoological Society of London (Sclater 1882 and Mitchell 1929; Scherren (1905) gives the date as 1828).

1828

February 21 — First extant reference to Gould in a letter, from Prideaux John Selby (1788-1867) to Sir William Jardine (1800-74) (Newton Library, Cambridge).

April 27 — The Zoological Society Gardens opened in Regents Park (Figure 28).

Figure 28 Gardens of the Zoological Society in Regent's Park, 1829 (from Scherren 1905).

PART THREE CHRONOLOGY OF THE LIFE AND WORKS OF JOHN GOULD

November 8	Thomas Bewick died at the age of 75 at Newcastle-upon-Tyne.
December 24	Gould stuffed and delivered four birds and five mammals for His Majesty, billing on 14 February 1829 (Royal Archives, Windsor Castle).

In 1828 *The Magazine of Natural History*, with Laudon as editor, was published as the 1828-29 volume. In 1840 it was amalgamated to form the *Annals and Magazine of Natural History*.

1829

January 5	Gould married Elizabeth Coxen (born at Ramsgate, 18 July 1804). They were both 24 years old.
October 7	A son, John Gould, was born but died in infancy.
October 17	'Messrs. Gould and Tomkins of the Zoological Society, are now dissecting the Giraffe which expired on Sunday last. We understand that when the skin is stuffed, His Majesty intends making it a present to the Zoological Society.' (From the *Windsor and Eton Express* of this date (Lambourne 1965).)
November 24	At a meeting of the Zoological Club of the Linnean Society, 'Mr. Yarrell, on behalf of Mr. Gould, exhibited a specimen of a Warbler, new to the British Fauna.' Gould's first published article was concerned with this discovery of *Sylvia tithys* in Britain (*Zool. Journal* 5:102-4).
November 30	The Zoological Club held its final meeting.

1830

	Early months. 'Stuffing Giraffe as per estimate...£148.10' from His Majesty to John Gould. A crane and two lemurs were also included in the bill (Royal Archives, Windsor Castle).
June 21	First extant letter from Gould to Lord Derby (1775-1851) from his 'home on Broad Street' (Wagstaffe and Rutherford 1954-55).
June 26	(King George IV died, and William IV with Queen Adelaide came to the throne. The new king was 65.)
September 1	First extant letter from Gould to Sir William Jardine. It was concerned with Mrs Gould preparing three drawings for Jardine for £1.16.0. (Roy. Scot. Mus.). Gould's address on this date was 20 Broad Street, Golden Square (now 56 Broadwick Street). He lived there until 1859 (Figures 29 & 30). [Correspondence from Gould or Edwin Prince to Jardine in the Royal Scottish Museum, in the Newton Library, Department of Zoology, University of Cambridge, and in the British Museum (Natural History) Zoology Department and General Library amount to a total of 88 letters. I have prepared typescripts of these for future publication.]
November 1	The first two parts of Edward Lear's *Illustrations of the family of Psittacidae, or Parrots* were published. Lear was 18.
December 12	First extant letter from Gould to William Swainson (1789-1855) (Linnean Society, London). Of this series of three letters the concluding one is dated 21 January 1837.
December 20	On or before this date, part no. 1 of *A century of birds hitherto unfigured from the Himalaya Mountains* was published with four coloured plates (Letter to Jardine, Roy. Scot. Mus.).

JOHN GOULD — THE BIRD MAN

December 21   A son, John Henry Gould was born.

In 1830 Gould was an 'Associate' of the Linnean Society (Sharpe 1893). The 'Proceedings of the Committee of science and correspondence of the Zoological Society of London' first appeared. In 1833 this journal continued as the *Proceedings of the Zoological Society of London*.

Edwin C. Prince began working for Gould eighteen years prior to 12 June 1848, thus, in 1830 (Preface, *Introduction to the Birds of Australia*).

1831

February	*Century of Birds* No. 2 scheduled.
March	*Century of Birds* No. 3 scheduled.
April	*Century of Birds* No. 4 scheduled.
May	*Century of Birds* No. 5 scheduled.
June	*Century of Birds* No. 6 scheduled.
July	*Century of Birds* No. 7 scheduled.
August	*Century of Birds* No. 8 scheduled.
August 28	Trip by Gould beginning on this date to Manchester, Jardine Hall in Dumfriesshire, Edinburgh, Glasgow, Trizell (P. J. Selby's home) and Newcastle (Letters, 17 August 1831 and 8 October 1831 to Jardine, Roy. Scot. Mus.).
September	*Century of Birds* No. 9 scheduled.

The first meeting of the British Association for the Advancement of Science was held at York in September 1831. Gould was one of 353 invited: 'Mr. Gould, member of the Zoological Society of London, exhibited select specimens of Birds figured and described in his work on the Ornithology of the Himalaya Mountains, and copies of this work were laid upon the table for inspection.'

The following paragraph is also included here for interest: 'Mr. R. Havell exhibited drawings of Birds for Mr. Audubon's great work on American Ornithology' (Report, 1831, Brit. Assoc.).

October	*Century of Birds* No. 10 scheduled.
November	*Century of Birds* No. 11 scheduled.
December	*Century of Birds* No. 12 scheduled.
December 27	H. M. S. *Beagle*, under the command of Captain Robert FitzRoy, and with Charles Darwin as 'Naturalist to the Expedition', set sail from Plymouth, England 'for the purpose of surveying the southern parts of America, and afterwards of circumnavigating the world'. Darwin was 22. They returned on 2 October 1836.

In 1831 *The Zoological Miscellany* was published by John Edward Gray until 1844.

1832

Gould visited Amsterdam, Rotterdam, Berne and Berlin with Edward Lear in 1831 or 1832 (Noakes 1969).

January	*Century of Birds* Nos. 13 and 14 probably published together.
March	*Century of Birds* concluding Nos. 15 to 20 probably published together as in the University of Kansas Spencer Library-Ellis copy.
March 23	(First British Reform Bill passed.)
June 1	*Birds of Europe* part 1, with twenty plates, scheduled.

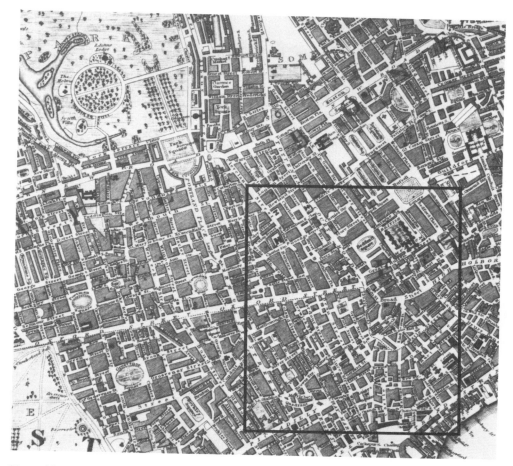

Figure 29 A section of London from 'actual survey 1832 ... United Kingdom Newspaper'.

Figure 30 Inset from 1832 map of London showing close-up of Broad Street, Golden Square, and The British Museum near Charlotte Street, Bedford Square.

September 1	*Birds of Europe* part 2 scheduled.
October 4	The Goulds' third child, Charles Gould, was born but died in infancy.
December 1	*Birds of Europe* part 3 scheduled.

1833

January 1	*The Zoological Magazine* was begun by Richard Owen, but the final issue was on 1 June 1833.
January 15	Gould was elected a Fellow of the Linnean Society (Obituary. 1881. *Proc. Linn. Soc. London*).
March 1	*Birds of Europe* part 4 scheduled.
April 26	Nicholas Coxen, father of Elizabeth Coxen Gould, died in London aged 67.
June 1	*Birds of Europe* part 5 scheduled.
September 1	*Birds of Europe* part 6 scheduled.
November 5	On or before this date was published *A monograph of the Ramphastidae, or family of Toucans*, part 1, twelve plates. Also on or before this date was published the letterpress written by N. A. Vigors for the *Century of Birds*.
December 1	*Birds of Europe* part 7 scheduled.

'Gould was appointed Superintendent of the Ornithological Department of the [Zoological Society] Museum, over which he presided for four years' (Scherren 1905).

Sir William Jardine began the publication of his forty-volume *The Naturalist's Library*. It was concluded in 1843.

1834

January 4	The Goulds' fourth child, Charles Gould, was born.
March 1	*Birds of Europe* part 8 scheduled.
June 1	*Birds of Europe* part 9 scheduled.
June 26	Beginning of correspondence and lists of exchanges or purchases between Gould and Hermann Schlegel (1804-84) and Coenraad Temminck (1770-1858) of the Rijksmuseum van Natuurlijke Historie of 'Leyden' (also known then as the 'Musée Royale des Pays-bas'). These letters were concluded on 1 December 1862.
July 8	Before this date *Ramphastidae* part 2 was published.
September 1	*Birds of Europe* part 10 scheduled.
October 16	(London fire destroyed Houses of Parliament and part of the City.)
December 1	*Birds of Europe* part 11 scheduled.

(In 1834 Samuel Morse invented the magnetic telegraph and Silas McCormick invented a practical reaper.)

1835

January	Gould visited Jardine at his winter residence 'The Holmes', St. Boswells, Roxburgh, Scotland (Letter from Jardine to Selby of 27 January 1835, Newton Library pers. corres. Christine Jackson.).
March 1	*Birds of Europe* part 12 scheduled.
April 1	Around this date was published *A monograph of the Trogonidae, or family of Trogons*, part 1, eleven plates.
June 1	*Birds of Europe* part 13 scheduled.

Plate 25 Drawing in pencil and watercolour, artist unknown, with many notes in hand of John Gould, of *Pteroglossus langsdorffi*, for *Ramphastidae*, plate 28, Kansas-Ellis-Gould MS 285.

Plate 26 Drawing in pencil and watercolour, attributed to John Gould, of *Pteroglossus mariae*, for *Ramphastidae*, 2nd edn, plate 30, Kansas-Ellis-Gould MS 956.

PART THREE CHRONOLOGY OF THE LIFE AND WORKS OF JOHN GOULD

July–September	Gould and his wife visited Temminck at Leyden, Ruppell at Frankfurt, and also Munich, Vienna, Salzburg and Switzerland (Letter, 28 September 1835, to Jardine, Newton Library).
September 1	*Birds of Europe* part 14 scheduled.
December 1	*Birds of Europe* part 15 scheduled.
December 24	'Her Majesty's ship *Sulphur*, accompanied by her consort the *Starling*, quitted the Coast of England on the 24th December, 1835, for the purpose of prosecuting surveys and scientific enquiries on the shores and among the islands of the Pacific Ocean.' Over six months were spent exploring the coast of North and Central America, from Alaska to Mexico, in 1839.
	In 1835 the *Transactions of the Zoological Society of London* were begun.

1836

February 29	On or before this date was published the third and final part of *Ramphastidae*.
	Around this date Gould helped move the offices and museum of the Zoological Society from 33 Bruton Street to 28 Leicester Square (Letter, this date, to Jardine, Newton Library).
March 1	*Birds of Europe* part 16 scheduled.
March 15	Around this date was published, *Trogonidae* part 2.
April 10	Eliza Gould born. She was the fifth child and first daughter. She later married John Muskett and they had a daughter, Helen, mother of Drs Alan and Geoffrey Edelsten and Mrs Doris Edelsten Twyman.
June 1	*Birds of Europe* part 17 scheduled.
September 1	*Birds of Europe* part 18 scheduled.
October 2	H. M. S. *Beagle* with Charles Darwin as 'Naturalist to the Expedition', arrived back in England after circumnavigating the globe.
October 31	A series of nine letters from Edward Lear (1812-88) to Gould began on this date and ended on 14 October 1863 (Houghton Library, Harvard).
December 1	*Birds of Europe* part 19 scheduled.

1837

January 2	*A synopsis of the birds of Australia, and the adjacent islands*, part 1, nineteen plates (of heads only), was scheduled for publication.
February 4	John Latham, notable ornithologist and physician, died at the age of 96.
March 1	*Birds of Europe* part 20 scheduled. Part 21 was not scheduled but appeared with part 20.
April	*Synopsis* part 2 scheduled.
June 20	(King William IV died and his niece, Victoria, became Queen of England.)
May–July	'I have been to Paris' (Letter, August 1837, Newton Library).
July	Probably in July *Birds of Europe* concluding part 22 was published, with twenty-eight plates.

August 28	Gould visited Jardine Hall. Jardine wrote

My brother ... is off to the moors this morning with Mr. Gould and I expect they will bring home 20 brace at least ... Gould has been here for four days, and leaves for Glasgow tomorrow morning partly on the way of business and partly on his way to Loch Awe where he is delivered to kill a big trout. He has been busy spinning some <u>Funny tackle</u> here and with perseverance has killed a few trout. He never saw black game [grouse?] alive before and was delighted. Missed his first shot at an old cock from perfect anxiety (he is a first rate shot) ... He returns from the west to Edinburg then to Newcastle and on to Liverpool where we shall see him. (Letter from Jardine to Selby, Newton Library, pers. corres. Christine Jackson.)

After this visit at Jardine Hall, Gould went on to Edinburgh and Liverpool, before returning to London.

August	*The birds of Australia and the adjacent islands* (the 'Cancelled Parts'), part 1, with ten plates, was scheduled. *Icones Avium or figures and descriptions of new and interesting species of birds from various parts of the globe*, part 1, ten plates, was also scheduled in this month.
September 26	On this date commences a series of seven letters from Dr George Bennett (1804-93) of Sydney, Australia, to Professor Owen (1804-92) in the Royal College of Surgeons Collection. All have pertinent comments concerning Gould, and particularly about his trip to Australia where Gould stayed with Bennett. The last letter of this series is dated 5 April 1840.
December 5	Louise Gould, the sixth child, was born.

In 1837 Gould presented 172 specimens of birds to the British Museum and 111 more were purchased from him. They were from various localities, but a few were Australian (British Museum (N.H.) 1906 History of the Collections).

In this year Gould left the service of the Zoological Society of London to prepare for his Australian trip (Scherren 1905, p. 53 and Mitchell 1929, p. 97).

The *Magazine of Zoology and Botany* was first published in 1837. Jardine was chief editor with P. J. Selby and Dr Johnston. With mergers this eventually in 1841 became the *Annals and Magazine of Natural History*.

1838

March 14	On or before this date was published the third and final part of *Trogonidae*. Also on or before this date was published *Birds of Australia and adjacent islands* ('Cancelled Parts'), concluding part 2.
April	*Synopsis* concluding parts 3 and 4 scheduled.
April 30	On this date Gould wrote a letter to Sir William Jardine with quite a detailed description of the ship *Parsee* on which he and his party were to travel to Australia (Collection: Newton Library, Cambridge University). For interest, the entire letter follows:

London 20 Broad St.
Golden Square
April 30th, 1838

Dear Sir William:

I received your letter of the 28th of March and in reply to which and your several inquiries I beg to say that I have spoken to Mr. Martin of our society who will be happy to send you an account of the Dog. You would I think do well to correspond with him. He is a very intelligent man and a very tolerable naturalist. Mr. Hawkins

PART THREE CHRONOLOGY OF THE LIFE AND WORKS OF JOHN GOULD

having no Town residence at present has requested me to allow any letter for him to be sent to my house. You will therefore direct to him there. I will endeavor to send you an account of the *Caprimulgidae* with drawings on the first part.

Your second letter with an order for forty pounds has this instant been put into my hands and the amount has been duly placed to your credit, pray accept my thanks for all past favours which I will endeavor to repay by thinking of you when in Australia. I shall be most happy to send you some seeds as well as the things you require in spirits. I am so busy that my brain is in a complete jumble. I will however detail a few of my plans. In the first place I am happy to say I have a most comfortable ship the "Parsee" although a small one 350 tons Barkew [Barque] with a commodious poop ['and carries' crossed out] no steerage passengers and not more than ten or twelve cabin P[assengers]. Messrs. Gordan the owner and the Captain both take an interest in my presents [sic]. The Captain's name is J. McKellar and the mates [sic] Murray so you see we cannot do without your countrymen and candidly I shall place my life in their hands with the utmost confidence. I have a Stern Cabin eleven feet by twelve principally for the comfort of Mrs. G. and one adjoining of a smaller size. I have had the former fitted up with Drawers for birds and books in hope to work going along and taking a quantity of birds with me as well as left over plates to cut up for box books etc.

My party consists of Mrs. G. with an attendant a woman who has accompanied Ladies twice to Calcutta, our little boy, two men one as a collector [John Gilbert], the other as a servant, and a nephew 14 years old who will settle there; in the next cabin is a Surgeon with his wife and family. You will say we have every comfort necessary to render so long and tedious a voyage as agreeable as it can be. We start in ten days and I firmly rely on the goodness of Providence to grant me a safe and prosperous passage. If permitted to return in safety I shall never regret the steps I am about to take. The voyage will certainly be a very expensive one but I doubt not my work will have a great sale in consequence and if so amply repay me for my exertion.

We go direct to Hobart Town. I have many reasons for visiting Van Diemens Land first—the climate is more like that of England and many of the Birds, I believe only pigeons may visit that island from the Continent, many are local species and others so nearly allied that an inspection of them in a state of nature is possible to ascertain their being specifically different. From V.D.L. I either go to Port Philip or Sydney and from there into the interior. Mrs. Gould's brothers have estates high up in the Hunter just under the Liverpool Range and within five miles of the Manuras [*Menuras*]. I then if possible send my assistant to the north of the Island and Swan River. New Zealand I hope to visit myself probably on my way home as we come round the Horn and thereby circumnavigate the globe!!! What an age.

May yourself, Lady J., and family every blessing and happiness, and believe me to remain

Yours Very Sincerely

John Gould

[This was addressed as follows:]
Sir William Jardine Bt.
Jardine Hall
Lockerby
Dumfriesshire
N. B. [North Britain]

May 8   First of a series of nine letters to or about Gould from George Bank, Governor John Hutt, C. Hirst[?], Sir John Franklin, Governor Sir George Gipps, Captain John Lort Stokes, Captain King, Captain George Grey, and Sir George Gawler. They are primarily concerned with introductions for Gould or discussions of Australian natural history. The last letter is dated 22 February 1842 (University of Tasmania Library, Hobart).

May 16   John Gould, at the age of 33, sailed from London in the barque *Parsee*, of 349 tons (356 tonnes), with his wife Elizabeth, son John Henry, nephew Henry Coxen, John Gilbert and two servants, James Benstead and Mary Watson (*Emu* 38:96).

A communication concerning the early part of this voyage was written by Gould on 20 June 1838, and is printed on that date.

May 20	'Off the Lizzard [Lizard Point, Cornwall, England].'
May 23	In the Bay of Biscay 'not more than forty miles from the British shores' where the ship was detained for eleven days.
May 24, 26, 27	See letter dated 20 June.
May 31	First of a series of thirty letters written by Gould's secretary, Edwin C. Prince, to Gould in Australia. These are long and very detailed (Mitchell Library, Sydney).
June 4	Off Oporto, Portugal.
June 7	'Off Cape Finnistore (coast of Spain).'
June 8, 9, 10	'Off Madeira and the Salvage Rocks.'
June 11	Landed at Santa Cruz, Island of Teneriffe where Gould travelled into the interior for parts of two days. (This note and those following after 7 July are from Gould's *Handbook* or from the *Proc. Zool. Soc. of London*, 8 October 1839 with letter from Gould of 10 May 1839.)
June 20	'We are now 7 degrees within the tropics.' On this date Gould wrote the following letter:

Latt 14 — 21.N. Long. 22 — 16.W
June 20th 1838

My Dear Sir

I feel assured that you as well as many of my scientific friends will be happy to hear that we are thus far safely on our Voyage and that the whole of our party are quite well. We are now 7 degrees within the tropics and if the trade winds in which we got off Madeira continue we may expect to cross the Equator in a week or ten days, about which time our Captn. informs us that we may expect to fall in with some homeward bound vessels which will enable me as well as the rest of our passengers to forward letters to our friends. Our passage at present may be considered a long one owing to a fortnights adverse winds at the commencement, the northeasterly wind which had prevailed so long previous to our leaving England having fairly blown itself out by the time we made the Land's End [Cornwall, England]. A dead calm for a day ensued which together with certain hazey [sic] appearances in the atmosphere with a depression of the murcury [sic] fully indicated to those accustomed to the sea that a change unfavourable to our equanimity was about to take place and these suspicions were fully verified, for the following day was herrolded [sic] in by a heavy gale from the southwest which lasted about 8 hours, two of which it blew tremendously. Our position at this time was critical being between the Lizard Point and the Scily [sic] Islands. We have however a tight little ship and what is more a very obedient one. Still we were on a sea [lee?] shore and our Captn showed some anxiety to get out to sea as a place of safety. The wind continued to blow in the same direction sometimes moderately and then a gale, with now and then a dead calm always accompanied by a heavy chopping cross sea, for 14 days, 12 of which we were in the Bay of Biscay. On the 4th of June however, while off Oporto we were gladdened with a change of wind which soon carried us to Teneriffe which Island we made on Monday the 11th. This bad weather had the effect of making us all good sailors and now we are enjoying fine weather and a fair wind. The past is quite forgotten and we are all cheerful and happy. During the bad weather we experienced a few disasters and events, but they were of little consequence. Our head rails were washed away and our water casks drove en masse from side to midships. Our Doctors Lady added a fine young Neptune to the ships company during some heavy weather while in the Bay. The mother and child have both done extremely well. I must say that I do not at present find the time at all tedious and I may say scarcely monotonous. In fact there has been so much to amuse that I have scarcely read a book and have written but little. There is always some new object presenting itself to view. Land birds are constantly visiting the ship and the pretty Stormy Petrels are our constant companions while the ocean itself contains an abundance to amuse and gratify one. The whales of which we saw several in Bay together with grampuses and porpoises are all novelties. The beautiful floating Moluscas (Portugese Men of War!) and the luminous species

PART THREE CHRONOLOGY OF THE LIFE AND WORKS OF JOHN GOULD

which bispangle the water at night are to a lover of Nature all objects of interest, not forgetting the most beautiful of all a shoal of Flying Fish. We are now just in the midst of them but the wind being directly off we have not been able to capture any. Their flight is more like that of a locust than anything else that have constantly not the power of adding a second impulse to their flight without first touching the water which sailors say is for the purpose of wetting their wings. I will refer to my Journal and give you a list of the Birds which I have seen on passage between the Land's End, this together with some observation on those Islands at Teneriffe, and if my friend Mr. Yarrell should think them worth being noticed at one of the Societies [sic] meetings or if Mr. Charlesworth would like to notice them in his Magazine, they are perfectly at liberty to do so.

May the 20th. Off the Lizzard [sic] wind S.W. and blowing hard. Common Goatsucker (*Caprimulgus Europaeus*) visited the ship and remained in company for more than an hour. It continued to fly round and close to the vessel without once settling. It was a fine old male and had in all probability passed the ['whole of the' crossed out] night on wing. Still its flight was performed with ease and vigour. It is likely that the Bird was performing a migration from the coast of Africa or Spain to England. Common Stormy Petrel (*Thal[assidroma] pelagica*) astern, in its flight and general appearance very like common Martin. The white rump is very conspicuous.

May 23. Female Yellow Wagtail (*Budytes flava*) flew on board and remained a few minutes and went off to sea again. Several petrels astern. Killed one which proved to be *Thal. Wilsoni*. This was in the Bay of Biscay and not more than 40 miles from the British Shores. It will consequently form an additional Bird for Mr. Yarrell for his work, and the plate which I have drawn from five examples killed since will be published in my supplementary plates on the European work at some future time. They differ considerably in their flight and actions from *Thal. pelagica*. Their structure is also different. These differences I have noted down and will appear in the description accompanying the plate. The species appears to enjoy a wide range, as I observed them almost daily from the time we first saw them untill [sic] we entered the tropics. The *Thal. pelagica* on the contrary appears to be a rare Bird toward the South as I only observed a single individual of the species after leaving the Bay of Biscay; this was off Madeira. I killed the Bird which proved to be a male, this was on June the 8th. I have not forgotten Mr. Eyton and will send him the body of one of those I have killed by the first opportunity that offers.

May 24. I saw several small Turtle Doves (*Columba Turta*), visited the ship, went off again immediately. At 4 o'clock in the afternoon surrounded by a large shoal of whales, they appeared about 50 feet in length and were called Black Whales by the sailors. One however was pronounced to be a Sperm.

May 26. Killed a Kestral (*Falco Tinnunculus*) from off the Rigging. Swallows paid visit. Grampuses around the ship.

May 27th. Short Eared Owl (*Strix brachyotus*) flew on Board during the night which was caught and kept alive for several days.

June 7th. Off Cape Finnistore (coast of Spain). Killed another petrel (*Thal. Wilsoni*). Saw Noddy Tern (*Sterna stoldida* [sic *stolida*]). The flight of this Bird differed considerably from any other terns I ever saw and resembled somewhat the true petrels (*Thal.*) in many of its actions.

June the 8 & 9 & 10th. Off Madeira and Salvage Rocks. Ship surrounded by hundreds of shearwaters, the great bulk of which was *Puffinus cinereus*. I did see however the smaller kind (*Puff. obscurus*). The petrels (*Thal. Wilsoni*) were also numerous.

Monday June 11th. Made the Island of Teneriffe by daybreak and came to an anchor in Santa Cruz Roads at three in the afternoon. Here I saw the first Gull since I left the Land's End. They were tolerably numerous in the Bay and appeared to be the Lesser Black Backed Larus. We were soon visited by the customs house officer and after a little ceremony permitted to land, which most of us did immediately having been informed however that the thermometer stood at 95 in the shade. I considered it best to leave Mrs. G. on board untill morning. I need scarcely say that cloths [sic] in a country like this is a burden not to be bourne [sic] and myself as well as the rest of our party dispensed with every article in this way that decency would allow off, but had we not to pay a visit to our Consul Mr. Bartlet.

[Letter draft from John Gould to the Chairman of the Scientific Committee of the Zoological Society of London, 20 June 1838, written by Gould but not signed (Collection: Mitchell Library, Library Council of New South Wales, MS A172.)]

June 20	(Audubon's *Birds of America* was completed.)
July	Charles Darwin's *Zoology of the Voyage H.M.S. Beagle, Birds*, Part III by Gould, Number I, ten plates, appeared.
July 7	The *Parsee* crossed the Equator.

July 20	They reached 26th degree of south latitude.
July 23	'On referring to my notes I find that it (*Phaebetria (Diomedea) fulginosa*, Sooty Albatross) first came under my notice on the 23rd of July 1838, in lat. 31°10′S., long. 34°[24?]W., when three examples were seen flying round the ship, which they continued to do until we doubled the Cape and entered the South Indian Ocean, on the 14th of August' (*Handbook* II:441).
July 24	Regarding *Diomedea exulans* (Wandering Albatross), Gould 'first hailed its presence on the 24th of July 1838, in lat. 30°38′ south, long. 20°43′ west, and from that day until my arrival at Storm Bay, Tasmania, it was constantly around the ship' (*Handbook* II:428).
	Also on this date Gould states (*Handbook* II:437) that he first observed the *Diomedea chlororhynchos* (Yellow-nosed Albatross).
July 27	From Gould's notes, when at lat. 26°54′S., long. 31°[11°?]25′W. he 'Saw the first Cape Petrel (*Daption capensis*), and from this date until we doubled the Cape of Good Hope it paid daily visits to the ship, sometimes in considerable numbers' (*Handbook* II:470).
August 12	'Off Cape Lagulhas [Agulhas] . . . saw first Black-bellied Storm Petrel' (*Handbook* II:479 states '1839' in error).
August 13	In London *Icones Avium*, concluding part 2, was published.
August 14	Entered the South Indian Ocean.
August 16	Gould first observed the Great Grey Petrel (*Adamastor cinerea*) 'in lat. 38°41′S., long. 36°30′W[E]' (*Handbook* II:447).
August 18	'The ship passed the islands of St. Paul and Amsterdam on August 18' (*Emu* 38:96). 'Off the island of St. Paul. Cape Petrels very numerous' (*Handbook* II:70).
August 21	'On the 21st it [Great Grey Petrel] was very numerous, and the day being nearly calm, I had a boat lowered and succeeded in killing several in lat. 39°23′S., long. 54°W [84°?E]. Its powers of flight are very great' (*Handbook* II:447).
September 8	'Off King George's Sound. Cape Petrels still very numerous' (*Handbook* II:70).
September 18	Reached Hobart, Tasmania (Figure 31) — then called Van Diemen's Land (*Emu* 38:96). [September 19 is the arrival date given in *Handbook* II:479.]

Figure 31 Hobart Town, Van Diemen's Land, Australia, in 1834 Government House is at the far right of the drawing. (from James Backhouse 1843, Author's Collection).

PART THREE CHRONOLOGY OF THE LIFE AND WORKS OF JOHN GOULD

The Goulds lived first on Davey Street then later moved into Government House with Sir John and Lady Franklin.

Gould's Australian journeys can be divided into eight explorations as follows:

I TASMANIA-HOBART AREA EXPLORATION (Figures 32 & 33)

Figure 32 Map of Van Diemen's Land, Australia in 1833. The Great Island in Bass Strait is Flinders Island. Note the road from Hobart Town to Launceston through Perth (Author's Collection).

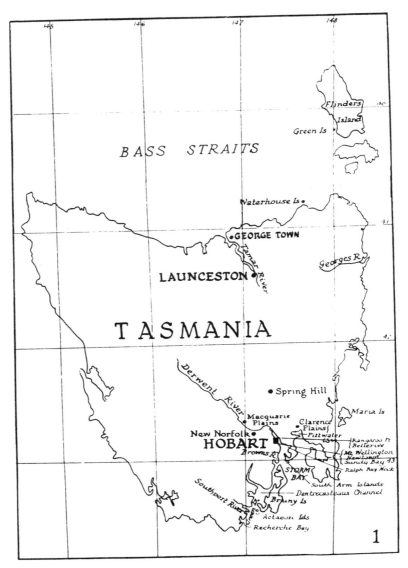

Figure 33 Map of Van Diemen's Land, now Tasmania, showing locations visited or mentioned by Gould (from *Emu* 38).

September	According to Hindwood (*Emu* 38:105) Gould did most of his Tasmanian collecting within eighty kilometres of Hobart. Some time was spent on or near the Derwent River, Macquarie Plains, and Green Island (Mackaness 1947, Franklin corres., p. 40).
October	'On the banks of the Derwent' observed nests of the White-fronted Heron (*Handbook* II:299).
December 4	In the 'Catalogue' of birds in the British Museum (vol. 3, 1877) there is listed a specimen of the Clinking Currawong, *Strepera arguta*, got by Gould at 'Norfolk, Van Diemen's Land' on this date (Whittell 1954).

PART THREE CHRONOLOGY OF THE LIFE AND WORKS OF JOHN GOULD

II RECHERCHE BAY EXPLORATION (Mackaness 1947, Franklin corres., item V, all from diary of Lady Jane Franklin, unless otherwise noted)

December 10 — Around this date Gould left Hobart with Lady Franklin and a small party travelling on the Government schooner. Their intention was to visit the south coast of Tasmania and Port Davey on the west coast. 'Stormy weather prevented the party from proceeding farther than Recherche Bay, where the boat was windbound for a fortnight' (*Emu* 38:96).

December 11 — Tuesday. Entrance to D'Entrecasteaux's Channel.
About 2 o'clock, when within about 4 miles of Green Island, for which we were tacking, Mr. Gould went off in a boat with the hope of reaching it before us, and of finding some penguin's eggs which he is much in want off [sic]. He had landed on the island before and killed penguins, quails, ducks, etc. On this occasion he did not succeed in finding any penguin's eggs, but came back with a live penguin, and with the eggs of various gulls. Some of these were too hard to be blown, in which case he cut with the point of a penknife or of a small knife adapted for the purpose, an oval-shaped piece of the shell out of the side, emptied the egg, and replaced the shell.

This was apparently the second visit for Gould to Green Island. (Also see *Handbook* II:365.)

December 12 — In 'Research Bay'.
Mr. Gould went off in the morning to the Acteon [Actaeon] Islands, which are about a mile distant from each other and at the distance of about 3½ or 4½ miles from Research Bay. The soil of them is sandy and much covered with scrub. He procured there 2 species of parrots he had not yet taken, an albatross, teal, gulls, and the eggs of the latter.

December 13 — Messrs. Gould and Gunn set off to-day to the head of the [Recherche] Bay to a plain, which appears to run up many miles into the country the hill called South Cape, which presents towards the bay a steep and particularly denuded surface ... Mr. Gould expected a rich harvest from the appearance of the plain, both as respect quadrupeds and birds, but it was remarkably destitute. He brought back, however, the nest of an emu wren, and a parrot which he had not killed before.

December 14 — All stayed on boat in Recherche Bay.
December 15 — Saturday. Gould wrote a letter to his wife in Hobart.
December 16 — Sunday. Tried for second time to get to Port Davey, but not enough wind for the schooner.
December 17 — Monday. 'It blew very hard, but not so much so as to prevent Messrs. Gould and Gunn going on shore. They visited the stream called Catamaran River in the Port du Nord.'
December 18 — Tuesday. Catamaran River area again.
December 20 — Thursday.
Mr. Gould was worn out by our reverses, regretted the loss of time, and this very afternoon had been declaring with many apologies that he must go back in the *Vansittart*. His good humour under his prolonged disappointment had never failed, but his time was precious to him. When he came to V.D.L. he intended to stop only a month. How should he get through his work if he went on in this way? Mr. Gould was to my very great regret in this disposition when Mr. Gunn, Elinor and I embarked again in the evening to visit the shores of that portion of the northern part of Research Bay where La Haye's garden was planted, and where Captain King had discovered signs of coal ... On our return to the schooner we found Mr. Gould no longer firm in his former determination to depart immediately, and on my telling him he must give me his hand in pledge that he would stay here and work longer if necessary, he, after a little hesitation, consented. As a little compensation I begged Captain King to let us remove either to Bruny, or to Muscle Bay, which would make very little difference when once the wind set in fair, and it was accordingly settled that at daylight we should sail for Muscle Bay.

Gould wrote a letter to his wife on this date, as follows:

Recherche Bay Thursday Night
20 Decr 38

My Dear Eliza,
Your letter of Sunday last was put into my hands this morning it having taken the Pilot with 5 men 4 days to reach this place so adverse has been the weather, you will therefore see we are still imprizoned [sic] here. We endeavored to get round the South West Cape on Sunday morning but was obliged quickly to retrace our steps. The Cutter Vansittart is here also and will take this letter to town. I would gladly have availed myself of the opportunity of taking a passage in her myself did I not find by so doing that the party would be entirely broken up and in all probability the trip to the westward abandoned altogether. I myself would I am sure hereafter regret not visiting this part of V.D.L., and under all circumstances I have agreed with Lady Franklin to wait one week longer after which to return if the wind should not come round to the eastward or northeast, either of these winds would take us to the desired place in a few hours.

We intend shifting our locality tomorrow morning; we are going into South Port 15 miles distance from this, (nearer home) this move is principally on my account for the purpose of affording me a new locality for my rambles.

I have this day killed rather an extraordinary Bird, and one that you will recollect, at least there is every probability of its being the same; you will remember the large Snow White Petrel (a variety of the black species) which followed us nearly all the way from the Cape to Tasman's Head and which I was so desirous of procuring without effecting my purpose.

I have also killed some beautiful albatrosses, petrels, ducks etc. and four nests with full compliment of eggs of the Black Oystercatcher, small and large gulls, a Teal etc. besides a nest with young of the Emu Wren and other minor things.

I am quite well and am happy to say Lady F. is much better and is quite well. I sincerely trust you are so also and that you are finally established at Government House. Tell Henry I hope he is good and very attentive to you and it is and will be his place to be your protector! in my absence. I am sorry you have not heard from Sydney, you cannot be long without letters, if they are not arrived already [letter torn]. The Marian Watson will return immediately. This is a cold cheerless place and a complete channel [?] house for whaling. The beach is almost everywhere strewn with their Bones, entire skeletons many of which are still putrid. You may therefore guess what we find to leeward.

I think I have not more to say.
God bless you my dear Eliza.

J. Gould

Hastily
P.S. I think we left some skeletons in water, if so, tell James to take them out merely and lay them on the table to dry.

[Letter from John Gould to Eliza Gould, 20 December 1838, written and signed by Gould. (Collection: Mitchell Library MS A172)]

December 21	Friday. Anchored in Muscle Bay. Gould and Thomas Smith, Assistant Police Magistrate at South Port, explored the bay and an inlet. 'I procured newly-hatched young [of the Black Swan] clothed in greyish white down at South Port River on the 31st [21st] of December' (*Handbook* II:348). According to Lady Franklin's letter No. VI (Mackaness) they were 'summoned back to Hobarton by Sir John', so they had to shorten this trip to only two weeks. [They would not have been at South Port River on December 31.]
1839	
	III NORTHERN TASMANIA AND BASS STRAITS ISLANDS EXPLORATION
January	In London appeared Darwin's *Zoology of the Beagle, Birds*, Number II.
January 2	In the 'Catalogue' of birds in the British Museum (vol. 3, 1877) there is listed a specimen of the Clinking Currawong (*Strepera arguta*), got

PART THREE CHRONOLOGY OF THE LIFE AND WORKS OF JOHN GOULD

	by Gould at 'Spring Hill [Inn], interior of Van Diemen's Land', on this date (Whittell, 1954).
January 5	At Launceston, Tasmania, Bill of Sale for shot, powder and caps (Mitchell Library).
January 7	At Launceston still.
January 8	Letter to his wife from 'George Town'. Gould and party had travelled overland from Hobart, through Perth and Launceston.

George Town
River Tamar
Jany 8th 1839,

My Dear Eliza

We reached this place about ten this morning and are at present wind bound but trust we shall be fairly under way for Flinders tomorrow morning. Your welcome letter found me at Launceston at which place I remained two days with Mr. Kerr, Captn Friend etc. I was not so much gratified with the interior of the country as I had anticipated but the people tell me I have seen it in an unfavorable season, in consequence of no rain having fallen for 3 months. The ground is literally parched and every blade of grass is yellow as sulphur, but independent of this the land in the interior is less rich and the fen of the country less fertile than any country I have ever traveled. On the whole the roads were good and also the Inns — the latter however are extravagantly high in their charges.

I collected a few birds on the way, among them a splendid Eagle which I killed at Perth. This place George Town is something like I should think Thenness[?] when that town was first established although the extensive docks of the latter render it now far more different.

I look forward with considerable pleasure to our arrival at Flinders whither I proceed purposely to see the remnants of the inhabitants which once peopled this fine Island over which they were Lords and Masters but now submissive creatures to the wiles of Englishmen. It will also be an interesting as well as expensive trip to me in other points. After Flinders you are aware I go to Kings and although I trust it will not be so it might take a month to accomplish the journey (that is if the winds prove contrary) I state this in order that you might not be alarmed in not hearing from me. You may depend if an opportunity should occur I will not fail to send a letter and I have to request in a [letter torn] fervor that you will post a letter to me at least once a <u>week</u> commencing from Monday next, with all particulars respecting your health etc etc directing them to P. Office Launceston until called for, and when I again reach this place I will write immediately. I am happy to hear that Henry was drawing and trust he will be attentive to you in every way. I should say give Sophy a Dollar, and I will think of the men when I return.

I wish you to send by the Launceston coach two <u>clean</u> parts, of course, (nos. one and two) of the Birds of Australia with some prospectuses which you will address to Captn Friend Post Office Launceston. You will either get the four nos of the Synopsis from the work room drawing portfolio, in Government House and send them with the larger copy.

You had better get Fisher to plain [sic] a very thin teak board the size of the large books for protection, and a common book cover from the work room will do for the protection of the other side, spreading the four parts of the Synopsis over and sealing the whole in brown paper. This parcel I wish <u>sent immediately</u>; James can pack it.

Must now conclude trusting my dear that every blessing will attend you and trusting I may find you well when I return — kiss dear Fitz [?] for me and believe me to remain Eliza

Yours ever affectionately
John Gould

[This letter is addressed:]
Mrs. Gould
at Mr. Fishers
Davey Street
Hobart Town

[Letter from John Gould to Eliza Gould, 8 January 1839, written and signed by John Gould. (Collection: Mitchell Library, MS A172)]

	Gould and party then visited the Bass Straits Islands of Flinders, Isabella, Green (*Handbook* II:219 and II:460) and Waterhouse.
January 12	On Isabella Island he killed a pair of Cape Barren Geese (*Handbook* II:350).
January 13	'I took five newly-laid eggs [of Black Swan] on Flinders' Island, in Bass's Straits on the 13th of January' (*Handbook* II:348) (Figure 34).

Figure 34 Flinders Island grass trees (from James Backhouse 1843, Author's Collection).

PART THREE CHRONOLOGY OF THE LIFE AND WORKS OF JOHN GOULD

January 20   Back at George Town, Gould wrote to his wife reporting the death of one of the crew on 'pulling the gun from the boat':

> George Town
> Sunday 20th Jany 1838 [1839]
>
> My Dear Eliza
>   I am now in Capt Friend's office and have only a short time to write before the post leaves. We arrived here yesterday from Flinders at which place I was especially gratified to [word not clear] my acquaintance with the Natives and other things, and I should have left the Island with a light heart and proceeded to Kings had [not] a fatal accident happened to one of the men who shot himself dead by uncautiously pulling the gun from the boat with the muzzle toward his chest, the cock of the gun caught the seat of the boat and all was over with the poor fellow in half a minute. I cannot tell you my Dear Eliza how great a shock I sustained. I have scarcely been myself since and I almost hate the sight of a gun. I have given up all idea now of going to Kings and shall make my way across the Island as quickly as I can making a call or two on the way. The inquest will be held in the morning. The man had every caution given him not a minute before to be careful with the gun, but his time was come as his poor shipmates say and with that they console themselves.
>   Pray give my kind remembrances to Sir John and Lady Franklin to whose kindness we owe so much. Tell Lady Franklin I have her little page with me, he is a most interesting little fellow, throws the spear and waddy with the utmost dexterity and extremely useful to me in the bush, an eye like a hawk discovers birds nests & eggs in a most astonishing manner.
>   Trust you and Dear Little Rif [?] are quite well, I remain
>
> My Dear Eliza
> Yours affectionately
> J. Gould
>
> P.S. I received both your letters
>
> [The letter is addressed:]
> Mrs. Gould
> Government House
> Hobart Town

[Letter from John Gould to Eliza Gould, 20 January 1839, written and signed by Gould (Collection: Mitchell Library, MS A172).]

January 21   Gould at Launceston. He wrote a letter of credit for John Gilbert (Gould's collector) from Messrs. Wm. Alexander & Co. for a proposed Swan River trip (Whittell 1954, Plate 20).

January 23   At Launceston still as shown on Bill for shot, caps, paper and powder (Mitchell Library).

On the return trip overland to Hobart, Gould possibly went to Maria Island off the east coast of Tasmania.

February 4   John Gilbert apparently remained in Launceston, because on 4 February he sailed by the *Comet* for Swan River, Western Australia, arriving on 6 March (Chisholm, *Emu* 39:162).

IV SYDNEY-MAITLAND-UPPER HUNTER-LIVERPOOL RANGES EXPLORATION (Figure 35)

February 15   'Gould and a servant left Hobart on the *Potentate* for Sydney, at which place he arrived on the 24th of the month' (*Emu* 38:97).

February 20   An account of his shooting a White-headed Petrel from the ship (*Handbook* II:451-2).

February 24   Arrived at Sydney, New South Wales. Stayed there only three days according to letter of 20 March to his wife (Mitchell Library).

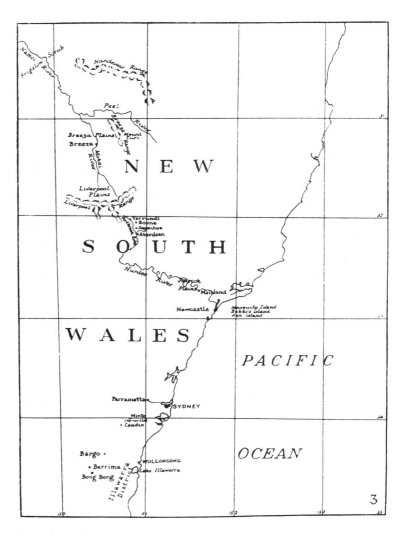

Figure 35 Map of New South Wales showing principal localities visited by Gould (*Emu* 38).

	Gould's staying in Sydney only three days is corroborated by a letter dated 25 February 1839 from Gould to Sir John Franklin (Scott Polar Research Inst.).
February 27	In the same letter to Sir John Franklin, Gould wrote 'On Wednesday [27th] I proceed to the Hunter River into the interior towards Liverpool plains there to commence in my favourite pursuits.'
February 28 [?]	Maitland. Stayed there one week according to a letter of 20 March to his wife. Then Gould proceeded to the Coxens' homestead, 'Dart Brook', at Yarrundi near the Hunter River (Figure 36).
March 8	First met with Australian Swift (*Cypselus Australis*) on this date on the Upper Hunter, with Stephen Coxen (*Ann. Nat. Hist.* 1840, 5:139-46).
March 14	Liverpool Range. Collected several Lyrebirds [see next entry].
March 20	Letter from Gould at 'Dart Brook' to his wife in Hobart as follows:

PART THREE CHRONOLOGY OF THE LIFE AND WORKS OF JOHN GOULD

Dart Brook
River Hunter
20th March, 39

My dear Eliza

I have received two letters from you and I am truly happy to find you are well. Rest assured my dear E that like yourself next to seeing you the greatest pleasure is receiving a letter. I staid in Sydney only 3 days after which I proceeded to the Hunter having previously sent off a letter to Stephen to say that I was coming, but you will consider me particularly unfortunate when I inform you that we crossed on the road, or rather on the river (Hunter), Stephen having started from home the day previous to the arrival of my letter and [was in] Maitland the day I was going up the river, the consequence was that I had to wait a whole week at Maitland before we met. He was most glad to see me and was especially kind, thinks I did perfectly right in leaving you at Hobart Town for the present. I have so much to say that I cannot tell you half in a letter. You will be happy to hear that I am in excellent health and the colony agrees with me well. I have already made an expedition to the Liverpool Range. On the top of one of the highest peaks I mounted, and had a good view of the plains below.

The birds are all molting consequently in a bad state being our excuse, so have killed many other things. We killed 9 *Manura* [sic] *superba* 3 of which I have in brine for dissection, 2 for skeletons, the rest in skin, a pair of which with several other fine birds I intend for Lady Franklin. Pray give my respects to Their Excellencies and remember me kindly to all. Tell dear little Henry for me I hope he is a good boy.

Stephen is doing well, better certainly than his neighbors. I find everything realized and more so than was stated. This is a pretty place far beyond what I anticipated. His <u>establishments</u> are very extensive and extremely well conducted. He has his failing however, I am sorry to say, the particulars of which I will explain when we meet. The drought is becoming very serious, no rain having fallen for nine months, in fact nothing but ruin presents itself. There is scarcely a blade of green grass in the colony. I could only have believed what I have seen.

Stephen of course is suffering much but the distress is general.

I intend leaving this the first moment I can for Sydney hence to *Hobart Town* to see you, although it will verily interfere with my pursuits. But how I am to get conveyed with my luggage to Maitland neither Stephen or I can devise, three teams of bullocks having left Warra on their way from the plains to this place eight days ago and have not yet arrived. Warra is 35 miles distant. When we heard of them last they were cutting down trees for them to feed upon which is very generally adopted. As to water every river is dried up, and men are constantly employed in sucking wells. Cattle and sheep are dying in every direction. All this makes Stephen fretful and peevish as well it might. We shall contrive some way however to get me down to Maitland town and the end of the week, when I shall make my way to Hobart Town the best way I can.

I trust my dear Eliza you are still well and that you received my letter. You have my sincere hope and prayers for a safe delivery out of your troubles, and which I fear not will be granted to you, and if I am not with you at the time you will have something to present to me when I do.

Charles is at the plains. I have [letter torn] he is quite well. Henry Coxen is well also but I am sorry to say not in the good graces of his uncle. What will become of him I know not. Stephen I think is sending him to the Cape. The first thing that got him into disgrace was that instead of delivering Mr. Blackett the Indescriber [?] clock he took it with him to Dart Brook, in consequence of which Mr. Blackett wrote a sharp note for it to be returned. Stephen is extremely strict with but all will not do. His [word not clear] puts him out of patience! Stephen's boys are at school. I shall see them when I return to Sydney. Trusting you are happy I conclude with the best wishes for your safety and remain my dear Eliza

Yours very affectionately
John Gould

P.S. I write this after dinner in a great hurry and with a bad pen.

[The letter is addressed to:]
Mrs. Gould
Government House
Hobart Town
V.D. Land

[Letter from John Gould to Eliza Gould, 20 March 1839, written and signed by J. Gould (Collection: Mitchell Library, Sydney).]

Figure 36 The Coxen homestead at Yarrundi, Upper Hunter River District, N.S.W. The small house, demolished about 1925, immediately above the white cross, is where Gould stayed during 1839–40 (Photograph taken by S. W. Jackson in 1907) (*Emu* 38).

	[There are four postmarks on the envelope, dated 21 March 1839 at Scone; 25 March 1839 at Sydney; 8 April 1839 Ship Letter at Sydney, and 20 April 1839 at Hobart Town[?].]
March 22	Gould saw a number of Australian Swifts 'hawking over a piece of cleared land at Yarrundi, on the Upper Hunter' (*Ann. Nat. Hist.* 1840 5:139-46).
April 4	At Maitland, on return to Sydney, shot a Little Whimbrel. In a letter dated 7 August 1839 Dr George Bennett of Sydney wrote to Professor Owen (Royal College of Surgeons) that 'Gould was with me in April leaving Mrs. G. at Hobart Town where she was delivered of a fine boy on the 8th [6th] of May.'
c. April 9	When in Sydney, Gould met the officers of H.M.S. *Beagle*, under the command of Commander John C. Wickham. Included were Surgeon Benjamin Bynoe, Lt. John Lort Stokes, Lt. James Emery and Purser John E. Dring. According to Whittell (1954, p. 102) it was at this time that Dr Bynoe and others of the *Beagle* agreed to let Gould have for naming and describing any birds they might collect on their surveying voyage.
April 11	Gould sailed from Sydney on the *Susannah Anne* to return to Hobart.
April 16	Gould in the Bass Strait area on the boat (*Handbook* II:474). From 6 March 1839 to the end of January 1840, John Gilbert was in the Perth area of Western Australia (*Emu* 39:162) (Figure 37).
May 6	Franklin Tasman Gould, a son, was born in Hobart. The Goulds were living at Government House with Sir John and Lady Franklin.

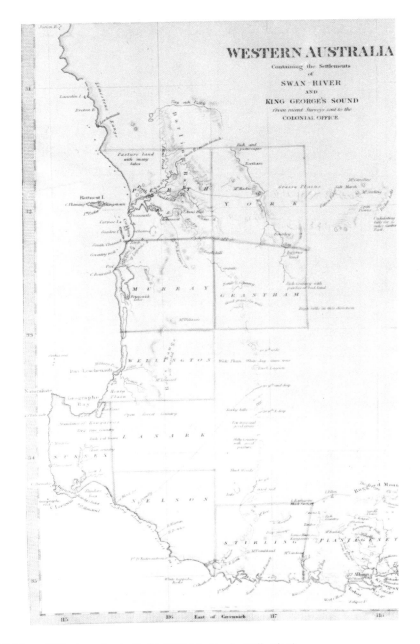

Figure 37 Map of Western Australia in 1833 (Author's Collection).

May 10   On this date Gould wrote a long letter to the Chairman of the Scientific Committee of the Zoological Society of London. In it he recounted his voyage, as related previously, in more detail in the incomplete draft letter dated 20 June 1838. Gould also described nineteen new species of birds. Thirteen of these he stated he had received from Dr Bynoe of the *Beagle*. This letter was read in London on 8 October 1839 and published in the *Proceedings of the Zoological Society* (7:139-45).

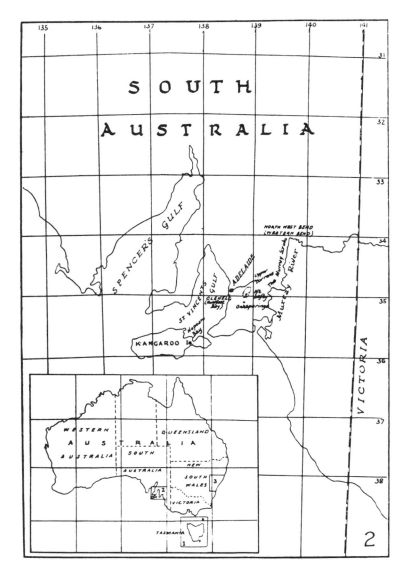

Figure 38 Map of South Australia showing principal localities visited by Gould (*Emu* 38).

V SOUTH AUSTRALIA-ADELAIDE-MURRAY RIVER EXPLORATION (Figure 38)

May 11 [?]	Gould sailed or went overland, from Hobart to George Town and Launceston. Then, probably on 18 May according to Hindwood (*Emu* 38:99), he sailed to Adelaide, South Australia, in the *Black Joke*.
June 12	According to the *South Australian* newspaper, Gould was in Adelaide on this date (Cleland 1937, *Emu* 36:203). 'The city of Adelaide [was] a chaotic jumble of sheds & mud-huts' (*Handbook* I:484).
June 17	Gould began a journey into the Murray Scrubs with Captain Sturt (see next item).

PART THREE CHRONOLOGY OF THE LIFE AND WORKS OF JOHN GOULD

June 22
Saturday

...the Surveyor-General [Captain Sturt] started on Monday [the 17th] on an expedition towards the Murray. We believe the principal object of his journey is to inspect the boundaries of the numerous special surveys in that direction...Mr. Gould, the celebrated South Australian Ornithologist, accompanies the Surveyor-General.

(Cleland 1937 *Emu* 36:203 from the South Australian Register, vol. II, No. 74.) [For a further reference to this trip see Mahoney 1928.]

There is also information on this trip in a published letter of 28 September 1839 written by Gould from Maitland:

Out of two and a half months visit to this part [slightly exaggerated], I spent five weeks entirely in the bush in the interior, partly on the ranges and partly on the belts of the Murray...[Some accounts follow on bird species, then a reference to Lake Alexandria and a] belt of dense dwarf *Eucalypti*, through the centre of which the river Murray winds its course...Through the kindness of Col. Gawler, the Governor, and Capt. Sturt, whom I accompanied into the interior on an especial expedition of survey, I was provided with horses, a cart, and a small company, with the view of reaching the Murray. Having with difficulty crossed the range over an entire new country, and penetrated to the centre of the dense *Eucalypti* scrub alluded to, in which I spent a night and part of two days without water for my horses, I was compelled, much to my regret, to beat a retreat back to the ranges, in the gullies of which I even found difficulty in obtaining water. During a week's stay under the ranges I made daily visits to this rich arboretum, which would have served me to investigate until this time without exhausting its treasures; but, alas! our provisions failing, we were obliged to retrace our steps (Letter to the Editor, *Ann. Nat. Hist.* 5:116-19).

June 23

In a letter from Dr Bennett of Sydney to Professor Owen (Royal College of Surgeons), Bennett wrote that Gould

proceeded to South Australia and a few days since I received a letter from him from the "Upper part of the Torrens["], S. Aust. dated June 23rd/39 in which he says "Capt. Sturt has just joined me on our way to the Murray" so I suppose he intends coming overland to Sydney, as he also observes that he shall be in Sydney not later than the middle of Augt. He also states in the same letter "I am much pleased with my visit to this portion of Australia which will be attended by very important results regarding the Geographical range of species besides many new ones."

[Gould did not go overland to Sydney but back to Adelaide, then to Hobart, then with his family to Sydney in early September.]

June 26

'On the range near the Upper Torrens' River he collected the first specimen of the Purple-gaped Honey-eater (*Emu* 38:111 and *Handbook* I:513).

July

In London appeared Darwin's *Zoology of the Beagle, Birds,* number III.

July 1

Gould killed a Bittern this day near the Murray River, above Gleeson's Station (*Handbook* II:31). After Gould returned to Adelaide, he crossed over to Kangaroo Island.

July 19

Gould sailed from Adelaide on the *Katherine Stewart Forbes* back to Hobart (*Emu* 38:100). Mrs Gould, while she had been in Van Diemen's Land, had been occupied in giving birth to a son, and also by drawing birds and plants. An example of a drawing by her of a plant is Figure 39.

VI EN ROUTE FROM HOBART TO YARRUNDI (WITH HIS WIFE AND TWO CHILDREN)

Mrs Gould wrote a diary from 20 [21] August to 30 September 1839. It is completely transcribed and corrected by Hindwood in the *Emu* (38:134-7).

Figure 39 Watercolour drawing of '*Richea dracophylla*', attributed to Mrs Gould, with 'V. D. Land' written in left lower corner (Collection of Dr and Mrs Geoffrey Edelsten).

August 20   The family left Hobart for Sydney in the *Mary Ann*.
September 3   The Goulds arrived in Sydney and stayed ten days with Dr George Bennett (*Emu* 38:100).
September 4, 7, 13   Dates of letters from Mrs Gould in Sydney to her mother and Mrs Mitchell in London which were published in full by Chisholm (Chisholm 1944, *Elizabeth Gould*).

PART THREE CHRONOLOGY OF THE LIFE AND WORKS OF JOHN GOULD

September 14	The Goulds left by steamer for Newcastle arriving on the 15th. Gould explored the neighbouring country, the Hunter River mouth, and Mosquito Island (*Handbook* I:467). Mrs Gould went over to Mosquito Island on the 17th with her husband. From Newcastle they went up the Hunter River by riverboat to Maitland.
September 23–29	In Maitland, according to Mrs Gould's diary.
September 28	Gould's party in Maitland.

> Nearly a fortnight was occupied in Sydney in preparing for the journey. My men proceed with the drays to the upper part of the Hunter, near the Liverpool range, but this being Saturday I follow on Monday, and from the slow travelling shall soon overtake the party.

September 29	(Letter of this date published in 1840 *Ann. Nat. Hist.* 5:116-19). They left by cart for Yarrundi and arrived on the 30th. This was the homestead of Mrs Gould's two brothers, Stephen and Charles Coxen.
October & early November	Gould pursued his studies in the nearby forests and the eastern slopes of the Liverpool Range, and also Aberdeen on the Dartbrook (*Handbook* II:240).
October 11	Gould was on the flats at Yarrundi on the upper Hunter (*Handbook* II:104).
October 23	Natty collected the eggs of the Whistling Eagle and the Spotted-sided Finch (*Emu* 38:100).
October 26	Gould collected in the 'cedar brushes of the Liverpool Range' (*Handbook* I:620).
November	In London appeared Darwin's *Zoology of the Beagle*, *Birds* by Gould, number IV.
November 2	In a letter to Sir John Franklin of this date (Scott Polar Research Inst.) Gould wrote from Yarrundi that he had just returned from the Liverpool Range:

> On leaving Sydney I went to the Hunter, where I found I had arrived at a good time, the birds having just commenced breeding. After spending a fortnight in the lowland brushes I proceeded to the upper districts and the Liverpool ranges whence I have just returned having made several discoveries of new species both of Birds and Quadrupeds, of the latter I believe I have two new kinds of Kangaroo... I have not yet succeeded in finding the breeding haunts of the *Manura* [sic]...
>
> I am now preparing for my journey into the interior directing my course by way of Namoi [River] and returning through New England etc.

VII LIVERPOOL RANGE-MOKAI RIVER-NAMOI RIVER-BRIGALOW SCRUB EXPLORATION

This was Gould's most ambitious and successful expedition through sparsely-settled country. There were seven in the party, including two 'faithful companions' Natty and Jemmy (*Emu* 38:100).

November 2	As seen from the above letter from Yarrundi, Gould was preparing for the journey into the interior of New South Wales.
December 2	Gould first saw the Flock Pigeon (Harlequin Bronzewing) on the banks of the Mokai River (*Emu* 38:107).
December 10	Dr Bennett of Sydney wrote to Professor Owen on this date (Royal College of Surgeons) that

> Gould is at present in the interior and Mrs. G. remains at her brother's farm at the Hunter River, they are in excellent health and Gould has already made a noble collection, numerous new species and some new forms with nests and eggs. I have just

	received for him, from the Commandant of Norfolk Island, a box containing the nest and eggs of most if not all the birds of Norfolk Island which will be a great acquisition ... Gould informed me that he preserved the bodies of all the Mammalia and birds in spirits together with skeletons for you, that will be indeed a glorious haul.
December 11	Gould shot the first specimen of the Tri-coloured Ephthianura (Crimson Chat) in the forest lands near the Peel River to the east of the Liverpool Range on this date (*Emu* 38:107; *Handbook* I:380).
c. December 16	'A fortnight after this [2 December] I descended about one hundred and fifty miles down the Namoi and while traversing the extensive plains, studded here and there with patches of trees that skirt the Nundawar range, I was suddenly startled by an immense flock of these birds [Harlequin Bronzewings]' (*Handbook* II:127-8).
December 23, 24	Gould was 'near Gunderheim on the lower Namoi' where he saw a Crested Bronze-wing and a Black-eared Cuckoo. (*Handbook* II:140 and I:622).

1840

January 2	Gould was on the banks of the Mokai where he took an Australian Egret (*Handbook* II:301).

VIII SYDNEY-WOLLONGONG-ILLAWARRA DISTRICT EXPLORATION

February-March	Gould visited Bong Bong, Berrima, and Camden in the Illawarra District (*Emu* 38:101, 109).
April 5	Dr Bennett's letter of this date to Professor Owen (Royal College of Surgeons) states
	I avail myself of the return of our friend Mr. Gould and his amiable lady to write to you and forward also by him the box you last sent me; I have filled the two bottles with contents mentioned further on, and have left him the two jars to fill with such specimens of Natural History, as he may be desirous of preserving during the homeward voyage. I am the more anxious to send the box by him as it contains the viscera of the *Alectura* and as he has a skeleton, I hope it may suffice to enable you to supply most part[s] of the Anatomy; the head and another body of an *Alectura* I hope to procure to forward to you another opportunity. Gould has made a splendid collection which you no doubt will have an opportunity of having laid before you at many meetings of the Zoological Society ... The bottle No. 2 contains the Body of the *Alectura Lathami* or Brush Turkey of the colonists ... This was the same specimen for so long a time domesticated at Mr. M'Leay's and was accidentally drowned in the well; seeing its shadow and mistaking it for a rival, he darted at it, precipitated himself in the water and was drowned. Not long before its death Mrs. Gould made a very accurate sketch of it which you will have an opportunity of seeing ... For further news I must refer you to Gould.
April 6	'We go on board [the *Kinnear*] today', wrote Gould to Sir John Franklin on this date (Scott Polar Research Inst.).
April 9	Gould, his wife, and two children sailed from Sydney in the *Kinnear* via the South Pacific Ocean and Cape Horn for London (*Emu* 38:101).
April 30	John Gilbert arrived in Sydney from Western Australia to find the Goulds had already left Sydney. He decided to go to Port Essington where he arrived on 12 July 1840.
April	Gould observed the Black-bellied Storm-Petrel and the Grey-backed Storm-Petrel in considerable numbers between the eastern coast of Australia and New Zealand (*Handbook* II:476, 479).

PART THREE CHRONOLOGY OF THE LIFE AND WORKS OF JOHN GOULD

May	'The Blue Petrel was very abundant off the north-east coast of New Zealand' (*Handbook* II:457).
May 6	There are references by Gould to the Sooty Albatross, Great Grey Petrel and the Cape Petrel at lat. 40°S., long. 154°W. in the South Pacific (*Handbook* II:442, 447, 470).
May 20	Near Cape Horn in the Pacific ocean in lat. 50°S., long. 90°W., there are references to the Sooty Albatross, Blue Petrel and Cape Petrel (*Handbook* II:442, 457, 470).
June 12	Gould and his party were in the Atlantic ocean on this date and he observed the Sooty Albatross, Great Grey Petrel, and the Blue Petrel at lat. 41°S., long. 34½°W. (*Handbook* II:442, 447, 458).
June 15	Gilbert left Sydney on the *Gilmore* for Port Essington, on the north-central coast of Australia (Chisholm *Emu* 39:162).
July 12	Gilbert arrived at Port Essington. He stayed there for eight months, collecting specimens.
Early August	John Gould, his wife, and two sons arrived back in England, having circumnavigated the globe (*Emu* 38:101). Eight months were spent in collecting and observing oceanic birds, four months going and four months on the return trip.
August 18	The ship landed in England. This date was obtained from a letter dated 23 August 1841 written by Eliza Macintyre to John Gould on the death of Mrs Gould (Collection: Mrs Grace Edelsten).
August 25	On this date Gould appeared before the Zoological Society of London and presented his information on the Bower-birds. The habits of these birds in their bowers became the parlour talk of England.
September 25	The first of a series of ten letters from Gould to Thomas Campbell Eyton (1809-80). These are now in the collection of the American Philosophical Society Library, Philadelphia. The last dated letter from Gould to Eyton is 2 April 1874. Three of the letters are not dated.
October 21	The first of a series of fourteen letters from Gould to Hugh E. Strickland (1811-53) ending 21 September [1849].
October 26	Nicholas Aylward Vigors died, aged 55. He had been an early supporter of Gould, very prominent in the Zoological Society of London, and wrote the letterpress for Gould's *Century of Birds*.
December 1	*Birds of Australia* part 1, seventeen plates, was scheduled for publication.

During this year, according to Sharpe (1893), Gould was elected a Fellow of the Royal Society. (Sclater (1882), however, gives the date as 1843, which is probably correct.)

Gould also became a full Fellow of the Zoological Society of London in 1840 (P. Chalmers Mitchell, 1929:97).

Gould made his first contribution of some species of mammals of Australia to the British Museum according to the listing in the 1906 History of the Collections. Between 1840 and 1849 the original Montagu House was demolished. The British Museum had been housed in this building from 1755. During the time the Montagu House was being torn down, the present structure for the British Museum was built, being completed in 1845.

1841

January 5	In a letter of this date to Jardine, Gould mentions having returned from Edinburgh, Jardine Hall, Aberdeen, Newcastle and Captain Mitford's home in Yorkshire (Newton Library).
February 2	The first of a series of nineteen letters from Gould to Prince Charles Lucian Bonaparte (1803-57) in the collection of the Bibliothèque Centrale du Muséum National d'Histoire Naturelle of Paris. In this collection in addition there is a letter to Gould from George Ransome of the Ipswich Museum. The final Gould letter in this series to Bonaparte was written in 1856. Bonaparte died in 1857.
March	Darwin's *Zoology of the Beagle, Birds* by Gould, concluding number V appeared.
March 1	*Birds of Australia* part 2 scheduled.
March 4	Gould went to Leiden, then to Paris and Rotterdam to research further species for his Kangaroo monograph.
March 17	John Gilbert left Port Essington in northern Australia for Singapore, where he departed on or about 29 April for London.
March 27	'Lord Northampton's evening party. Richard [Owen] very tired, and thought he would not go... after some dinner R. felt better... He was glad afterwards he went, for Prince Albert was there, and Mr. Gould brought his pretty singing New South Wales parrots' (Owen 1894).
May 21	'I have just taken a cottage at Egham on the Banks of the Thames (Figure 40) from this time untill [sic] September in order to give Mrs. G. and our little folks the benefit of country air, I shall also get some fishing although I must be in London nearly half my time.' (Letter to Jardine, Roy. Scot. Mus.)
June 1	*Birds of Australia* part 3 scheduled.
July 31	Before this date *J. Gould's monographie der Ramphastiden*, parts 1 and 2, with ten plates each, was published by the Sturm brothers in Nuremberg.
August 10	Sarah (Sai) Gould was born, their eighth child, and the sixth who lived past infancy.
August 15	Elizabeth Coxen Gould, John Gould's wife, died of 'puerperel [sic] fever' at Egham, aged 37 (Figure 41).
September 1	*Birds of Australia* part 4 scheduled.
September 30 [?]	John Gilbert arrived in London from Australia.
October 1	Before this date was published *A monograph of the Macropodidae, or family of kangaroos*, part 1, fifteen plates.
November 15	'Went to see Gould's birds — not to be imagined till seen... After dinner this evening Mrs. Darwin, Mr. Gould, and his brother came here for some music.' [Whose brother, Owen's or Gould's?] (Owen, 1894.)
December	John Gould's sister, Sarah Wilson, died at the age of 25.
December 1	*Birds of Australia* part 5 scheduled.

During this year the British Museum made the first purchase of Gould's Australian birds, 332 specimens from South Australia (Brit. Mus. (N.H.) 1906).

Figure 40 An 1846 map of the Thames River from Oxford to Kew. Egham is No. 19 on the map. (from the *Illustrated London News*, Author's Collection). ▶

THE ILLUSTRATED LONDON NEWS.

[August 15, 1846.

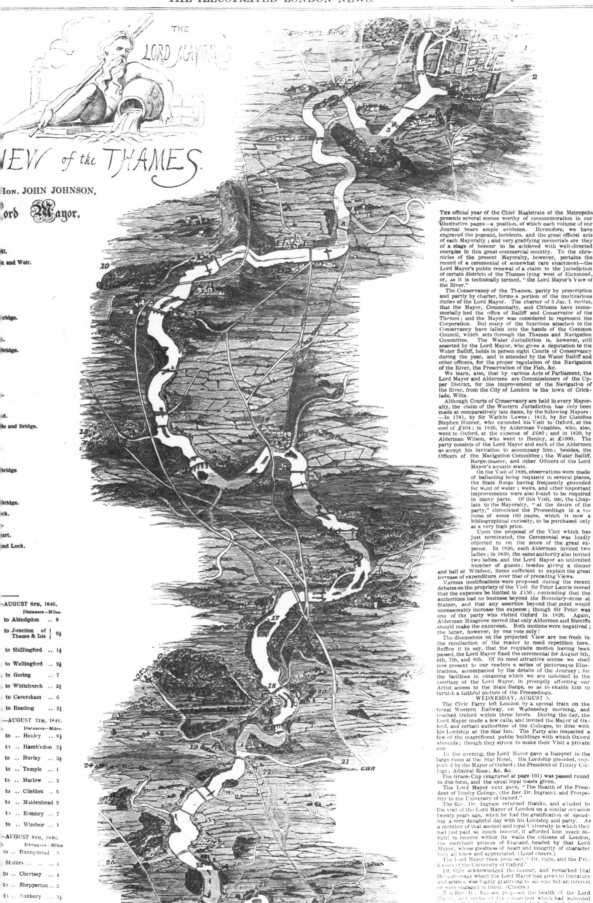

THE official year of the Chief Magistrate of the Metropolis presents several scenes worthy of commemoration in our illustrative pages—a position, of which each volume of our Journal bears ample evidence. Heretofore, we have engraved the pageant, incidents, and the great official acts of each Mayoralty; and very gratifying memorials are they of a stage of honour to be achieved with well-directed energies in this great commercial country. To the chronicles of the present Mayoralty, however, pertains the record of a ceremonial of somewhat rare enactment—the Lord Mayor's public renewal of a claim to the jurisdiction of certain districts of the Thames lying west of Richmond, or, as it is technically termed, "the Lord Mayor's View of the River."

The Conservancy of the Thames, partly by prescription and partly by charter, forms a portion of the multifarious duties of the Lord Mayor. The charter of 3 Jac. I. recites, that the Mayor, Commonalty, and Citizens have immemorially had the office of Bailiff and Conservator of the Thames; and the Mayor was considered to represent the Corporation. But many of the functions attached to the Conservancy have fallen into the hands of the Common Council, which acts through the Thames and Navigation Committee. The Water Jurisdiction is, however, still asserted by the Lord Mayor, who gives a deputation to the Water Bailiff, holds in person eight Courts of Conservancy during the year, and is attended by the Water Bailiff and other officers, for the proper regulation of the Navigation of the River, the Preservation of the Fish, &c.

We learn, also, that by various Acts of Parliament, the Lord Mayor and Aldermen are Commissioners of the Upper District, for the improvement of the Navigation of the River, from the City of London to the town of Cricklade, Wilts.

Although Courts of Conservancy are held in every Mayoralty, the claim of the Western Jurisdiction has only been made at comparatively late dates, by the following Mayors:—In 1781, by Sir Watkin Lewes; 1812, by Sir Claudius Stephen Hunter, who extended his Visit to Oxford, at the cost of £404; in 1826, by Alderman Venables, who, also, went to Oxford, at the expense of £680; and in 1839, by Alderman Wilson, who went to Henley, at £1000. The party consists of the Lord Mayor and such of the Aldermen as accept his invitation to accompany him; besides, the Officers of the Navigation Committee; the Water Bailiff, Barge-master, and other Officers of the Lord Mayor's aquatic state.

On the Visit of 1826, observations were made of ballasting being requisite in several places, the State Barge having frequently grounded for want of water; weirs, and other important improvements were also found to be required in many parts. Of this Visit, too, the Chaplain to the Mayoralty, "at the desire of the party," chronicled the Proceedings in a volume of some 160 pages, which is now a bibliographical curiosity, to be purchased only at a very high price.

Upon the proposal of the Visit which has just terminated, the Ceremonial was loudly objected to on the score of the great expense. In 1826, each Alderman invited two ladies; in 1839, the same authority also invited two ladies, and the Lord Mayor an unlimited number of guests; besides giving a dinner and ball at Windsor, items sufficient to explain the great increase of expenditure over that of preceding Views.

Various modifications were proposed during the recent debates on the propriety of the Visit. Sir Peter Laurie moved that the expenses be limited to £150; contending that the authorities had no business beyond the Boundary-stone at Staines, and that any assertion beyond that point would unreasonably increase the expense; though Sir Peter was one of the party who visited Oxford in 1826. Again, Alderman Musgrove moved that only Aldermen and Sheriffs should make the excursion. Both motions were negatived; the latter, however, by one vote only!

The discussions on the projected View are too fresh in the recollection of the reader to need repetition here. Suffice it to say, that the requisite motion having been passed, the Lord Mayor fixed the ceremonial for August 5th, 6th, 7th, and 8th. Of its most attractive scenes we shall now present to our readers a series of picturesque Illustrations, accompanied by the details of the Journey; for the facilities in obtaining which we are indebted to the courtesy of the Lord Mayor, in promptly affording our Artist access to the State Barge, so as to enable him to furnish a faithful picture of the Proceedings.

WEDNESDAY, AUGUST 5.

The Civic Party left London by a special train on the Great Western Railway, on Wednesday morning, and reached Oxford within three hours. During the day, the Lord Mayor made a few calls, and invited the Mayor of Oxford, and certain authorities of the Colleges, to dine with his Lordship at the Star Inn. The Party also inspected a few of the magnificent public buildings with which Oxford abounds; though they strove to make their Visit a private one.

In the evening, the Lord Mayor gave a Banquet in the large room at the Star Hotel. His Lordship presided, supported by the Mayor of Oxford; the President of Trinity College; Admiral Ross; &c. &c.

The Grace-Cup (engraved at page 101) was passed round in due form, and the usual loyal toasts given.

The Lord Mayor next gave, "The Health of the President of Trinity College, (the Rev. Dr. Ingram), and Prosperity to the University of Oxford."

The Rev. Dr. Ingram returned thanks, and alluded to the visit of the Lord Mayor of London on a similar occasion twenty years ago, when he had the gratification of spending a very delightful day with his Lordship and party. As a member of that ancient and loyal University to which they had just paid so much honour, it afforded him much delight to receive within its walls the citizens of London, the merchant princes of England, headed by that Lord Mayor, whose goodness of heart and integrity of character they all knew and appreciated. (Loud cheers.)

The Lord Mayor then proposed, "Dr. Ogle, and the Professors of the University of Oxford."

Dr. Ogle acknowledged the honour, and remarked that the patronage which the Lord Mayor had given to literature and science was highly gratifying to all who felt an interest or were engaged in them. (Cheers.)

The Rev. D. Ingram proposed the health of the Lord Mayor, and spoke of the connection which had subsisted between the cities of London and Oxford, which he dated a far back as the year 891. These two important cities, it was spoke of the Thames, had, in days gone by, been associated in many ways, and that association, he doubted not, had been beneficial to both. The still closer communication by railway and the advancement of art would unite them still more, so that the connection between Oxford and London might be important to the whole country. He hoped that the stream which ran through the two cities would

Figure 41 Elizabeth Coxen Gould, wife of John Gould (Courtesy of Dr Geoffrey Edelsten).

Gould was elected Corresponding Member of the Royal Academy at Torin, Italy (Salvadori 1881).

Henry Constantine Richter began working as the main artist for Gould after Elizabeth Gould's death. According to Christine Jackson (pers. corres. 31 March 1976) he was a bachelor and son of the famous artist Henry Richter, who died in 1857.

1842

February 2   Gilbert left London for his second trip to Australia, arriving in Fremantle, Western Australia, on 17 July. There was a stayover of five weeks at the Cape.

March 1   *Birds of Australia* part 6 scheduled.

June 1   *Birds of Australia* part 7 scheduled.

PART THREE CHRONOLOGY OF THE LIFE AND WORKS OF JOHN GOULD

June 23	The British Association meeting at Manchester adopted Hugh Strickland's Code of Nomenclature based on the rule of priority founded upon the twelfth edition of Linnaeus's 1766 'Systema naturae'.
July 19	H.M.S. *Sulphur* returned to England after a voyage of over five years, primarily devoted to exploration of the Pacific islands and shores.
August 13	First of a series of six letters from George Grey (1812-98), Governor of South Australia, ending on 10 January 1843 (Chisholm 1938 *Emu* 38:216-26).
September 1	*Birds of Australia* part 8 scheduled.
November 12	On or before this date, part 3 of the German *Ramphastiden* was published.
December 1	*Birds of Australia* part 9 scheduled.
December	*Macropodidae*, concluding part 2 appeared.

1843

March 1	*Birds of Australia* part 10 scheduled.
June 1	*Birds of Australia* part 11 scheduled.
August 8	Gould was on the Continent for a journey to Paris, Brussels, Frankfurt, Nuremberg, Leipzig, Berlin, Hanover, Amsterdam and Leiden. (Letter, Roy. Scot. Mus.)
September 1	*Birds of Australia* part 12 scheduled.
October	Richard B. Hind's *Zoology of the voyage of H.M.S. Sulphur*, Number III, *Birds*, by Gould, part 1 appeared.
December 1	*Birds of Australia* part 13 scheduled.
December 20	Gilbert left Western Australia for Adelaide, and Launceston, en route to Sydney.

Gould was elected a Fellow of the Royal Society in 1843, according to Sclater (1882). *Zoologist* journal first published. In 1916 merged into *British Birds*.

1844

January	Hind's *Voyage of the Sulphur, Birds*, part 2, appeared.
January 30	Gilbert arrived in Sydney. From February to August he collected in New South Wales, to as far north as Darling Downs of the present state of Queensland.
March 1	*Birds of Australia* part 14 scheduled.
June 1	*Birds of Australia* part 15 scheduled.
September 1	*Birds of Australia* part 16 scheduled.
September 18	Gilbert begins his diary on the Leichhardt Expedition. The explorers were to go from Moreton Bay, Brisbane to Port Essington. (Figure 42)
September 26	Gould presented a paper on the partridges of America at the British Association meeting in York.
October 26	Near this date was published *A monograph of the Odontophorinae, or partridges of America*, part 1, ten plates.
December 1	*Birds of Australia* part 17 scheduled.

In 1844 Hermann Schlegel adapted the use of trinomial nomenclature for bird sub-species.

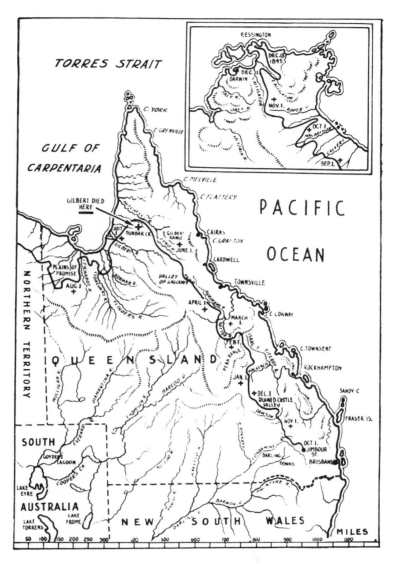

Figure 42 Map showing route, dates and major discoveries of Leichhardt's Expedition en route to Port Essington in 1844 to 1845 (Chisholm 1941).

1845

March 1	*Birds of Australia* part 18 scheduled.
May 1	*Mammals of Australia*, part 1, fifteen plates scheduled for publication.
June 1	*Birds of Australia* part 19 scheduled.
June 28	Gilbert was speared by Aborigines in a night attack on Leichhardt's camp near the Gulf of Carpentaria (Figure 43).
September 1	*Birds of Australia* part 20 scheduled.
September 4	Stephen Coxen died in Sydney at the age of 47 from self-administered poison because of financial reverses. He was a brother of Elizabeth Coxen Gould. (Chisholm, *Emu* 51:78-9)
December 1	*Birds of Australia* part 21 scheduled.

PART THREE CHRONOLOGY OF THE LIFE AND WORKS OF JOHN GOULD

Figure 43 Mural tablet in memory of John Gilbert in St. James's Church of England, in Sydney. The date of death is incorrect and should read '28th of June' (*Emu* 38).

December 5	Gould was visiting at Jardine Hall (Letter, Roy. Scot. Mus.).
	During this year the new building for the British Museum was completed in Bloomsbury.

1846

February 10	Edward Lear's *A Book of Nonsense* was published.
March 1	*Birds of Australia* part 22 scheduled.
April 1	Before this date was published *Odontophorinae*, part 2.
June 1	*Birds of Australia* part 23 scheduled.
June–July	Gould was on the Continent with his son Henry. They visited the south of France, Switzerland and part of Germany. They had returned by 30 July (Letters of July 24 & 30. Roy. Scot. Mus.).

September 1	*Birds of Australia* part 24 scheduled.
September 10	'Mr. Gould exhibited several new species of humming-birds from the Andes' at the British Association meeting in Southampton.
December 1	*Birds of Australia* part 25 scheduled.

1847

February 11	John Gould's grandmother, Ann Gould, died at Ilminster aged 90.
March 1	*Birds of Australia* part 26 scheduled.
May 1	The cornerstone was laid for the Smithsonian Institution in Washington.
May 8	Letter from John Gould to Edward Wilson offering his Australian bird collection to him for £800, or for £1,000 with eggs included. The collection, with eggs, was purchased by Wilson, then sent to his brother, Dr Thomas B. Wilson of Philadelphia, who in turn presented it to the Academy of Natural Sciences (ANSP Collection 364). (*See also* June 1849 when the collection arrived in Philadelphia.)
June 1	*Birds of Australia* part 27 scheduled.
June 11	Sir John Franklin perished in Canada on his ill-fated expedition for the Northwest Passage.
September 1	*Birds of Australia* part 28 scheduled.
November 22	Richard Bowdler Sharpe was born in London.
December 1	*Birds of Australia* part 29 scheduled.

In 1847 the German *Monographie der Ramphastiden* concluding part 4 was published by Sturm brothers.
(In 1847-48 an influenza epidemic in London killed 15 000.)

1848

February	Joseph Wolf left Germany and took up residence in London (Mullens & Swann 1917).
March 1	*Birds of Australia* part 30 scheduled.
June 1	*Birds of Australia* part 31 scheduled.
June	Gould on 'a trout excursion' to Chemise, forty kilometres northwest of London. (Jardine, 1849, *Contributions to Ornithology* 2:137)
August 1	*An Introduction to the birds of Australia*, post octavo, published.
September 1	*Birds of Australia* part 32 scheduled.
December 1	*Birds of Australia* concluding parts 33, 34, 35 and 36 are all dated 1 December 1848. This was considerably ahead of proposed scheduling.

In 1848 Jardine's *Contributions to Ornithology*, a year-book, was first published, and continued until 1852.

1849

May 12	Before this date *A Monograph of the Trochilidae, or family of humming-birds*, part 1, fifteen plates, was published.
June	The collection of birds upon which Gould has based his *Birds of Australia* arrived in Philadelphia, at the Academy of Natural Sciences by this date. The 1848 [1847] specimens were purchased from Gould between May and July 1847 by Edward Wilson of London for his brother Dr Thomas B. Wilson of Philadelphia, who in turn gave the collection to the Academy. Verreaux Freres of Paris mounted the collection. Unfortunately they removed Gould's original labels, but

PART THREE CHRONOLOGY OF THE LIFE AND WORKS OF JOHN GOULD

	copied onto the stands an abbreviated transcription of the original data (Meyer de Schauensee 1957).
September 1	*Mammals of Australia* part 2 scheduled.
December 5	Before this date was published *The birds of Asia*, part 1, seventeen plates.
December 12–13	Gould attended the meeting to commemorate the second anniversary of the opening of the Ipswich Museum. On Wednesday evening, the 12th, Professor Owen gave a lecture on 'Gigantic Birds of New Zealand'. The meeting was held on the next day. (Abstracted from local newspapers by H. Mandel, Ipswich Museum, and from a letter from Gould to Bonaparte, of 5 December 1849 (Mus. Nat'l. d'Hist. Nat., Paris.))

In 1849 Gould was portrayed by T. H. Maguire (Figure 44). A bronze plaque by J. A. Heyman was undoubtedly made from the Maguire portrait (Figure 45).

1850

July 1	*Birds of Asia* part 2 scheduled.
July 30	Gould apparently read papers at the 20th meeting of the British Association at Edinburgh (Jardine's *Contributions to Ornithology* 1850, 3:86). [There are no references to this in the 1850 Report of the British Association.]
October 24	'Note from Mr. Gould to ask us to step round and see the skin of a *notornis* which has been sent him' (Owen 1894).

Figure 44 John Gould at the age of 45 years, from an engraving by T. H. Maguire, 1849, published by George Ransome of the Ipswich Museum (Author's Collection).

Figure 45 A bronze plaque of John Gould by J. A. Heyman, undoubtedly taken from the Maguire portrait. Where is this bronze now? In Australia? (Barrett 1938).

November 18   Before this date was published *Odontophorinae*, concluding part 3.
December 16   Hugh Strickland, of Apperley Green, Tewkesbury, wrote to T. C. Eyton on this date (Amer. Philosophical Soc., Phila.) 'We have had a flying visit of a few hours from Gould today. He brought with him his drawing of the "Balaeniceps", an extraordinary bird from Central Africa, like a Cancroma but the size of a large crane. He has gone on this evening to Birmingham, and thence to Lincoln.'

1851
January 27   John James Audubon died in New York.
March 15   *Birds of Australia, Supplement* part 1, sixteen plates, was scheduled.

Figure 46 Plan of the Zoological Society Gardens in 1851. The building for Gould's humming-bird display is shown at lower right centre (from Henry Scherren, 1905).

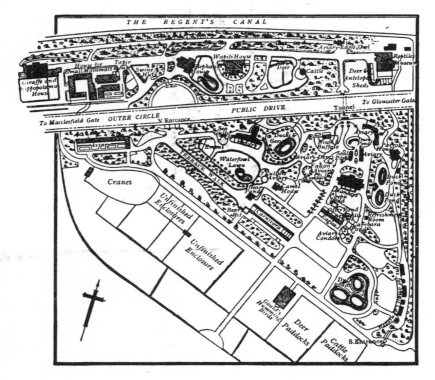

May 1   The Great Exhibition opened in Hyde Park, closing on 15 October 1851. *See also* Royal Commission 1851-52 for information on the Gould exhibit of 'the luminous and metallic colouring of the *Trochilidae*'.
May 17   This is the date Gould began keeping records of the gate receipts for his humming-bird exhibition at the Zoological Gardens. Chalmers Mitchell wrote that 'Gould lent to the [Zoological] Society his collection of mounted hummingbirds and these were exhibited in a special building erected near the Lion House' (Figures 46 to 49). According to Mitchell this exhibit 'attracted over 75 000 visitors and without doubt added to the popularity of the Gardens in the year of the Great Exhibition' (C. Mitchell 1929, p. 102).

Plate 27 Drawing in pencil and watercolour, artist unknown, of *Garrulus taivanus*, for *Birds of Asia*, V: 58, Kansas-Ellis-Gould MS 763.

Plate 28 Drawing in pencil and watercolour, signed by Wolf, of *Galloperdix lunulosa*, for *Birds of Asia*, VI: 69, Kansas-Ellis-Gould MS 224.

PART THREE CHRONOLOGY OF THE LIFE AND WORKS OF JOHN GOULD

Figure 47 Interior view of the Humming-bird house (*Illustrated London News*).

Figure 48 Pencil sketch, possibly by John Gould, of the interior of his Humming-bird house. Notice the similarity to Figure 47, except for reversal of image (University of Kansas Spencer Library Ellis-Gould MS 507).

                For Gould's letters to the Zoological Society of London concerning this humming-bird exhibition *see* Gould, John. *Correspondence.* 1851.

June 2   *Birds of Asia* part 3 scheduled.
June 10   Queen Victoria visited Gould's humming-bird exhibit at the Zoological Gardens. An account follows (C. Mitchell 1929):

Figure 49 Exterior view of Gould's Humming-bird house in the Zoological Society Gardens (*Illustrated London News*, 12 June 1852, University of Kansas, Watson Library).

This collection was one of the great attractions during the year of the Great Exhibition. On June 10 the Queen and Prince Albert, accompanied by the Princesses, the Duke and Duchess of Saxe-Coburg, and Duke Ernest of Würtemberg, visited the Gardens. The "Times" of the following day said:

'Her Majesty occupied a considerable period of her visit in inspecting the celebrated collection of humming birds which has been placed in the Garden by Mr. Gould. The admirable manner in which this beautiful group is illustrated, and the extreme rarity of several of the species, have rendered the building in which they are contained a most important addition to the previous attractions of the establishment, and supplied in the only possible manner a great *desideratum* in the ornithological part of the Society's collection. The visitors who have repaired to the Gardens for the purpose of examining the humming birds include the most distinguished names in science and in art, as well as in rank, and they have universally expressed their surprise and admiration at the unexpected extent of the species, the peculiar forms of their plumage, and the intense brilliancy of colour for which they are remarkable above every other part of the animal kingdom.' [For Queen Victoria's record of this visit, *see* Queen Victoria. *Manuscript*. 1851.]

Mitchell concludes:

At the end of the season this house was taken down and re-erected in what is now the Middle Garden, where, till the end of 1852, it served to contain the humming-bird collection.

June 30   The Thirteenth Earl of Derby, Edward Smith Stanley, died at Knowsley Hall, aged 77.

November 1   *Trochilidae* part 2 scheduled.

November 8   This is the date of cessation of Gould's records for the gate receipts at his humming-bird exhibition. The admission charge had been sixpence. According to these records in the Zoological Society Library,

PART THREE CHRONOLOGY OF THE LIFE AND WORKS OF JOHN GOULD

the income from visitors from 17 May to 8 November was £1589.15.1.

December 1   *Mammals of Australia* part 3 scheduled.

In 1851 W. Hart began working as an artist for Gould
Figure 50 shows an 1851 view of London.
(Gold was discovered in Australia in 1851.)

Figure 50 'A balloon view of London showing Broad Street, the British Museum, Golden Square, Russell Square, Charlotte Street, and the Bedford Square area. [as seen from Hampstead] . . . May 1st 1851, By Banks & Co. . . .'

Figure 51 Drawing in pencil, signed 'J. Gould Jan. 7, '52', of unidentified bird (unpublished?), Kansas-Ellis-Gould MS 478. This bird is very similar to the upper figure in the pencil drawing by John Gould in the Smithsonian Institute National Museum of Natural History (Mansuetti 1953).

1852

January 7	Drawing in pencil by John Gould on this date (Figure 51).
May 1	*Trochilidae* part 3 scheduled.
May 1	*A monograph of the Ramphastidae, or family of Toucans* (second edition), part 1, fourteen plates, was scheduled for publication.
July 1	Frederick Strange, a natural history collector, was in London on this date. Gould assisted him in selling some specimens from Australia (Letter to Jardine, British Museum (N.H.)).
September 11	'I spent 10 days with Rendel and Fowler killing plenty of grouse and ptarmigan in the midst of fine scenery' in the Highlands of Scotland (Letter to Jardine, 1852 [?] Brit. Mus. (N.H.)).

PART THREE CHRONOLOGY OF THE LIFE AND WORKS OF JOHN GOULD

October 1	*Trochilidae* part 4 scheduled.
November 1	*Birds of Asia* part 4 scheduled.
December 1	*Mammals of Australia* part 4 scheduled.

1853

May 1	*Trochilidae* part 5 scheduled.
July 13	Gould was given a Diploma by the German Ornithological Society signed by Martin Lichtenstein, Zander, Ludwig Brehm, Dr J. F. Naumann, Von Homeyer, and Eduard Baldamus, Secretary (Figure 52).

Figure 52 Diploma of the German Ornithological Society granted to Gould in 1853 (*Emu* 38).

September 7-14	Gould attended the British Association meeting at Hull and presented a paper 'On a new species of *Cometes*; a genus of Humming-Birds'.
September 14	On the way home from the British Association meeting, Hugh Edwin Strickland took a side trip to examine some railroad cuttings for geological specimens and was killed by an express train. He was 42.
October 1	*Trochilidae* part 6 scheduled.
October 1	*Birds of Asia* part 5 scheduled.
November 1	*Mammals of Australia* part 5 scheduled.
December 31, Saturday	Edward Forbes wrote to J. S. Henslow, 4 January 1854 (Collection: Royal Botanic Gardens, Kew, copied for me by Philip Rehbock).

I was in London on Saturday for an odd purpose.'Waterhouse Hawkins, the maker of the gigantic beasts at the Crystal Palace, having completed his Iguanodon of the natural size resolved to end the year by giving a dinner party in the monster's belly. The party consisted of himself, Owen, Prestwick, Gould and me — with some of the Crystal Palace Directors, sundry representatives of the press, and the skilled workmen who built up the beasts — very fine specimens they were too. You may fancy what a jolly feast we had — one to remember (Figure 53).

DINNER IN THE IGUANODON MODEL, AT THE CRYSTAL PALACE, SYDENHAM.

Sir W. Stewart, Bart.; with, in the middle ground, an old Scotch fir wood and forester's lodge.

Should the winter of 1854 prove to be as severe a one as has been predicted, and as it threatens to be, we should not be much surprised to see a number of Curling Clubs formed on this side of the Tweed.

THE CRYSTAL PALACE, AT SYDENHAM.

In our Number of last week we gave a whole-page Illustration of Mr. B. Waterhouse Hawkins's Model-room, or Studio, at the Crystal Palace, Sydenham, where he is constructing his gigantic restorations of the Extinct Inhabitants of the Ancient World. We then had only the opportunity briefly to allude to this novel and great undertaking; and repeated the speculations of enthusiastic discoverers of antediluvian remains which are now, with anatomical severity, being reconstructed and restored to a state of life-like nature by Mr. Waterhouse Hawkins; to whose talents and knowledge this department of the great educational scheme of the Directors of the Crystal Palace Company is now confided; and with how much credit to their judgment was most agreeably exemplified on Saturday evening last (the last day of the year 1853), when Mr. W. Hawkins, with the concurrence of the Directors, invited a number of his scientific friends and supporters to dine with him in the body of one of his largest models, called the Iguanodon, which occupies so conspicuous a place in our Illustration of last week. In the mould of this colossal work of art—for as such it must deservedly rank very high—Mr. Hawkins conceived the idea of bringing together those great names whose high position in the science of palæontology and geology would form the best guarantee for the severe truthfulness of his works; and, at the same time, show to the public the high tone of criticism and knowledge which the Directors of this truly national undertaking require those officers to sustain to whom they confide the carrying out of any important part of their plan which so particularly bears on the education of the people.

To carry out this extraordinary idea, cards were issued at the beginning of last week—and such cards! as startling as the invitation they bore: "Mr. B. Waterhouse Hawkins solicits the honour of Professor ——'s company at dinner, *in the Iguanodon*, on the 31st of December, 1853, at four p.m." The incredible request was written on the wing of a Pterodactyle, spread before a most graphic etching of the Iguanodon, with his socially-loaded stomach, so practicably and easily filled, as to tempt all to whom it was possible to accept, at such short notice, this singular invitation. Many have to regret the rapidity of executing this novel idea, at a season when almost all have a plurality of engagements. Nevertheless, Mr. Hawkins had one-and-twenty guests around him in the body of the Iguanodon on Saturday last; at the head of whom, most appropriately, and in the head of the gigantic animal, sat Professor Owen, supported by Professor E. Forbes; Mr. Prestwick, the geologist; Mr. Gould, the celebrated ornithologist; and the Directors and officers of the Company.

The dinner, which was luxurious and elegantly served, being ended, the usual routine of loyal toasts were duly given and responded to—allusion being gracefully made by Mr. Francis Fuller, Managing Director, to the great interest evinced and approbation expressed by H.M. the Queen and H.R.H. the Prince, on their recent visit to the extraordinary works by which the company were surrounded.

Professor Owen then took occasion to explain, in his lucid and powerful manner, the means and careful study by which Mr. Hawkins had prepared his models, and had attained his present truthful success; Professor Owen adding that it had been a source of great pleasure to him to aid so important an undertaking, by assisting with his instruction and direction a gentleman who possessed the rarely-united capabilities of an anatomist, a naturalist, and a practical artist, with a docility and eagerness for the truth which ensured Mr. Hawkins's careful restorations the highest point of knowledge which had been attained up to the present period. The learned Professor then briefly commented upon the course of reasoning by which Cuvier, and other comparative anatomists, were enabled to build up the various animals of which but small remains were at first presented to their anxious study; but which, when afterwards increased, served to develop [and confirm their confident conceptions—instancing the Megalosaurus, the Iguanodon, and Dinornis as striking examples.

Professor Forbes also bore testimony to the truthful care and study with which these great models were produced by Mr. Hawkins, and which would render them trustworthy lessons to the world at large in a branch of science which had hitherto been found too vast and abstruse to call in the aid of art to illustrate its wonderful truths.

After several appropriate toasts, this agreeable party of philosophers returned to London by rail, evidently well pleased with the modern hospitality of the Iguanodon, whose ancient sides there is no reason to suppose had ever before been shaken with philosophic mirth.

The numbers attending the Museum of Ornamental Art at Marlborough House, during December, were—19,630 persons on the public days, free; 475 persons on the students' days, and admitted at 6d. each; besides the registered students of the classes and schools—an increase of 5,867 over last year.

GIGANTIC BIRD OF NEW ZEALAND.

In the fine Museum of the Royal College of Surgeons, in Lincoln's-inn-fields, is a most interesting illustration of the pitch to which comparative anatomy has reached in this country; the result of an immense induction of particulars in this noble science. Such is the Skeleton of the *Dinornis* of New Zealand, which the visitor will immediately recognise on the left side of the old Museum, having the skeleton of O'Brien, the Irish giant, on its right. The means by which the Museum obtained this valuable acquisition is thus graphically described in Mr. Samuel Warren's truthful and eloquent lecture on "The Intellectual and Moral Development of the Present Age:"—

In the year 1839, Professor Owen was sitting alone in his study, when a shabbily-dressed man made his appearance, announcing that he had got a great curiosity which he had brought from New Zealand, and wished to dispose of it to him. Any one in London can now see the article in question, for it is deposited in the Museum of the College of Surgeons, in Lincoln's-inn-fields. It has the appearance of an old marrow-bone, about six inches in length, and rather more than two inches in thickness, *with both extremities broken off*: and Professor Owen considered that, to what-

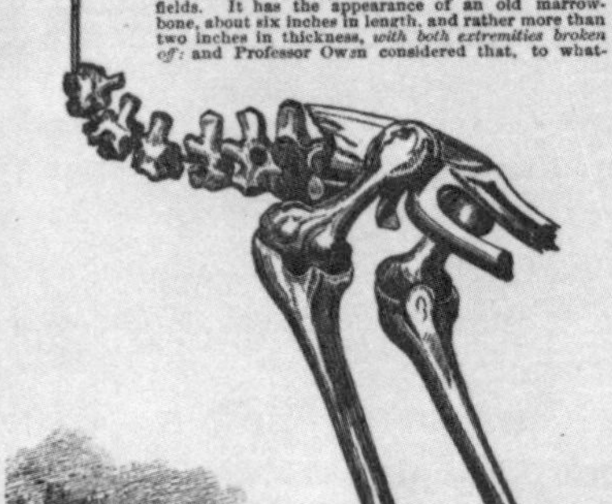

SKELETON OF THE DINORNIS, IN THE MUSEUM OF THE ROYAL COLLEGE OF SURGEONS,

ever animal it might have belonged, the fragment must have lain in the earth for centuries. At first he considered this same marrow bone to have belonged to an ox—at all events, to a quadruped; for the wall or rim of the bone was six times as thick as the bone of any bird, even the ostrich. He compared it with the bones in the skeleton of an ox, a horse, a camel, a tapir, and every quadruped apparently possessing a bone of that size and configuration; but it corresponded with none. On this he very narrowly examined the surface of the bony rim, and at length became satisfied that this monstrous fragment must have belonged to *a bird!*—to one at least *as* large as an ostrich, but of a totally different species; and quently one never before heard of, as an ostrich was by far the bigge known. From the difference in the *strength* of the bone, the ostrich unable to fly, so must have been unable this unknown bird: and anatomist came to the conclusion that this old shapeless bone indica former existence, in New Zealand, of some huge bird, at least as g an ostrich, but of a far heavier and more sluggish kind. Prof. Ow confident* of the validity of his conclusions, but could communica confidence to no one else; and, notwithstanding attempts to d him from committing his views to the public, he printed his deduct the "Transactions of the Zoological Society" for the year 1839, fortunately, they remain on record as conclusive evidence of the his having then made this guess, so to speak, in the dark. He cau bone, however, to be engraved; and having sent a hundred copies engraving to New Zealand, in the hopes of their being distribu leading to interesting results, he patiently waited for three yea till the year 1842, when he received intelligence from Dr. Buckl Oxford, that a great box, just arrived from New Zealand, consi himself, was on its way, unopened, to Professor Owen; who found with bones, palpably of a bird, one of which was three feet in leng much more than double the size of any bone in the ostrich! And the contents of this box the Professor was positively enabled to ar almost the entire skeleton of a huge wingless bird, between t eleven feet in height, its bony structure in strict conformity with th ment in question; and that skeleton may be at any time seen at t seum of the College of Surgeons, towering over, and nearly tw height of the skeleton of an ostrich; and at its feet is lying the o from which alone consummate anatomical science had deduced astounding reality: the existence of an enormous extinct creature bird kind, in an island where previously no bird had been known larger than a pheasant or a common fowl!

* The paper on which he even sketched the outline of the unknown bird, is n hands of an accomplished naturalist in London—Mr. Broderip.

The distinguished artist, Don Rafael Benjumea, has h honour of a private audience of her Majesty the Queen of Spain liver to her Majesty a splendid historical picture of the presenta the Princess Royal in the Royal palace after the birth took place painting has been executed by the artist, by special command Majesty, and has taken the author two years to complete.

South Sea Company.—On Thursday the half-yearly m of this company was held at the South Sea House, when a divi $1\frac{1}{2}$ per cent for the half-year was declared. After some discu was resolved that a special meeting should be called, to take in sideration the proposition for obtaining an Act of Parliament to c the company as a trust company.

Asylum for Female Orphans.—On Thursday the qu court of this charity was held at the Asylum, Westminster-ro Wild, Esq., in the chair; when it was reported that the school w quite full, the number of children in the Asylum being 160, a general health, notwithstanding the inclemency of the weathe satisfactory. Since the last meeting, £500, being part of the pro the triennial dinner, at which his Royal Highness the Duke of Ca presided, had been invested in the Funds. The report was adopt a special vote of thanks given to his Royal Highness for the gr evinced by him in the success of the institution.

Royal National Institution for the Preservati Life from Shipwreck.—On Thursday last a meeting of the Committee of this old and valuable institution, was held at the John-street, Adelphi; Mr. Alderman Thompson, chairman of the presiding. The gold medallion of the institution was voted gallant Captain Ludlow, of the American ship *Monmouth*, miration of his noble and humane conduct to the crew a sengers of the unfortunate emigrant ship *Meridian*. The medal of the Society was also presented to B. Harringto W. Waters, first and second coxswains of the *Southwold* life-b their frequent services in saving life in the life-boats of tha The thanks of the committee, on vellum, were voted to the R Williams, jun., and to Captain Hedenscht, of the Peninsular and Steam Navigation Company's Services, for their intrepidity in sav A reward of £4 10s. was granted to the crew of the *Boulmer*, n boat, for saving the crew, consisting of seven men, of the brig *Nicol*, of Perth. The boat was placed on this station by the Northumberland, K.G., and liberally presented by him, as presi the institution. Other rewards having been voted, the commi cided that a public dinner should be held, early in March next, the funds of the society, and for the purpose of bringing the tru volent and national objects of the institution prominently before th It was reported that four new life-boats were nearly ready to be the coast. Colonel Tulloch, R.A., as successor to the late Colquhoun, R.A., having been elected a member of the committ proceedings closed.

A Fire broke out in Doctor's-commons at half-pa on Thursday morning, destroying the extensive manufactory of Hodgkinson and Burnside, envelope makers, and the premises Coombs, builder—these buildings lying between the Church of S and the College of Advocates, both of which narrowly esca struction; but the residence of Mr. Pritchard, adjoining the Hall, with some partial damage from the heat of the flames. Had the pened a few hours earlier, the library of the College (only separat feet from the burning premises), and the whole range of the build cluding the Will-office, might have become a heap of ruins. The is estimated at from £3000 to £4000.

PART THREE CHRONOLOGY OF THE LIFE AND WORKS OF JOHN GOULD

In 1853 parts 2 and 3 of *Ramphastidae*, second edition, were undoubtedly published.

(1853-56 was the period of the Crimean War.)

1854

January — Gould left England around the 22nd and travelled with his son, Dr John Henry Gould, to the island of Malta in the Mediterranean. Henry left his father there and proceeded on 2 February *en route* to India to assume a post as an Assistant-Surgeon in the Honourable East India Company's Service. Gould published privately twelve letters from his son beginning on 8 February 1854 written as he travelled down the Red Sea, then later from Bombay and Kurrachee [Karachi, Pakistan], with the last letter dated 7 November 1854 (Letters courtesy of Dr Geoffrey Edelsten).

May 1 — *Trochilidae* part 7 scheduled.

May 1 — *Ramphastidae*, second edition, concluding part 4 scheduled.

July 1 — *Birds of Asia* part 6 scheduled.

August 19 to Sept. 30 — These are the dates of an intensive study by Dr John Snow during a cholera epidemic in the neighbourhood of the Broad street pump. In ten days over 500 persons died. Dr Snow published a map of the area showing the houses in which the many hundreds of deaths

Figure 54 Map of the Broad Street pump area from Dr Snow's 1855 article on cholera. A black bar marks each death in a house during the epidemic of cholera from 19 August to 30 September 1854. Gould's home was near the 'X' Mark. (Photograph by Larry Brown).

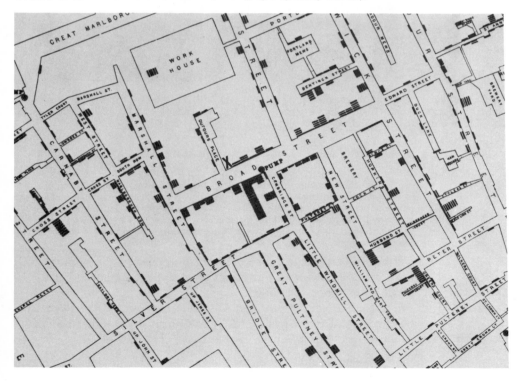

Figure 53 Dinner given at the Crystal Palace by Waterhouse Hawkins set in the belly of an Iguanodon model. ◄

135

Figure 55 Broad Street, mid-nineteenth century. According to L. E. James Helyar, these homes were across the street from Gould's home (from Barton's 1959 article on Dr Snow, courtesy St. Bartholomew's Hospital, London).

occurred (Snow 1855 and 1936) (Figures 54 & 55). Dr Snow wrote that

> Mr. Gould, the eminent ornithologist, lives near the pump in Broad Street, and was in the habit of drinking the water. He was out of town at the commencement of the outbreak of cholera, but came home, on Saturday morning, 2nd September [1854], and sent for some water almost immediately, when he was much surprised to find that it had an offensive smell, although perfectly transparent and fresh from the pump. He did not drink any of it.

Further quotations from Snow's article appear in the bibliography. Barton (1959) reviewed Dr Snow's work in this epidemic.

October 1	*Trochilidae* part 8 scheduled.
October 14	Frederick Strange, one of Gould's collectors, was killed by natives on the Second Percy Island off Cape Capricorn, Queensland.
December 1	*Mammals of Australia* part 6 scheduled.

In this year Gould was elected to membership in the Athenaeum Club as one 'distinguished in science'. (Pers. corres. from the Secretary, 28 October 1974.)

PART THREE CHRONOLOGY OF THE LIFE AND WORKS OF JOHN GOULD

1855

January 1	*Supplement to the first edition of a monograph on the Ramphastidae, or family of toucans*, parts 1 and 2 complete, twenty plates, scheduled.
April 2	*Birds of Asia* part 7 scheduled.
May 1	*Trochilidae* part 9 scheduled.
July 1	*Mammals of Australia* part 7 scheduled.
September 1	*Birds of Australia Supplement*, part 2 scheduled.
September 1	*Trochilidae* part 10 scheduled.
October 4	Dr John Henry Gould, aged 24, oldest son of John Gould, died at Bombay of a 'fever'.
October 15	First of a series of sixteen or so letters from Lizzy (Elizabeth) Gould to her sisters. The latest letter is dated 3 February 1859 (Courtesy of Mrs Alan Edelsten).
December 5	William Swainson died in New Zealand.

(In 1855 the Metropolis Local Management Act was passed. A clause was inserted that gave the Metropolitan Board of Works of London comprehensive powers in regard to the naming and numbering of the streets. Numbering of houses prior to that time had been 'consecutive'. It was then changed to the 'odd and even' principle — odd on one side of the street, even on the other. Also over 1500 streets which had duplicate names were renamed. The task took years to accomplish (Cunningham 1927, p. xxi)).

1856

May 1	*Birds of Asia* part 8 scheduled.
May 1	*Trochilidae* part 11 scheduled.
May 3	In a letter from Gould's daughter Lizzy to Louisa this date (collection of Mrs Alan Edelsten) Lizzy wrote

We have had the Prince of Wales and his brother here today... They were principally delighted with the birds Papa shewed them in the office and which they could handle. They took away with them two humming bird's nests and eggs... The street was in an uproar.

June	Joseph Wolf and Gould 'sailed for Norway on board the yacht of Mr. Bidder, formerly The Calculating Boy, and then Mr. Dresser tells me, an eminent consulting engineer. Landing at Christiania, Wolf and Gould went on alone, with an interpreter' (Palmer 1895, p. 77). (Figures 56 & 57)
July 1	With Wolf at 'the celebrated Snee Haetten range of mountains' in Norway (Palmer 1895, p. 78).
August 1	Gould wrote to Alfred Newton (1829-1907) that he had just returned from a 'lengthened tour of Norway, part of the time in company with Mr. Woolley [John Wolley]'. This letter is the first of a series of five letters between Gould and Newton, concluded on 26 March 1861 (Newton Library, Cambridge).
September 1	*Trochilidae* part 12 scheduled.
October 1	*Mammals of Australia* part 8 scheduled.

1857

May 1	*Birds of Asia* part 9 scheduled.
May 1	*Trochilidae* part 13 scheduled.

Figure 56 Watercolour signed 'J. Wolf 1856' of Ptarmigan in summer plumage, probably done while with Gould in Norway (Courtesy of Dr and Mrs Geoffrey Edelsten).

Figure 57 Watercolour signed 'J. Wolf 1857' of King Duck [Eider], similar to *Birds of Great Britain*, V: 27, designated on plate as by J. Gould & H. C. Richter (Courtesy of Dr and Mrs Geoffrey Edelsten).

PART THREE CHRONOLOGY OF THE LIFE AND WORKS OF JOHN GOULD

May 2	John Gould and his son Charles departed from Liverpool on the *Asia* at 3 p.m. on this Saturday for New York City.
May 15	According to the New York *Evening Post*, the British Mail Steamer *Asia* had arrived in the morning. The boat had departed from Liverpool at 3 p.m. Saturday 2 May. On page 2 of the newspaper was a note 'Passengers by the Asia...Gould, Gould, Jr.'
May 16	A letter of this date soon after their arrival in New York City follows (Courtesy of Alec H. Chisholm of Sydney, 1954):

> New York.
> 16th May 57
>
> My dearest children,
> I take the earliest opportunity after arriving to say we are well, in a day or two either Charles or myself will write more fully. Indeed it is just probable that our second letter may arrive first. The entrance to New York from the sea is truly beautiful and from what I have seen of the place (I mean the city) I am not a little astonished and amused. I trust in a day or two to change this turmoil for quieter scenes in a state of nature. Trusting my dearest children that you are all well and that a kind providence will ever protect you. I will ask you to kindly remember me to Miss Yates, Mr. Prince, and all enquiring friends, and believe me to remain, my dear children,
>
> Your ever affectionate father,
> John Gould

May 19	Gould and son were in Philadelphia. 'Hall of the Academy ordinary meeting. Present Messrs. Leidy, A. H. Smith, Corse, W. S. Wilson, J. D. Sergeant, Draper, J. A. Meigs, Woodhouse, Schafhirt, Morris, Rand and Mr. John Gould, Correspondent. Dr. T. B. Wilson in the chair.' From the Academy of Natural Sciences of Philadelphia, minutes of meeting, 1857 (Collection 502, meeting of 19 May 1857).
May 21	It was on the 21st of May, 1857, that my earnest day thoughts and not infrequent night-dreams of thirty years were realized by the sight of a Humming Bird. The period of my visit to America being somewhat early in the season my attempts to discover a living 'Hummer' in the neighborhood of New York during the second week of May were futile, and it was not until I arrived at the more southern city of Philadelphia that my wish was gratified by the sight of a single male in the celebrated Bartram's Gardens, whither I was conducted by my friend Mr. Wm. L. Baily, from whom I also received many other kind attentions.

The almost total absence of Humming Birds around Philadelphia proved to me that I was still too early for them, the lateness of the season of 1857 having retarded their movement. I therefore, determined to proceed further south to Washington, where in the gardens of the Capitol, I had the pleasure of meeting with them in great numbers; in lieu of the single individual in Bartram's Gardens, I was now gratified by the sight of from fifty to sixty on a single tree. [Gould's *Trochilidae*, vol. 3, pl. 131, the account of the Ruby-throated Humming-bird.]

While in Philadelphia it appears that Gould had a portrait taken by J. E. McClees (Figures 58 & 59).

May 23	This note appeared in the Washington *Evening Star*: 'Arrivals at the Hotels...Willards' Hotel...Mr. Gould, son, Eng.'
June 3	A letter of this date from Gould in Toronto to Spencer Fullerton Baird (1823-87) of the Smithsonian Institution, Washington, begins a series of fourteen back and forth communications. The dates of the American correspondence are as follows:

3 June [May in error] 1857 from Gould in Toronto to Baird concerning Dr Brewer's paper on 'North American Oology', sets of books sold in Buffalo and Toronto, and a request for a few live hummingbirds to be sent to Gould when he arrives in New York City.

Figure 58 Portrait of John Gould probably taken by 'J. E. McClees Photograph Establishment' in Philadelphia in 1857. This 90 x 55mm wide photograph appears to be the basis for a watercolour portrait that is the Frontispiece in volume three of Baily's unpublished *Illustrations of the Trochilidae*, 1855-8 (Courtesy of the Library, The Academy of Natural Sciences of Philadelphia).

Figure 59 Preliminary watercolour portrait of John Gould, painted in 1857, by 'Jas. E. McClees, Photograph Establishment' in Philadelphia. *See* [Gould, John.] Portrait. 1857 (Courtesy of the Library, The Academy of Natural Sciences of Philadelphia).

June 11	(To maintain continuity, a letter on this date from Boston from Charles Gould to his family in London will follow after this Baird exchange.)
June 18	From Gould at the Brevoort Hotel in New York City to Baird concerning desiderata of Baird, a Dr Rae of Montreal, and sending off a box of books to the Library of Congress.
June 19	From Gould in New York City to Baird requesting a specimen of *Trochilus alexanderi*, with mention of John Cassin and George Lawrence.
June 20	From Baird to Gould in New York City concerning Baird wanting a good skin of the Barren Ground Reindeer and Musk-Ox, and not having a specimen of *T. alexanderi*.
June 22	From Gould in New York City to Baird promising to attempt to get those two mammals from McClintock's expedition, and reporting that Gould was leaving New York for home in two days.

The balance of the correspondence to Baird is on Gould's return to London. It consists of four letters from Gould to Baird, one draft from Baird, probably for Joseph Henry, to Gould thanking him for a collection of 160 specimens of American birds given to the Smithsonian, and four drafts of letters from Baird to Gould, the last one dated 31 January 1864. [This ends the Gould-Baird interchange of letters.]

June 11	Letter from Charles Gould in Boston to the family at home. Charles was travelling with his father. Since the letter details their American

PART THREE CHRONOLOGY OF THE LIFE AND WORKS OF JOHN GOULD

trip so completely, it will be published in full here, through the courtesy of Dr Geoffrey Edelsten, Gould's great-grandson.

June 11, 1/57 Boston

My dear Sarah

If I remember rightly I am in debt to you on the score of letters, and therefore proceed to wipe off part of it. As you observe from the date etc. we have made the most of our time. In fact we have been dashing over the country at such a rate that I have scarcely had time to think of you all at home, much less to write. In fact my log book from which I expected great things is as yet uncommenced, the books almost unopened, the shell compeller, that extraordinary compound of tin and wire gauze, is rusting from inactivity and the leather fossil bag, after lingering for some time in ignominious obscurity at the bottom of my carpetbag waiting for better days to come, was finally used up as packing material a short time back at Cleveland. Thus much as a sort of apology for not writing more frequently, and next to acknowledge letters No. I, II, III from Mr. Prince with their enclosure from the misses, No. II reached us before No. I containing letters from Louisa and Eliza, both of course very interesting, and that of the first rather enigmatical, she says that you have all been to the polite time [?], what can she mean?? in a latter letter she says the weather is fine. What new article has she been purchasing I wonder.

I dare say you will be wondering what sort of route we have taken, I will tell you.

1. from *New York* to Philadelphia
   Philadelphia — Washington
      (through Baltimore).
   Washington — Altoona (in the Allegheny Mountains)
   Altoona — Cleveland (on Lake Erie)
   Cleveland — Buffalo
   Buffalo - Niagara. There ought to have been a flourish of trumpets at this point — imagine it —
   Niagara — Toronto
   Toronto — Montreal
   Montreal — Portland (coming through the White Mountains)
   Portland — Boston.

I quite forgot what I told you about in my first letter so that I must run the risk of repetition, I believe I said something about N.Y. (The Empire City, or Commercial Emporium) and Phila (The City of brotherly love). At Washington, (city of magnificent distances) pater familias dined with Lord ['Elgin' crossed out] Napier and spent the evening with the President [Buchanan], the small boy was left at home upon each occasion, and consequently adopted for a short time the profession of mud larker, on the banks (the very muddy banks) of the Potomac collecting however Melauria's in place of halfpennies, and of course humming the appropriate air of "Shells of Ocean".

It was here that we first met with humming birds and coloured gentlemen in any abundance. The first buzzing by dozens round horse chestnut trees in the gardens of the Capitol, the latter especially numerous in and about the hotel. The weather was exceptionally hot, so I found that it was much better to keep them "men and brother" at a distance, and at all events to get to windward of them. N.B. also a great consumption of cobblers. I suppose you will have seen in the Times an account of the row which took place lately in Washington between plug-uglys, and some other political body, I forgot its name. This happened a few days after we had left — 6 or 7 men were killed, and 17 wounded — only — much too trifling to cause any excitement here.

From W[ashington] to Cleveland, from Cl[eveland] to Buffalo (rubie H B), Buffalo to Ni____ag____a____ra____. I don't like anything sentimental, therefore do not expect raptures and all that sort of thing. Suffice it to say I was *not* disappointed. It is in every respect as vast and glorious as represented. The only humbug about the falls, is the going underneath. I had heard so much about it that I expected this would be attended with some difficulty, and that there would be something to see — all hum — neither the one nor the other.

There being no H[umming] B[irds] in the neighbourhood, and no savants Mr. Gould found the beauty of the falls alone insufficient to attract him more than a few hours, so off we started again before I had hardly got a glimpse of them, which was the more aggravating in as much as he had been saying for the last fortnight, how glad he should be to get to the falls where he might be quiet for two or three days — (Sic vita Gouldi Johnni) — I do not mean to give you even the suspicion of a hint of the sundry bracelets, etc. we bought after a good deal of small talk, with a young lady at a kind of Niagaran edition of the Lowther bazaar.

From Niagara we went partly by rail and partly by boat to Toronto, which is a very large place, and like most American cities very uninteresting. The only novelty about the place, is the use of planks for foot paths and roadways. There is hardly a stone in the place, half the town consists of plank houses. I got a few land and fresh water shells here and had some pleasant days collecting on a kind of sand bank or peninsula running out into the lake. Tell Frank there were no butterflies, no insects of any kind with the exception of a few very small butter [flies].

We went to one or two dinner parties and to an evening party at Government house. The hotels here very dirty and wretched.

On reaching Montreal you find quite a different state of things. It is a much older town, and has not those dreadfull [sic] monotonous streets, all straight and crossing one another at right angles. The people are more than half French, though English is mostly spoken. While here we stayed with Mr. Hodges one of the engineers of the Grand Trunk R.R. He was a fellow passenger of ours in the Asia. From Montreal to Portland.

From P[ortland] to Boston

Your affectionate brother
Charles Gould

An interesting note concerning Gould in Toronto occurs in the September 1857 *Canadian Journal of Industry, Science and Art* (2:328) written by Dr James Bovell of Toronto: 'During the present summer we were visited by Mr. John Gould, the distinguished Naturalist, whose chief object in his tour through Canada was for the purpose of studying the habits and manners of the species of *Trochilus* frequenting this portion of the North American Continent.'

In a report to the Montreal Natural History Society dated 29 June 1857 written by William D'Urban, Sub-Curator of the Society's Museum is the following: 'At the beginning of this month the celebrated Ornithologist Mr. Gould F.R.S. . . . paid Montreal a flying visit of 3 or 4 days . . . He brought his son with him . . . ' (*See also* Montreal Nat. Hist. Soc. Minutes 1857).

July 14 In the *Proceedings of the Zoological Society* of London under this date is the statement 'Mr. Gould having returned from a visit to the United States whither he had proceeded for the purpose of studying the habits and manners of the species of *Trochilus* frequenting that portion of the American continent, detailed some of the results of his observations.'

July 29 Charles Lucian Bonaparte of France and Italy died, aged 54.
September 1 *Trochilidae* part 14 scheduled.
October 1 *Mammals of Australia* part 9 scheduled.
December 7 First of a series of letters from Gould to Professor or Sir Frederick McCoy, (1823-99) Director of the National Museum of Victoria. McCoy requested from Gould 'an entire set of generic types of birds' of the world for his museum. In these thirty letters concluding on 15 April 1876, is the record of over 4000 bird specimens sent to McCoy. The latter in turn subscribed for and sold in Australia a few sets of Gould's published works.

In 1857 Gould was on the Council of the Zoological Society when the museum collection of the Society was transferred to the British Museum at Bloomsbury. Gould 'made an offer which was accepted, to buy the bird' skins not otherwise disposed of for the sum of £40' (Mitchell, 1929, p. 104).

PART THREE CHRONOLOGY OF THE LIFE AND WORKS OF JOHN GOULD

1858

January 30	Coenraad Jacob Temminck, a famous Dutch ornithologist and friend of Gould, died, aged 80. He was Director of the Rijksmuseum, Leiden.
May 1	*Trochilidae* part 15 scheduled.
June 1	*Birds of Asia* part 10 scheduled.
July 1	On this date Charles Darwin and Alfred Russell Wallace had their two short papers on evolution by natural selection read at a meeting of the Linnean Society by the Secretary. Neither author was present.
August 5	(First Atlantic cable completed. Queen Victoria and President Buchanan exchanged greetings. But the cable soon failed. It was successful in 1866.)
September 1	*Trochilidae* part 16 scheduled.
October 1	*Mammals of Australia* part 10 scheduled.
	Gould's son Franklin, who was born in Tasmania and educated at King's College, graduated in 1858 at the age of 19, with B.A. from the University of London (Dr Geoffrey Edelsten family scrapbook).

A monograph of the Trogonidae, or family of Trogons, second edition, part 1, thirteen plates, was published in 1858.

1859

January	*The Ibis* organ of the British Ornithologists Union, first published as a quarterly journal.
January 11	'You will soon see my son [Charles] the Geologist who I believe will go to Tasmania, taking Melbourne in his way on a visit to Mr. Selwyn. He has been recommended and I believe his appointment settled to undertake a Geological Survey of Tasmania' (Letter to McCoy, Nat'l. Mus. Victoria, Melbourne).
January 24	'I was shooting at Somerleyton at the time', so writes Gould in a letter to the Editor of the 1859 *Ibis* 1(1):206-7.
May 1	*Birds of Asia* part 11 scheduled.
May 1	*Trochilidae* part 17 scheduled.
September 1	*Mammals of Australia* part 11 scheduled.
September 1	*Birds of Australia, Supplement*, part 3 scheduled.
September 1	*Trochilidae* part 18 scheduled.
September 14-21	Gould gave three papers at the British Association meeting in Aberdeen on pheasants, birds of paradise and a new species not mentioned in the Report.
September 17	Around this date begins a series of seventy-eight letters from John Gould to his oldest daughter Eliza, or Lizzie. Most of the letters are undated. The last definite date in this group of letters is 1 December 1869 (Courtesy of Mrs Alan Edelsten).
October 17	Gould's residence for a short time was at 94 Gt. Russell Street, Bloomsbury W.C. (Letter to McCoy, Nat'l. Mus. Victoria). Thus Gould and his family moved from 20 Broad Street, Golden Square in the fall of 1859. He was living on Broad Street on 14 April 1859 according to a letter of this date to McCoy. The Gould family had resided at 20 Broad Street for twenty-nine years, since 1830.
November 24	The first edition of *On the origin of species* by Charles Darwin was published.

Figure 60 A gallery of portraits of John Gould when he was in his fifties and sixties.
A From Whittell (1954), of a portrait in possession of Chisholm. **B** From Dr Geoffrey Edelsten, photographed by W. Walker. Similar setting to C. **C** From Barrett (1928), of a portrait in possession of Mr E. A. Vidler, whose grandfather was Dr George Bennett. **D** From Mrs Doris Twyman, great-granddaughter of John Gould. **E** From Mrs Doris Twyman. **F** From Reeve (1863–67), photographed by Ernest Edwards.

December 16   Gould's address on a letter of this date to McCoy was 26 Charlotte Street, Bedford Square, W.C. No further residential moves for Gould and his family were to take place. 26 Charlotte Street is now 23 Bloomsbury Street (Cunningham 1927, p. 55).

1860
May 1   *Trochilidae* part 19 scheduled.
June 1   *Birds of Asia* part 12 scheduled.
June 30   The famous meeting between Bishop Samuel Wilberforce and T. H. Huxley on Darwin's evolution theories occurred at the British Association meeting at Oxford.
September 1   *Trochilidae* part 20 scheduled.
November 1   *Mammals of Australia* part 12 scheduled.

(In 1860 Burke and Wills made the first south-to-north crossing of the Australian Continent from Victoria to the Gulf of Carpentaria.)

1861
April 12   (In the United States the war between the states began on this date, and continued until 9 April 1865.)
May 1   *Birds of Asia* part 13 scheduled.
May 1   *Trochilidae* part 21 scheduled.
July 1   *Trochilidae* part 22 scheduled.

PART THREE CHRONOLOGY OF THE LIFE AND WORKS OF JOHN GOULD

September 1	*Trochilidae* concluding parts 23, 24 and 25 were published.
September 1	*An Introduction to the Trochilidae, or family of humming-birds*, post octavo, was published.
September 4–11	A paper by Charles Gould entitled 'Results of the Geological Survey of Tasmania' was abstracted as presented at the British Association meeting in Manchester.
December 14	(Prince Albert, Consort, husband of Queen Victoria, died.)
December	The John McDouall Stuart expedition began, which eventually crossed the continent to Darwin, on the north coast. F. G. Waterhouse collected some birds for Gould on this journey.

1862

April 26	Gould 'killed' a male Thames trout at Cookham on this date, according to notes on a drawing [by Hart?] of a fish in the British Museum (Natural History).
May 1	*Birds of Asia* part 14 scheduled.
May 15	Gould 'killed' a male Colne trout on this date in Mercer's pool, West Drayton, according to notes on another drawing. Both of these drawings were presented to the British Museum (N.H.) by Mrs Alan Edelsten.
October 1	*The Birds of Great Britain* parts 1 and 2, with fifteen plates each, were published.
	In 1862 Richard Bowdler Sharpe (1847-1909) first became acquainted with Gould in the neighbourhood of Cookham, on the Thames. Sharpe was to become one of the world's most eminent ornithologists. In 1893 he wrote an *Analytical Index* to Gould's works.
	Figure 60 is a gallery of portraits of John Gould taken around this time.

1863

May 1	*Mammals of Australia* concluding part 13 scheduled.
May 13	On or bedore this date was published *An introduction to the mammals of Australia*.
June 1	*Birds of Asia* part 15 scheduled.
September	'Owen, accompanied by John Gould was the guest of the Duke of Northumberland at Olnwick Castle. Afterwards they went to Lord Tankerville's where, among other things, they inspected the Chillingham cattle' (Owen 1894, vol. 2, p. 140).
September 23	First of twelve letters from Gould to Frederick George Waterhouse (1815-98), who was the first Curator of the South Australian (Institute) Museum (Sutton, 1929, S. Aust. Ornith.).

In 1863 *Birds of Great Britain* parts 3 and 4 were published.
(In 1863 the first part of the London underground railway (tube) was completed, which finally relieved some of the street-traffic congestion. An epidemic of scarlet fever in England killed over 30 000.)

1864

April 1	*Birds of Asia* part 16 scheduled.

In 1864 *Birds of Great Britain* parts 5 and 6 published.
Zoological Record first published and continues to date.

1865

April 1	*Birds of Asia* part 17 scheduled.
June 11	Gould purchased a Great Auk's egg for Mr G. Dawson Rowley for £33 (Parkin 1911).
September 1	*Handbook to the birds of Australia*, volume one, was published.
October 25	First of a series of twelve letters from Gould to Edward P. Ramsay (1842-1916) of Sydney, concluded on 29 January 1869 (Hindwood, 1938, *Emu* 38:206-12).
December 2	*Handbook* volume two published.

In 1865 Gould's son, Franklin, became a member of the Royal College of Surgeons in London, and in 1866 graduated M.D. from the University of Edinburgh (Dr Geoffrey Edelsten family scrapbook).

Birds of Great Britain, parts 7 and 8 were published.

1866

April 1	*Birds of Asia* part 18 scheduled.
August 1	*Birds of Great Britain* part 9 scheduled.
September 1	*Birds of Great Britain* part 10 scheduled.
December 24	'I shall be away from home next month on a visit to Lord Falmouth' (Letter of this date to McCoy, Nat'l. Mus. Victoria).

1867

March 27	Prideaux John Selby died, aged 79. He was a friend of Gould's and a close associate of Jardine's with whom he published several ornithological works.
May 1	*Birds of Asia* part 19 scheduled.
August 1	*Birds of Great Britain* part 11 scheduled.
September 1	*Birds of Great Britain* part 12 scheduled.
December 1	*Birds of Australia, Supplement* part 4 scheduled.

(The Reform Bill of 1867 allowed many working men to vote for the first time.)

1868

April 1	*Birds of Asia* part 20 scheduled.
July 11	First letter from Gould to George N. Lawrence (1806-95) of New York City. This began a series of thirty-one letters ending on 1 January 1878. They are in the American Museum of Natural History Library, New York City. I have made typescripts of these letters of which twenty-two are from Gould to Lawrence, five from Lawrence to Gould, two from Gould's secretary Amelia Prince to Lawrence, one from A. S. Bickmore to Gould, and one from Lawrence to J. Henry.
August 1	*Birds of Great Britain* part 13 scheduled.
September 1	*Birds of Great Britain* part 14 scheduled.

(In 1868 public hangings outside Newgate prison came to an end.)

1869

March 1	*Trogonidae*, second edition, part 2 scheduled.
April 1	*Birds of Asia* part 21 scheduled.

PART THREE CHRONOLOGY OF THE LIFE AND WORKS OF JOHN GOULD

July 16	Gould inscribed one of his *Introduction to Trochilidae* to Sir James Walker of the Bahamas. Probably included in the book was a sheet with some elaborate instructions on obtaining and preserving bird specimens (Harvard University, Houghton Library Collection).
August 1	*Birds of Australia, Supplement* concluding part 5 scheduled.
August 1	*Birds of Great Britain* part 15 scheduled.
September 1	*Birds of Great Britain* part 16 scheduled.
September 7	Eliza Gould, aged 33 married John Muskett (Figure 65).
November 16	(The Suez Canal opened.)
December	Samuel White (1835-80) of Adelaide, and his new wife visited England in December 1869. According to his son's account

> A great deal of time was spent with John Gould, who gave a little dinner on one occasion in honour of the Australian ornithologist, and at this gathering many Old World men met my father. I have been told days and nights were spent discussing ornithological subjects and a great and lasting friendship sprang up between the great author of "The Birds of Australia" and the greatest field ornithologist Australia ever possessed. (White 1914-19)

1870

March 1	*Birds of Asia* part 22 scheduled.
August 1	*Birds of Great Britain* part 17 scheduled.
September 1	*Birds of Great Britain* part 18 scheduled.

1871

January 21	Letter from Gould to Sir James Walker of the Bahamas thanking him for his assistance in obtaining a specimen of humming-bird (Houghton Library, Harvard).
March 1	*Birds of Asia* part 23 scheduled.
August 1	*Birds of Great Britain* part 19 scheduled.
September 1	*Birds of Great Britain* part 20 scheduled.

1872

March 1	*Birds of Asia* part 24 scheduled.
May 6	George Robert Gray died. He had been in charge of the British Museum's bird collection since 1831.
August 1	*Birds of Great Britain* part 21 scheduled.
September 1	*Birds of Great Britain* part 22 scheduled. From this part on, W. Hart 'commenced to draw the plates on stone' (Sharpe, 1893).

1873

March 1	*Birds of Asia* part 25 scheduled.
March 19	Dr Franklin Gould died and was buried in the Red Sea. He had travelled to India with Earl Grosvenor, eldest son of the Marquis of Westminster, as his personal medical attendant. In this capacity, on a voyage from Bombay to Suez, Franklin developed a very high fever and after several days' illness died (Figure 61). (Medical report MS. courtesy of Dr Geoffrey Edelsten.)
July 2	First of a series of thirty-two letters from Gould's son Charles to his father (two only), and to his sisters Louisa and Sarah. Figure 62 is a portrait of Charles. The first is from a survey camp in New South Wales. The later letters, beginning 16 December 1880 are from Shanghai, Hongkong,

Figure 61 Dr Franklin Gould (Courtesy of Dr Geoffrey Edelsten).

Figure 62 Charles Gould, the geologist and author (Courtesy of Dr Geoffrey Edelsten and Mrs Doris Twyman).

	Singapore, Bordeaux (on his way home to stop an impending marriage of his widowed sister Eliza), Buenos Aires respectively, and finally twelve letters from Montevideo ending on 13 August 1892. Charles died on 15 April 1893 in Montevideo, aged 59.
August 1	*Birds of Great Britain* part 23 scheduled.
September 1	*Birds of Great Britain* part 24 scheduled.
November 1	*An introduction to the birds of Great Britain*, post octavo, scheduled.
December 3	On or before this date was published the concluding part 25 of *Birds of Great Britain*.

1874

August 1	*Birds of Asia* part 26 scheduled.
November 21	Sir William Jardine died, aged 74. The most extensive Gould correspondence is with Jardine, and Gould had visited at Jardine Hall in Scotland on many occasions.

1875

March 1	*Birds of Asia* part 27 scheduled.
March	Portraits of Gould were made by Messrs. Maull and Fox (Figure 63).
May 17	Charles Coxen, brother of Elizabeth Coxen Gould died, aged 67, in Queensland, Australia. There were no children of his marriage. Charles Coxen, Sylvester Diggles and John Thomas Cockerell formed 'a small local group of working ornithologists' of the Brisbane area (Whittell 1954 II:156). [The Coxen genealogy gives the date of death as 1875, but Whittell says 1876.]
June	*Trogonidae*, second edition, part 3 scheduled.

PART THREE CHRONOLOGY OF THE LIFE AND WORKS OF JOHN GOULD

Figure 63 a & b Portraits of John Gould in March 1875 by Messrs. Maull & Fox (Figure 63a from Dr Geoffrey Edelsten, and 63b from Mrs Doris Twyman).

July 18   Lady Jane Franklin died. She was buried in Kensal Green Cemetery, London, where John and Elizabeth Gould are also buried.

September   *Trogonidae*, second edition, concluding part 4 scheduled for publication.

December 1   *The birds of New Guinea*, part 1, thirteen plates, scheduled.

In 1875 Edwin C. Prince died. He had been Gould's highly-esteemed and devoted secretary and business manager for 45 years (Figure 64).

1876

January 1   *Birds of New Guinea* part 2 scheduled.

April 15   The first extant letter written by Amelia A. Prince as secretary for Gould was to Professor McCoy (Nat'l Mus. Victoria).

May 1   *Birds of New Guinea* part 3 scheduled.

July 1   *Birds of Asia* part 28 scheduled.

(In this year Alexander Graham Bell invented the telephone.)

1877

January 1   *Birds of New Guinea* part 4 scheduled.

January 8   John Gould's daughter, Eliza Muskett (Figure 65), gave birth to a daughter Helen Catherine Elizabeth Muskett. On 10 August 1898 Helen (Figure 65) married Ernest Alfred Edelsten and they had five children, but one died in infancy. Doris Edelsten (Twyman) was born 29 July 1899, Geoffrey Edelsten was born 20 May 1903, Alan Edelsten was born 15 January 1907 and Ronald Edelsten was born 14 September 1912.

Figure 64 Edwin C. Prince, secretary and business manager for Gould and Company (Chisholm 1939, *Victorian Naturalist*, **56**:22).

Figure 66 Portrait in oils by Robertson of John Gould, around 1878 (Courtesy of Dr Geoffrey Edelsten).

April 1	*Birds of Asia* part 29 scheduled.
June 1	*Birds of New Guinea*, part 5 scheduled.
October 1	*Birds of Asia* part 30 scheduled.
Autumn	Count Tommaso Salvadori (1835-1923) of Italy visited Gould and 'admired the stupendous collection of Humming Birds, which was then placed in a gallery expressly built in the house which he inhabited' (Salvatori 1881).

1878

February 1	*Birds of New Guinea* part 6 scheduled.
April 25	John Gould's will was signed. Figure 66 is a portrait in oils painted in 1878.
June 1	*Birds of New Guinea* part 7 scheduled.
June 4	The first meeting of the Committee of the British Ornithologists Union to revise the nomenclature of British birds.
June 21	First of a series of short letters between Gould and R. Bowdler Sharpe, concluded on 8 September 1880 (Sharpe Collection; Blacker-Wood Library, McGill University, Montreal).
October 1	*Birds of New Guinea* part 8 scheduled.

1879

March 1	*Birds of New Guinea* part 9 scheduled.
July 1	*Birds of Asia* part 31 scheduled.
September 1	*Birds of New Guinea* part 10 scheduled.

(In this year Thomas Edison invented the incandescent lamp.)

1880

February 1	*Birds of New Guinea* part 11 scheduled.

PART THREE CHRONOLOGY OF THE LIFE AND WORKS OF JOHN GOULD

Figure 65 Eliza Gould Muskett family portraits (A, B & D from Dr Geoffrey Edelsten, and C from Mrs Doris Twyman).
A Eliza Gould Muskett **B** John Muskett **C** Eliza Gould Muskett (later Moon). **D** Helen Catherine Elizabeth Muskett (later Edelsten).

July 1	*Birds of Asia* part 32 scheduled.
August 1	*A Monograph of the Trochilidae, or family of humming-birds, Supplement*, part 1, twelve plates, was published.
September 8	Letter from Gould to Sharpe written in a shaky hand. 'In great pain. My opiate leaves me... I write this and all other notes on birds on my bed, so excuse them' (Wood Library, McGill University).
October 1	*Monograph of the Pittidae*, part 1, ten plates, was published.
October 25	Thomas Chalmers Eyton died, a colleague and friend of Gould who had 'a spacious museum of birds and bird skeletons' at Eyton Hall in Shropshire (Whittell II:233).
October 27	A Codicil to John Gould's will was written.

1881

February 3	John Gould died. He was buried in Kensal Green Cemetery in London.

In 1881 the following works were published: *Birds of New Guinea*, part 12; *Pittidae*, part 2 (text only); and *Supplement to Trochilidae*, part 2.

The British Museum (N.H.) purchased '6315 skins of birds, being the private collection of the late John Gould' (Brit. Mus. (N.H.) 1906).

1882

February — *Birds of Asia* part 33 scheduled.

April 19 — Charles Darwin died.

Gould's humming-bird collection was purchased by the British Museum (N.H.) and consisted of over 5000 specimens (*Obituary*, Sclater 1882).

In 1882 *Birds of New Guinea*, part 13 was published.

1883

January — *Birds of Asia* part 34 scheduled.

August — The move of the Zoological Department from the British Museum at Bloomsbury to the new Natural History building at South Kensington was completed.

August — *Birds of Asia* concluding part 35 scheduled.

In 1883 the following works were published: *Supplement to Trochilidae*, part 3; *Birds of New Guinea*, parts 14 and 15; and *Introduction to Gould's birds of Asia*, folio size.

1884

In this year the British Museum (N.H.) purchased 579 eggs from various localities from the collection of the late John Gould, Esq. In 1885 an additional 1157 eggs were purchased (Brit. Mus. (N.H.) 1906).

Birds of New Guinea parts 16, 17 and 18 were published in 1884. *The Auk*, organ of the American Ornithologists Union was begun as a quarterly publication.

(In 1884 Waterman invented the fountain pen.)

1885

Supplement to Trochilidae part 4 and *Birds of New Guinea* parts 19 and 20 were published.

The American Ornithologists Union, mainly through the efforts of J. A. Allen and Elliott Coues, adopted a stricter law of priority based on the 1758 tenth edition of Linnaeus's *Systema naturae*.

1886

Birds of New Guinea parts 21 and 22 were published.

1887

Supplement to Trochilidae part 5 and *Birds of New Guinea* part 23 were published.

1888

January 29 — Edward Lear died at San Remo.

Birds of New Guinea concluding parts 24 and 25 were published.

Figure 67 illustrates the specially constructed cabinet Gould had built to house all his folio and other volumes.

PART THREE CHRONOLOGY OF THE LIFE AND WORKS OF JOHN GOULD

Figure 67 Gould's personal, specially constructed cabinet to house all his imperial folios and other volumes (Courtesy of Dr Geoffrey Edelsten).

1893

April 15 — Gould's last son, Charles Gould, the surveyor and author of *Mythical Monsters* died at Montevideo, Uruguay.

Louisa Gould lived until 1894, Eliza Gould Muskett Moon until 1918, and Sarah (Sai) Gould until 1926. Figure 68 is a portrait of the Goulds' three daughters when they were in their twenties, in the 1860s.

1909

October — 'The Gould League of Bird Lovers' was organized in Melbourne, Victoria (Figure 69).

1910

October — The New South Wales branch of 'The Gould League of Bird Lovers' was inaugurated at Wellington.

* * *

By 1881, the year of John Gould's death, England had become a powerful industrial nation with extensive colonial ties. The population had tripled since Gould's birth in 1804. (The population in the United States had increased fifteen-fold in the same period.) The streets of London and most towns were then, in the 1880s, lit by gas. Transportation was still by horse-drawn carriage in England, but the number of roads had proliferated immeasurably and were smooth and well drained. Steam locomotives travelled on miles of railroads. Steamships had cut the time of ocean travel from four months by sail from England to Australia to one-fourth the time by steamship.

Figure 68 Photograph of John and Elizabeth Gould's daughters, Eliza, Louisa and Sarah undoubtedly taken when they were in their twenties, or around the late 1850s or early 1860s (Courtesy of Dr G. Edelsten and Mrs Doris Twyman).

Farming was accomplished with the help of iron ploughs drawn by horses, and by the use of mechanical reapers. The telegraph, the Atlantic cable, the typewriter, and the telephone were all in use. Photography began with the daguerreotype in 1839, but the latter was obsolete by 1857, having been replaced by the wet collodion method.

Breech-loading shot-guns were now made of fine hardened steel; the hammerless guns were beginning to replace the hammer type, and the gunpowder was smokeless.

PART THREE CHRONOLOGY OF THE LIFE AND WORKS OF JOHN GOULD

Figure 69 First certificate of The Gould League of Bird Lovers of Victoria (*Emu* 38).

The binocular 'telescope' had not yet been invented, nor had the fountain pen.

Medicine had made considerable advances by 1881. Anaesthesia was now established for surgical procedure. Surgery itself was usually successful, particularly since antisepsis based on the elimination of contaminating bacteria had been promulgated by Lister in 1865. More and more diseases were being found to be caused by infective organisms. Semmelweis in 1847 associated uncleanliness on the part of the doctor or midwife with the cause of puerperal or child-bed fever. Mrs Gould died of this complication of pregnancy in 1841.

Pasteur propounded the germ theory as a cause of infection in 1857. But the acceptance of his and others' findings and theories of bacterial causation for disease came slowly.

By the 1880s the field of ornithology was leaving the era of morphology and systematics. Stresemann (1975, p. 269) writes that in 1812 there were almost 3800 forms of birds. By 1871 there were over 11 000 forms, which by 1946 had jumped to 28 500 forms. But 'ornithological systematics, in order to accomplish its most important task, is more and more compelled to ally itself with biologic research' (p. 279), and the biologists must closely study behavioural characteristics. These and many other avenues of scientific investigation were opening up for the twentieth-century ornithologist.

Part Four

Bibliography of John Gould, his family and associates

Here is a bibliography of an ornithologist written by neither a bibliographer nor an ornithologist. This statement will be obvious to both.

I am especially concerned that proper emphasis may not have been given to a particular book or article in the bibliography. An ornithologist could feel that certain important scientific points were missed, while a bibliographer could see fault in my selections and in my simple collations. But these words were written over many years, sometimes when I may have felt like waxing poetic on a given subject that stimulated me at the moment or, at other times, a particular subject would only elicit a cursory comment.

It should be particularly noted in this bibliography that certain important books or journals relative to Gouldiana have been completely indexed for references to John Gould, his family, and John Gilbert. Selective indexing has been done for his artists, Lear, Richter, Hart and Wolf, and also for some of his scientific associates and collectors.

At first this time-consuming indexing was instituted for my own edification. It began with Whittell's 1954 compendium of *The Literature of Australian Birds*. Here I found so many references to Gould in the biographies of individuals and in the listed articles written by a multitude of authors, that, for any sensible order, and for future use with Gould's correspondence and published text material in his works, it became a necessity to compile an index of these references. In a few other major ornithological bibliographies and journals it was also noted that there were relevant but hidden Gouldiana references, which for any serious student deserved notation by indexing. Thus evolved quite a complete index of Gouldiana which I felt might also serve others interested in this field of Victorian-age natural history. The works that are quite completely indexed include:

Anker, Jean. 1938. *Bird books and bird art.*

British Museum (Natural History). 1904-1917. *The history of the collections contained in the Natural History Departments of the British Museum*, 3 vols.

Burmester, C. A. 1974-77. *National Library of Australia, guide to the collections*, 2 vols.

Coues, Elliott. 1878-80. [Ornithological bibliography.]

Journal. *Contributions to ornithology.*

Journal. *The Emu.* 1938.

Journal. *The Ibis.*

Journal. *Proceedings of the Zoological Society of London.*

Journal. *Transactions of the Zoological Society of London.*

Mackaness, George. 1947. *Some private correspondence of Sir John and Lady Jane Franklin.*

Palmer, A. H. 1895. *The life of Joseph Wolf, animal painter.*

Phillips, V. T. and Phillips, M. E. 1963. *Guide to the manuscript collections in the Academy of Natural Sciences of Philadelphia.*

Ripley, S. D. and Scribner, L. L. 1961. *Ornithological books in the Yale University Library.*

Sharpe, R. Bowdler. 1893. *An analytical index to the works of the late John Gould, F.R.S.*

Sotheran, Ltd, Henry. 1934. No. 9 'Piccadilly Notes'.

Whittell, H. M. 1954. *The Literature of Australian Birds.*

Wood, Casey. 1931. *An introduction to the literature of vertebrate zoology.*

Zimmer, John Todd. 1926. *Catalogue of the Edward E. Ayer ornithological library.*

Extensive references and data on Gould's published works appear in Anker 1838, Ripley and Scribner 1901, Wood 1931, and Zimmer 1926. As I realized that copies of these bibliographies may not be readily accessible, the sections on Gould's works are reproduced in their entirety in the bibliography. The courtesy of the institutions allowing this reprinting is gratefully acknowledged.

A   ACADEMY OF NATURAL SCIENCES OF PHILADELPHIA, *Minutes of Meeting* 1857. Meeting of 19 May 1857. Library, Acad. Nat. Sciences, Philadelphia, Collection 502.

In the minute book is this record: 'Hall of the Academy ordinary meeting. Present Messrs. Leidy, A. H. Smith, Corse, W. S. Wilson, J. D. Sergeant, Draper, J. A. Meigs, Woodhouse, Schafhirt, Morris, Rand and Mr. John Gould, Correspondent. Dr. T. B. Wilson in the chair...'

ADAMS, ANDREW LEITH. 1873. *Field and forest rambles ... of Eastern Canada.* London, Henry S. King.

The title-vignette is an engraved illustration of the Ruby-throated Humming-bird. In the Preface Adams states 'I am indebted to the masterhand of my distinguished friend J. Gould, F.R.S., for the appropriate vignette on the title-page.' To my knowledge, this is the only vignette done by Gould for another author.

ADAMS, HENRY GARDINER. 1856. *Humming Birds.* London, Groombridge.

There is a three-page section devoted to Gould, recording the admiration with which Gould's works were held during the mid-nineteenth century. The short account of Gould's life printed here is one of the earliest.

A section from pages 53 to 70 is entitled 'My Humming Birds' and is a personal account by C. W. Webber of America.

Eight hand-coloured plates are bound at the front of the volume, and depict sixteen humming-birds.

The Ellis–Kansas Collection also has these eight coloured plates unbound, with a title-page undated and otherwise different from the 1856 copy.

AGASSIZ, LOUIS. 1848-54. *Bibliographia zoologiae et geologiae, corrected, enlarged, edited by H. E. Strickland.* 4 vols. London, Ray Society.

Volume three contains a listing of 122 papers and folio works by John Gould to 1851. Some reviews of his works are also listed.

The Johnson Reprint Corporation, New York and London, has this set available in the series 'The Sources of Science', No. 20, 1968.

Plate 29 Three steps in the development of a 'Gould' lithograph of a bower-bird.
A The rough preliminary watercolour sketch undoubtedly drawn by John Gould.
B A fine watercolour rendition of the rough sketch by one of Gould's artists, possibly H. C. Richter.
C The hand-coloured finished lithograph. In the right upper corner are the pencilled words 'Australia/ no stone kept' (Author's Collection).

Plate 30 Drawing in pencil, watercolour and chalk, artist unknown for fine drawing, but with rough sketch work undoubtedly by John Gould, of *Uria troile*, for *Birds of Great Britain*, V: 48, Kansas-Ellis-Gould MS 871.

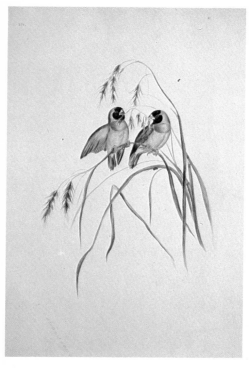

Plate 31 Drawing in watercolour, attributed to H. C. Richter, of Gouldian Finch, for *Birds of Australia*, III: 86, Blacker-Wood, McGill.

Plate 32 Drawing in watercolour, signed E. Lear 1832 June 28, of White Stork. for *Birds of Europe*, IV: 283, Blacker-Wood Library, McGill.

PART FOUR BIBLIOGRAPHY OF JOHN GOULD, HIS FAMILY AND ASSOCIATES

ALBERTIS, LUIGI MARIA D'. 1880. *New Guinea: what I did and what I saw.* 2 vols. London, S. Low, Marston, Searle, and Rivington.

In volume one there are four chromo-lithograph plates of birds, three of which have 'Gould and Hart/March, 1879' on the plate. These face pages 82, 104, 114 and 123.

Italian and American editions of this book were also published.

ALEXANDER, W. B. 1921. 'The Tubinares (Petrels and Albatrosses) in the Gould Collection at Philadelphia'. *Emu* 20:224-7.

In a visit to Philadelphia, Alexander, through the kindness of Dr Witmer Stone, examined the 'famous Gould collection of Australian birds'. 'I was a good deal surprised to find in what an excellent state of preservation the majority of the specimens are.'

Alexander then copied from Verreaux's list the specimens of *Tubinares* from Gould's collection and commented on a few of them. Twelve specimens were noted to be types.

ALEXANDER, W. B. 1953. 'Ornithological illustration'. *Endeavor* 12:144-53.

Written by the librarian of the Edward Grey Institute of Field Ornithology, University of Oxford, this is an excellent article, beautifully illustrated mainly in colour, covering bird art from Palaeolithic rock-paintings to Fuertes and Brooks. Gould's art is represented by a figure in colour of a Mandarin Duck from his *Birds of Asia*, and a long paragraph is devoted to him.

Lear is also mentioned, and his Red and Blue Macaw is shown in colour.

Alexander felt that Joseph Wolf 'far surpassed all previous bird-painters, and has remained unequalled in his capacity to make birds appear alive'. Wolf's watercolour of Tengmalm's Owl is used as his example of bird art for the article. This watercolour, dated 1867, is in the possession of Alexander.

ALLEN, GLOVER M. 1924. *An introduction to the study of birds.* New England Bird Banding Association.

A compilation of ten lectures by Dr Allen resulted in this 118-page book. [My copy is unbound.] Gould's collection of Australian birds, which is now in the Academy of Natural Sciences in Philadelphia, is mentioned on page 12.

AMERICAN ORNITHOLOGISTS' UNION. 1933. *Fifty years progress of American ornithology 1883-1933.* Lancaster, Pa., A.O.U.

Frank Chapman and T. S. Palmer collected fourteen articles by prominent ornithologists on various subjects for this volume. While Gould is mentioned only once on page 134 with reference to his collection of Australian birds that went to the Academy of Natural Sciences of Philadelphia, this series of articles is very relevant for the end of Gould's period. Many references are made to work or works prior to the founding of the A.O.U. in 1883.

Especially relevant is the article by James L. Peters on study collections in North America, and the article by George Miksch Sutton on American bird art.

ANKER, JEAN. 1938. *Bird books and bird art.* Copenhagen, Levin and Munksgaard.

A comprehensive survey of bird books, including Gould's works, in the University Library at Copenhagen. There is no Gould manuscript material listed.

The collation is preceded by a brief historical survey of ornithological books and authors, including several pages on or about Gould.

The volume is indexed, but not completely. There are references to Gould on page xviii, and a coloured plate of a humming-bird (plate VII) facing page 56. Further references are to be found on pages 58, 60, 61-4, 66, 97, 98, 112, 115, 122, 123, *128-31*, 132, 137, 141, 142, 162, 163, 176, 195, 199, 201, 208, 214, 215, 226, 227, 228, and 234.

Hart is referred to on pages 60, 107, 117, 130, 131, 195 and 235.

Richter is referred to on pages 60, 129, 130, 131, 215, and 241.

The entire sections pertaining to Gould's published works in this bibliography are reprinted here (Figure 70). The courtesy of the University Library in Copenhagen in allowing this reproduction is gratefully acknowledged.

ANONYMOUS. 1830. 'Remarks on some of the advantages and disadvantages of periodical works on natural history'. *Mag. Nat. Hist.* 3:297-308.

An interesting and relevant commentary on the issuance over long periods of time of parts or numbers of major works, which on completion were to be bound in volumes. Gould did just this.

The criticisms by this anonymous purchaser may well have prompted Gould to set up publication schedules from which, because of his self-discipline and that of his co-workers, he seldom wavered.

ANONYMOUS. 1921. 'The "Times" on John Gould'. *Emu* 21:114-25.

The London *Times* of 3 September 1851 printed a three-column article of critical appreciation of John Gould and his works. Apparently Alex Chisholm saw the value of reprinting this for Australian readers in 1921, so it was published in *The Emu*.

The *Times* article is a glowing one and favourably compares Gould with Latham, Catesby, Selby, Audubon and others. Many flowery quotes could be taken from this article.

Also, in *The Emu*, the Gould portrait by T. H. Maguire appears, with a page of an incomplete letter from Gould to 'A. Denny' dated 15 April 1851.

The portrait, letter and *Times* clipping were all included with a set of Gould's *Birds of Australia* purchased by the Queensland Government and now in the Public Library of Brisbane.

ARMSTRONG, EDWARD A. 1975. *The life and lore of the bird.* New York, Crown.

This lavishly illustrated book contains three references to Gould. On page 139 is the statement 'budgerigars were brought from Australia by John Gould in 1840'. On page 224 is a short paragraph about Gould. On page 229 is an uncoloured illustration of an original drawing of a Red-winged Lory, the property of the Blacker-Wood Library of McGill University.

ASHBY, EDWIN. 1919. 'Some of Gould's types'. *Emu* 18:311-12.

Ashby discusses his examination of two Gould types in Philadelphia, namely (*Gilbertornis*) *Pachycephala rufogularis*, Gould and *Platycercus adelaidae*, Gould. [See also the article by White (*Emu* 19:72) concerning this report.]

ASHBY, EDWIN. 1924. 'Gould's description of the Brown Warbler (*Gerygone fusca*)'. *Emu* 24:154.

Ashby states that the tail feathers of *Gerygone fusca* as described by Gould (*Handbook*, vol. 1, species 156, p. 268) should be *brown* not white.

Figure 70 Pages concerning Gould's works from Jean Anker. 1938. *Bird books and bird art.* ▶

PART FOUR BIBLIOGRAPHY OF JOHN GOULD, HIS FAMILY AND ASSOCIATES

XXXII,* XXXVII,* XXXIX,* XL,* XLII,* XLVII,* XXIX,** XXXVII,** XXIX,** missing pl. XXIX*). Vol. IV. - - < pp. 526—595: Birds >. 4 pl. (numb. XXV,* XXVII, XXX, XXXV).

The first edition of this popular work by the famous author appeared in 1774 and was succeeded by a number of editions in which the account was improved and enlarged by means of corrections, notes, and additions.

The appendix to the present edition (Vol. IV, pp. 429—654) contains a survey of the animal kingdom with explanations of technical terms and an outline of the Cuvierian and other systems. The birds are arranged according to Temminck's system of classification.

The engravings were executed by R. Scott, many birds being figured on each of the ornithological plates.

Several bird books with coloured plates, based entirely or partially on Goldsmith's work, have appeared, for instance a work in two volumes in 1815 under the title 'The natural history of birds; from the works of Oliver Goldsmith', etc., with altogether 152 coloured plates, and a volume in 1838 with the title 'Goldsmith's history of British and foreign birds' with 46 coloured plates.

GOULD, J.

168. [1831—] 1832. *A century of birds from the Himalaya Mountains. By John Gould. London. pp. [XI] + [143]. 80 col. pl. (numb. I—LXXX in text). large fol.*

The first of the long series of Gould's pictorial works in large folio.

The leaf of text accompanying each, or in a few cases two or three of the plates, contains the diagnosis in Latin of the bird or birds figured on the plate, an indication of the measures, and some few other data of interest, such as notes on the history, habitat, and habits of the species. The greater part of this text was apparently written by N. A. Vigors, who is thanked in an advertisement in the following words, 'by him not only the nomenclature, but also the accompanying letterpress descriptions were liberally contributed'.

The plates contain altogether 102 figures, two of which, however, represent birds figured twice; thus the 100 birds referred to in the title are actually figured.

The plates (drawn from nature and on stone by E. Gould) from sketches by J. Gould, are executed in lithography coloured by hand (printed by C. Hullmandel). In the present copy the backgrounds of the plates are coloured; however, the work also occurs with these backgrounds uncoloured.

169. [1832—] 1837. *The birds of Europe. By John Gould. Published by the author. London. 5 vols. 448 col. pl. (numb. consecutively in 'List of plates'). large fol.* — Vol. I. Raptores. *pp. XII + [II] + 4 + [I] + [99]. 50 col. pl. (Nos. 1—50).* Vol. II. Insessores. *pp. [IV] + [197]. 99 col. pl. (Nos. 51—149).* Vol. III. Insessores. *pp. [IV] + [185]. 93 col. pl. (Nos. 150-242).* Vol. IV. Rasores. Grallatores. *pp. [IV] + [205]. 103 col. pl. (Nos. 243—345).* Vol. V. Natatores. *pp. [IV] + [205]. 103 col. pl. (Nos. 346—449; Figs. 447 and 448 on one plate).*

This work was published in 22 parts, the issue of which was commenced on June 1, 1832. Each volume contains a 'List of plates' in which the plates of the particular volume are numbered.

In the introduction (pp. IX—XII) questions relative to migration are briefly dealt with, and an outline is given of the plan of the work.

The brief text which accompanies each plate by way of explanation gives a characterization of the genera, while the individual form is treated with regard to its geographical distribution, habitat, habits, and other facts of interest, and is briefly described.

The plates are executed in lithography coloured by hand. The greater number were drawn and lithographed by Mrs. E. Gould from sketches and designs by the author, taken from nature. The remainder of the plates were drawn and lithographed by E. Lear. All the plates were coloured under the direction of Mr. Bayfield. Finally, the printing was done by C. Hullmandel.

170. [1833—] 1834 [—35]. *A monograph of the Ramphastidæ, or family of toucans. By John Gould. Published by the author. London. pp. [XX] + [74]. 34 pl. (33 col.). large fol.*

This work — Gould's first monograph — was issued in three parts. The introduction contains a general review of the history of the group and of the geographical distribution, habits, and classification of the birds. The explanatory text accompanying each of the plates gives in Latin the specific characters of the form figured, and also the description, measurements, synonyms, and, in a different type, information about the iconography, special characters, habitat, etc., of the birds.

The plates (lithographs coloured by hand, printed by C. Hullmandel) were mostly drawn and lithographed by J. & E. Gould, though some were executed by E. Lear.

The work has a final four-page chapter with the title 'Observations on the anatomy of the toucan' by Richard Owen, illustrated with figures on an uncoloured plate drawn and lithographed by G. Scharf.

A new edition of this work, with 52 plates, was issued in 1852—54 (in three parts?), while a supplement was published in 1855 (180).

A German edition (four parts, 38 plates, folio) was issued at Nürnberg in 1841—47.

171. [1836?—] 1838. *A monograph of the Trogonidæ, or family of trogons. By John Gould. Published by the author. London. pp. [V] + VII + [III] + [71]. 36 col. pl. (numb. 1—36 in 'List of plates'). large fol.*

Published in three parts, the first of which possibly appeared in 1836.

The present work is Gould's second monograph, in which altogether 34 species are treated, Gould having added to the 22 species known hitherto 12 others new to science.

The introduction, dealing with general conditions relative to the group, is succeeded by a 'Synopsis specierum'. The text accompanying each plate contains a diagnosis in Latin and also a description, measurements, and synonyms, of the bird figured on the particular plate, to which is added some information on the history, habitat, and iconography of the bird.

The plates (drawn from nature and on stone by J. & E. Gould) are executed in lithography coloured by hand (printed by C. Hullmandel).

Another, revised, edition of this work (4 parts, 47 col. pl., large fol.) was issued in 1858—75.

172. 1837—38. *Icones avium, or figures and descriptions of new and interesting species of birds from various parts of the globe. By John Gould. Forming a supplement to his previous works. Published by the author. London. 2 parts. 18 col. pl. large fol.* — Part I. 1837. *pp. [19]. 10 col. pl.* Part II. Monograph of the Caprimulgidæ. Part I. 1838. *pp. [15]. 8 col. pl.*

The present two parts were the only ones of this work to appear, their covers being dated August, 1837, and August, 1838, respectively, after which publication was stopped owing to Gould's journey to Australia.

Each plate is accompanied by a leaf of text, containing a diagnosis in Latin of the form figured, description, synonyms, and information about the history and geographic distribution of the bird, etc.

The plates (printed by C. Hullmandel) are executed in lithography coloured by hand (drawn from nature & on stone by J. & E. Gould).

173. [1838—] 1841. *The zoology of the voyage of H. M. S. Beagle, under the command of Captain Fitzroy ... 1832—1836. Published with the approval of the Lords Commissioners of her Majesty's Treasury. Edited and superintended by Charles Darwin. Part III. Birds, by John Gould. Smith, Elder and Co. Lon-*

JOHN GOULD — THE BIRD MAN

don. pp. [VII] + II + 3—156 + [8]. 50 col. pl. (numb. 1—50). 4to.

The ornithological section of the report on the zoological collections from the famous voyage round the world of the 'Beagle', in which Darwin participated as zoologist, and during which he conceived the idea of his theory of evolution. The whole zoological report (five parts, Part III : Birds) appeared in 1838—43, and was issued in numbers, of which 3, 6, 9, 11, and 15 contained the ornithological section (863).

The subtitle of this section runs as follows : 'Birds, described by John Gould, with a notice of their habits and ranges, by Charles Darwin, and with an anatomical appendix, by T. C. Eyton'. In the advertisement Darwin states that Gould had prepared the descriptions of the new species and supplied those already known with names, but that his manuscript was so incomplete that G. R. Gray had to come to his assistance with information as to some parts of the general arrangement and the use of proper generic terms, while Darwin himself enlarged some of Gould's descriptions which were too brief.

The main part of the work (pp. 3—146) is taken up with a description of the comprehensive material, after which follows (pp. 147—156) the appendix by Eyton, while the last eight pages contain an 'Index to the species'.

The plates (lithographs coloured by hand) are made from sketches by J. Gould, and executed on stone by Mrs. E. Gould.

174. [1840—] 1848. The birds of Australia. By John Gould. Published by the author. London. 7 vols. 600 col. pl. (numb. in the lists of plates). large fol. — Vol. I. pp. [XVIII] + V—CII + 13 + [I] + [71]. text-figs. 36 col. pl. (Nos. 1—36). Vol. II. pp. [IV] + [207]. 104 col. pl. (Nos. 1—104). Vol. III. pp. [IV] + [193]. 97 col. pl. (Nos. 1—97). Vol. IV. pp. [IV] + [207]. 104 col. pl. (Nos. 1—104). Vol. V. pp. [IV] + [183]. 92 col. pl. (Nos. 1—92). Vol. VI. pp. [IV] + [163]. 82 col. pl. (Nos. 1—82). Vol. VII. pp. [IV] + [169]. 85 col. pl. (Nos. 1—85).

In 1837—38 Gould issued two parts of a work with the title 'The birds of Australia, and the adjacent islands' (20 plates, text, large fol.). Owing to lack of material, however, he did not go on with the work, but went to Australia to collect the material on which the present work is based. This was published in 36 parts, the first of which was dated December 1, 1840, the last December 1, 1848.

Vol. I, in addition to the material concerning individual forms, contains a general index to the whole work, a preface (pp. V—XII) with information about the origin of the work, and a long introduction (pp. XIII—CII) with a general account of the avifauna of Australia, consisting chiefly of a synoptic table with references to the volumes in which the respective plates are to be found. This is followed by a table of the range or distribution of the species (pp. 1—13), in which the species are arranged in the same order as in the work itself. These parts of Vol. I (preface and introduction with the table of the range) were also issued separately in 1848, 8vo, under the title 'An introduction to the birds of Australia'.

Each volume contains a 'List of plates'. The treatment of the subject is otherwise as in Gould's other works in large folio, a brief explanatory text accompanying each plate.

The plates (J. Gould and H. C. Richter del. et lith., or J. & E. Gould del.) are executed in lithography coloured by hand (Hullmandel & Walton, or C. Hullmandel, imp.).

A supplement to the work was issued by Gould in 1851—69 (179), the text of the main work and the supplement being published in 1865 in an enlarged form under the title 'Handbook to the birds of Australia' (London, 2 vols., 8vo).

175. 1843—44. The zoology of the voyage of H. M. S. Sulphur, under the command of Captain Sir Edward Belcher ... 1836—42. Published under the authority of the Lords Commissioners of the Admiralty. Edited and superintended by Richard Brinsley Hinds. No. III (IV). Birds, by John Gould. Smith, Elder and Co. London. 2 parts. 16 col. pl. fol. — Part I. 1843. pp. 37—44. 8 col. pl. (numb. 19—26). Part. II. 1844. pp. 45—50. 8 col. pl. (numb. 27—34).

This expedition made studies and scientific investigations on the shores and among the islands of the Pacific Ocean. The zoological results were published in 12 parts in 1843—45.

The present section on birds appeared as Nos. III and IV, as stated in the still extant covers of the parts of the present copy, which are dated October, 1843, and January, 1844, respectively. The above title is taken from these covers. A total of 16 species is dealt with, all of which are figured and eleven of which, described as new, have also been treated by Gould in the Proceedings of the Zoological Society of London, Parts X and XI, 1842—43.

The majority of the plates (lithographs coloured by hand, printed by C. Hullmandel) are signed : drawn by J. Gould, on stone by B. Waterhouse Hawkins.

176. [1844—] 1850. A monograph of the Odontophorinæ, or partridges of America. By John Gould. Published by the author. London. pp. 23 + [1] + [63]. 32 col. pl. large fol.

Published in three parts appearing in succession in 1844, 1846, and 1850 respectively. Altogether 35 species are described, of which three are more closely treated in the introduction. This gives a general account of the group, especially as regards classification.

Each of the 32 forms figured is treated in detail in the text, which accompanies the individual plates and contains a brief diagnosis in Latin, description — sometimes of both male and female —, measurements, and synonyms, and also data of the appearance, habits, and habitats, of the birds, and information about the figures on the plates.

The plates (J. Gould and H. C. Richter del. et lith.) are executed in lithography coloured by hand (Hullmandel & Walton, or C. Hullmandel, imp.).

177. [1849—] 1861. A monograph of the Trochilidæ, or family of humming-birds. By John Gould. Published by the author. London. 5 vols. 360 col. pl. (numb. in the lists of plates). large fol. — Vol. I. pp. [IX] + V—CXXVII + [81]. 41 col. pl. (Nos. 1—41). Vol. II. pp. [III] + [149]. 75 col. pl. (Nos. 42—116). Vol. III. pp. [III] + [175]. 87 col. pl. (Nos. 117—203). Vol. IV. pp. [III] + [159]. 80 col. pl. (Nos. 204—283). Vol. V. pp. [III] + [153]. 77 col. pl. Nos. 284—360).

This work was published in 25 parts. Each volume contains a 'List of plates' giving the numbers of the plates of the particular volume.

In addition to the matter concerning the birds figured Vol. I contains a preface (pp. V—VIII), dated September 1, 1861, and a long introduction (pp. IX—CXXVII) with a general review of the group, including a synopsis of the classification and a review of the individual forms, including also nomenclature. The introductory chapters conclude with lists of generic and specific names. These introductory chapters (preface, introduction, and list of names) were issued separately in 1861 under the title 'An introduction to the Trochilidæ, or family of humming birds' (London, 8vo).

The explanatory text about the birds figured contains synonyms, history, and, in addition to other information of interest, a brief description.

The plates (J. Gould and H. C. Richter, del. et lith.) are executed in lithography coloured by hand (Hullmandel & Walton (some few : Walter & Cohn) imp.), while Mr. Bayfield, who is thanked in the preface, seems to have been concerned in the colouring.

A supplement to the present work was issued under the same title as this work in 1880—87 (182).

1850—53. Contributions to ornithology ... See *Jardine, W.*, 1848—53.

178. 1850—83. The birds of Asia. By John Gould. Dedicated to the honourable East India Company. Published by the author. *London. 7 vols. 530 col. pl. (numb. in the lists of plates). large fol.* — Vol. I. *pp.* [*VII*] + 9 + [*1*] + [*151*]. 76 *col. pl. (Nos. 1—76).* Vol. II. *pp.* [*IV*] + [*149*]. 75 *col. pl. (Nos. 1—75).* Vol. III. *pp.* [*IV*] + [*155*]. 78 *col. pl. (Nos. 1—78).* Vol. IV. *pp.* [*IV*] + [*143*]. 72 *col. pl. (Nos. 1—72).* Vol. V. *pp.* [*IV*] + [*165*]. 83 *col. pl. (Nos. 1—83).* Vol. VI. *pp.* [*IV*] + [*149*]. 75 *col. pl. (Nos. 1—75).* Vol. VII. *pp.* [*IV*] + [*141*]. 71 *col. pl. (Nos. 1—71).*

This work was published in 35 parts, of which the last three were issued after Gould's death by R. Bowdler Sharpe, whose contributions are marked with the initials 'R. B. S.'. He also wrote the preface and the introduction to the work. Each volume contains a 'List of plates' which states in which part the particular plate is published and when that part appeared.

The introduction contains a general review of the history of Asiatic ornithology during the time the work was published.

As in Gould's other works, each plate is accompanied by a leaf of descriptive text giving synonymy, history, description, and other information of interest regarding the form figured on the particular plate.

The plates, executed in lithography coloured by hand, were drawn and lithographed by J. Gould in collaboration with W. Hart or H. C. Richter, or by the latter in collaboration with J. Wolf, or — in the parts issued after Gould's death — by W. Hart. The plates were printed by Hullmandel & Walton, T. Walter, or Walter & Cohn.

Plates from this work were reproduced in Gould's unfinished paper 'A monograph of the Pittidæ', etc. (London 1880, 1 part, 10 (13 ?) col pl., large folio).

179.   [1851—] 1869. The birds of Australia. By John Gould. Supplement. Published by the author. *London. 5 parts. 81 col. pl. (numb. 1—81 in 'List of plates'). large fol.* — Part I. 1851. *pp.* [*31*]. 16 *col. pl.* Part II. 1855. *pp.* [*35*]. 17 *col. pl.* Part III. 1859. *pp.* [*29*]. 16 *col. pl.* Part IV. 1867. *pp.* [*31*]. 16 *col. pl.* Part V. 1869. *pp.* [*31*] + *IV* + [*4*]. 16 *col. pl.*

In 1840—48 Gould had issued his work 'The birds of Australia' (174), to which the present treatise forms a supplement. The five parts, which are in the original covers, deal with a number of the forms belonging to the area and discovered after the conclusion of the main work cited above.

The plates are executed in lithography coloured by hand (J. Gould & H. C. Richter, del. et lith.; Parts I—III: Hullmandel & Walton, imp, Parts IV—V: Walter, imp.).

The text of this work was re-issued in 1865, as stated in the note on the main work (174).

180.   1855. Supplement to the first edition of A monograph of the Ramphastidæ, or family of toucans. By John Gould. Published by the author. *London. pp. 7—26 + [41]. 20 col. pl. large fol.*

This supplement was issued in two parts. The work to which it belongs appeared in 1833—35 (170), a new and enlarged edition being issued in 1852—54. The supplement, which was intended to balance the two editions, therefore deals with the forms which are found in the second edition but not in the first.

The introduction — a reprint of the introduction to the second edition of the work — treats of the group in general, principally as regards history and habits — illustrated by quotations from various authors — and classification.

The text accompanying the plates, as in the main work, gives under each form the specific character in Latin and English, measurements, and other data of the birds.

The plates (Gould & Richter, del. (or del. et lith.)) are executed in lithography coloured by hand (Hullmandel & Walton, imp.).

181.   1875—88. The birds of New Guinea and the adjacent Papuan Islands, including many new species recently discovered in Australia. By John Gould. Completed after the author's death by R. Bowdler Sharpe.

Henry Sotheran & Co. *London. 5 vols. 320 col. pl. (numb. in the 'Contents'). large fol.* — Vol. I. *pp.* [*III*] + *III* + [*1*] + [*109*]. 56 *col. pl. (Nos. 1—56).* Vol. II. *pp.* [*III*] + [*115*]. 58 *col. pl. (Nos. 1—58).* Vol. III. *pp.* [*IV*] + [*143*]. 72 *col. pl. (Nos. 1—72).* Vol. IV. *pp.* [*III*] + [*117*]. 59 *col. pl. (Nos. 1—59).* Vol. V. *pp.* [*IV*] + [*149*]. 75 *col. pl. (Nos. 1—75).*

The present copy of this work consists of the 25 parts in which it was issued, the contents of which are distributed among the individual volumes as stated above.

The individual parts may be briefly described as follows:—
Part I, 1875, (Dec. 1), *pp.* [*25*]. 13 *col. pl.*;
Part II, 1876, (Jan. 1), *pp.* [*25*]. 13 *col. pl.*;
Part III, 1876, (May 1), *pp.* [*25*]. 13 *col. pl.*;
Part IV, 1877, (Jan. 1), *pp.* [*25*]. 13 *col. pl.*;
Part V, 1877, (June 1), *pp.* [*25*]. 13 *col. pl.*;
Part VI, 1878, (Febr. 1), *pp.* [*25*]. 13 *col. pl.*;
Part VII, 1878, (June 1), *pp.* [*25*]. 13 *col. pl.*;
Part VIII, 1878, (Oct. 1), *pp.* [*23*]. 12 *col. pl.*;
Part IX, 1879, (March 1), *pp.* [*23*]. 12 *col. pl.*;
Part X, 1879, (Sept.), *pp.* [*23*]. 13 *col. pl.*;
Part XI, 1880, (Febr.), *pp.* [*25*]. 13 *col. pl.*;
Part XII, 1881, *pp.* [*25*]. 13 *col. pl.*;
Part XIII, 1882, *pp.* [*25*]. 13 *col. pl.*;
Part XIV, 1883, *pp.* [*25*]. 13 *col. pl.*;
Part XV, 1883, *pp.* [*25*]. 13 *col. pl.*;
Part XVI, 1884, *pp.* [*25*]. 13 *col. pl.*;
Part XVII, 1884, *pp.* [*25*]. 13 *col. pl.*;
Part XVIII, 1884, *pp.* [*25*]. 13 *col. pl.*;
Part XIX, 1885, *pp.* [*25*]. 13 *col. pl.*;
Part XX, 1885, *pp.* [*25*]. 13 *col. pl.*;
Part XXI, 1886, *pp.* [*25*]. 13 *col. pl.*;
Part XXII, 1886, *pp.* [*25*]. 13 *col. pl.*;
Part XXIII, 1887, *pp.* [*25*]. 13 *col. pl.*;
Part XXIV, 1888, *pp.* [*25*]. 13 *col. pl.*;
Part XXV, 1888, *pp.* [*19*] + *title-pages, preface, introduction, and contents.* 10 *col. pl.*

After Gould's death the work was continued and finished by R. Bowdler Sharpe who is responsible for the last thirteen parts, as is evidenced by the initials (R. B. S.) appended at the foot of each article. Sharpe further prepared the preface and the sketch of the zoological work carried out in New Guinea and the Moluccas, which constitutes the introduction (pp. I—III).

The explanatory text which accompanies each plate contains synonymy, description, and other information about the birds figured on the plate.

The plates are executed in lithography coloured by hand, those in Parts I—XII, some in Part XIII, and a few in the subsequent parts, being signed: J. Gould & W. Hart, del. et lith., while the remainder of the plates are signed: W. Hart del. et lith. The printing of the plates in Parts I—XV and of some in Part XVI was done by Walter, Mintern Bros. printing the rest.

182.   [1880—] 1887. A monograph of the Trochilidæ, or family of humming-birds. By John Gould. Completed after the author's death by R. Bowdler Sharpe. Supplement. Henry Sotheran & Co. *London. pp.* [*VII*] + [*199*]. 58 *col. pl. large fol.*

In 1849—61 Gould issued his work 'A monograph of the Trochilidæ' (177), to which the present forms a supplement; it was issued in five parts, which appeared in 1880, 1881, 1883, 1885, and 1887 respectively.

After Gould's death, which occurred before the publication of Part II, the continuation of the work was entrusted to Sharpe, whose contributions are marked with his initials (R. B. S.). In the preface Sharpe thanks Osbert Salvin 'for his advice in planning the present work and for having supervised the proofs of each Part'. A large number of species are included in the supplement without accompanying plates, Sharpe having included descriptions of all the species of Trochilidæ discovered since 1861, when Gould's monograph was finished.

The plates are executed in lithography coloured by hand. Most of the drawings for the plates were executed while Gould was alive, while the finishing of the plates was entrusted to

W. Hart. The majority of them are therefore signed: J. Gould & W. Hart del. et. lith., some of the supplementary plates being signed: W. Hart del. et lith. The plates were printed by Walter or Mintern Bros.

GRAESER, K.

183. *1904.* Der Zug der Vögel. Eine entwicklungsgeschichtliche Studie von Kurt Graeser. Hermann Walther. *Berlin. pp. 96. text-figs. 5 col. pl. 8vo.*

An attempt to explain the origin of the migration of the birds on the basis of the theory of evolution by assuming the development of a migratory instinct in all species and its decline or disappearance in some of them.

The plates (three-colour prints, by J. S. Preuss, Berlin) were executed from original water-colours by E. Bade.

The second edition of the book appeared in 1905, the third edition in 1911.

GRANT, W. R. O.- See Ogilvie-Grant, W. R.

GRANVIK, S. H.

184. *1923.* Journal für Ornithologie. 71. Jahrgang. <-- Sonderheft. Contributions to the knowledge of the East African ornithology. Birds collected by the Swedish Mount Elgon Expedition 1920. By Hugo Granvik >. Deutsche Ornithologische Gesellschaft. *Berlin. pp. [III] + 280 + [1]. text-figs. 10 pl. (numb. 1—10; 5 col. (numb. 1—5)). 1 map. (numb. 11). 8vo.*

A report, issued on February 15, 1923, on the ornithological results of the Swedish expedition, cited in the title, carried out under the leadership of S. A. Lovén.

The first section, the general part, deals with the itinerary of the expedition, system and nomenclature, the route of the expedition, and the ornithology of Mount Elgon and surrounding regions.

The systematic part, in which the birds are arranged in accordance with the system of classification of Reichenow's work 'Die Vögel Afrikas' (416), discusses the material collected, which consists of 1517 specimens, distributed over 330 forms, ten of which are described as new.

The coloured plates (three-colour prints) were executed from drawings by Erich Schröder.

GRAY, G. R.

1838—41. The zoology of the voyage of H. M. S. Beagle ... 1832—1826 ... Part III. Birds. *See Gould, J.*

185. *1844—75.* The zoology of the voyage of H. M. S. Erebus & Terror, under the command of Captain Sir James Clark Ross ... 1839—43. By authority of the Lords Commissioners of the Admiralty. Edited by John Richardson and John Edward Gray. Vol. I. Mammalia, birds. - - Birds. By George Robert Gray and R. Bowdler Sharpe. E. W. Janson. *London. pp. 39. 37 col. pl. (numb. 1—35, 1*, 11*, 20*, XXI*, missing Nos. 12 and 22). 4to.*

The special title-page for the ornithological section of the report of this expedition is dated 1846—75. The section is divided into two parts, of which Part I, 'Birds of New Zealand', pp. 1—20, was written by G. R. Gray. Pp. 1—8 of this part were published in 1844, while pp. 9—20 were published in 1845 (328, VII, p. 484). The appendix (pp. 21—39), which appeared in 1875, was written by R. B. Sharpe. It ends with a list of the plates originally issued and those now issued. In this list numbers 12 and 22 are missing. The plates are executed in lithography coloured by hand. The plates in the latter series, however, are marked: Wolf del. et lit., printed by Hullmandel and Walton, while the former are marked: C. Hullmandel's Patent Lithotint.

186. *1873.* Jottings during the cruise of H. M. S. Curaçoa among the South Sea Islands in 1865. By Julius L. Brenchley. <-- *pp. 353—394:* Birds. By G. R. Gray >. Longmans, Green, and Co. *London. 21 col. pl. (numb. 1—21). 8vo.*

This work, the ornithological section of which is cited here, may be briefly described as follows: pp. XXVIII + 487, text-figs., 60 pl. (44 col.).

The section on birds was prepared on the basis of material collected by Brenchley during the cruise of the frigate 'Curaçoa' in the western Pacific. Only birds new to science or particularly rare are noted.

The bird portraits were drawn and lithographed by J. Smit, and the plates (lithographs coloured by hand) impressed by Mintern Bros.

GRAY, J. E.

1827—29. The animal kingdom ... by the baron Cuvier ... Vols. VI—VIII. The class aves ... The additional species inserted in the text of Cuvier by John Edward Gray. *See Cuvier, G. L. C. F. D.*

187. *1830—35.* Illustrations of Indian zoology; chiefly selected from the collection of Major-General Hardwicke. By John Edward Gray. Treuttel, Wurtz, Treuttel, jun., and Richter. *London. 2 vols. 90 col. pl. birds. large fol.*

This work was published in altogether 20 parts, of which, however, only the first twelve are found in the present copy, which may be briefly described as follows, the ornithological plates only being mentioned:

Vol. I, 1830—32, pp. (VI). front. 58 col. pl. birds
    (numb. 14—71 in the list of plates).
Part XI (1832) 4 col. pl. birds (numb. V—VIII on cover).
Part XII (1832) 4 col. pl. birds (numb. VI—IX on cover).

The list of subscribers is missing in Vol. I.

A summary of the approximate date of publication and the contents of each part was given by Zimmer (589, pp. 272—273) from N. B. Kinnear (773). In its complete form the work contains 202 coloured plates for, as planned, each part was to contain 10 plates. The plates themselves are not provided with numbers but are numbered in the 'Directions for arranging the plates' appended to each volume. Vol. I is composed of Parts I—X, Vol. II of Parts XI—XX.

After the appearance of Part XII publication was interrupted for more than a year. In the present copy Parts XI and XII are found in their original covers, whose title runs as follows: 'Illustrations of Indian Zoology, consisting of coloured plates of new or hitherto unfigured Indian animals, from the collection of Major-General Hardwicke. Selected and arranged by John Edward Gray. Dedicated, by permission, to the honourable court of directors of the East India Company'. On the back of the cover there is a 'Prospectus', in which among other announcements is the statement that 'shortly will be published ... Part I, containing the Mammalia of Prodromus Faunæ Indicæ, or a synopsis of Indian vertebrated animals, containing the generic and specific characters of all recorded Indian Vertebrated Animals ... This Synopsis will form a Text to the »Illustrations«', and 'will be complete in four parts'. However, this work was never issued. The Prospectus furthermore gives information about the material possessed by General Hardwicke, on the basis of which the present work was prepared, and of which are mentioned 'Drawings made upon the spot, and chiefly from living specimens of Animals ... executed by English and native Artists, constantly employed for this purpose, under his [Hardwicke's] own immediate superintendence'. This collection of water-colour drawings was bequeathed by Hardwicke to the British Museum; it consists of 32 volumes (552, II, p. 903; 677, I, pp. 37—38; II, pp. 169—170; II. Appendix, p. 54). The figures of birds, of which 76 are said to be due to T. W. Lewin, amount to a total of 1735 which, however, do not all represent Indian birds.

Not all of the plates in the present work were, however, executed from drawings in the collection mentioned above;

PART FOUR   BIBLIOGRAPHY OF JOHN GOULD, HIS FAMILY AND ASSOCIATES

AUROUSSEAU, M. 1967. *The letters of F. W. Ludwig Leichhardt.* 3 vols. Cambridge University Press for the Hakluyt Society.

This is an exceptional work. Aurousseau publishes 205 letters from Leichhardt to his family and relatives, the Nicholson family, his fellow explorers, scientists, Australian officials, Australian newspapers, and many other persons. A considerable number of these letters are translated from the German.

John Gilbert was with Leichhardt on his expedition to Port Essington, and was killed by natives 28 June 1845.

In these letters and in Aurousseau's notes and comments there is quite a bit of relevant information on Gilbert and Gould, with a wealth of material relating to Australia at the time of Gould and Gilbert's journeys there.

In the introduction to volume one, Aurousseau discusses a controversy that is still continuing over whether Leichhardt was as accomplished an explorer as Australians were led to believe in the mid-1800s. Part of this controversy against Leichhardt had as a proponent Alex Chisholm. Gilbert had written in his journal a few disparaging sentences against Leichhardt. With these unflattering appraisals and many more from others, Chisholm in his book *Strange New Journey* debunks Leichhardt as a competent explorer. Aurousseau says that the letters by Leichhardt 'speak for himself' and support Aurousseau's contention that Leichhardt demonstrated the 'steady, serious purpose of a man' who was 'one of the earliest geographers, in the modern sense'.

B

BACKHOUSE, JAMES. 1843. *A narrative of a visit to the Australian colonies.* London, Hamilton, Adams & Co.

James Backhouse was a member of the Society of Friends of York. 'For the purpose of discharging a religious duty', he and a friend, G. W. Walker, travelled to Australia in 1832 and ministered to the inhabitants and to the prisoner population.

In addition to their religious experiences, this narrative contains a wealth of information on the people, including the Aborigines and transported prisoners, the topography, towns, climate, plants, birds and animals noted during their travels made over a period of six years. They left Australia in 1838, the same year Gould and his family arrived.

BAILY, WILLIAM LOYD. 1855-58. *Illustrations of the Trochilidae or Humming Birds.* Philadelphia. Not published.

This unpublished work is in four volumes as follows:

1855 Part 1st	12 col. plates
1855 Part 2nd	12 col. plates
1857 Part 3rd	18 col. plates
1858 Part 4th	14 col. plates
	56 total

The volumes are listed as 'Collection of the Acad. Nat. Sc. Phil., #11.'

The four volumes were presented to the ANSP in March 1946 by descendants of W. L. Baily.

Also present are letters from Charles M. B. Cadwalader; from Crawford H. Greenewalt (30 April 1959) on evaluation of Baily's versus Gould's drawings; and a long note about these four volumes apparently written by the donors.

PART 1ST   1855. Photograph of Baily.
Twelve hand-drawn and coloured plates by Baily and J. Collins — very few plates have the artist's name appended. Gould is mentioned in the Preface: 'For the most reliable information relating to some obscure species I am indebted mostly to my friend John Gould of London, whose excellent work now in publication has proved a valuable assistant [sic] and from which I have borrowed largely.' End of Preface.

PART 2ND   1855. Haunts and Habits of the Humming Birds, 14 pages.

Page 1: 'Prince Lucian Bonaparte and John Gould have both been the means of furnishing the world much valuable information relating to this interesting group (of Humming Birds).

'... at the present time specimens have been procured and sent to Europe of upwards of three hundred and twenty distinct and well defined species.

'This result is in great measure owing to the energetic exertions of John Gould of London whose collectors have distributed themselves throughout the continents of America in search of new varieties.'

Page 8: An account of Polytmus of Jamaica 'taken from Martin's Humming Birds of Gould's collection'.

Page 14: 'I cannot but admire the indefatigable zeal manifested by Gould in defining the limits, and arranging the species of each of the subgenera, and adopting a nomenclature, suited to their pecularities, and to whose zeal and deep research his beautiful work on the Trochilidae must ever stand as a monument.' End of this preface.

Plate 3: In left lower corner ... 'From the original of J. Gould by W. L. Baily and J. Collins'.

PART 3RD   1857.
The frontispiece is a brown tinted photograph portrait of John Gould embellished with hand drawing. The portrait is signed in ink by John Gould. An attached note states that this frontispiece was a compliment to J. Gould because of their 'intimate correspondence during their contemporary preparation of their Humming Bird books, notably when Mr. Baily gratuously [sic] furnished Mr. Gould with his remarkable method of portraying the iridescense on the plumage of humming birds, which he had perfected'.

'Their intimacy was increased when Mr. Gould visited America and called upon Mr. Baily, who guided him to Bartram's Gardens, and showed him his first living humming bird, in which Mr. Gould expressed most enthusiastic interest.'

Plate 10: Delattre's Sabre wing - 'The specimen of the male bird from which the drawing was made was presented to me by my friend John Gould of London.'

PART 4TH   1858.
Eleven colour plates are listed but there are fourteen, three without letterpress.

Plate 7: Named after Gould *Lesbia Gouldi* by Mr Loddiges of London at meeting of the Zoological Society of London 10 January, 1832. Gould said he would 'prefer the designation of Sylphia' because its 'motion's light and sylph-like'.

BAILY, WM. L. 1919. 'William Loyd Baily, Sr.' *Cassinia* No. 23: 1-13.
Baily had evinced considerable interest in birds and had water-coloured a set of humming-birds that were bound in four volumes. These now reside in the library of the Academy of Natural Sciences of Philadelphia (Baily, W. L. 1855-58). Baily died at the age of 32.

Much of this article is devoted to the polite contention by Baily's nephew that Gould used Baily's technique to incorporate gold leaf in the colouring of Gould's

PART FOUR BIBLIOGRAPHY OF JOHN GOULD, HIS FAMILY AND ASSOCIATES

humming-bird drawings without giving credit to Baily. A letter dated 19 October, 1854 from Gould to Baily is published which does acknowledge that Baily suggested an improvement in the colouring of the humming-bird prints. But the first part of Gould's Trochilidae had appeared in 1849 and eight parts had been published by 19 October, 1854.

It is conceivable that Gould may have profited from a suggestion for improvement by Baily, but in the Exposition of 1851, Gould displayed his own process using iridescent gold-leaf for humming-bird drawings (*see* Royal Commission 1851-52).

Three letters from Gould to Baily are printed in this article, along with part of the Preface to Gould's Trochilidae where Gould describes his visit with Baily in Philadelphia and his first view of a living humming-bird. The dates of the letters are 19 October 1854, 8 August 1855 and 4 June 1856.

BAIRD, SPENCER F. 1858. *Reports of exploration and surveys...for a railroad from the Mississippi River to the Pacific Ocean. Birds.* Vol. IX. Washington, Nicholson.

Prepared by Baird with John Cassin and George N. Lawrence. The names are given of many collectors on the various survey expeditions and other naturalists who supplied specimens. On page xv is the statement 'A collection of about 150 species received from Mr. John Gould, of London, contains many rare birds from the northwest and Arctic regions, (some of them types of the "Fauna Boreali-Americana") as well as others from Mexico and Guatemala. The latter have proved of great value for comparison with closely allied species of the United States.'

Gould's name appears rather frequently as the donor of the specimen. The collectors listed for Gould's specimens are John Taylor for birds from Mexico, David Douglas for birds from California, A. Sallé for the type specimen of *Paris meridionalis* from El Jacalo, Mexico, Capt. Beechey from N.W. America, and Capt. Belcher.

BARCLAY-SMITH, PHYLLIS. 1939. *British birds on lake, river and stream.* Harmondsworth, Middlesex, England, Penguin.

This small quarto book contains thirty-one pages of text. Of these text pages, four are devoted to a résumé of Gould's life and works. There is also a portrait of him courtesy of the Trustees of the British Museum. Coloured plates of sixteen British birds from Gould's work complete this small quarto volume.

BARCLAY-SMITH, PHYLLIS. 1945. *Garden Birds.* London, King Penguin Books.

A small octavo book with twenty coloured plates from Gould's *Birds of Great Britain.* On page 18 is a short biographical note about Gould.

BARLOW, NORA. 1963. 'Darwin's ornithological notes'. London, *Bulletin of the British Museum (N.H.) Historical Series,* vol.2, no. 7, pp. 203-78.

In the Darwin papers at the Cambridge University Library, item 29 has the general title: 'MS. notes made on board H.M.S. Beagle, 1832-6', and section 29 (ii) has the sub-title 'Birds'. Barlow has compared these notes with other MS. items, including the 1839 and 1845 editions of *The Voyage of the Beagle.*

There are references to John Gould on pages 206, 271, 272, 273 and 278. The latter references relate to Gould naming a small South American ostrich discovered by Darwin *Rhea darwinii.*

BARRETT, CHARLES. 1928. 'Notable naturalists'. I. - John Gould. *Victorian Nat.* 45:42.

A short article that states that Gould was the 'father of Australian ornithology' and that he 'was both scientific and popular in his writings'.

Barrett says the photograph in the article was loaned by Mr E. A. Vidler of Melbourne, whose mother, Amelia, was a daughter of Dr George Bennett. Erroneously he states that Amelia 'was Gould's only grandchild'. [Could this have been a misprint for 'godchild'?]

BARRETT, CHARLES. 1938. *The Bird Man.* Melbourne, Whitcombe and Tombs.

This small octavo book of fifty-two pages contains a popular account of Gould's life and works. Many interesting facts are stated, but corroborative references are missing.

BARRETT, CHARLES. 1947. *An Australian animal book.* Rev. ed. Melbourne, Cumberlege.

A popular account of Australian wildlife. Coloured plates 'after Gould' face pages 44, 60, 76, 96, 104, 240, 300, and the frontispiece. There are also many quotes from Gould's works.

BARRETT, CHARLES. 1949. *Parrots of Australasia.* Melbourne, Seward.

A popular account of parrots that contains quite a bit of historical information and many quotations from Gould's *Birds of Australia.* A coloured plate of the Golden-shouldered Parrot, after Gould, faces page 73.

Most valuable are good black-and-white photographs of habitat areas.

BARRETT, CHARLES. 1955. *An Australian animal book.* 2nd ed. London, Oxford University Press.

This contains many quotations from the writings of Gould and Gilbert.

BARTON, M. T. 1959. *The life and works of John Snow,* 2 parts. St. Bartholomew's Hospital Jour., Sept. pp. 238-45; Oct. pp. 271-6.

Dr Geoffrey Edelsten brought this article to my attention. He had told me that Gould moved from Broad Street, Golden Square because of the 1854 cholera epidemic that had its origin from a pump on Broad Street which delivered water to the area. The pump had been contaminated with the cholera *vibrio.* Dr Edelsten said he remembered an article on this subject in the St. Bartholomew's Hospital Journal, so he obtained it for me. In fact, however, Gould and his family moved from the Broad Street address in 1859, five years after Snow's study. Undoubtedly, the cholera epidemic could have been a factor in deciding on the move.

The interesting story of Dr Snow's work on this cholera epidemic is presented so succinctly in the first part of Barton's article that it will be abstracted here. A further value of this article is the excellent portrayal of the living conditions in London in the mid-1800s.

John Snow arrived in London in October 1836 at the age of 23. He had, since the age of 14, been an apprentice to various surgeons in the Newcastle area. After further study at the Hunterian School of Medicine in Great Windmill Street and at Westminster Hospital 'he qualified M.R.C.S., and Licentiate of the Apothecaries Society in 1838 and set up as a practitioner in Soho'. He qualified as an M.D. in 1844. The Soho area encompasses Broad Street, Golden Square where Gould lived from 1830 to 1859. The extensive quotation from Barton's article follows:

John Snow's experiences among the miners of Killingworth in the cholera epidemic of 1832 must have left a deep impression on his mind. He was (when not yet 19), sent out to treat a disease of startling rapidity and high mortality with a pack of empirical remedies and common sense. Richardson says "in

PART FOUR BIBLIOGRAPHY OF JOHN GOULD, HIS FAMILY AND ASSOCIATES

this labour he was indefatigable and his exertions were crowned with great success". Newcastle had the third highest mortality of any town in the country: the epidemic ran for a month resulting in 801 deaths and in the neighbouring colliery villages there were over 250 dead. The legacy of this terrible month may well have been to leave with Snow a determination to discover the real cause of the disease.

In the following years Snow came to London and gained his medical qualifications, building up a general medical practice and, with the discovery of anaesthesia, vigorously developing his knowledge and understanding of the actions and administration of anaesthetics. It was 16 years after Snow's Killingworth days that cholera broke out again. Having made its ominous way across Europe cholera came to England in the autumn of 1848 and in the summer of 1849 it raged throughout the country, causing 53,000 deaths in England and Wales and over 14,000 deaths in London. During the epidemic the Lancet published a spate of letters announcing innumerable cures for and causes for cholera. One cynical writer capped them all by collecting the remedies advocated, fifty in all, ranging from cayenne pepper, croton oil and chloroform to the reeking skin of a fresh slaughtered sheep.

Despite his being absorbed in anaesthesia Snow turned his mind to cholera. Already he was mentioning his theories by the autumn of 1848 to recognized authorities on collected evidence to support his views. In August 1849 he published 'On the Mode of Communication of Cholera' a pamphlet of 31 pages. Snow claimed that cholera was a communicable disease affecting the alimentary [sic] system and the general symptoms were all attributable to the loss of fluid in the body. The cholera poison was to be found in the vomit and excreta of infected persons, and, as well as by personal contact and the eating of contaminated uncooked food, Snow considered that cholera was spread by the emptying of sewers into the drinking water of the community. Infection was by swallowing the poison which probably consisted of "organised particles" not capable of indefinite division. He stated his opinions: "Not as matters of certainty, but as containing a greater amount of probability in their favour than any other in the present state of our knowledge". Snow supports his conclusions by quoting, among others, cases of personal infection and a limited but very fatal outbreak in a terrace of houses in Wandsworth. Here all the evidence pointed to the water supply being the agent of infection. Snow realized at the time that his researches were supported by insufficient evidence. He concludes "It would have been more satisfying to the author to have given the subject a more extensive examination and only to have published his opinions in case he could bring forward such mass of evidence in their support as would have commanded ready and almost universal assent; but being preoccupied with another subject (i.e. anaesthesia) he could only either leave the enquiry or bring it forward in its present state and he has considered it his duty to adopt the latter course..."

Our knowledge of Snow's early views on the transmission of disease is limited. Current theories mostly favoured the idea of effluvia emanating from the sick and the effects of damp atmosphere or obnoxious smells. The causes of a disease were divided into predisposing and exciting. Predisposing causes of cholera might be the place you lived in, exciting causes the state of the weather. By 1838 Snow saw reason to believe 'typhus' was contagious and in 1839 he disagreed with the 'malaria' theory that the products of animal decomposition caused ague. The question of contagion had secured as many opponents as supporters, and its history is beset with controversies. Ignorant of the actual causes, doctors did not know which way to turn in the light of conflicting evidence. Frascaturo in 1546 and Kircher in 1658 had suggested the basis of contagion and the existence of micro-organisms. Sydenham in the latter part of the 17th century wrote of pestilential particles and infection from carriers, but obscured the idea with theories of miasma, disturbances of the humours and climate and atmospheric influences. Webster in 1799 cites many cases against the theory of contagion. In 1840 Henle suggested a germ theory of contagion. Thus, though the works of Hunter and Jenner show that there is a communicable poison, the extreme diversity of opinion, and the often inexplicable circumstances surrounding cases of disease, only served to confuse the issue. And in the living conditions of the day there was so much that supported the 'malaria' theory. In 1842 the Report on the Sanitary Conditions of the Labouring population of Great Britain gave examples of the most frightful squalor: in Inverness there were very few houses with a W.C. or privy and only 2 or 3 public privies for the great bulk of inhabitants: in London the night soil overflowed from cess pools to a depth of three feet in two houses. In Manchester there was so much rubbish and mud in the streets that the ambulances could not get to the houses to take the sick away: in Glasgow there were dung heaps behind the tenements instead of lavatories and the excrement was a profitable source of manure. A revealing cri du coeur was published by "The Times" in 1849:

THE EDITOR OF THE TIMES PAPER.

Sur,

May we beg and beseach your proteckshion and power, We are Sur, as it may be, livin in a Willderniss, so far as the rest of London knows anything of us, or as the rich and great people care about. We live in muck and filthe. We aint got no privez, no dust bins, no drains, no water-splies, and no drain or suer in the hole place. The Suer Company, in Greek St., Soho Square, all great, rich and powerfool men, take no notice watsomedever of our cumplaints. The Stenche of a Gully-hole is disgustin. We all of us suffer, and numbers are ill, and if the Colera comes Lord help us.

Some gentlemens comed yesterday, and we thought they was comishoners from the Suer Company, but they was complaining of the noosance and stenche our lanes and corts was to them in New Oxforde Street. They was much surprized to see the seller in No. 12, Carrier St., in our lane, where a child was dyin from fever, and would not believe that Sixty persons sleep in it every night. This here seller

you couldent swing a cat in, and the rent is five shilling a week; but theare are greate many sich deare sellars. Sur, we hope you will let us have our complaints put into your hinfluenshall paper, and make these landlords of our houses and these comishoners (the friends we spose of the landlords) make our houses decent for Christions to live in.

Preaye Sir com and see us, for we are livin like piggs, and it aint faire we shoulde be so ill treted.

We are your respeckful servents in Church Lane, Carrier St., and the other corts.

Teusday, Juley 3, 1849.

The drawback to the acceptance of the contagion theory was the effectiveness of general sanitation measures in cutting down disease. Chadwick, one of the great sanitarians of the century showed that drainage and sanitary improvements stopped an epidemic of dysentery in Cork and generally reduced the incidence of disease. The care of public health proceeded under the banner of sanitary reform. Conditions were so bad that people could not visualise anything so theoretical as contagious organisms. It was in fact the right cure, but the cause was not accepted.

The effect of Snow's tract on Cholera was minimal. He had 125 copies printed at his own expense, of which only 30 copies were sold, at a shilling each. The Lancet commented that "the arguments adduced by the author against emanations causing the disease are by no means conclusive" and "Dr. Snow's exclusive views must be received with great limitation". In the report on the Epidemic Cholera of 1848-49 by the General Board of Health no mention is made of Snow's views and the blame for the epidemic is laid on atmospheric pollution aggravated by the poor sanitary conditions ...

Snow was not content to let his theories be propagated solely by his pamphlet and in 1849 the London Medical Gazette and the Medical Times of November 1851 published his views. Subsequently Snow was elected Orator of the Medical Society of London, with which the Westminster Society had joined in 1849, and his oration, delivered in March 1853, deals mainly with his reasons for considering cholera among others a communicable disease. He infers that the communicable cause of disease "resembles a species of living being", despite the fact that we cannot see it, from the manner of its growth and reproduction, and he brings forward evidence and reasons against the rival theories of the day. The society were so impressed that the oration was published "by request of the society". Again he wrote an article for the Medical Times, published in October 1853 and in February the next year he presented to the Epidemiological Society a rational mode of treatment of Cholera based on his views on the communication and pathology of the disease. When Cholera again broke out in England in the summer of 1854 Snow had given the medical profession ample opportunity to pay heed to his opinions.

THE EPIDEMIC OF 1854-55

Snow's theory by this time lacked only that mass of evidence necessary to support it. This epidemic supplied it in horrible abundance, as a result of which Snow published in November 1885 [1855] a second edition of his pamphlet 'much enlarged'. It was in fact a clearly-reasoned book that presented a great number of cases illustrating the communication of cholera on small and large scale. Well-marshalled facts drive home his theories and his arguments refute the conflicting theories of the day. Snow states the pathology of the disease and reasons that "the morbid matter of cholera having the property of reproducing its own kind, must necessarily have some sort of structure, most likely that of a cell". It was not until 1863 that Pasteur's work established the connexion of micro-organisms and disease. Snow lays down principles of personal and public hygiene that are essential in the prevention of the spread of disease. The outstanding ability of Snow is however shown in the conduct of his researches and the evaluation of the facts in two centres of the epidemic. One was a strictly localised outbreak of devastating severity, centred on the Broad Street pump near Golden Square, Soho. The other was "an experiment on the grandest scale" in the parts of London South of the Thames. Here the inhabitants received water from two different companies, one of which supplied relatively pure water and the other water grossly contaminated with London sewage. Snow uses the statistics of cholera mortality in these districts to provide the most convincing proof of his theory.

THE BROAD STREET PUMP

Snow had already started his researches in South London when a disastrous outbreak of cholera occurred in Soho. Broad Street (Figure 40) was the centre of an area partly residential and partly commercial. There were the houses of middle-class tradesmen and houses of a much poorer sort: a workhouse with five hundred inmates, a percussion-cap factory employing 200 people and a brewery employing 70; there were shops, coffee houses, public houses and many small workshops in the area. As soon as Snow heard of the outbreak he suspected the pump in Broad Street: this was a popular pump supplying many in the area and preferred by some to a nearer water supply. It was used in the eating houses in the neighbourhood, and a tub of it was kept for the workmen in the factory. There was a piped supply from the Grand Junction and New River companies but this was intermittent, being turned on for only a few hours each day. Householders had to store it in tubs open very often to the weather, never cleaned out and in the hot summer this supply must have been unpleasant compared with the pump water.

During the last few days of August 1854 there were one or two deaths occurring each day in the area.

PART FOUR BIBLIOGRAPHY OF JOHN GOULD, HIS FAMILY AND ASSOCIATES

Suddenly the mortality shot up: on Sept. 1st, 70 people died; 127 died the next day and over 300 in the following week. As soon as he heard of the outbreak Snow suspected the pump. From his knowledge of cholera Snow worked out the date of the probable infection of the well and personally investigated the 83 deaths in the area registered since that date. In only fourteen cases did his researches not definitely incriminate the pump. That, and the extreme severity of the outbreak in such a limited area was enough to convince Snow that action must be taken to close the pump. In the evening of the 7th September Snow explained his reasons to the Board of Guardians of the parish and the next day the pump handle was removed. It was an action of such beautiful simplicity that it would be satisfying to record that the epidemic came to a dramatic close. But in fact the cholera mortality had declined with as much rapidity as it had at first risen. The population had deserted the area and of course the cholera *vibrio* does not have an indefinite life in water. Snow himself does not make any claim for proof resulting from his initiative, but the simplicity of his action has seized the imagination of subsequent writers, including Richardson, who attribute the ending of the outbreak to the removal of the pump handle. The greatness of Snow's work lies in his investigation, not in the prevention of any one outbreak. His work in connection with the Broad Street pump continued with the collection of supporting evidence all pointing to the source of infection being the pump. Perhaps the most striking case is that of the lady living in Hampstead who had Broad Street water sent out to her every day. She and a niece who was visiting her drank the water. The lady died from cholera and the niece who returned to her home in Islington, was attacked with cholera and also died. Neither at Hampstead nor at Islington was there any cholera at the time.

Snow prepared a map showing the places of residence of all those who died from cholera, including those of the sick who were removed to the Middlesex hospital where Florence Nightingale was working. The black marks cluster incriminatingly around the Broad Street pump.

Following the outbreak a local Committee of Enquiry was set up on to which Snow was co-opted. After Sir Benjamin Hall, the president of the General Board of Health, had refused his help, the Committee circulated a questionnaire, with little success. Then, following Snow's methods, the interviewing team undertook a house to house visit, armed with a list of twenty four questions, drawn up by Snow and Dr. Lankester, the Chairman. This had a better result, and the committee prepared and published a report of the outbreak. Snow submitted an individual report that followed the lines of his previous investigations. One of the interviewers was a Rev. H. Whitehead who also submitted an individual report. Whitehead was the curate of a neighbouring church, who showed much courage and initiative in the cholera epidemic in London. He was at first sceptical of Snow's theory and wrote to him to tell him so. He instituted his own enquiry in Broad Street itself and found the evidence so conclusive that he did not hesitate to confirm Snow's results.

Snow had had the pump opened at the time of the outbreak but could find no definite proof of a source of contamination. Whitehead, whilst going through the Registrar General's returns stumbled on the case which gave rise to the epidemic. Two days before the violent outbreak an infant was attacked with choleraic diarrhoea in No. 40 Broad Street: she died four days later. Whitehead found that the mother washed the nappies and tipped the water into a cesspool which lay within feet of the pump well. Again the well was opened and this time there was clear evidence of the contamination of the water through defective brickwork.

This ends the lengthy but poignant quotation from Barton's article.

In 1855 Dr John Snow published a small book entitled 'On the mode of communication of cholera'. In this book is a map 'showing the deaths from cholera in Broad Street, Golden Square, and the neighbourhood, from 19th August to 30th September 1854'. 'A black mark or bar for each death is placed in the situation of the house in which the fatal attack took place'. This map is illustrated in Figure 39.

For more information on this 1855 book on cholera see Snow (1936). There is a reference to 'John Gould, the eminent ornithologist' in Snow's book. He did not drink the pump water.

BASTIN, JOHN. 1970. 'The first Prospectus of the Zoological Society of London: new light on the Society's origin'. *J. Soc. Biblphy. Nat. Hist.* 5:369-88.

Bastin ably presents the case for the Zoological Society of London being the brain child of Sir Stamford Raffles. Bastin also refutes the inference by both Henry Scherren (1905) and P. C. Mitchell (1929) that 1824 was the date for the post organizational meeting of the Society and that a Prospectus was issued in that year. The correct date for both these matters is 1825.

Bastin, again contrary to Mitchell's thoughts, felt that there were definite links between the members of the Zoological Club and those forming the Zoological Society.

BASTIN, JOHN. 1973. 'A further note on the origins of the Zoological Society of London'. *J. Soc. Biblphy. Nat. Hist.* 6:236-41.

Additional letters and reports are presented to substantiate Bastin's earlier view (1970) that the first Prospectus of the Zoological Society was issued in March 1825, and that Sir Stamford Raffles was the leading promoter and founder of the Society.

BATESON, CHARLES. 1972. *Australian shipwrecks 1622-1850.* Sydney, Reed.

Terry Callen of North Stockton, N.S.W. has very kindly supplied me with answers to my questions about the vessel *Parsee* and others used by the Goulds when in Australia.

Here is a reference supplied by Mr Callen from Bateson's book concerning the shipwreck of the *Parsee* soon after the Gould's had landed at Hobart Town.

> The barque *PARSEE* bound from Hobart, Tasmania, to Adelaide, South Australia, was totally wrecked on Troubridge Shoal in St. Vincent's Gulf, Sth Aust., on Nov. 17, 1838. No lives were lost. The passengers and crew landed on Torrens Island. The master, Cpt J. McKellar, claimed that he had shaped course to lie along the east side of the gulf after passing Cape Jervis in the evening but was carried onto the shoal by currents. She struck heavily and the main and foremast went by the board. William Cleveland, who visited the wreck on NOV. 26, said she was a complete wreck and everything on board was damaged. The *PARSEE* was a barque of 349 tons, built in Dumbarton, Scotland, in 1831, and owned by J. Gordon of London. She had recently arrived at Hobart Town from London.

BECKER, BERNARD H. 1874. *Scientific London.* London, Henry S. King. (Reprint 1968 by Frank Cass & Co., London.)

After some historical facts on the London learned societies of the mid-1800s, the author relates a few personal comments on his own visits to these societies and his attendance at lectures. Of interest are his comments on The Royal Society, the Department of Science and Art which includes the British Museum (Natural History), the British Association for the Advancement of Science and the Royal Geographical Society.

BENNETT, EDWARD TURNER. 1831. *The gardens and menagerie of the Zoological Society delineated . . . birds.* London, John Sharpe.

A volume that is contemporary with Gould's employment as 'Curator and Preserver' at the newly formed Zoological Society of London. The Society sponsored the Gardens.

This volume on birds is the second volume of a set, the first volume being concerned with the quadrupeds at the Gardens.

BENNETT, GEORGE. 1834. *Wanderings in New South Wales, Batavia, Pedir Coast, Singapore, and China; being a journal of a naturalist in those countries, during 1832, 1833 and 1834.* 2 vols. London, Bentley.

Bennett returned to England after these wanderings and undoubtedly met Gould when he made presentations at the meetings of the Zoological Society of London. They were to become good friends, and the Gould family were to stay at his home in Sydney in 1839.

A wealth of information is presented in these two volumes on many aspects of Australia, but unfortunately the work is not indexed.

BENNETT, GEORGE. *Correspondence. 1837-1840. Correspondence from Dr. Bennett to Professor Owen.* Collection: Royal College of Surgeons of England.

Through the courtesy of E. H. Cornelius, Librarian, seven letters were copied written by Dr Bennett of Sydney, to Professor Owen which contained comments about John

PART FOUR BIBLIOGRAPHY OF JOHN GOULD, HIS FAMILY AND ASSOCIATES

Gould. Three were written before Gould arrived in Australia and the remaining four contain some comments on Gould during his visit to Australia.

The dates of these letters are 26 September 1837, 5 March 1838, 7 May 1838, 27 December 1838, 7 August 1839, 10 December 1839 and 5 April 1840.

BENNETT, GEORGE. 1860. *Gatherings of a naturalist in Australia... London, Van Voorst.*

There are many comments or quotations from Gould, especially in the ornithological sections. This is particularly true in the chapter on 'The Mooruk or Cassowary of New Britain...' Dr Bennett's letter to Gould of 10 September 1857 with the account of this new bird is printed in full. Gould named the bird *Casuarius Bennetti.*

There are three fine hand-coloured plates by J. Wolf. The book is not indexed.

BONAPARTE, CHARLES LUCIAN. 1838. *A geographical and comparative list of the birds of Europe and North America.* London, Van Voorst.

An interesting comparative listing of European and North American bird species.

Throughout the list, I have quoted as Types of the Species under consideration, the figures of the great works of Mr. Gould and M. Audubon on the Ornithology of the two regions, as they must be considered the standard works on the subject. The merit of M. Audubon's work yields only to the size of his book; while Mr. Gould's work on the Birds of Europe, inferior in size to that of M. Audubon's, is the most beautiful work on Ornithology that has ever appeared in this or any other country. The latter Ornithologist has expressed his intention of including in his Supplement all the European Species indicated in this list that have not yet appeared in his work, together with such others as his unrivalled industry may add to the existing Fauna.

A generally favourable review of this small sixty-seven page volume appeared in June 1838 *Ann. Nat. Hist.* 1:318-19.

BORELL, ADREY E. Manuscript. 1974. Ralph Ellis, Jr. 1908-45. Memories by Adrey E. Borell. Collection: author.

As the result of a suggestion by me, Adrey Borell wrote a most interesting anecdotal remembrance of Ralph Ellis. The Ellis Collection at the University of Kansas Spencer Library contains the largest collection of original Gouldiana in the world.

Borell had been hired by the Ellis family to be a companion to the difficult-dispositioned Ralph. Adrey also prepared the bird and mammal skins Ralph collected. Their association extended over a six-year period.

This reminiscence is typed on legal-sized paper, double-spaced, on twelve pages.

BORLAND, HUGH A. 1941. The Gilbert water-shed. C.P. [newspaper?] date 3 April 1941, from the North Queensland Cutting Book, No. 1, p. 8, courtesy of M. I. Walker and C. G. Sheehan of the John Oxley Library, Brisbane.

This is an interesting account of the Gilbert River and the parallel Mitchell River, also named by Leichhardt. The Gilbert River flows into the Gulf of Carpentaria, Queensland. After Gilbert's death, Leichhardt gave the name of Gilbert to the next big river his expedition came upon.

BOUBIER, MAURICE. 1925. *L'Evolution de l'Ornithologie.* Paris, Librairie Felix Alcan.

There is a reference to Darwin's voyage on the Beagle on p. 122 stating that Gould and G. R. Gray determined the new species of birds brought back by Darwin. Pages 144 to 146 relate to Gould with the statement that 'John Gould was one of the most remarkable ornithologists of the 19th century'.

On page 219 is the statement that Gould was 'one of the first explorers of Australia'.

BRENNAN, M. T. & MORGAN, P. J. 1977. 'Lord Edward Smith Stanley, 1775-1851, XIIIth Earl of Derby: a review of his biological collections and their importance'. *BCG, Newsletter* No. 6 of the Biology Curators Group, July, pp. 20-8.

A preliminary paper 'intended to show the scope and wealth of the [Derby] collection and the role of Lord Derby in the development of natural history during the early and middle nineteenth century'.

In spite of the bombing and fire of 1941, around 20 000 of Lord Derby's bird skins are still extant.

Brennan and Morgan recount some of the history of the acquisition of this valuable collection. Derby received most specimens by purchase from Messrs. Leadbeater, who in turn had received their specimens, often duplicates of types, from early expeditions and collectors. Second in importance for source of specimens was John Gould who supplied at least 543 skins to Lord Derby between April 1830 to 27 February 1852.

A wealth of manuscript catalogues, lists, letters, paintings and books exist in the Merseyside County Museum, City of Liverpool Libraries and Knowsley Hall, and are in the process of being catalogued and analysed.

BRIDSON, GAVIN D. R. 1976. 'The treatment of plates in bibliographical description'. *J. Soc. Biblphy. Nat. Hist.* 7:469-88.

A valuable treatise on an important subject for consideration by the bibliographers. Many good references to adjunctive works relevant to the Gouldian era of print-making.

[Gould was his own publisher. This meant that in addition to being author and artist himself, he also employed in his home shop additional more competent artists, who were also lithographers and transferred the original drawing to the stone. He also employed the colourists. After the plates and text were printed, then Gould and his 'publishing company' sold and distributed the parts as separately issued.]

BRITISH ASSOCIATION FOR THE ADVANCEMENT OF SCIENCE. 1835–. *Reports*, 1831–. London, John Murray.

The first meeting of the British Association was held at York in September 1831. For this first meeting 353 tickets were issued and 300 scientists attended.

'Mr. Gould, Member of the Zoological Society of London, exhibited select specimens of Birds figured and described in his work on the Ornithology of the Himalaya Mountains, and copies of this work were laid upon the table for inspection.

'Mr. R. Havell exhibited drawings of Birds for Mr. Audubon's great work on American Ornithology'.

These two paragraphs appear on page 91 of the 'First Report-1831' published in 1835. Interestingly Havell wrote Audubon about this encounter with Gould (Ford 1964: 292).

The second meeting was in Oxford in 1832, and the third in Cambridge in 1833 was attended by 900 members. From this meeting appeared 'Lithographed Signatures of the Members who met at Cambridge in 1833... 4to. Price 4s'. I have not seen a copy of this.

The 1834 meeting was in Edinburgh, the 1835 meeting at Dublin, and the 1836 meeting at Bristol.

The Report of the 1837 meeting at Liverpool contains the first listing of almost 1000 members. Under Life Member is 'Gould, J., 20 Broad St. Golden Square, London'. The Earl of Derby and Jardine are also listed. On page xx is 'The following reports and monographs were requested... on the Caprimulgidae, by Mr. Gould'.

PART FOUR BIBLIOGRAPHY OF JOHN GOULD, HIS FAMILY AND ASSOCIATES

The 1838 meeting was at Newcastle in August. On page xix of the Report is 'The following Reports and Continuations of Reports have been undertaken to be drawn up at the request of the Association... On the Caprimulgidae, by N. [sic] Gould, F.L.S.'

Sir William Jardine was the President of Section D, Zoology and Botany. There is nothing in this volume on Gould's work on the *Caprimulgidae* being presented, as Gould had requested of Jardine, before he left for Australia.

The 1839 meeting was held at Birmingham in August.

The tenth meeting was held at Glasgow in August 1840. On page xxii is 'The following Reports and Continuations of Reports have been undertaken to be drawn up at the request of the Association. On the *Caprimulgidae* by N. [sic] Gould, F.L.S.'.

The eleventh meeting was at Plymouth in July 1841 and on page xix was the same announcement as in 1840 — 'by J. Gould'. On p. 331 was 'Notices of progress have been received in regard to the *Report on Caprimulgidae,* by Mr. Gould'.

The twelfth meeting was held in 1842 at Manchester. On page 69 of the Transactions of the Sections is a paper by Dr Richardson, 'On a Specimen of *Machaerium Subducens* from Port Essington, New Holland, Belonging to the Collection Made by Mr. Gilbert, Mr. Gould's Assistant'.

The thirteenth meeting was held in 1843 at Cork.

The fourteenth meeting was held at York in September, 1844. In this volume is the article, 'Report on the recent Progress and present State of Ornithology' by H. E. Strickland on pages 170-221. This article is reprinted completely in Jardine's 1858 book on the life of Strickland.

On pages 61 to 62 of the Transactions of the Sections is a paper by Gould on 'A Monograph of the Sub-family Odontophorinae, or Partridges of America'. A long paragraph summarizes Gould's presentation on this subject. [See Gould, John. Manuscript [1844] for the original handwritten paper in the McGill University Collection.]

The fifteenth meeting was held in 1845 at Cambridge in June.

The sixteenth meeting was held in 1846 in Southampton in September. On page 79 of the Transactions is a sentence 'Mr. Gould exhibited several new species of humming-birds from the Andes'.

The seventeenth meeting was held at Oxford in June 1847, the eighteenth meeting at Swansea in August 1848, the nineteenth meeting at Birmingham in September 1849, and the twentieth meeting at Edinburgh in July and August, 1850. There was no listing of Gould being in attendance at these meetings, but in Jardine's 1850 *Contributions to Ornithology* on page 86 is the note that 'Gould read papers at the 20th meeting of the British Association at Edinburgh, July 30, 1850'. I was unable to find any reference to this presentation in the published Report.

The twenty-first meeting was at Ipswich in July 1851, twenty-second at Belfast in September 1852, and the twenty-third at Hull in September 1853.

In the Report of this 1853 meeting on page 68 of the Transactions of the Sections is a paragraph listing a paper by Gould 'On a New Species of *Cometes;* a Genus of Humming-Birds'.

The twenty-fourth meeting was at Liverpool in September 1854, next at Glasgow in September 1855, Cheltenham in August 1856, Dublin in September 1857, and Leeds in September 1858.

The twenty-ninth meeting was at Aberdeen in September 1859 and Gould is listed on pages 148 to 149 of the Transactions as giving three papers, namely 'On the Varieties and Species of New Pheasants recently introduced into England', 'Several species of Birds of Paradise', and 'On some New Species of Birds'.

The thirtieth meeting of the Association was held at Oxford in June and July 1860, and the thirty-first meeting was at Manchester in September 1861. In the geology section transactions on pages 112 to 113 is a paper listed by C. Gould, B.A., F.G.S. entitled 'Results of the Geological Survey of Tasmania'. Obviously this is by John Gould's son Charles. Who read it, or was it read? Charles never returned to England after he left for Tasmania.

The thirty-second meeting was at Cambridge in October 1862, next at Newcastle-upon-Tyne August and September 1863, next at Bath in September 1864.

The thirty-fifth meeting of the Association was at Birmingham in September 1865. In the Report of this meeting is a very complete and valuable listing of members with addresses. No list had been printed for several years. Gould was listed as a 'Life Member not entitled to the Annual Report' as follows: 'Gould, John, F.R.S., F.L.S., F.R.G.S., F.Z.S., 26 Charlotte-street, Bedford-square, London'.

Nottingham was the city for the thirty-sixth meeting in August 1866, Dundee in September 1867, Norwich in August 1868, Exeter in August 1869, Liverpool in September 1870, Edinburgh in August 1871, and Brighton in August 1872. In the list of members in this Report of the forty-second meeting, Gould's name was printed in capital letters indicating he was a 'Member of the General Committee'. This committee passed on research projects and allocated funds if needed.

The listing continues of the meetings because Gould probably still attended until the late 1870s. The forty-third meeting was at Bradford in September 1873, Belfast in August 1874, Bristol in August 1875, Glasgow in September 1876, Plymouth in August 1877, Dublin in August 1878, Sheffield in August 1879, Swansea in September 1880, and York in August and September 1881.

Gould's name did not appear on that 1881 fifty-first meeting Report. Gould was mentioned, however, in the Presidential Address for the D Section by Richard Owen. In an historical review of Owen's attempts to get a new natural history museum on another site from the existing British Museum, and on the unheard-of site size of at least eight acres (three hectares), he related that his idea was opposed by 'other scientific gentlemen among them Professor Huxley, Professor Maskelyne, Mr. Waterhouse, Dr. Gray, Sir Roderick Murchison, Mr. Thomas Bell, Dr. Sclater, Mr. Gould, and Sir Benjamin Brodie'. Owen does say that the group, along with Parliamentary members, were really surprised when he put forth the large acreage. Fortunately, in a few years, Owen's plan, on five acres (two hectares), later expanded, was realized and we now have the British Museum (Natural History) on Cromwell Road, separated from the British Museum itself at no 'impediment to the studies of the scientific visitor'.

BRITISH MUSEUM. 1874-98 *Catalogue of the birds in the British Museum.* 27 vols. London, Trustees.

Zimmer in 1926 stated that this monumental work is 'unquestionably the most important work on systematic ornithology that has ever been published'. It was originally intended that R. B. Sharpe would prepare this work, but because of the magnitude of the project other specialists in various groups were called in to write certain sections.

As related in the British Museum (Natural History) 1906 'History of the collections' most of Gould's bird skins and eggs (the notable exception being his Australian type skins) were either given to or purchased by the British Museum. Thus, there is much in these twenty-seven volumes pertaining to the Gould collections.

In hand is the sixteenth volume with the section on *Upupae* and *Trochili* by Osbert Salvin, and the section on *Coraciae* by Ernst Hartert. Obviously there are many

references to Gould because of his great interest in the humming-birds and caprimulgidae.

BRITISH MUSEUM (NATURAL HISTORY). Guide. 1885. *A Guide to the Gould collection of humming-birds.* 4th ed. London, British Museum (N.H.).

A twenty-two-page booklet which contains seven pages of text by Albert Gunther concerning the history of the Gould humming-bird collection. The balance of the booklet is a listing by R. Bowdler Sharpe of the mounted species of humming-birds in the sixty-two cases comprising the exhibition.

Presumably other editions are similar but I have only examined this fourth edition. In the Library of the British Museum (N.H.) is the first edition of 1881, the second of 1883, the fourth of 1885, and another edition of 1889.

BRITISH MUSEUM 1901-12. *Catalogue of the collection of the birds' eggs in the British Museum.* 5 vols. London, Trustees.

Except for the eggs of Gould's Australian collection, the great majority of his egg collection was given to or purchased by the British Museum. As related in the 1906 'History of the Collections', the majority were purchased after Gould's death.

BRITISH MUSEUM 1961. *General Catalogue of Printing Books.* London, Trustees British Museum.

Volume 89, Columns 726 and 727 list thirty-one Gould items including his works and works about him.

BRITISH MUSEUM (NATURAL HISTORY.) 1903-40. *Catalogue of the Books, Manuscripts, Maps and Drawings in the British Museum (Natural History).* 8 vols. London, Trustees of the British Museum (N.H.)

In volume two on pages 701 and 702 are listed thirty-two items of works by and about John Gould. Missing is the 1855 edition of *Ramphastidae, An Introduction to the Mammals of Australia, An Introduction to the Birds of Great Britain,* and the large folio-size separate issue of the introduction to the *Birds of Asia.*

BRITISH MUSEUM (NATURAL HISTORY). 1904-12. *The history of the collections contained in the Natural History Departments of the British Museum.* 3 vols. London, Trustees of the British Museum (N.H.).

A most interesting and valuable historical record of this renowned museum and its acquisitions. Of particular value is the listing of and commentary on those who gave, traded, or sold specimens to the various departments of the British Museum (N.H.), since this list necessarily includes most of the world's naturalists, collectors, museums, and natural history dealers.

Volume 1 includes an Introduction and General Sketch of the several libraries in the British Museum (N.H.) with an accessions list. Then follow sections on the departments of Botany, Geology, and Minerals. There are no references to John Gould or son Charles, the geologist, in this volume.

Volume 2 contains the accounts and accessions of many other departments including Mammals and Birds.

Obviously, many of Gould's dealers and associates grace the pages. The dealers include Argent, W. Cutler, the Edward Gerrards, Messrs. Leadbeater, Maison Vereaux and others. Associates include Joseph Baker, William Briggs, Henry Denny, Dr Dieffenbach, Dr Robert McCormick, Frederic Moore, Captain Stokes, Frederick

Strange, Captain Charles Sturt, Edward Wilson, plus of course Sir William Jardine, T. C. Eyton, John MacGillivray, John H. Gurney, R. B. Sharpe, etc.

It is unfortunate that none of the volumes is indexed. The Gould references in volume 2 are as follows: under the collection of mammals, written by Oldfield Thomas: pages 6, 7, 8, *35*, 36, and 64; under the bird collection section, written by R. B. Sharpe: pages 130, 131, 246, 247, 248, 251, 252, 253, 255, 257, 258, 260, 263, 265, 266, 299, 304, 320, 327, 330, 335, *373-375*, 376, 377, 382, 395, 403, 421, 422, 424, 438, 477 (Sharpe), 493, 511, 514, and 515; under the collection of reptiles and batrachians: page 518.

Volume 3 is a thin volume concerned with the development from 1856 on when the Zoological Department became separated from the Natural History Department. Subjects covered include the growth of the collections, lists and catalogues and the staff.

On page 27 is mentioned the purchase of the 'Gould collection of Falcons' during the years 1870 to 1874.

The years 1875 to 1883 were concerned with the building and subsequent move of the Natural History section to the new building at South Kensington separate from the British Museum at Bloomsbury.

The acquisition of the bird collection of the late John Gould occupies page 45:

It consisted of 12,395 specimens and contained a large number of the objects described and figured in his various great works. Of special value in this acquisition was the series of Humming Birds, which consisted of 5,378 specimens, partly preserved as skins, partly mounted and grouped in the cases which they had been shown to the public at the first International Exhibition in 1851. They were the favourites of their former owner, and were brought together by forty years' careful collecting, regardless of expense. The price paid for the entire collection was £3000.

Gould is also mentioned on pages 44 and 49. The issuance of the 'Guide to the Gould Collection of Humming-Birds' is listed on page 63 and a new issue on page 71. The acquisition during 1886-87 of many of Gould's works, along with that of Audubon, as part of the Marquis of Tweeddale's Collection, is referred to on page 70.

BRITISH MUSEUM (NATURAL HISTORY). Guide. 1910. *Guide to the gallery of birds in the Department of Zoology, British Museum (Natural History)*, 2nd ed. London, British Museum.

The first and second editions were both prepared by W. R. Ogilvie Grant.

In the section of the book on the descriptions of the exhibits of the nesting of British birds, the Bullfinch case on page 142 is stated to be from the 'Gould Collection'.

BROWN, RICHARD & BRETT, STANLEY. 1977. *The London bookshop*. Part two. London, Private Libraries Association.

'Being Part Two of a Pictorial Record of the Antiquarian Book Trade: Portraits & Premises', so states the subtitle. There is no preface.

Eight bookshops are covered in this profusely illustrated small volume. Of interest is the six-page section on Henry Sotheran, Ltd of 2, 3, 4, & 5 Sackville Street. The firm was established around 1760. In 1845 was begun a monthly *Price Current of Literature*.

The following is stated: 'When they [Sotheran] purchased the rights and stock of the naturalist John Gould, they completed his publishing programme offering the complete work "elegantly bound" for one thousand pounds the set'.

In the article are three photographs of the Sotheran establishment and one of Mr Leslie G. Simpson. The present manager of the firm, Mr Simpson, kindly presented me with this copy of part two on the London bookshops.

The first part of *The London bookshop* was published in 1971.

BRYANT, C. E. 1938. 'Gould Commemorative Issue'. *Emu* 38: 89-90.

Bryant, as Editor, wrote this short introduction to the valuable Gould Commemorative Issue of *Emu*. He apologizes for the errors that he knows exist in dates and other information, but, with multiple authorship and the too recent discovery of the wealth of Gould material by Chisholm, this was unavoidable. (*See* Journal. *The Emu*. 1938.)

BRYANT, C. E. 1938. 'Gould miscellanea and some anecdotes'. *Emu* 38:226-31.

As an expression of the maxim 'Hear, also, the other side', Bryant provides some anecdotes not always complimentary of Gould. These are taken primarily from Sharpe's *Analytical Index* and Palmer's *The Life of Joseph Wolf*.

BRYANT, C. E. (ed.). 1938. 'The Gould League of Bird Lovers'. *Emu* 38:240-4.

Information about the Gould League chapters in Victoria, Queensland and New South Wales is presented.

BUCHANAN, HANDASYDE. 1979. *Nature into art*. New York City, Mayflower.

There are extensive references to John Gould and his associates, Lear and Wolf, in this elegant book. Two Gould plates are uncoloured and one coloured.

Here are two quotes from page 131: 'For sheer numbers of very good books Gould is unchallenged', and 'Hullmandel printed Gould's books, and they represent some of the best examples of lithography ever produced'.

BUCKMAN, THOMAS R. (ed.). 1961. R. V., nine eventful years. Lawrence, University of Kansas Libraries.

Mr Buckman, Alexandra Mason and staff have compiled an index to the first twenty-six issues of 'Books and Libraries' as a tribute to Robert Vosper, who was transferring to California. The twenty-six issues of this library publication covered the period from December 1952 to May 1961.

There are numerous references to the Ralph Ellis Collection. My article on Gould is indexed as being in copy 12: 2-3.

BUCKMAN, THOMAS R. 1964. *A guide to the collections, Department of Special Collections, University of Kansas Libraries*. Lawrence, University of Kansas Libraries.

This thirty-two-page booklet contains many references to the Ellis ornithology collection (pp. 14, 15, 16, 22, 28, & 29) and to the Gould volumes and manuscripts therein.

BUCKMAN, THOMAS R. 1966. *Bibliography and natural history*. Lawrence, University of Kansas Libraries.

A collection of essays presented on the above subject at a Conference at the University of Kansas in June, 1964. The articles by Mengel and by Zeitlin will be commented on further since they have more to do with ornithology and Gould. The other papers have primarily a botanical slant.

BUERSCHAPER, PETER. 1975. *Animals in art.* Toronto, Royal Ontario Museum.

The Royal Ontario Museum published this beautifully illustrated fifty-page booklet in conjunction with 'An International Exhibition of Wildlife Art' at the museum in the fall of 1975.

Gould is represented by Item 119 listed as 'Rosebreasted Cockatoo, watercolour, 50.8 x 38.1' which is illustrated in black and white. Item 120 is a hand-coloured lithograph of *Notornis mantelli.* Both were loaned by the Blacker-Wood Library, McGill University.

Lear has two watercolours listed, Item 169 White Stork 1831, and Item 170 Osprey 1831, lent from the Blacker-Wood Library, McGill University.

Josef Wolf is represented by Item 308 Guinea Fowl (Verreaux's) in charcoal (illustrated), and Item 309 Lammergeier & Ibex in oil.

BUFFALO SOCIETY OF NATURAL SCIENCES. 1938. *Seventy-five years. A history of the Buffalo Society of Natural Sciences 1861-1936.* Buffalo, New York, Buffalo Society of Nat. Sci.

In the chapter on 'The Two Libraries' (page 51) is this comment: 'In 1865 a notable work was presented by Everard Palmer - John Gould's monumental "Monograph of the Trochilidae, or Family of Humming Birds" (1861) in five volumes'. Henry S. Sprague reported that this was a great addition to the library, 'only 275 copies being in existence ... probably not more than thirty copies in the United States'.

In 1875 a further report on the library listed Gould's 'Birds of Great Britain', 'Mammals of Australia', 'Partridges of America', Toucans (1854), and Trogons (1875). They were obtained through 'liberal donations of Everard Palmer, Esq. [and] the late Coleman T. Robinson'.

BURMESTER, C. A. 1974-77. *National Library of Australia, guide to the collections.* 2 vols. Canberra, National Library of Australia.

'This *Guide* is made up of a series of notes on elements of the National Library's collections.'

Volume 1 has references to John Gould on page 197 under the section on Silvester Diggles (1817-80), naturalist, musician and artist; on page 207 under Joseph R. Elsey (1834-57) surgeon, explorer and naturalist; on page 407 under Mathews Collection. In this latter collection are some MS. materials and watercolours relating to Gould. These are more extensively catalogued in the National Library (1966) checklist to the Mathews ornithological collection. The estensive Rex Nan Kivell Collection is listed on page 491 and contains some Gouldiana items but they are not specifically mentioned.

In volume 2 on pages 181 to 183 is a section devoted to John Gould, naturalist, written in March 1974. Following a short review of his life and works, a list of the National Library's holdings of Gould's manuscripts is printed:

[MS 587] Letters (29) to various people and some other papers
[MS 454] Letters (10) to E. P. Ramsay (1865-9) of Dobroyde near Sydney
[MS 1217] Letter from J. R. Elsey relating to birds collected on A. C. Gregory's North Australia expedition 1855-6
[MS 1465] 24 Photostat copies of letters from Gould to E. P. Ramsay, 1866-9
  Typescript copies of letters from J. Gilbert to Gould 1839-42
  Typescript copies of 18 letters from Elizabeth Gould and 3 letters from Lady Franklin to Elizabeth Gould
  Correspondence (1938-41) between Australian ornithologists concerning historical research on the work of John Gould
[MS1579] Handbook to the Birds of Australia with a letter from Gould to W. Butler inserted
[MS 1755] Letters (4) concerning purchase of pattern plates of Birds of Australia for proposed new edition, 1888-91

PART FOUR BIBLIOGRAPHY OF JOHN GOULD, HIS FAMILY AND ASSOCIATES

Article by T. Iredale on transfer of plates to National Library
   Microfilm Copies
[Mfm G 746] John MacGillivray's Notebooks entitled 'Voyage of H.M.S. Rattesnake: Correspondence 1847-9' which includes letters sent to Gould 1848-9
[Mfm G 743] Stanley family papers: notes by Owen Stanley on Gould-Gilbert-Stanley relations
   Of Gould's illustrations the National Library has 80 pattern plates of The birds of Australia; 179 pattern plates of Mammals of Australia; a number of original water colour drawings of birds attributed to John Gould and to Elizabeth Gould respectively. These are filed in the Pictorial Collection together with photographs of portraits of Gould.

On pages 374 to 375 Gould is mentioned in the section on Parrots. John Gilbert is referred to on page 290 in the section on Leichhardt.

CALABY, J. H. 1971. 'The current status of Australian Macropodidae'. *Aust. Zool.* 16:17-29.

This paper is part of an entire conference devoted to the Australian kangaroos. I am listing it here because of its contemporary relevance to Gould's pioneer work on kangaroos.

CAMERON, RODERICK. 1971. *Australian history and horizons.* New York, Columbia University Press.

A full-page colour plate of Gould's Black cockatoo appears on page 176. Interesting chapters have to do with the course of the long four-to-five-month voyages from England to Australia via the African Cape of Good Hope, the exportation from New South Wales in the 1830s of large amounts of wool and whale products, and the discovery of gold in the early 1850s.

CAMPBELL, A. J. 1900. *Nests and eggs of Australian birds.* Sheffield, Pawson and Brailsford.

Campbell dedicates this book to Gould and Gilbert 'who together performed such great and good pioneer work in Australian ornithology'. Note is made of Gould's stated intention to publish a book on the oology of Australia.

   In the Introduction is a lengthy quote from Mrs Robert Brockman of Guildford, Western Australia, concerning Gilbert's disposition and work habits. Gilbert visited almost two months in this neighbourhood, and her house was his headquarters. The section on Gould is paraphrased from Sharpe's *Analytical Index*.

CAMPBELL, A. J. 1938 'John Gould amongst Tasmanian birds'. *Emu* 38: 138-41.

A short article concerned with the birds of Tasmania discovered or figured by Gould.

CARUS, JULIUS VICTOR & ENGELMANN, WILHELM. 1861. *Bibliotheca zoologica.* 2 vols. Leipzig, W. Englemann.

Another monumental compilation of scientific papers and volumes published on all forms of zoology. The section on Aves is in volume two, from page 1112 to page 1260.

   Gould's papers and volumes are listed on several pages under specific sections, such as those related to general collections, specific areas of the world and, finally, under the specific headings of various genera and species.

CASSIN, JOHN. 1847. Annual meeting Dec. 28, 1847, report of ornithological department. *Proc. Acad. Nat. Sciences of Philadelphia* 3:347-8.

Cassin reports on the recent gifts by the Wilson brothers to the Academy:

One other collection, which is to be delivered to this Society early in the coming year, remains to be noticed, and that is Mr. John Gould's collection of the birds of Australia.

When I inform the Society that this collection contains specimens of all the known Australian birds, except five species, and of the nests and eggs of a large number, its peculiar value will be immediately understood. I may be excused for remarking, however, that Mr. Gould's collection acquires additional interest from the consideration that it contains the original specimens from which many of the numerous species described by him were first characterized, and that the specimens comprised in his collection are those from which the drawings were made for his latest and splendid work, "The Birds of Australia".

The number of specimens now contained in the collection and those which will be received in the course of the ensuing year, according to arrangements now completed, may be estimated as follows:

Duke of Rivoli's collection	12,500	specimens
M. Bourcier's collection	1,000	”
Mr. Edward Wilson's collections	4,000	”
Mr. Gambel's and Mr. Cassin's collection	1,000	”
Mr. Gould's collection	2,000	”
Former collections of the Academy	2,500	”
	23,000	

CATALOGUE. 1932. *The library of an Ohio collector, sale number 3960.* New York City, American Art Association.

Item 149 is a 'complete set of the folio works of John Gould... over 3000 beautiful coloured plates... 71 volumes and parts' which sold, according to a pencilled note, for $2600.00. The original edition of *Ramphastidae,* the suppressed *Birds of Australia and adjacent islands* ('Cancelled Parts'), and the monograph *Pittidae* are not included.

This catalogue is item Ellis Aves D268 at the University of Kansas Spencer Library.

CAYLEY, NEVILLE W. 1932. *Australian finches in bush and aviary.* Sydney, Angus & Robertson.

This is a book on aviculture but it contains many notes on the finches in their natural habitat. Gould is quoted on many pages.

CAYLEY, NEVILLE W. 1938. 'John Gould as an illustrator'. *Emu* 38:167-72.

As is true of all the articles in this Gould Commemorative Issue, this one should also be read and digested in its entirety. Cayley writes on the lithographic process, especially as it relates to Gould's works, and also on Gould's role of artist and his responsibility for the finished plates.

He states: 'Amongst writers on natural science John Gould stands alone. The amount and quality of his work are unsurpassed, his writings will ever be a source of reference and inspiration, and the beautiful illustrations remain unchallenged.'

Cayley makes some comparisons regarding craftsmanship and artistry among Gould, Audubon, Fuertes, and Thorburn.

CAYLEY, NEVILLE W. 1958. *What bird is that?* 2nd ed. Sydney, Angus & Robertson.

In the Preface of this book is a short history of the Gould League of Bird Lovers written by Alec Chisholm.

CHALMERS-HUNT, J. M. 1976. *Natural history auctions 1700-1972.* London, Sotheby Parke Bernet.

'A register of sales in the British Isles'. Clive Simson wrote a section on 'ornithology' on page 29 which contains several comments by David Evans on prices of Gould's works around 1973.

PART FOUR BIBLIOGRAPHY OF JOHN GOULD, HIS FAMILY AND ASSOCIATES

On page 114 is a sale entry under the date of 4 May 1881: 'Gould, (John) F.R.S. ornithological lib. Puttick 301 lots 2 + 18pp. BML [Brit. Mus. Library Reading Room] Camb. [Balfour Library, Camb.] p. n.'

The abbreviations 'p' means price, and 'n' means that names of buyers are included in these copies of the catalogue.

CHANCELLOR, JOHN. 1978. *Audubon.* New York, Viking Press.

A profusely illustrated volume filled with facts. The contemporary illustrations of hunting scenes, Audubon's homes, individuals and family, and some original watercolour drawings greatly enhance the value of this work on Audubon.

Gould is mentioned on four occasions. On page 199 Chancellor writes that Audubon 'renewed his attacks on contemporary English bird books [around 1835], sparing only John Gould'. Further, on the next page, in letters to Bachman, Audubon wrote that 'Gould's *wife* makes his drawings on stone - she is a plain fine woman, and although these works are not quite up to Nature, both deserve great credit'. In another place Audubon wrote about the pedigree dog Gould sent to Bachman, and further on, how Swainson and Bonaparte had dropped him.

A black-and-white illustration by Gould of an Asian pheasant is on page 201. On page 205 is a plate (No. 424) of several finches and buntings. 'The top figure, a male house finch, was based on a specimen that belonged to John Gould.'

CHILDS, JOHN LEWIS. 1917. *Catalogue of the North American Natural History Library of John Lewis Childs.* Floral Park, New York, private.

Three works by Gould are listed: 'Partridges', 'Icones Avium' and 'Trochilidae', the latter in six volumes.

CHISHOLM, ALEXANDER HUGH. 1938. 'Gould, the Birdman', letter to the Editor. London, *The Times,* 30 June 1938.

In the Gould family news clipping book is this letter to the Editor written by Chisholm requesting information on the Gould descendants, if any. Chisholm was amply rewarded in his search.

CHISHOLM, A. H. 1938. 'Out of the Past: Gould Material Discovered'. *Vic. Nat.* 55:95-102.

Mr Chisholm wrote this article while still in London after discovering the wealth of Gouldiana owned by the descendants of John Gould. He reviews some of the family history from Gould's children to the present, and relates some notes taken from letters written in Australia by Mrs Gould. Four letters are reported completely, two from Lady Jane Franklin to Mrs Gould (undated and 15 July 1840); one from John Gould to his wife, dated 20 January 1839, and one from Mrs Gould to Mrs Mitchell in London, dated 28 May 1839.

A colour plate of the Buff-sided Scrub Robin and a portrait of John Gould seated accompany the article.

CHISHOLM, A. H. 1938. 'John Gilbert and some letters to Gould'. *Emu* 38:186-99.

Four of the fifteen Gilbert-to-Gould letters, uncovered by Chisholm when he went to England in 1938, are printed. These letters were with other valuable Gouldiana held by the Gould descendants, his sole grandchild, Mrs Helen Edelsten, and her two sons, Drs Geoffrey and Alan Edelsten.

Gilbert's very important role as Gould's primary collector of Australian birds is appreciated and emphasized by Chisholm.

CHISHOLM, A. H. 1938. 'Some letters from George Grey to John Gould'. *Emu* 38:216-26.

These six letters to Gould, printed verbatim, cover the period from 13 August 1842 to 10 January 1843. Grey was Governor of South Australia at that time. He wrote quite detailed information on the habits and nesting of the birds he sent to Gould.

These letters had been in the possession of Gould's great-grandsons, Drs Geoffrey and Alan Edelsten of England, but in 1938 they gave them to Alec Chisholm who deposited them in the Mitchell Library, Sydney.

CHISHOLM, A. H. 1939. 'The Story of Eliza Gould'. *Vic. Nat.* 56:22-5.

The letters written by Mrs Gould from Australia are alluded to, and one dated 8 October 1838, written to her mother, is printed in full.

There are photographs of Henry W. Coxen, a nephew of John Gould, and a full page one of E. C. Prince, Gould's very able secretary.

CHISHOLM, A. H. 1939. 'Charles Gould to John Gould'. *Vic. Nat.* 56: 99-101.

A letter is printed written 21 October 1861 from Hamilton, Tasmania, by Charles Gould to his father in London. Charles relates some notes on birds and on a Tasmanian 'tiger'.

CHISHOLM, A. H. 1940. 'The mother of Australian bird study'. *Wild Life*, February, p. 32.

A photocopy of this article was very kindly sent to me by Mrs Alan (Grace) Edelsten. In 1938 Alec Chisholm visited with Mrs Edelsten and her husband Alan, great-grandson of John Gould.

In the article Chisholm recounts his uncovering of the John Gould and John Gilbert memorabilia in England in 1938. Two photographs of portraits of Mrs Gould accompany the article.

CHISHOLM, A. H. 1940. 'The Story of John Gilbert'. *Emu* 39:156-76.

A fascinating and quite comprehensive story of what was known about John Gilbert, who was Gould's most valuable and outstanding collector. The dates are given for Gilbert's movements in Australia during the 5½ years he spent in that country. The majority of the material is from Gilbert's diary written during the nine months while he was on Leichhardt's expedition of 1844-45. Gilbert was killed by Aborigines 28 June 1845 while on this expedition.

Chisholm mentions that the dates given in this article correct numerous errors given in previous articles, including those in the 1938 Gould Centenary issue of *The Emu*.

CHISHOLM, A. H. 1941. *Strange new world. The adventures of John Gilbert and Ludwig Leichhardt*. Sydney, Angus & Robertson.

Chisholm has written an absorbing account of Gould's Gilbert and of Leichhardt. The basis for the book is the collection of letters from Gilbert to Gould, and Gilbert's journal found by Chisholm in 1938 in the hands of Gould's descendants in England.

This book is one of the important Gouldiana documents with many references to Gould, especially in his relationship to Gilbert. We must be very thankful that Leichhardt saved Gilbert's journal after Gilbert was killed by Aborigines, and that he then sent it to Gould in London.

The indexing of references to Gould seems to be complete.

PART FOUR BIBLIOGRAPHY OF JOHN GOULD, HIS FAMILY AND ASSOCIATES

CHISHOLM, A. H. Review. 1942. 'Strange new world, by A. H. Chisholm'. *Emu* 41: 312.

This is a short review by P. C. M. [Mitchell] of Chisholm's book about John Gilbert and the first Leichhardt Expedition.

CHISHOLM, A. H. 1941. 'Mrs. John Gould and Her Relatives'. *Emu* 40: 336-54.

A very comprehensive article on Mrs Gould and her family including brothers, children, nephew and later descendants. There is a figure of Mrs Gould holding a cockatiel, photographs of portraits of Charles Coxen and Henry Coxen, and lastly a photograph of a page of a letter from Mrs Gould to her mother, dated 9 March 1839.

Practically all of this material is also in Chisholm's 1944 book *The Story of Elizabeth Gould*. The only figure appearing in the book, however, is the portrait of Mrs Gould.

CHISHOLM, A. H. 1942. 'John Gould's Stolen Birds'. *Vic. Nat.* 58: 131-3.

Some valuable bird skins were stolen from Gould in England around September, 1845. This is a short article on the subject with an illustration of the handbill for a £20 Reward.

Mention is also made in the article of the theft of Captain Charles Sturt's collection of watercolour drawings of Parrots which had been admired by Gould when he was in Australia.

CHISHOLM, A. H. 1942. 'John Gould's Australian Prospectus'. *Emu* 42:74-84.

This is a detailed commentary on two copies of a Prospectus, printed in Sydney, Australia for Gould's *Birds of Australia*. The two copies were found by Chisholm in 1938 in a box of Gould's personal papers in the possession of Gould's granddaughter, Mrs Helen Edelsten of Hampshire, England.

Chisholm comments on Gould's tremendous 'mental and physical energy' and also on some of his financial arrangements with his agent, John Fairfax, and with his collector, John Gilbert.

A list is given also of eleven other published articles that are based on the material recovered by Chisholm when he visited the Gould relatives in England in 1938.

CHISHOLM, A. H. 1944. *The Story of Elizabeth Gould*. Melbourne, The Hawthorn Press.

This 74-page book contains extensive material on Mrs Gould and her relatives, a portrait of Mrs Gould, eight letters written to her mother and four letters written to Mrs Mitchell when Mrs Gould was in Australia. These letters were written between 8 October 1838 and 13 September 1839.

Essentially the same text, minus the twelve letters, is in Chisholm's article in the April 1941 *Emu* entitled 'Mrs. John Gould and her relatives'. The 8 October 1838 letter is also reproduced in the *Victorian Naturalist* of June 1939.

CHISHOLM, A. H. 1944. 'An interesting old note-book'. *Emu* 43: 281-8.

In 1902 the widow of Charles Coxen presented Dr Arthur Malaher of Queensland with a notebook she thought previously belonged to John Gilbert. Upon examining the bulky volume (230 x 190mm), Chisholm concluded that it was really the notebook of Eli Waller, an early naturalist of Australia, particularly Queensland, who had a taxidermy shop in Brisbane. Waller was a friend of Charles Coxen and Silvester

Diggles. Waller and his son, who had considerable artistic ability, sent Gould quite a few bird-skins and paintings. [The unsorted box (Gould, John. Manuscript. [Several Dates.] Autograph...) in the British Museum (Natural History) might contain relevant source material in regard to this Gould-Waller association.]

CHISHOLM, A. H. 1944-45. 'Birds in the Gilbert Diary'. *Emu* 44: 131-50; 183-200.

This two-part article, except for the second and last paragraphs, is identical with the pamphlet published by Chisholm entitled 'An explorer and his birds'. The type set for each page is not even altered.

CHISHOLM, A. H. [1945 ?]. *An Explorer and His Birds, John Gilbert's Discoveries in 1844-45.* Melbourne, Brown, Prior, Anderson Printers.

This paper-bound booklet of thirty-eight pages was probably published in 1945 or 1946 since the material 'was first published, in slightly different form, as "Birds of the Gilbert Diary", in the Emu in October, 1944, and January, 1945'.

There are four black-and-white plates, one of the diary, and three of Australian scenes. A map of the Leichhardt expedition's journey is also included. Then follows a rather complete listing of the birds seen and discovered by Gilbert. This list is taken from Gilbert's diary which was eventually sent to Gould after Gilbert's death.

CHISHOLM, A. H. 1951. 'Gouldian tragedies'. *Emu* 51: 78-9.

In the *Sydney Morning Herald* of 5 September 1844 is a report on the suicide by poison of Stephen Coxen. It had previously been stated that he died 'probably about 1845, through falling overboard from a vessel between Newcastle and Sydney'.

Stephen Coxen emigrated to New South Wales in 1827 and developed the property 'Yarrundi' on the Dartbrook, a tributary of the Hunter River. He was the brother of Gould's wife, Elizabeth. Some five years later Stephen was joined by his brother Charles, who became both a pastoralist and ornithologist. Gould visited them in 1839 when he was in Australia.

CHISHOLM, A. H. 1951. 'Skymaster bears famous birdman's name'. *Sydney Morning Herald.* 27 October 1951.

In the Gould family clippings collection in the home of Dr Geoffrey Edelsten is this newspaper article concerning an Australian airplane christened 'John Gould'. The name of John Gilbert was already on a Skymaster plane of Trans-Australia Airlines, as were those of other early Australian explorers.

Chisholm relates in this article Gould's travels in Australia from 1838 to 1840 and then mentions his contact with the Edelsten family in England in 1938.

CHISHOLM, A. H. 1952. 'John Gould's "missing" egg paintings'. *Emu* 52: 301-5.

In 1951 Dr Geoffrey Edelsten of England, a great-grandson of John Gould, added to previous memorabilia of Gould which he had given to Chisholm in 1938, by giving him a bound volume entitled 'Bird Eggs/Gould'. The volume contains seventy-two plates of eighty-seven species of birds. Many are signed 'J.H.G.' or 'J.H.G. del' and appear to be the result of the artistry of John Gould's eldest son John Henry. The plates were never published. (*See* [Gould, John.] Original drawing. Birds Eggs.)

Information is also given concerning John Gould's birthplace of Lyme Regis in Dorset, his birth date of 14 September 1804 and his baptism date of 7 October 1804.

PART FOUR BIBLIOGRAPHY OF JOHN GOULD, HIS FAMILY AND ASSOCIATES

CHISHOLM, A. H. 1955. *Strange new world.* Sydney, Angus & Robertson.

The first 1941 edition is revised and re-illustrated. Chisholm mentions some of the changes in his Preface. There are more illustrations in the 1941 edition and they are also more interesting than those in this second edition.

Chisholm's debunking of Leichhardt appears to me to be even more intense in this second edition.

The indexing of references to Gould is complete.

CHISHOLM, A. H. 1955 'Dr. Henry Gould, writer and artist'. *Vic. Nat.* 71: 182-5.

A sixteen-page booklet containing five letters from Dr John Henry Gould to his father is the subject of this article. The first date is 8 February 1854 from the Red Sea, and the fifth letter is dated 17 April 1854 from Kotree opposite Hyderabad. Chisholm gives some short excerpts from the five letters.

In the final paragraphs of the article, Chisholm reports on a collection of drawings of birds' eggs, now in Australia, presented by Dr Geoffrey Edelsten. These were also the work of John Henry Gould.

[Obviously, these letters are the first five that were in a booklet of thirty-two pages copied for me by Dr Geoffrey Edelsten. In this larger booklet a total of eleven letters are printed. See [Gould, John Henry] 1854. Letters . . .]

CHISHOLM, A. H. 1956. *Bird Wonders of Australia.* 4th ed. Sydney, Angus & Robertson.

The references to Gould in the Index are complete. An American edition was published by the Michigan State University Press in 1958.

CHISHOLM, A. H. 1964. 'Some early letters in Australian ornithology'. *Emu* 63: 287-96; 373-82.

Concerning a detailed criticism of the series of papers by Wagstaffe and Rutherford (1954-55) on the Lord Derby letters of John Latham, Gould, John Gilbert, Charles Sturt and John Edward Gray.

Chisholm states that 'the compilers would have been well advised to have sent copies of the Gould-Gilbert correspondence to Australia, for checking and annotating, before publication'. [A reasonable statement. Hope I remember to do this when the volume on Gould's correspondence is ready.]

Detailed comments and some corrections of the transcriptions of the Lord Derby letters are made by Chisholm. He was pleased that three more of Gilbert's letters had been 'discovered', but wondered where others might be found. [I suggest that the Box of Manuscripts in the British Museum (N.H.) is the likely repository for additional Gilbert letters and those of many other Gould correspondents.]

CHISHOLM, A. H. 1964. 'Elizabeth Gould. Some "new" letters'. *J. Roy. Australian Hist. Soc.* 48: 321-36.

This article contains a complete transcript of three letters written by John Gould's wife, Elizabeth. The first one was written to her mother apparently near the end of 1827, before she married. She was a governess for a nine-year-old girl and had the qualifications to teach her French, Latin, and music. The letter refers to her own family living on Broad Street.

The two other letters were cross written on the same sheet while the Goulds were in Australia, one to Elizabeth's mother and the other to a close family friend, Mrs

Mitchell. The date was 6 December 1839, from 'Yarrundi', the Dartbrook property of Elizabeth's brothers, Stephen and Charles Coxen. John Gould had gone collecting to the Liverpool Plains and the Namoi.

CHISHOLM, A. H. 1973. 'Earth is our home'. *Sydney Morning Herald,* 13 January.

In the family book of clippings in the possession of Dr Geoffrey Edelsten, seen when we were in Winchester in 1974, is a clipping of this article.

The article concerns 'The First and Greatest Ambassador with which Australia has been blest' namely John Gould. Chisholm says that Gould's eight large illustrated volumes on 'The Birds of Australia' recently sold at auction for $A20,000.

He then details some of the information on the organization of the Gould League of Bird Lovers which was founded in Melbourne, Victoria in 1909. The organization is primarily made up of school children who pledged to take a kindly interest in their country's bird life and in general conservation. The League's pledge has recently been broadened to read 'Earth is our home and I promise to keep it beautiful by learning to understand and conserve its soil, air, water, natural beauty, and all its living things.'

Chisholm then goes on to give a considerable amount of information about his discovery in England of the letters and diaries pertaining to Gould's and Gilbert's exploration of Australia from 1838 to 1845.

The special purpose of the article, however, was to bring attention to the fact that John Gould's relative, Mrs Grace Edelsten, was visiting in Australia. Her late husband, Dr Alan Edelsten, with his brother, Dr Geoffrey Edelsten, in 1938 provided Chisholm with some extremely valuable and unique historical and ornithological material from Gould's and Gilbert's Australian trip. Chisholm in turn gave this material to the Mitchell Library in Sydney. Mrs Alan Edelsten was entertained at the Mitchell Library, visited History House, attended a meeting of the Council of the Gould League, and addressed children of the large new public school at Belrose.

Finally, there is the information that Mrs Edelsten's daughter Helen, now Mrs Guy Mitchell, with her husband and children, have become established on the land in Western Australia.

CHISHOLM, A. H. 1973. *Strange journey.* Melbourne, Rigby.

Except for a prefatory three-page 'Notes written in 1973' this is a line-for-line reprint of Chisholm's 1955 *Strange new world.*

In the 'Preface to the second edition' the words 'Strange New World' have been replaced with 'the present book' and 'the present edition'. The map of Gilbert's journeys placed inside the covers is not in this 1973 edition.

The 'Notes Written in 1973' by Chisholm present more current evidence for Chisholm's contention that Leichhardt was a poor bushman and leader.

CHISHOLM, A. H. & SERVENTY, V. 1973. *John Gould's the birds of Australia.* Melbourne, Lansdowne.

This is a beautiful book with ninety-six of Gould's plates from *Birds of Australia.* The introduction by Alec Chisholm is quite extensive and contains much material from the 1938 *Emu* Commemorative Issue, including three maps, two portraits of Gould, and Elizabeth Gould's original sketch of the Spotted Owl. It is erroneously stated that Gould visited America in 1851, when it was 1857.

A short note taken from Gould's original text accompanies each coloured plate, which is followed by a contemporary note on distribution, food and nesting of each bird portrayed.

PART FOUR BIBLIOGRAPHY OF JOHN GOULD, HIS FAMILY AND ASSOCIATES

CHISHOLM, A. H. 1964. *Strange journey.* Adelaide, Rigby.

According to correspondence from Mr Chisholm (18 July 1976), this book is a reprint of his *Strange New World*, but he did not say of which edition.

[CHISHOLM, A. H.] Obituary. 1977. *Emu:* 77: 232-5.

The following is an extract from the Obituary:

> A close link with the early history of the Royal Australasian Ornithologists Union was broken with the death of Alexander Hugh Chisholm, OBE, journalist, author, ornithologist and historian, on 10 July 1977, at the age of 87. With his passing, those who knew him best felt that it was the end of an era, as he was intimately associated with RAOU pioneers such as A. J. Campbell, W. H. D. LeSouef, J. A. Leach, C. F. Belcher, A. H. E. Mattingley and C. L. Barrett. Alec Chisholm had been actively connected with the Royal Australasian Ornithologists Union for seventy years, having first become a member in 1907 and before his death he was the Union's longest surviving member. He succeeded W. B. Alexander as editor of the Emu in April 1926 and continued in that office until April 1928; became President of the Union during 1939-40; and was elected a Fellow in 1941.

A. R. McGill gives an account of his youth, aborted education, early days as a newspaper reporter, and his early interest in natural history, particularly birds.

During a visit to England in 1938 he met Dr Alan Edelsten, a great-grandson of John Gould. From this association Chisholm was to bring to Australia many letters from or to John Gould, and John Gilbert, and also the long-lost Gilbert diary written on the Leichhardt expedition. Since then Chisholm published several articles and books on this Gouldiana.

He is survived by a daughter.

CHITTENDEN, CAROL. 1975. Index of manuscript Gouldiana in the Ellis Collection at the University of Kansas Libraries. Typescript.

This index to the original drawings by Gould and his artists evolved over a period of four years. Ralph Ellis had purchased this Gould collection of original drawings from Henry Sotheran Ltd, of London, in 1937 (Sauer (1976) Index of Ellis Archives, p. 133) and it was left to the University of Kansas on his death.

Chittenden numbers 1286 items, listing in columns the species of the bird or mammal, if determined, secondly, the medium of the drawing and notes, and thirdly, the location in Gould's works of the published drawing, if this was determined. Over fifty per cent of the drawings were identified by Chittenden.

Around 1971 the Spencer Library at the University of Kansas purchased two volumes containing an additional 237 original drawings, mainly by John Gould and a few by Mrs Gould, purportedly done for *Birds of Australia*. Chittenden similarly catalogued these drawings.

The total listing comes to 1523 items, all original drawings, but for ten or so extraneous items such as manila folders, etc.

CHITTENDEN, CAROL. 1976. 'On birds and bird men: Gouldiana at the University of Kansas'. Books and Libraries at the University of Kansas 13 (part 2): 1-5.

The author has spent more hours with the Ellis Collection of Gouldiana at the University of Kansas than any other individual. In the process of cataloguing the over 1500 original drawings by Gould and his artists in the Ellis Collection she has some relevant opinions on Gould, his methods and his associates.

Comparisons between Audubon and Gould are touched upon also in this article.

As she states, only fifty per cent of these original drawings have been identified in relationship to a corresponding published lithograph, so much more remains to be done with this collection. The sheer bulkiness of this material makes any such study very time-consuming and very fatiguing.

CLEEVELY, R. J. 1976. 'The Sowerbys and their publications in the light of the manuscript material in the British Museum (Natural History)'. *J. Soc. Biblphy. Nat. Hist.* 7(4): 343-68.

The Sowerby's were publishers and illustrators of natural history books. They were also naturalists in their own right. This article is relevant to the Gould interests because it covers the years from 1782 to 1890 and relates to many of the publishing problems of the nineteenth century.

Gould may well have asked the Sowerby's to publish his works, but when he was turned down, fortunately for his own interests, he began to publish the works himself. Being his own publisher proved to be a very profitable venture for Gould. He also had direct control over the artists, lithographers, printers, colourists, and sales, all of which emanated (except for the printing) from his own home.

In reading this article one sees the publishing problems in a proper perspective. There are references to Edward Lear, to Richard Taylor, whose company did the printing of Gould's works, and to Samuel Highley, another publisher.

CLELAND, J. BURTON. 1937. 'The history of ornithology in South Australia'. *Emu* 36:197-221; 296-312; 37:33-47.

Gould was in South Australia from 'May or early June [1839], and visiting Onkaparinga, the Murray scrub and Kangaroo Island' leaving by August. In part 1 of this series of articles, Gould's visit is recorded on pages 201 to 204, but there are also additional references to him.

Other naturalists discussed in this article include Captain Charles Sturt, Edward John Eyre, and John McDouall Stuart.

CLELAND, J. B. 1939. 'Gouldiana'. *Emu* 38: 536.

Cleland presents convincing evidence that the 'un-named collector' who wrote Item 6, 1841 of Hindwoods 'Gouldiana' article (*Emu* 38: 237) entitled 'Voyage from London...' is without doubt Sir George Grey.

[*See also* Somerville 1939.]

COLP, RALPH, JR. 1977. 'Charles Darwin and the Galapagos'. *N.Y. State J. Medicine* 77: 262-7.

Dr Colp attempts in this article to answer the question 'When and how did the Galapagos affect Darwin's theory of evolution?'

The descriptions of the Galapagos finches and mocking-birds by John Gould at the 10 January 1837 meeting of the Zoological Society of London is reported. Gould obviously provided Darwin with important scientific data regarding distinct modifications of the forms of finches on the different islands. But it took Darwin several years to formulate his theory of evolution by adaptation.

A letter from Darwin to Gould is printed in the article. It concerns Gould's role in preparing the bird section of Darwin's *The zoology of the voyage of H.M.S. Beagle*.

CONIGRAVE, C. PRICE. 1938. 'The "Gilbert Country" in Western Australia'. *Emu* 38: 233-5.

Conigrave relates personal memories of the areas of Western Australia transversed by Gilbert in the 1840s.

CORNING, HOWARD. 1930. *Letters of John James Audubon, 1826-40.* 2 vols. Boston, Club of odd volumes. (A facsimile reprint in one volume was published in 1969 in New York by Kraus Reprint Co.)

PART FOUR BIBLIOGRAPHY OF JOHN GOULD, HIS FAMILY AND ASSOCIATES

It is great to have these valuable letters published, but there are no annotations and, what's worse, no index!

The following quotations regarding Gould are from these letters:

To Robert Havell, Jr, from York, 13 September 1830, volume I, page 116:

Mr. Gould must forward the Box to Manchester, by canal directed to the Secretary of the Natural History Society there... he may have a sett uncoloured for 30 shillings a number, the price I have sold one at Manchester, and you must see that he pays you cash down, as I fear that he is not over and above able to purchase such a work on credit...

[Gould apparently was already at work acting as an agent of sorts for authors and publishers.]

To Robert Havell, Jr, from 'New Castle Tyne'. 30 September 1830, volume I, page 119: '... let me know if Mr. Gould has actually sent the box to Manchester and when it went, as my promise is out with the Society there...'

To Reverend John Bachman from London 25 August 1834, volume II, page 29:

Since here I have read, Selby's & Temminck's Works, but they are I am sorry to say not from Nature. not a word could I find in them but what was compilation... I could not even be told at what time the Golden Eagle laid her eggs in Europe! Sir Wam Jardine is published an enormous quantity of trash all compilation, and takes the undue liberty of giving figures from my Work and those of all others who may best suit his views... Mr. Gould is publishing the Birds of England, of Europe &c &c &c... in all sorts of ways... Swainson has 17 volumes in his head and on papers half finished... You could not pass a Bookseller's shop from the extreme "West end" to Wapping without seeing New books on Zoology in every window...

To Reverend Bachman from London, 20 April 1835, volume II, pages 67-70:

Here there are at present three Works publishing on the Birds of Europe... one by Mr. Gould and the others by no one knows who... at least I do not know... Works on the Birds of all the World are innumerable... Cheap as dirt and more dirty than dirt... Sir William Jardine will encumber the whole of God's creation with stuff as little like the objects of the Creator's formations as the moon is unto cheese... but who cares? as long as these miscellanies bring forth 5 shillings per Vol to the pocket bag of the one who produces them as a Hen that hatches Duck Eggs, and whom I have no doubt is as much surprised to see his progenitor go to Market as the Hen is at seeing her Webbed brood take to the Water...

... I have agreed to exchange a copy of my Work with Mr. Gould, for his publications... and have by me, 13 Numbers of his Birds of Europe... his century of Hymealan Birds... and his monograph of Trogons... I have also purchased a monograph of Parrots from a Mr. Lear... When you & I are old men... how pleasing it will be to us to look at these together to quiz them all, and pass our Veto upon them!

Gould is a man of great industry... has the advantage of the Zoological Society's Museums, Gardens &c... and is in correspondance with Temminck, Jardine, Selby, James Wilson and the rest of the Scientific Gentry... his Wife makes his drawings on stone... she is a plain fine woman, and although these works are not quite up to Nature, both deserve great credit.

To Robert Havell, Jr, from Edinburgh, 25 September 1835, volume II, page 93: 'has Gould delivered you any numbers of his works for me? if not ask him for what he has published... I am told that he is now in London'.

To Reverend Bachman from Edinburgh, 1 December 1835, volume II, page 103:

I wish you would write me about the Beautiful Dog I sent you, for I should like to offer your thanks to Mr. Gould who gave it to me for you, and hoped that you might be pleased with it. he assured me that it is of the very best blood in England, and her Whelps would point as well as she Naturally! You ought to raise several from the best dog you can procure...

This is the only account I have seen referring to Gould having a dog, but then Gould enjoyed his 'shooting' trips, in addition to fishing.

To Reverend Bachman from London, 14 August 1837, volume II, page 176, after Audubon's return to England:

London is just as I left it, a Vast Artificial area, as well covered with humbug, as are our Pine Lands and old fields with Broom grass.... Swainson is publishing his incomprehensible Works... Gould has just finished his Birds of Europe and now will go on with those of Australia. Yarrell is publishing the British birds quarto size... and about one thousand other niny tiny Works are in progress to assist in the mass of confusion already scattered over the World...

To 'dear friend' (undoubtedly Bachman), dated 4 October 1837, volume II, pages 185-6: 'Gould is publishing the Birds of Australia from stuffed Skins! Next to that a monograph of the Caprimulgae, and next to that the next things he can get.'

The letter continues on 8 October: '[Charles Bonaparte] dislikes the Humbug of the Nobles...he came a third time to us...he is almost constantly with Gould, at the Zoological Museum and Indeed every where, where there are bird skins. Me thinks that he is over anxious to pump me...'

To Reverend Bachman from London, 14 April 1838, volume II, page 202:

Mr. Gould the author of the Birds of Europe is about leaving this Country for New holland, or as it is now called Australia.... he takes his Wife and Bairns with him, a Waggon the size of a Squatters Cabin and all such apparatus as will imcumber him not a little...he has never travelled in the Woods, never salted his rump Stakes with Gun Powder and how he will take to it, will be a "sin to Crockett".

To Robert Havell from Edinburgh, 13 March 1839, volume II, page 212:

I have written to Mr. Baker but have had no answer from him and as he is very frequently absent from England, I would advise you to call on Mr. Gould "Secretary"! and try to find from him whether he is at home or not.

To Robert Havell from Edinburgh, 30 June 1839, volume II, page 222:

'Respecting the letter of Mr. Prince, I have only to say that no answer to it is neccessary. I have no numbers for Charles Bonaparte and no 5th Vol. of Biogs for Mr. Gould. Let these gentlemen purchase or procure what they want where they can.

In addition to these comments about Gould, Audubon's letters throw an interesting perspective on all the naturalists of those years, and especially on Sir William Jardine, Charles Waterton, William Swainson, George Ord, William MacGillivray, Charles Bonaparte and Robert Jameson.

COTTON, CATHERINE DRUMMOND. 1938. *Ludwig Leichhardt and the great South Land.* Sydney, Angus & Robertson.

The John Oxley Library, Brisbane provided me with photocopies of the pages of this book that contain references to John Gilbert. They are on pages 190-1, 194-7, 202-7, 212, 218-21, 226-7 and 236-7.

COUES, ELLIOTT. 1878-80. [Ornithological bibliography.] Washington, Government Printing Office.

Linda Hall Library has the four parts of Coues's American and British bibliographic instalments bound in one volume. The four articles originally appeared in different United States government publications.

The first part is a 'List of Faunal publications relating to North American ornithology'. Gould is listed under the date of 1834 for one article published in two journals, *Proc. Zool. Soc.* 2: 14-15 and *Lond. and Edinb. Philos. Mag.* 5: 72-3.

The second instalment is an extension of the first, also on the Americas. Gould is listed under 1834 for an article 'Sur des Oiseaux de l'Amerique Meridionale' from *L'Institut* 1834, 2: 252. This probably is the same material as in the article in instalment one under 1834. Further articles by Gould appear under the years 1837, 1839, 1841 (on 'The Zoology of the Voyage of the Beagle'), 1855 (2 ref.), 1856, 1857, and 1859.

Instalment three is a bibliography of ornithology arranged according to the families of birds. There are many, many Gould references to published articles in this section and also excellent collations of his major works. This is especially extensive for Gould's monograph on the humming-birds. It is quite monumental really, since Coues lists all the humming-bird species illustrated in Gould's work, first as they appeared in the separate twenty-five parts, and secondly, as the plates were to be bound in the volumes!

PART FOUR BIBLIOGRAPHY OF JOHN GOULD, HIS FAMILY AND ASSOCIATES

In an 'Addendum to Trochilidae' Coues then proceeds to write an historical sketch of workers and works on 'Hummers', the synonymy of which developed into a 'malignant epidemic which we may call the genus-itch'. [As a dermatologist this has a familiar ring.] Coues's writing in this section is especially witty.

The fourth instalment of this ornithological bibliography 'being a list of faunal publications relating to British birds' may have been prompted by the gracious letter to Coues mailed from England on 14 May 1879 and signed by an imposing list of British naturalists. They wrote to request Coues to write a bibliography of European ornithology, as he had done for American ornithology. For this letter, see Manuscript letter to Elliott Coues (1879) from the Smithsonian Institution, Division of Birds, Washington.

Since Gould wrote very few articles on British birds, Coues lists him only for the years 1831, 1835, 1864 and 1869. Under the years 1862-73 he of course refers to Gould's folio work *The Birds of Great Britain*.

[COUES, ELLIOTT.] Correspondence. 1879. [Manuscript letter to Elliott Coues.] Collection: Division of birds, Smithsonian Institution, Washington, D.C.

A very relevant manuscript from the ornithological and bibliographical standpoint. Thirty-eight of the most respected naturalists of Great Britain (including John Gould, naturally) got together and signed a statement to Elliott Coues. 'We wish to place on record our gratitude to you and to the Surgeon General of the U.S. Army for the scientific work you have performed' in your compilation of American ornithological bibliography and 'hope that time and means will be found you to prosecute in Europe' a similar undertaking. I have paraphrased the letter by way of explanation, but through the courtesy of Secretary Ripley of the Smithsonian Institution and Dr George Watson of the division of birds, it is printed in full:

To Elliott Coues Esquire, Assistant Surgeon, United States Army.
We the undersigned beg leave to express our high appreciation of the 'Bibliographical Appendix' to your recent work 'Birds of the Colorado Valley' being No. 11 of the Miscellaneous Publications of the United States Geological Survey of the Territories under the charge of Dr. Hayden. And at the same time we also wish to place on record our gratitude to that gentleman and to the authorities of the Department to which you are attached, for the liberality they have shown in granting you permission to stay at Washington for the completion of this and other important works upon which you have now been so long and so usefully engaged.

The want of indexes to the ever increasing mass of Zoological literature has long been felt by all workers in every department of that science; but the enormous labour of compilation has hitherto deterred many from undertaking a task so appalling. It is with no small satisfaction that we recognize your readiness to devote yourself to work of this nature. Moreover we feel justified in hoping that should the instalment now published in the volume above be enlarged in a similar manner so as to include a complete Bibliography of Ornithology this branch of science will possess an index to its writings perhaps more complete as to its scope and contents than any kindred subject of similar extent.

An undertaking of this sort is beset with formidable difficulties; not only is its extent enormous and the works relating to the subject are widely scattered through many Libraries public and private; but the qualifications of a good Bibliographer are not easily to be found united in one person. His application and industry must be untiring and he must be thoroughly conversant with the art of Bibliography. In addition to these requirements, in a case like the present an equally thorough knowledge of the subject under consideration is indispensable. You happily combine all these qualifications; your industry has long been approved, your knowledge of books is evident from what you have now put before us, your knowledge of Ornithology has long been known to us. We can well believe that the libraries of your own country are better stored than any others with works relating to the Ornithology of North America and that therefore the 'List of faunal publications relating to North American Ornithology' could be nowhere better prepared than in Washington; but when the ornithological literature of the whole world has to be examined it seems to us almost indispensable that the older libraries of Europe, and especially of England, France, Italy, Germany and Holland should be consulted if one of the chief merits of your work is to be maintained viz:- the consultation at first hand by yourself of every work mentioned therein.

This brings us to one of the chief objects of this memorial which is to express our sincere hope that time and means will be found you to prosecute in Europe the great undertaking you have commenced so well and bring it to a successful conclusion. Should the authorities who preside over the Department

to which you belong — and especially the Surgeon General of the U.S. Army — who have hitherto so liberally granted you facilities for the scientific work you have performed be disposed to furnish you with these means of perfecting your undertaking we are convinced it will reflect great credit to them and the country to which you belong. We on our part, so far as England is concerned, are ready not only to welcome a brother Ornithologist but also to render you every assistance in our power.

W. H. Flower, F.R.S. etc.
   President of the Zoological Society of London
F. W. Manley [?] Sec. R. C.
Charles Darwin, F.R.S.
St. Geo. Mivart, F.R.S., Sec.L.S.
Alfred R. Wallace
A. Günther, F.R.S.
   Keeper of the Department of Zoology,
   British Museum
Philip Lutley Sclater, M.A., Th.Dr., F.R.S.
   Secretary to the Zoological Society of London
Alfred Newton, F.R.S., V.P.Z.S., Professor of
   Zoology in the University of Cambridge
H. B. Tristram, F.R.S.
Osbert Salvin, M.A., F.R.S., Editor of
   "The Ibis"
F. DuCane Godman, Secretary of the
   British Ornithologists' Union
Henry Seebohm
Edward R. Alston
R. Bowdler Sharpe, British Museum
H. E. Dresser
J. E. Harting, F.L.S. Editor of "The Zoologist"
H. H. Godwin-Austen, Lt. Colonel
W. H. Hudleston
E. W. H. Holdsworth
J. H. Gurney, President of the Norwich Museum
H. J. Elwes
John Van Voorst
Wm. Borrer
John Cordeaux
W. B. Tegetmeier
Chas. W. Shepherd
C. Bygrave Wharton
Charles A. Wright, F.L.S.
L. Howard Irby
G. E. Shelley
Henry T. Wharton, Met. Oxon.
H. W. Feilden — late Nat. Arctic Ex. 1875-76
H. S. Marks, R. A.
A. H. Garrod, N.A., F.R.S., Prosector to the
   Zoological Society
W. R. Parker [?] F.R.S., F.Z.S., etc.
John Gould, F.R.S., etc.
Hy Stevenson, F.L.S., Hon. Secty. Norwich Museum
Howard Laundry [?] autograph affixed by request
   in letter of June 7, 1879

[The envelope is dated 14 May 1879 and:] "Received May 26, 79/Elliott Coues."

COWTAN, ROBERT. 1872. *Memories of the British Museum.* London, Bentley.

A more proper title would be 'Memories of the Department of Books of the British Museum'.

A humbly written account of an imposing institution by one of its underpaid but devoted Assistants. The almost monasterial life of a librarian of Victorian times is quite well portrayed in this interesting but somewhat tedious book.

There is no index. A John Cleave is mentioned, on page 13, as being an accountant of the Library, apparently around 1872. A John Cleave was named in John Gould's will of 1838 written before Gould left for Australia.

PART FOUR BIBLIOGRAPHY OF JOHN GOULD, HIS FAMILY AND ASSOCIATES

John Gould's works are referred to on page 183 in relationship to the hardship imposed upon authors of sumptuous works being required to deliver free copies to the British Museum Library and four other libraries.

CUNNINGHAM, GEORGE H. 1927. *London.* London, Dent.

This thick book is 'a comprehensive survey of the history, tradition, and historical associations of buildings and monuments arranged under streets in alphabetical order'. Needless to say, it is quite an interesting and valuable book.

On page 55, is the notation 'Bloomsbury Street, Bedford Square (Formerly Charlotte Street): 23, formerly 26 Charlotte Street; ... John Gould, ornithologist, died here in 1881'. The British Museum is in the Bedford Square area.

Nothing is noted in this book for Gould's earlier residence at 20 Broad Street, Golden Square. William Blake, the engraver and mystic, lived at 28 Broad Street until the early 1800s (p. 79). Golden Square itself was built up during the years from 1688 to 1700 (p. 264), but Cunningham states that now (1927) it has been almost entirely rebuilt.

DALL, WILLIAM HEALEY. 1915. *Spencer Fullerton Baird, a biography.* Philadelphia, Lippincott.

D

A prominent part of this excellent biography is the printing and collation of many letters between Baird and contemporary ornithologists and naturalists. Gould's correspondence with Baird is not mentioned, probably because it was so uninteresting — undoubtedly important and necessary — but uninteresting. Several Audubon and Cassin letters grace the book chosen undoubtedly because Audubon was Audubon, and from Cassin, well, because Cassin was so humorous, with other qualities.

DANCE, S. PETER. 1978. *The art of natural history.* London, Country Life Books.

A lavishly illustrated book with an erudite discourse on animal art and artists. The index is quite complete.

Part I discusses zoological art in a chronological way.

John Gould, because of his total output of bird volumes, 'did more to make men and women bird conscious' than Audubon. Joseph Wolf is thought of highly with 'originality and vision'.

In Part II Dance is even more the editorialist. He decries anthropomorphism and felt Audubon was a prime proponent of it.

DARWIN, CHARLES. 1859. *On the origin of species.* New York 1963, Heritage Press edition.

The first edition of this monumental work was published on 24 November 1859.

There are three references to Gould. On page 107 of this edition in the chapter on Laws of Variation 'Mr. Gould believes that birds of the same species are more brightly coloured under a clear atmosphere, than when living near the coast or on islands, and Wollaston is convinced that residence near the sea affects the colours of insects'.

Page 220 in the chapter on Instinct has the following:

In the case of the European cuckoo, the offspring of the foster-parents are commonly ejected from the nest within three days after the cuckoo is hatched; and as the latter at this age is in a most helpless condition, Mr. Gould was formerly inclined to believe that the act of ejection was performed by the foster-parents themselves. But he has now received a trustworthy account of a young cuckoo which was actually seen, whilst still blind and not able even to hold up its own head, in the act of ejecting its foster-brothers. One of these was replaced in the nest by the observer, and was again thrown out. With respect to the

means by which this strange and odious instinct was acquired, if it were of great importance for the young cuckoo, as is probably the case, to receive as much food as possible soon after birth, I can see no special difficulty in its having gradually acquired, during successive generations, the blind desire, the strength, and structure necessary for the work of ejection; for those young cuckoos which had such habits and structure best developed would be the most securely reared.

Finally on page 371 in the chapter on Geographical Distribution Darwin states:

The relation between the power and extent of migration in certain species, either at the present or at some former period, and the existence at remote points of the world of closely-allied species, is shown in another and more general way. Mr. Gould remarked to me long ago, that in those genera of birds which range over the world, many of the species have very wide ranges. I can hardly doubt that this rule is generally true, though difficult of proof.

DARWIN, CHARLES. 1887. *The descent of man, and selection in relation to sex.* New edition. New York, Appleton.

Darwin calls this volume the 'second edition' of his work which was originally published in 1871.

An index of the importance of an individual, albeit of dubious value, is a statement concerning the number of lines his or hers biography occupies in *Who's Who*. Along this line of reasoning, the number of lines devoted to references about Gould in the index of this book by Darwin is only exceeded by the number of lines for A. R. Wallace, E. Blyth, and Professor R. Owen. And in a somewhat cursory examination of this book, I found thirteen references to Gould in the book that are missing in the Index. *Trivia et trivia*! But the information does shed some light on Darwin's opinion of Gould's observations on birds and marsupials.

The references to Gould not appearing in the Index, are on pages 360, 388, 398, 402, 406, 453, 457, 476, 477, 479, 481, 485 and 492.

DARWIN, FRANCIS. 1893. *The life and letters of Charles Darwin.* 2 vols. New York, Appleton.

In these two volumes there are two references to Gould. On page 386 of volume 1, in a letter dated 11 January 1844 from Darwin to Sir J. D. Hooker, the botanist, is the following:

In discussion with Mr. Gould, I found that in most of the genera of birds which range over the whole or greater part of the world, the individual species have wider ranges, thus the Owl is mundane, and many of the species have very wide ranges . . . I do not suppose that the converse holds, viz. . . . that when a species has a wide range, its genus also ranges wide.

On page 533 is a list of works of Charles Darwin. Under 'Zoology of the voyage of H.M.S. "Beagle", part III, Birds, by John Gould' follows the editorial statement, 'An "Advertisement" (2pp.) states that in consequence of Mr. Gould's having left England for Australia, many descriptions were supplied by Mr. G. R. Gray of the British Museum. 4to. London, 1841'.

DAVIDSON, ANGUS. 1938. *Edward Lear.* London, Murray.

A comprehensive biography of Lear illustrated with several of his watercolours in black and white. Gould is mentioned on page 23 as employing Lear 'to make drawings for his *Indian Pheasants* [sic], *Toucans* and *Birds of Europe,* and they had travelled together to Rotterdam, Berne, Berlin and other places, in the course of their collaboration'.

[DICKENS, CHARLES.] 1851. *Household Words.* A weekly journal. No. 65, pp. 289-91, June 21. London.

An article entitled 'The tresses of the day star' is concerned with John Gould's humming-bird exhibit at the Zoological Gardens: 'The collection of Mr. Gould, . . . is comprised in twenty-four cases. His materials for a history of Humming Birds extend

to about three hundred and twenty species ... The most vivid colours of the painter's pallette cannot duplicate their ever-varying tints.'

DICKISON, D. J. 1938. 'A résumé of Gould's major works'. *Emu* 38: 118-31.

As the title states this article is a quite complete commentary on all of the books published by Gould. It contains information of dates of publication (not of the parts however), number of subscribers, prices as issued, and current (1938) availability and prices.

Dickison states that 3325 folio-sized coloured plates were published, but that several were used more than once.

DICKISON, D. J. 1938. 'Gould's illustrations'. *Vic. Nat.* 55: 94.

A short five-paragraph article relates aspects of Gould's publishing ventures. 'Gould, himself, sketched many of the plates' for *The Birds of Australia*. Dickison states that Gould's folio works contain over 3300 coloured plates, some being duplicated in another publication.

DIGGLES, SILVESTER. [1866-1870], *The ornithology of Australia*. Brisbane, Pugh.

This copy, Ellis Aves G10, is in the original nineteen parts. Parts twenty and twenty-one are missing.

From a 'Prospectus' of this new work on the back of part one is the statement that this work was produced because 'copies of Mr. Gould's ornithology are so difficult to procure'.

DIGGLES, SILVESTER. 1877. *Companion to Gould's Handbook; or Synopsis of the Birds of Australia*. Brisbane, Thorne and Greenwall.

As stated in the 'Original Prospectus' at the beginning of the volume, Diggles's purpose in publishing this work was to provide an inexpensive book with coloured plates of the birds of Australia since Gould's volumes 'are so difficult to procure, and their price is so far beyond the means of any but the most wealthy'.

Originally issued from 1866 to 1870 in 21 parts with six plates each, the 1877 edition is a re-arranged re-issue in two volumes, imperial quarto. A complete copy is a rarity. My copy is bound in one volume, the plates except for one are uncoloured and a few are missing.

There are numerous references from Gould's *Birds of Australia*, and also to personal notes on bird habits from Charles Coxen, Gould's brother-in-law, who was a friend of Diggles.

DIXON, JOAN M. (ed.). 1973. *Kangaroos John Gould.* New York, Doubleday.

This folio-sized book is a reproduction of the coloured plates and text, with minor changes, of Volume II of Gould's *Mammals of Australia* published from 1845 to 1860. In addition, Joan Dixon has added pertinent notes regarding present nomenclature and present known range of the various species.

Frith and Calaby (1969. *Kangaroos*. Melbourne, Cheshire) stated that 'Gould became the foremost authority on Australian birds and mammals, and most specimens collected by official exploring and surveying expeditions, and by others, were sent to him for study. At the end of the period of his active interest in the Australian fauna, few species of macropodids remained to be discovered'.

DIXON, JOAN M. (ed.). 1973. *Kangaroos. John Gould.* New York, Doubleday. Melbourne, Macmillan.

This is the second volume of Australian mammals published 'with modern commentaries' by Joan M. Dixon. This contemporary volume is a reprinting, with additions, of Gould's *Mammals of Australia,* volume I, 'first published in 1863'.

The reprinting is quite complete and includes the Dedication, List of Subscribers, General Index, Preface, Introduction, Text and sixty coloured plates. Ms. Dixon includes some introductory notes, the current accepted scientific name, and indexes.

DIXON, JOAN M. (ed.). 1976. *Placental mammals of Australia* [by] *John Gould.* South Melbourne, Macmillan.

The Gould plates and text appearing in this edition were first published in 1863 as volume III of the *Mammals of Australia.* The format is similar to the first two volumes of her set.

DIXON, JOAN M. (ed.). 1977. *Gould's mammals.* London, David and Charles.

This edition contains a selection of sixty colour plates taken from Macmillan's (the Australian publisher) three-volume reproduction of Gould's *Mammals of Australia.* Joan M. Dixon's updated commentaries also accompany these plates.

DRUMMOND, JAMES. 1844. 'Letter to Sir W. J. Hooker relating to Swan River botany'. *London Journal of Botany* 3: 300-14.

Extracts are made from three letters written by James Drummond to Hooker dated September 1842, October 1842 and January 1843.

The first letter written from Hawthornden Farm, concerns a trip taken with John Gilbert some eighty kilometres north to Lakes Dalarn and Maradine, so called by the natives. Drummond collected plants and Gilbert some birds, mammals and shells.

The October 1842 letter was also written from Hawthornden Farm, Toodjay Valley and gives an account of a trip some eighty kilometres to the east to the Wangan Hills. Gilbert found the nests and eggs of 'an extraordinary *Gallinaceous bird* which breeds there' which deposits its eggs in large mounds (Mallee-Fowl).

The January 1843 letter is shorter and tells about another trip with Gilbert, this time to the south of Perth to Mount William.

E   EISELEY, LOREN C. 1959. 'Charles Darwin, Edward Blyth, and the theory of natural selection'. *Proc. Amer. Phil. Soc.* 103: 94-158.

A most erudite and interesting paper showing how with intelligent deductions and sleuthing Eiseley attempts to prove 'that Darwin made unacknowledged use of Blyth's work' which was published in *The Magazine of Natural History* beginning in 1835.

Regarding Gould, on page 157 of this article, there is the statement that Gould referred to Blyth as 'one of the first zoologists of his time, and the founder of the study of that science in India'. Gould made use of Blyth's extensive bird collections for his *Birds of Asia.*

EKAMA, C. 1885-88. *Catalogue de la bibliothèque de la fondation Teyler.* Vol. 1. Harlem, Heritiers Loosjes.

In the section on birds is a listing of Gould's works with rather extensive and glowing commentary on the major works.

ELLIOT, DANIEL GIRAUD. 1878-79. 'A classification and synopsis of the Trochilidae'. *Smithsonian Contributions to Knowledge,* No. 317. Washington, Smithsonian Institution.

An important reference giving the characteristics for the genera of humming-birds that Elliot acknowledged. Gould and Elliot were working with a difficult taxonomical subject.

ELLIOT, DANIEL GIRAUD. 1879. 'List of described species of humming birds'. Smithsonian miscellaneous collections 334. Washington, Smithsonian Institution.

A list of 120 genera and 427 species of humming-birds, referrable to Elliot's larger 1878 work 'A classification and Synopsis of the Trochilidae'. Gould was the author of the names of 37 of the genera and of 102 of the species listed.

ELLIS, RALPH. 1948 and 1949. In the Supreme Court of the State of Kansas. In the matter of the estate of Ralph Ellis, deceased.

The following matters of record of the Appeal of this case from the District Court of Douglas County, Kansas are published in seven pamphlets.

Volumes 1, 2 and 3 are the 'Joint abstract of the appellant and appellees for use on the various appeals involved herein'. These are contained in 304 pages with an Index in volume 3. They are dated 22 December 1948. The remaining four pamphlets consist of the following:

'Brief of the Board of Regents of the State of Kansas, University of Kansas and the State of Kansas', no date; 'Opening Brief of Irene S. Ellis, respondent, appellant and appellee', dated 4 April 1949; 'Reply brief of Irene S. Ellis, respondent, appellant and appellee' dated 1949; 'Petition of Irene S. Ellis, respondent, for a rehearing', dated 26 October 1949.

These court records provide the fascinating details of the life of Ralph Ellis, his parents, their financial status, his accumulation of an extensive ornithological library with the largest collection extant of original drawings for John Gould's publications, Ellis's two wives, his medical and emotional problems, his home in Berkeley, California, the movement of his library to the University of Kansas around March 1945, and his death on 17 December 1945.

Ralph and his wife had signed an agreement to leave his library to the University of Kansas, but upon his death legal proceedings were instituted by his wife, primarily based upon her statement that through fear for what her husband might do, she had signed the agreement. Simply stated, the Kansas Supreme Court decided in the favour of the University of Kansas, and the Ellis collection is now in the Spencer Research Library as its main holding.

ENGELMANN, WILHELM. 1846. *Bibliotheca historico-naturalis... 1700-1846*. Leipzig, W. Engelmann.

The volumes published by Gould to date (1846) are listed, with no commentary. Prices of the sets are given.

This work was continued as 'Bibliotheca zoologica... 1846-1860' by J. V. Carus and Engelmann published in 1861, and as 'Bibliotheca zoologica II... 1861-1880' by O. Taschenberg published in 1886.

ENWRIGHT, W. J. 1939. Gould Commemoration *Emu*. *Emu* 38: 426-7.

Here are four short notes in reference to articles in the Gould Commemorative issue of Emu (1838). Enwright describes the town of Maitland in more depth; he comments on James Backhouse [1843] having crossed the Hunter River and how he described it more completely than Mrs Gould; he writes concerning a W. Allen of Wingham who knew Gould, and a John Hopson.

EVERETT, MICHAEL. 1978. *The birds of paradise.* New York, Putnam's.

A beautiful book with fifty-eight full-page colour plates. Most are taken from R. B. Sharpe's *Monograph of the Paradiseidae and Ptilonorhychidae,* which contained plates by Gould, Hart and Keulemans. Additional colour plates for this book by Everett are by Peter Hayman.

A full-page portrait (by Ransome-Maguire) of Gould graces the beginning of this volume.

EYRE, EDWARD JOHN. 1845. *Journals of expeditions of discovery into central Australia... in the years 1840-1...* 2 vols. London, T. and W. Boone. (Australiana Facsimile Editions No. 7, Adelaide, Libraries Board of South Australia, 1964)

This is a valuable account of Aboriginal Australia with fine illustrations. The dates of 1840 to 1841 cover the immediate period after Gould's Australian venture.

A 'List of Birds/known to inhabit southern Australia,/by John Gould, Esq. F.R.S.' is in the Appendix from pages 440 to 448 of volume 1.

EYTON, THOMAS CAMPBELL. 1867. *Osteologia avium; or, a sketch of the osteology of birds.* 2 vols. Wellington, R. Hobson.

The title-page of this volume by Gould's colleague states that this work is 'To be had of Mr. Prince, at Mr. J. Gould's, Charlotte Street/Bedford Square, London/' (Zimmer 1926).

This work (and the next one) by Eyton are the only ones with the reference printed on the title-page of the fact that Gould and Prince were acting as agents or brokers for publications by others. In Gould's correspondence, however, are many references to this fact. Gould wrote about, attempted to sell, and, if successful, sent billings to individuals and museums for some of the works of Bonaparte, Wolf, Jardine, the Zoological Society of London, and as shown here, Eyton.

EYTON, THOMAS CAMPBELL. 1869. *Supplement to osteologia avium...* Wellington, R. Hobson.

On the title is 'To be had of Mr. Prince, at Mr. J. Gould's, Charlotte Street,/Bedford Square, London/'. See my comment on the preceding work by Eyton.

[EYTON, THOMAS CAMPBELL.] Review. 1838. 'A monograph of the Anatidae or Duck Tribe by Thomas Campbell Eyton. 1838.' *Ann. Nat. Hist.* 1: 473-8.

Of the illustrations of birds in the second part of this work 'six... are beautifully figured by Mr. Gould'.

F FERGUSON, JOHN ALEXANDER. 1941-69. *Bibliography of Australia.* 7 vols. Sydney, Angus & Robertson.

One of the monumental bibliographic compilations. All of Gould's works on Australia, plus three Prospectus are listed and collated:

Volume 2
  Item 1. 1836, pp. 258-9. Prospectus for 'A Synopsis of the
    Birds of Australia, and adjacent islands'. 2 pages, no date. 2127.
  Item 2. 1837, p. 317. A Synopsis of the Birds of Australia... 2271.
    Part II was not issued in January 1837, but in April. A Gould bibliography is appended.
  Item 3. 1837, pp. 317-18. The Birds of Australia... Suppressed parts. 2272. The size is imperial folio, not elephant folio.

PART FOUR BIBLIOGRAPHY OF JOHN GOULD, HIS FAMILY AND ASSOCIATES

Volume 3
   Item 4. 1840, p. 105. Prospectus for 'The Birds of Australia'. 4 pages, March 6, 1840. 2990.
   Page 2 has the names of Subscribers at Home and in the Australasian Colonies. Issued in Sydney.
   Item 5, 1841, pp. 184-5. Monograph of the Macropodidae... 3197.
   Item 6. 1841, p. 185. Prospectus for 'The Birds of Australia'. 4 pages, 3198. Issued in London.

Volume 4
   Item 7. 1848, p. 189. An Introduction to the Birds of Australia. 4772.
   Item 8. 1848, pp. 190-1. The Birds of Australia, 7 vols. 4773.
      Mathew's (1920-25) bibliographic data is reprinted. A long paragraph is included on prices of sets from 1922 to 1953.

Volume 5
   Item 9. Pp. 1100-1101. An Introduction to the Mammals of Australia. 10030.
   Item 10. P. 1101. Handbook to the Birds of Australia. 10031.
   Item 11. The Mammals of Australia, 3 vols. 10032.
   Item 12. The Birds of Australia, Supplement. 10032a.
   Item 13. The Birds of New Guinea... 5 vols. 10033.

FISHER, JAMES. 1942. *The Birds of Britain.* London, Collins.

This forty-nine page book contains a good concise review of British ornithology and is illustrated with twelve coloured plates and twenty-six black and white. Included are three coloured plates from Gould's works (Kestrel, Goldfinches, Lapwing and young) and six black-and-white Gould illustrations. On page 40 Fisher refers to 'Gould's superb plates' and states 'Since the days of Gould and Wolf there has been only one really important bird artist in Britain, Archibald Thorburn; Thorburn was just about as good as Gould...'
   The bird section in Turner's 1946 *Nature in Britain* contains Fisher's identical text but there are fewer Gould illustrations.

FISHER, JAMES. 1966. *The Shell bird book.* London, Ebury Press and Michael Joseph.

This is a fascinating and learned book with much information on the history of British ornithology. There are five drawings from Gould's works by Wolf, Lear, Richter, and Hart on pages 44, 49, 91, 97 and 107. Gould is mentioned on pages 68 and 69 with a section devoted to the 'impresario' on pages 214 to 216. Lear is portrayed on page 195 and Wolf on page 215.

FITTER, R. S. R. [No date] *British birds in colour.* London, Odhams.

This book contains seven articles on bird lore, illustrated with many black-and-white photographs and 108 colour plates. These colour plates are all from Gould's *Birds of Great Britain.* Some of the plates are full page in size but in some instances two or four plates are on a page. The colour reproduction is quite acceptable.
   In the Introduction, on pages 6 and 7, is a listing of the Gould plates and a short biographical note on Gould.

FITZPATRICK, KATHLEEN. 1949. *Sir John Franklin in Tasmania 1837-1843.* Melbourne, Melbourne University Press.

Several letters in the Royal Society of Tasmania, Hobart, and at the Scott Polar Research Institute, Cambridge (Lefroy Bequest) are quoted which contain references to Gould.

The references to Gould are as follows: On page 40 'some people thought Gould, the great ornithologist, a vain man, but that for her [Lady Franklin's] part she could not see "why a bird fancier should come all the way to the Antipodes in pursuit of his peculiar fame and not think the better of himself for it" '.

On page 192 is a comment about Gould's *Birds of Australia* being in Sir John Franklin's library.

Pages 200-2 relate to Gould and especially the trip he took with Lady Franklin to the Recherche Bay area. Of the five letters referred to, four of these letters were published by Mackaness, 1947, and by Chisholm, 1944, *Elizabeth Gould*. The fifth letter is in the Lefroy Bequest at the Scott Polar Institute.

FITZPATRICK, KATHLEEN. 1958. *Australian explorers*. London and New York, Oxford University Press.

Quotes from journals or correspondence of the early explorers of Australia beginning with Blaxland, Lawson and Wentworth in 1813, to the Giles expedition of 1874-76. Leichhardt's exploration is included.

FLEXNER, MARION W. 1947. 'John Gould and company'. *Antiques Magazine*. 51: 386-9.

This is a detailed interesting article with information taken from Sharpe's *Analytical Index,* the 1938 *Emu* Commemorative Issue, and Sotheran's 'Piccadilly Notes No. 9'. There are eight black-and-white illustrations of Gould's bird lithographs, and the cover of the magazine has an additional print illustrated.

FORD, ALICE. 1964. *John James Audubon.* Norman, Oklahoma, University of Oklahoma Press.

This detailed book on the life of Audubon contains seven references to Gould. On page 246 is a note that Gould had commissioned Henry Ward to collect specimens for him in America. On page 256 is a further note on Ward squandering the money paid him in advance by Gould on marriage.

On page 292 is a reference from a letter from Robert Havell, Jr. to Audubon on 30 November 1831 saying that 'Mr. Gould has been very successful in his [book] of Birds [of the Himalaya Mountains], which has caused him to think not a little of himself'.

On page 308 is a mention of Victor Audubon, in 1833 apparently, canvassing France, Holland, and western Germany 'to head off a similar subscription tour by the hugely successful bird painter John Gould'.

On page 335 is a note of an 1836 visit to Audubon when he and his family lived on Wimpole Street, London, by 'Gould, rich and renowned' with his wife 'a skillful lithographer of his birds'.

On page 347 is mention that in 1837 Gould's and Yarrell's ornithological productions 'were the topic of the hour', with the intimation that Audubon considered these to be artificial works.

On page 355 is a paragraph about 'Gould's preparations for a family expedition to Australia'. Audubon saw this as an absurd move 'by one who had never so much as learned to "salt steaks with gunpowder" '. Ford further states that Gould's rumoured annual income was $10 000 but that Gould still was unable to continue his subscription for Audubon's Folio. 'The family had scant admiration for the art of this man whom

Victor, for one, rated a scamp for having aped "The Birds of America" '.
More complete quotations from Audubon's letters are in Corning (1930).

FORSHAW, J. M., FULLAGAR, P. J. & HARRIS, J. I. 1976. 'Specimens of the Night Parrot in museums throughout the world'. *Emu* 76: 120-6.
This article lists the twenty-two specimens of Night Parrot in Museums. Gould described the type specimen in 1861 which is named *Geopsittacus occidentalis,* Gould.

The history of a specimen in the Merseyside County Museum, formerly Liverpool or Derby Museum is detailed. Apparently it was collected by John Stuart, and sent by Charles Sturt to Gould in 1847. Gould incorrectly identified it as the Ground Parrot and sent it on to Lord Derby.

FRASER, LOUIS. 1846-48. *Zoologia typica, or figures of new and rare mammals and birds described in the Proceedings, or exhibited in the collections of the Zoological Society of London.* London, published by the author.
The date on the title-page is 1849. Gould is listed as a Subscriber to this work. Fraser's intent was to have a continuing publication that would contain drawings of birds and mammals not previously figured in the *Proceedings of the Zoological Society.* The plan was not continued beyond 1848.

The birds depicted on plates 40, 64, and 65 illustrate three papers by Gould published in the Proceedings.

FREEMAN, R. B. 1965. *The works of Charles Darwin.* London, Dawson.
There is a detailed description of the dates of publication of Part III, *Birds,* of Darwin's *The Zoology of the voyage of H.M.S. Beagle.* Gould was the author of this part, assisted by G. R. Gray. Part III was issued in five numbers. Number I appeared in July 1838, and Number V in March 1841. There was a total of fifty plates of birds in the five numbers.

FREEMAN, R. B. 1977. *The works of Charles Darwin.* 2nd ed. London, Dawson.
The unbound and bound issues of Darwin's *The zoology of the voyage of H.M.S. Beagle...* are collated as in the first edition of Freeman's book.

Items 1643 (not 1603 as in the index) and 1644 refer to comments by Darwin in the *Proc. Zool. Soc. London* 1837 Part V following descriptions and presentations of bird specimens by Gould.

FRIES, WALDEMAR H. 1973. *The double elephant folio.* Chicago, American Library Association.
This is a truly monumental work. Fries has attempted, and admirably succeeded in most instances, to trace the Audubon folio sets as published to their present repository.

Obviously there is also much information on the process of the painting, printing, colouring, and distribution of these large bird engravings. For instance, on page 79, the drawings of waterbirds for volume three 'had become a family project at the Bachman home. Audubon wrote to Victor (his son) that "John (Bachman), myself and Miss Maria (sister of John Bachman's wife) are drawing constantly, to finish all the unfinished Drawings now on hand".... Maria Martin's work for Audubon was to continue after the naturalists had left'.

Audubon was in England and Scotland for varying periods from 1826 to 1839 and there were many contacts among Audubon and his sons with the naturalists of England and Europe. Some of these encounters and Audubon's account of these encounters are valuable for an understanding of this exciting time.

Gould is mentioned on page 88 from an Audubon letter of 25 August 1834, as 'publishing the Birds of England, of Europe etc. etc. etc. — in all sorts of ways —'. Audubon goes on in a critical manner to deride the publications of Selby, Temminck, Jardine ('an enormous trash all compilation'), and Swainson. Yarrell he praised for a 'beautiful work'.

According to Fries, in the Spring of 1835 Audubon agreed to exchange a copy of his work 'with Mr. Gould, for his publications — and have by me, 13 numbers of his Birds of Europe — his Century of Hymealan [sic] Birds — and his monograph of Trogans — I have also purchased a monograph of Parrots from a Mr. Lear'.

These Gould references are in letters written by Audubon now housed in the Houghton Library, Harvard (Collection bMS Am 1482) dated 25 August 1834 (vol. 2, p. 29) and 20 April 1835 (vol. 2, p. 69).

Gould is listed as a subscriber in vol. 4 of the 'Ornithological Biography' published in November 1838 (p. 444 of Fries book).

More complete quotations from Audubon's letters are in Corning (1930).

FROGGATT, WALTER W. 1921. *Some useful Australian birds.* Sydney, Gullick.

Sixty-one coloured plates copied from Gould's *Birds of Australia* illustrate this book. These same plates appeared earlier in a series of papers in the *Agricultural Gazette* of New South Wales issued from 1896 to 1905.

Gould's works are referred to on pages 3, 7, 14, 15 and 17 of the Introduction, and throughout the text pages.

G [GALLATIN, FREDERIC, JR.]. 1908. *Catalogue of a collection of books on ornithology in the library of Frederic Gallatin, Jr.* New York, privately printed.

'The books in this library comprise the complete or most important ornithological writings of... Gould' and thirty-six other authors who are listed alphabetically. No important commentary is given on the works of Gould listed.

[GILBERT, JOHN]. [No date] 'List of Swan River Birds'. Collection: Balfour and Newton Library, Department of Zoology, University of Cambridge.

This list was not examined carefully by me when I visited Cambridge in 1974. So Mr Ron Hughes, librarian, kindly sent me a photocopy. He states that 'it consists of one large sheet folded to make four pages'.

Under the heading is a long alphabetical list of the scientific names of birds arranged roughly according to species and groups, written in a rather careful hand. Many corrections and additions are noted in the listing.

There are also three other shorter lists. The headings of these lists are 'Birds seen at the Swan but not obtained' with the number '14' at the end of the list, 'List of Skeletons Prepared' with the notation at the end of the list of '73 specimens', and a final list of 'Aboriginal Terms at the Swan[?]'.

There is no question but that this 'List of Swan River Birds' was written by John Gilbert. The handwriting here is identical with that in his letters to Gould in the Mitchell Library of the State Library of New South Wales in Sydney.

A long notation by Gould is also present on page 3.

GILBERT, JOHN. Manuscript. 1842-1845. [John Gilbert's journal.] Collection: Mitchell Library, State Library of New South Wales, Sydney.

This is a two-volume set. Volume 1 is oblong, 6 x 3 inches (152 x 76mm), with a leather cover with inner pockets and a brass clip. Volume 2 is oblong, 7 x 4½ inches

(178 × 114mm) with no cover. The writing in the first half of volume 1 is almost illegible. The part headed 'Port Essington Expedition' begins in the middle of the first volume and fills volume 2.

There is a 226-page typescript of the section on the Port Essington expedition (ML-A.2587).

Gilbert's important journal should be published.

GILBERT, JOHN. Manuscript. [Not dated.] [Four volumes of Gilbert items.] Collection: Queensland Museum, Fortitude Valley, Queensland.

In response to a request for information on John Gilbert items, Dr D. J. Robinson, Senior Curator, sent me the following information on the four volumes in their collection.

This material should be compared with the Gilbert items in the National Library of Australia at Canberra (MS587), and the manuscript material in the Newton Library, Cambridge University.

Whittell (1951 *Emu* 51: 17-29) reports on two of these three items, namely on the two-volume 'Ornithological Notes–Gilbert', and on the altered volume of Jardine's 'Marsupials'.

Gilbert, John

2 volumes "Ornithological Notes – Gilbert". Believed to be based on pages from Gould's "Synopsis of the Birds of Australia and the Adjacent Islands"; interleaved with notes by Gilbert.

Pagination is by Gilbert; introductory distribution list of 623 birds, compiled by Gilbert, used as an index by insertion of page numbers.

1st vol. - pages 268

2nd vol. - pages 269-693.

Gilbert, John

A single volume bearing a handwritten endorsement "The Naturalist's Library. Jardine's. Vol. XXIV by G. R. Waterhouse", has been interleaved by Gilbert for the insertion of his own notes and comments.

The printed text appears on pages which are numbered 117-309 (inclusive). Spine title of volume appears to be "Naturalist's Library. Mammalia. Vol. XI. Marsupials".

Gilbert, John

A volume which has title written on spine. "J. Gould's Ornithological Notes - Instructions to Collectors". (illustrated).

Apparently, this volume came to the Queensland Museum through the hands of Mr. H. C. Coxen, a great nephew of Mrs. John Gould. It contains an annotation by A. H. Chisholm, dated 7th October, 1940, which apportions authorship of various sections of work.

Of the contents, Gilbert, according to Chisholm, was responsible for:-

Distribution list of birds of Australia (17pp.)

Linnaean index (24pp.)

Distribution list for 142 mammals.

Text accompanying illustrations of birds (28pp.).

Illustrations cut from some other publication and pasted in.

Index, which does not appear related to rest of work, attached in the front section of book

According to Chisholm's note, Gould's secretary, E. C. Prince, was responsible for 8 pages of instructions which follow immediately after Gilbert's 17-page distribution list.

[GILBERT, JOHN.] Anonymous. 1945. 'John Gilbert: centenary of his death'. *Australian Museum Magazine.* (December) pp. 403-4.

A service honouring the memory of John Gilbert on the centenary of his death was held at St. James' Church, Sydney on 28 June 1945. A mural tablet to John Gilbert is in this church. Among those present were the Reverend E. J. Davidson, rector, K. A. Hindwood, President of the Royal Australasian Ornithologists' Union and C. Price Conigrave, of the Royal Australian Historical Society who gave an address.

GILL, THEODORE. 1900. 'Esthetic birds: the bower birds of Australia and New Guinea'. *Osprey*, 4: 67-71.

Gould's statements on the Bower Birds that appeared in his *Handbook* are repeated here.

GILL, THEODORE. 1902. 'Biographical notice of John Cassin'. *Osprey* 1: 50-3.

Cassin, in Philadelphia, had access to the extensive ornithological collection fostered by his friend Dr Thomas B. Wilson and his brother, Mr Edward Wilson. The Wilsons were responsible for the gift of Gould's types of his Australian collection to the Philadelphia Academy of Natural Sciences.

Cassin had a most substantial knowledge of birds of the world. He was born in 1813 and died in 1869.

GILLIARD, E. THOMAS. 1969. *Birds of paradise and bower birds.* Garden City, New York, Natural History Press.

A comprehensive book on two fascinating groups of birds. Gilliard's evolutionary hypotheses are thought-provoking.

The many black-and-white illustrations are taken from Sharpe's 1891-98 monograph of the birds of paradise and bower birds (utilizing Gould's and Hart's drawings) and from Elliott's similar monograph of 1873 illustrated by Wolf and Smit. There are many references to Gould and Sharpe.

GOULD, C. 1861. 'Results of the geological survey of Tasmania'. Report, British Assoc. for the Advancement of Science, *Transactions*. London, Murray, pp. 112-13.

This is obviously by Charles Gould because he did participate in a geological survey of Tasmania around this date.

GOULD, CHARLES. 1886. *Mythical Monsters.* London, W. H. Allen & Company.

An exceedingly scholarly work wherein John Gould's son proposes 'that many of the so-called mythical animals, which throughout long ages and in all nations have been the fertile subjects of fiction and fable, come legitimately within the scope of plain matter-of-fact Natural History, and that they may be considered, not as the outcome of exuberant fancy, but as creatures which really once existed'.

Charles Gould, the surveyor, had become a student of language, mythology and natural history. His erudition was anticipatory. He supported his theories with many references indicating a wide breadth of reading and knowledge.

Chapters include 'on the extinction of species', 'antiquity of man', 'the deluge not a myth', 'the dragon', 'the sea-serpent', 'the unicorn' and 'the Chinese phoenix'.

John Gould is mentioned only cursorily on pages 43, 66 and 77.

[GOULD, CHARLES.] Obituary. 1899. *Geol. Soc. Quart. Jour.* 55: lxii. London.

'Charles Gould was a son of Dr. [sic] Gould, the well-known naturalist. He was educated at the Royal School of Mines in 1854-56, becoming Associate in Mining, Metallurgy, and Geology, and winning the Duke of Cornwall's Scholarship in 1854, the Royal Scholarship and the Edward Forbes Medal in 1856.

'He joined the Geological Survey of England in 1857, about three months after I [William Whitaker, President] had done so, and worked in the Weald, his notes being incorporated in Topley's memoir on that tract. A year or two later he left the English Geological Survey to join that of Tasmania, on which country he published several Reports and papers, his work there ending about 1874 or 1875.

Plate 33 Drawing in watercolour, attributed to Mrs E. Gould, species not labelled [Bohemian Waxwing?], unpublished, Kansas-Ellis-Gould MS 1033.

Plate 34 Drawing in ink and watercolour, attributed to Mrs E. Gould, of *Coccothraustes icterioides*, for *Century of Birds*, plate 45, Kansas-Ellis-Gould MS 995.

PART FOUR BIBLIOGRAPHY OF JOHN GOULD, HIS FAMILY AND ASSOCIATES

'Gould was elected a Fellow of this Society in 1859, and gave us two papers on Fossil Crustacea. He died at Montevideo (Uruguay) on April 15th, 1895, but no report of his death reached the officers of this Society until lately'.

GOULD, ELIZA. Manuscript. 1885. *See* MOON, ELIZA GOULD MUSKETT. Manuscript.

GOULD, ELIZABETH. Manuscript. 1839. 'Mrs. Elizabeth Gould/Diary/20 Aug. – 29 Sept. 1839'. Collection: Mitchell Library, Sydney, item A1763.

Mrs Gould's diary is bound with the 1840-1846 'Letterbook of J. Gould'. The diary is ten pages long.

On the first page is an annotation 'Sotheran 14-8-34', undoubtedly indicating it was purchased from Henry Sotheran, Ltd, London in August 1934. This diary is listed in the [1934] 'Piccadilly Notes' No. 9 Catalogue issued by Henry Sotheran as item 1438 on page 97 as follows: 'Gould, (Mrs.) Part of her MS. Diary while in Australia, from Aug. 30th [sic] to Sep. 26th [sic], 1839, giving some impressions of Sydney, Newcastle, Maitland, local scenery, bird life, etc., 10 pp., 4 to., £10.10s.' The dates mentioned are in error.

The diary in its entirety is transcribed by Hindwood (1938 *Emu* 38: 134-7) with very helpful annotations. In going over my copy kindly supplied by the Mitchell Library, I would make only minor changes except for one item. On page 136 where Hindwood has note 17, the '22' he inserts in the sentence obviously refers to the date '22' for the next paragraph. Thus '22 Left Maitland about...' supplies a questionable date.

Enright (*Emu* 38: 426) also comments on this diary.

[GOULD, ELIZABETH.] Death Certificate. 1841. General Register House, Somerset House, London.

A copy of this death certificate for Mrs Gould was sent to me by Mrs Tess Kloot, Archivist, Royal Australasian Ornithologists' Union, Melbourne.

Here are the pertinent facts on this certificate: 'Application Number P.A.S./061135/7C/F... Registration District Windsor... 1841 Death in the Subdistrict of Egham in the Counties of Berks and Surrey... No. 490... Fifteen of August 1841 Egham Surrey, Elizabeth Gould, female, 37 Years, Wife of John Gould Gentleman, cause of death Puerperel [sic] Fever, John Gould present at the death, Egham, Surrey, registered Seventeenth of August 1841, Charles Brown, Registrar... DA 560892'.

GOULD, FRANKLIN. Manuscript. 1866-73. 'Correspondence and journals'. Collection: Duke of Westminster.

Mrs Grace Edelsten, widow of Dr Alan Edelsten, one of John Gould's great-grandsons, presented these manuscripts to the Duke of Westminster. The items were copied for me by the Eaton Estate office, through Mr T. H. Carter. They were sent to me divided into three sections.

Section one is a preliminary compilation by Dr Franklin Gould, son of John Gould, of material for a journal of his trip to India as the Physician for Lord Belgrave, later Lord Grosvenor, son of Lord and Lady Westminster.

The first page is dated November 1872 and Franklin presents an introductory paragraph prior to the onset of their trip on November 9, from Dover. 'We are going to India, China and America and hope to be back by the end of next August.' This preliminary journal covers their expedition through to 20 February 1873 when they

are in Benares, India. Franklin writes interestingly about the people, the towns, the customs and especially about the vegetation and wildlife.

Section two is Dr Gould's more finished journal covering a shorter period from 9 November 1872 when preparing to leave England, to 14 December 1872, when they were near Colombo on the island of Ceylon.

Section three consists of fifty-one letters to and from Dr Franklin Gould or his family. The greatest number were written by Franklin to his father. The earliest dated letter is 9 November 1866 from Walmer Castle where Franklin was with Lord and Lady Granville.

Beginning with a letter dated 28 December 1867 from Mentone [France], Franklin becomes associated with the Lord Grosvenor, later Westminster, family as the physician for their son, Lord Belgrave, who has 'tiresome fainting fits', but only very rarely. Franklin accompanies him to Eton College. Later letters are dated from Reay Forest, Sutherland and from Eaton, Chester.

On 15 November 1872 Franklin writes his first letter to his father en route to India, from Bologna, Italy. This excursion apparently was mainly for hunting in India and consisted of Belgrave, now Lord Grosvenor, Franklin, and two or three others.

Franklin was apprehensive concerning the health of his ward, Lord Grosvenor. His fears were justified, but not only for Grosvenor. In India 'Disentry' seriously affected Grosvenor and required many 'night watches' on the part of Franklin to nurse him back to health. In the process, Franklin wrote he had an 'inflammatory sore throat'. Franklin then became the patient, and en route home from a shortened journey, Franklin developed high fever and died on ship in the Red Sea on 19 March 1873.

Item 40 is the original and a copy of an account of Dr Franklin Gould's illness written by a physician on board the P. & O. S.S. *Behar*, Dr John Henry Sylvester, en route from Bombay to Suez. It was written on 20 March 1873, the day after Franklin's death.

Item 41 is an original and a copy of an account by Wm. Ford Creeney, M.A., who administered Holy Communion to Dr Gould and attended further to him on board the S.S. *Behar*.

The remaining letters are primarily from Lord and Lady Westminster to John Gould and the family expressing their sympathy and concern on Franklin's death.

This interesting, revealing and poignant series of journals and letters should eventually be reproduced.

GOULD, JOHN.

The bibliographical material on John Gould follows. For ease of reference, the items appear under these headings:

Announcement (of his works)
Anonymous (notes on Gould)
Biography
Correspondence (and statements of account)
Lithographic stones
Manuscript (with notebooks and his will)
Newspaper items
Obituary
Original drawing(s)
Portrait
Prospectus (for Gould's works) with List of Subscribers

Reviews (of Gould's works)
Specimens
Stamp (postage, published in Australia)
Miscellaneous items including facsimile sets.

[GOULD, JOHN.] Announcement. 1832. [Concerning *The Birds of Europe*.] *Mag. Nat. Hist.* 5: 190-1.

This announcement consists of five paragraphs from a prospectus or 'sketch of his means and terms'.

This issue of the magazine was dated March 1832.

[GOULD, JOHN.] Announcement. 1833. [Concerning *The Birds of Europe*.] *Mag. Nat. Hist.* 6: 509.

'Of Gould's "Birds of Europe" the sixth part is just published.' This notice appeared in the November 1833 issue.

[GOULD, JOHN.] Announcement. 1838. [Note on Gould's proposed trip to Australia.] *Mag. Nat. Hist.* 2 (new series): 107.

Sir William Jardine undoubtedly wrote this note. It is quoted here in toto:

Mr. Gould, the eminent ornithologist, has determined on leaving England in the spring for Australia. He purposes spending 2 years there; his object being to render the splendid work which he has commenced, on the birds of that portion of the globe, a contribution to science which shall furnish naturalists, not merely with faithful delineations of the numerous and interesting forms of that country, but also contain a history of the habits and general economy of the subjects represented. That no interruption to the progress of the work may occur, all the materials necessary for continuing its publication will be carried out with him. The outlines of the subjects will be made with his own hand, and the lithography, as in all his previous illustrations, will be executed by Mrs. Gould, who will accompany him in his travels.

GOULD, JOHN. Announcement. 1840. 'Notice to subscribers'.

A single sheet in the University of Kansas Spencer Library–Ellis Collection.

Ralph Ellis wrote this notation on the sheet: 'From Pt. I, John Gould/Birds of Australia, 1840.'

'The paper covers heretofore employed for the present Work not having been found a sufficient defence to the Plates, the Author has determined upon issuing it in boarded Parts, containing seventeen Plates instead of ten, and at intervals of three instead of two months...' There was to be a 'slight difference in the price per Plate' also.

[GOULD, JOHN.] Anonymous. 1881. [Note on the collection of birds of the late John Gould.] *Nature* (March 24) 23: 491.

The note in *Nature* is as follows:

Writing to the Times on Friday last, Mr. Sclater calls attention to the fact that the collection of birds of the late John Gould, the ornithologist, have been offered to the trustees of the British Museum for 3000 pounds, and expressed the hope that there will be no difficulty on the part of the Treasury in sanctioning about 1500 mounted and 3800 unmounted specimens of hummingbirds, being the types from which the descriptions and figures in the celebrated "Monograph of the Trochilidae" were taken. There are besides, 7000 other skins of various groups, amongst which are splendid series of the families of Toucans, Trogons, Birds of Paradise, and Pittas.

[GOULD, JOHN.] Anonymous. 1881. [Note on Gould.] Printing Times and Lithographer. 15 April 1881, p. 81.

Salvadori (1881) gives this note as a reference for his statement that 'After Gould's death the copyright of those works not previously sold was acquired by Messrs. H. Sotheran & Co. for the sum of about 125,000 lire'.

[GOULD, JOHN.] Biography. 1885-1900. *Dictionary of National Biography,* Vol. 8, pp. 287-8. Oxford University Press.

JOHN GOULD — THE BIRD MAN

A very complete biography which contains the statement that 'without any advantages of education and position Gould achieved a great success by perseverance and love of his subject. He united in himself the qualities of a good naturalist, artist, and man of business'. This was written by G. T. Bettany.

[GOULD, JOHN.] Biography. 1949. *Dictionary of Australian biography.* Sydney, Angus & Robertson, vol. 1, pp. 358-9.

The dates of '1818 to 1824' are given for Gould's employment as a gardener under his father in the Royal Gardens at Windsor, but no reference for these dates is listed.

'In 1909 the Gould League of Bird Lovers was founded in Australia. Thirty years later it had a membership of 250,000, largely school children.'

The rest of the biography is repetitious from Sharpe's *Analytical Index*.

[GOULD, JOHN.] Biography. 1972. *Dictionary of scientific biography.* C. C. Gillispie, Editor in Chief. New York, Scribner's.

Diana M. Simpkins has written a short history and bibliography of Gould.

GOULD, JOHN. Correspondence.

Under this heading follows a listing of the major collections of manuscript Gouldiana correspondence. Some family and smaller collections are mentioned in the chronology.

Typescript copies have been made of the majority of these letters, bills, receipts, notes, etc. John Gould's handwriting has been very difficult to decipher.

Again I acknowledge with appreciation the many institutions and individuals who made copies for me of correspondence in their collections, and also their permission to publish quotations. I hope to publish a second volume with this correspondence at hand and additional correspondence that undoubtedly will be uncovered.

GOULD, JOHN. Correspondence. 1830-37. [Correspondence from John Gould to William Swainson.] Collection: Linnean Society, London.

Three letters make up this collection. The first letter dated 12 December 1830 concerns lending or selling some of Gould's collection of American birds to Swainson, a study of the generic characters of birds by Gould, and the sending to Swainson 'in a few days... my first number' apparently of Gould's *Century of Birds.*

The second letter is dated 30 April 1834 and is concerned with some new species of Toucans and Trogons.

The letter dated 21 January 1837 from Gould to Swainson is concerned with Swainson accepting the first part of Gould's *Synopsis,* a comment that Mrs Gould's brothers are in Australia and collect for him, and the letter closes with the statement 'I would hope to visit the southern hemisphere'. [The earliest record of Gould's interest in travelling to Australia appeared in a letter to Jardine dated 2 November 1836.]

GOULD, JOHN. Correspondence. 1838-74. [Correspondence between T. C. Eyton, John Gould, Sir William Jardine, Edward Wilson, H. E. Strickland, Rev. J. S. Henslow, Philip Sclater, John MacGillivray and Hugh Cuming.] Collection: American Philosophical Society Library, Philadelphia.

Here is a group of twenty letters in the T. C. Eyton collection. Eight are from Gould to Eyton plus one from Gould to Mrs Maria Emma Gray and one to Rev. J. S. Henslow. The remaining ten letters contain comments about Gould.

PART FOUR BIBLIOGRAPHY OF JOHN GOULD, HIS FAMILY AND ASSOCIATES

The first letter is dated 8 December 1838 from Jardine to Eyton, and the concluding one is dated 2 April 1874 from Gould to Eyton.

The contents are concerned with the sale and exchange of specimens to Eyton, plus some more personal remarks, as on Gould having his rooms redecorated, get-togethers, etc.

GOULD, JOHN. Correspondence. [1838-1891.] [Various manuscripts, letters, and miscellaneous items.] Collection: National Library, Canberra.

Photocopies of this Gouldiana collection were sent to me in January 1979 in eleven folders.

Folder 1 consists of a letter from John Gould to W. Buller dated 30 December 1873, a letter from Gould to Richard Owen on 9 May 1842, and finally four typed pages referring to the Owen Stanley-John Gilbert problem mentioned in Alec Chisholm's 1941 *Strange New World.*

Folder 2 consists of MS 1217 which is a twenty-page letter from Dr J. R. Elsey, Jr. to John Gould dated 29 September 1857. This is a poignant letter in several respects. Elsey relates his very difficult experiences with A. C. Gregory's expedition across northern Australia of 1855-56, and then he states that at present he has 'haemoptysis' [tuberculosis]. He was to die of this in January 1858. The letter is replete with descriptions of many birds found on the Gregory expeditions.

Folder 3 contains copies of MS 454 which consists of thirteen or so items relating to E. P. Ramsay-Gould correspondence. Hindwood (1938 *Emu* 38: 206-12) wrote about and commented on these letters and receipts. Two of the letters were printed in full in *Emu*. Ramsay was twenty-two years of age when this correspondence began in October 1865.

In the folder sent me, there are three additional items not mentioned by Hindwood. Five sheets relate to Ramsay in 1884 being given Power of Attorney by Henry Sotheran, owner of the copyright of the late John Gould's works, to sue 'Mr. Bronowski' [sic] if necessary.

Folder 4 contains six miscellaneous items labelled MS 1755. Four items are typed copies of letters covering the period from 17 August 1888 to 7 May 1891 concerned with the purchase from Henry Sotheran of the pattern plates for Gould's *Birds of Australia* with the *Supplement,* and for the *Mammals of Australia.* The purchase was instigated by E. P. Ramsay and consummated by the Department of Public Instruction of New South Wales. Specimen plates by Mr Broinowski are referred to in the 7 May 1891 letter. Item 5 in this folder is a typed copy of the article by Tom Iredale entitled 'Again Gould' (1951 *Aust. Zoologist* 11: 316) which refers to the pattern plates for Gould's *Birds of Australia Supplement* and for his *Mammals of Australia.* They are now housed in the National Library at Canberra. Item 6 is a photocopy of Allan McEvey's 1968 article on 'Collections of John Gould Manuscripts and Drawings'.

Pertinent to a further consideration of the pattern plates, McEvey mentions the ones for the Supplement (80 plates) and for the Mammals (156 only examined) as being in the Library at Canberra, but then adds that those for the original seven volumes of Gould's *Birds of Australia* are in the Mitchell Library, Sydney. He does not give the number of plates in this latter collection.

Folders 5 through 11 contain items labelled MS 587.

Folder 5 contains a letter from Edwin Prince to John Gould in Australia of 21 December 1838 with a postscript by Mrs Mitchell, mainly concerning the health of the family in London and 'Money is very scarce'; a letter entitled 'No. 8 Duplicate'

from Prince to John Gould, 4 January 1839, concerned business affairs, taxidermy jobs in the work room, and noted that Louisa, the Gould's daughter, had been very ill; a receipt from Jons Piele [?] and Co. to Captain Orand [?] 7 January, 1842 regarding a deposit from Mr Gilbert for passage apparently from England back to Swan River and Perth; a business statement from Robert Kerr of Kerr Alexander & Co. from Launceston with dates of January 21st to 25th and February 1st and 4th, 1839; a one-page item with many separate notations in the hand of John Gould, which presumably are his instructions to John Gilbert in late January regarding items to collect in Western Australia, beginning with 'Gain all possible information respecting the identification of the Cuculidae' and ending with 'Ascertain about the Harriers'; a receipt dated 4 February 1839 for £10 for Gilbert's passage on the *Comet* to Swan River; a list 'Contents of Box by Madras' with postmark dated June 28, 1839 [?] sent to Prince, quite apparently from Gilbert with many natural history items, birds, forefeet of Kangaroos, box of skulls, '2 Waddies and a Walking Stick', etc.; a list of 'Contents of Box by Shepherd' with postmark of 2 May, 1840, sent to Prince obviously by Gilbert with many natural history items including '175 skin of birds/95 skeletons, 17 quadrupeds skins/9 Reptile Skins/1 Fish Skin' plus insects, shells, eggs and nests; a rough draft of letter to 'Bennett' [Dr George Bennett] from Gould undated [on or soon after December 1840] stating 'My Dear Friend, I have at length the pleasure of shipping the first part of my work on to Sidney [sic] ... for your acceptance'; a letter from Gould to Lieut. Friend, R.N., 4 January 1841 stating that the first part has been shipped to Hobart Town; draft letters from Gould to Lady Franklin, and Sir John Franklin, both dated 4 January 1841, concerning the shipment to Hobart Town of the first part of his work for their acceptance, with grateful thanks for their kindness to him and Mrs Gould; a letter from William Ince to John Gould, 20 January 1842 offering 'sincere congratulations on your having been elected an F.R.S. ... you may, I doubt not, have heard of the pleasing intelligence, but probably you are not aware that your name immediately succeeded that of Capt. Belcher O.B., R.N. who had 17 Black & only 31 White Balls ... you had but 5 of the former & 45 of the latter ... Sir Benj. Heywood Bart. had 8 Black ... & Edw's Lolly [?] Jun'r. 5'; a post card from Louisa Gould to R. B. Sharpe, 22 October 1891 stating 'Dear Mr. Sharpe, In the Westminster Review No. LXIX. April 1841 there is a long notice of my Father [John Gould] & his earlier works with illustrations & about his visit to Australia. This was soon after his return. We have the journal of Gilbert with whom my Father was connected in his work. Is it of any value to any one. Hoping you are all well. Believe me, Yours truly, Louisa Gould, 3 Steder Grand [?] N.W., Oct. 22nd.' [Gilbert's journal remained with the Gould family until 1938 when Chisholm re-discovered it]; a draft of letter from John Gould to John Gilbert, 14 July 1842 'No. 1' concerning revision of Gould's instructions to Gilbert since Mr Fairfax of Sydney had been so remiss in sending the subscribers payments, requesting Gilbert to go to Sydney at intervals, noting that Mr Priess of Germany had returned with an extensive collection from the Swan River area, giving further instruction for collecting mammals, seeds, etc., and 'Bye the bye I dreamed last night that I was among the Menuras in N.S.W.', a rough list in Gould's hand, apparently for Gilbert, undated: 'The following is a list of some of the birds of New S. Wales whose nidification or number and colour of their eggs is wholly unknown to me. You will thus pay particular attention to these and offering some rewards in money etc. for any of those marked with stars'; a shorter list probably in Gould's hand, undated, 'Desiderata of Eggs and Birds in Van D. Land'; an envelope from H.M.S. *Beagle* to John Gould in London postmarked 25 September 1843; a broadsheet of 23 September 1845, advertising £20 reward for

PART FOUR BIBLIOGRAPHY OF JOHN GOULD, HIS FAMILY AND ASSOCIATES

recovery of a number of skins of valuable birds apparently stolen from Gould [See Chisholm 1942]; a letter from James D. Somerville to A. H. Chisholm, of 14 November 1941, with comments on the enclosed typewritten copies of two letters from Gould to J. B. Harvey, one dated 27 December 1840, and the second undated [but obviously soon after Gould's return to London in August 1840], requesting specimens from Harvey [the originals of these letters are in the State Library of South Australia, Adelaide]; and a small card with a note to Charles Gould concerning the address of Count Strzelecki.

Folder 6 consists of MS 587 continued which contains ten letters from Gould to Gilbert dated 16 November 1840, 5 April 1841, 28 November 1842 [labelled No. 2 of the series written during Gilbert's second trip to Australia; No. 1 is in Folder 5], 18 April 1844, 15 May 1844, 22 May 1844, 1 June 1844, 1 July 1844 [?], 24 August 1844 [one copy in Gould's hand and a second copy in Prince's hand], and 20 December 1844. These valuable letters need to be printed completely. They have much to do with Gould's instructions to Gilbert, Gould's comments on the contents of the letters received from Gilbert, his growing concern for Gilbert, etc. etc.

Folder 7 (MS 587) is a letter from I. Agnew to Charles Gould, 14 May 1874, concerning meetings of the Royal Society, Landseer's paintings, Gould's *Handbook,* and chit chat. Charles was apparently in Tasmania, and Agnew on the mainland of Australia.

Folder 8 (MS 587) consists of an incomplete letter from Lady Jane Franklin to Elizabeth Gould, 15 January 1840, asking 'can you not again come over to us with the two boys and make this your home till your husband sends for you or fetches you away ... I almost envied you to hear of your living in tents on the Hunter ... the two French ships of Discovery here', commments on Lady Jane's household, and 'our little natural history society has its meetings in this house every Monday fortnight'. [This letter was partially printed in Chisholm's book *Elizabeth Gould* (1944)].

Folder 9 (MS 587) contains three letters from Ronald Gunn of Launceston, Tasmania to Charles Gould, dated 1 May 1867 [?], 26 May 1867 and 29 September 1869. Gunn discusses survey points in north-west Tasmania, offers comments on maps of the area, and answers survey questions from Charles Gould.

Folder 10 (MS 587) consists of two short letters from Richard Owen. The first is to John Gould, 4 December 1844: 'Will you join our little musical party here on Saturday evening at 8?' The second letter is to a Lord dated 8 November 1866 concerning Dr Franklin Gould being chosen as Medical Attendant for Sir [?] and Lady Georgina Millarton [?] and 'the high opinion I have been led to form of the moral, professional and scientific qualities and attainments of Dr. Franklin'.

Folder 11 (MS 587) consists of six short letters from John Gould to his daughter, Lizzie. None is dated. But two envelopes are available, dated 29 August 1864 and 30 November 1868. A third one is dated 7 July with an 1856 postmark, from Trondhjem, Norway when Gould travelled there with Joseph Wolf.

GOULD, JOHN. Correspondence. 1839. 'Extract of a Letter from Mr. Gould, the Ornithologist, dated June 30, City of Adelaide, South Australia.' *Mag. Nat. Hist.* 3 (New Series): 568.

This letter was addressed to Mr Prince, Broad Street Golden Square and follows:

I wish it were in my power to give you a faithful picture of this famed city of two years standing. People live in tents, and customs are so different from what they have been used to, that I really wonder how they reconcile themselves to their new mode of life. On the whole, however, I think South Australia may be considered as flourishing, and its condition will ultimately be prosperous.

The Zoology here, from what I have already seen, is likely to be of a most interesting description, totally different in its nature from that of Sydney, but probably approaching nearer in its character to the productions found beyond the Liverpool range, or what is more properly called the interior of New South Wales. — John Gould.

[GOULD, JOHN.] Correspondence. 1840-42. [Miscellaneous letters.] Collection: State Library of South Australia, Adelaide.

Four letters are in this collection. Two letters are the originals from John Gould to J. B. Harvey of Port Lincoln. This first is undated but probably was written soon after Gould's return to London on 1 August 1840, requesting specimens of birds and mammals. The second letter is from Gould to Harvey dated 27 December 1840 also requesting specimens.

The third item is a letter from Harvey to Governor George Grey dated 28 June 1841 asking for a list of specimens already sent to Gould from South Australia. This letter was printed by Cleland in 1937 (*Emu* 36: 201-04).

The fourth letter is from J. Newman of Port Adelaide to Governor Grey dated 23 November 1842, concerning the receipt of five cases of specimens addressed to J. Gould which will be shipped on the *Taglioni*.

[GOULD, JOHN] Correspondence. 1840-46. 'Letterbook of J. Gould in connection/with the shipping and sale of his Birds of Australia/not in Gould's handwriting/Diary of Mrs. Gould, Au. 20-Sep. 29, 1839/is bound in this volume.' Collection: Mitchell Library, Sydney, item A1763.

This bound volume contains copies of letters regarding sales and shipments of Gould's works to his Australian friends, purchasers and agents. These copies were written by Gould's secretaries, mainly by Edwin C. Prince.

The letterbook begins with correspondence dated 26 December 1840, from Gould to Mr Hack of Adelaide. Gould requested him to be his agent in South Australia for a commission of 10 per cent of sales.

On the first page of Mr Hack's letter is an annotation on the edge 'Sotheran 19-5-31 d/c[?]/e'. Undoubtedly this letterbook was purchased by the Mitchell Library, or a donor, from Henry Sotheran, Ltd, London in May 1931.

There are forty-nine letters in this file. The date of the final letter is 15 December 1846. See Hindwood (1938 *Emu* 38: 211) for a footnote about this letterbook.
Notes on the letterbook item A1763:

1. Copy of letter from John Gould to Mr Hack, Adelaide, 26 December 1840, written and signed by Edwin Prince, concerning offer of 10 per cent commission to Mr Hack for sales and delivery of Gould's *Birds of Australia* in South Australia.

2 Extract of letter written by Edwin Prince to Capt. Sturt, Adelaide, 26 December 1840, requesting payment for *Birds of Europe*.

3 Copy of letter from Gould to Rev. T. J. Ewing, 29 December 1840, written by Prince asking him to get an agent for Gould in Hobart Town, Van Diemen's Land. It also contains a detailed accounting of several of Gould's works sold in Van Diemen's Land and the works of others (Ruppell, Jardine, Darwin, Andrew Smith, Bonaparte, and the Zoological Society Proceedings) for which Gould was acting as an agent.

4 Copy of note to Mr Lewis of Hobart Town, Van Diemen's Land, written by Prince, requesting his undertaking job of agent for Gould at 10 per cent commission on sales. This note was enclosed in the previous letter.

5 Copy of letter from Gould to Rev. T. J. Ewing, 4 January 1841, written by Prince, concerning the fact that the first part of Gould's *Birds of Australia*, and other 'books, skins of birds, etc. are shipped on board the *Winrick*'.

PART FOUR BIBLIOGRAPHY OF JOHN GOULD, HIS FAMILY AND ASSOCIATES

6 Copy of letter from Gould to John Fairfax, Sydney, 27 January 1841, written by Prince, concerning shipping of two cases of seventy copies of the first part of Gould's *Birds of Australia* plus several other works, for delivery to subscribers and friends. A complete set of the *Birds of Europe* for the 'Australian Library' was also enclosed.

7 Copy of letter from Gould to Rev. T. Ewing, 29 March 1841, written by Prince concerning shipment of a case with part 2 of the *Birds of Australia* for the 'various subscribers in V.D.L'.

8 Copy of letter from Gould to Rev. Ewing, 5 April 1841, written by Prince, concerning collecting money from Lt. Friend of George Town, V.D.L.

9 Copy of letter from Gould to Mr Fairfax, 5 April 1841, written by Prince, concerning shipment of seventy copies of part 2. 'I long to hear from you.' Also concern over a financial crisis in the Colony.

10 Copy of extract of letter from Gould to Rev. Ewing 26 June 1841, written by Prince, concerning shipment of part 3, and receipt from Ewing of some eggs and nests.

11 Copy of letter from E. C. Prince to Rev. Ewing 5 July 1841 concerning shipment of '12 Cop. Pt. 3 Birds of Australia as before' to Van Diemen's Land subscribers.

12 Copy of letter from E. C. Prince to J. Fairfax 7 July 1841, concerning shipment of case of seventy copies of part 3 to Sydney.

13 Copy of extract of letter from Gould to Governor John Hutt, Western Australia, dated 28 July 1841, but not sent until 20 September 1841, written by Prince, concerning first three parts of *Birds of Australia* and part 1 of *Macropodidae*.

14 Copy of letter from E. C. Prince to J. Fairfax of Sydney, 15 December 1841, concerning receipt of first letters from Fairfax, and shipment of seventy copies of part 4 of the *Birds of Australia* and five copies of the first part of *Macropodidae*.

15 Copy of letter from E. C. Prince to Rev. Ewing, 17 January 1842, concerning Gilbert's return to England and imminent return to Australia, Mrs Gould's death, shipment of twelve copies of parts 4 and 5 of *Birds of Australia*, five copies of part 1 of the *Macropodidae*, and other books ordered by Ewing.

16 Copy of letter from E. C. Prince to J. Fairfax, 20 January 1842, concerning shipment of seventy copies of part 5 of *Birds of Australia,* and bill for £1153.11 to close of 1841.

17 Copy of letter from E. C. Prince to J. Fairfax 23 April 1842, concerning shipment of 70 copies of part 6 of *Birds of Australia*, and imploring him to remit and write.

18 Copy of letter from E. C. Prince to Rev. Ewing, 29 April 1842, concerning Ewing agreeing to take the Van Diemen's Land agency himself; a printed list of subscribers enclosed with the shipment of twelve copies of part 6 of *Birds of Australia* and other books; part 2 of *Macropodidae* coming out the first of May; the naming of another bird after Rev. Ewing; and a request for monies received to be sent to London.

19 Copy of letter from Gould to Messrs. Hack, Watson & Co. of Adelaide, 14 June 1842, written by Prince, acknowledging their agreement to 'become agents for the sale and delivery of my works in South Australia'. Parts were to be delivered to the Governor, Capt. Sturt, Mr Fortnum, and Mr Harvey (gift in exchange for specimens).

20 Copy of letter from Gould to J. Fairfax of Sydney, 14 June 1842, written by Prince, concerning the death of Mrs Gould and Gould's anxiety over his not having heard from Fairfax acknowledging the receipt of the parts sent and the remittances due.

21 Copy of letter from Edwin Prince to Rev. Ewing, 22 July 1842, concerning fourteen copies of part 7 of *Birds of Australia*, five copies of part 2 of *Macropodidae* plus many other items, such as personal books for Ewing, Lady Franklin, Lieut. Breton and Lieut. Friend.

22 Copy of letter from Gould to J. Fairfax, 10 October 1842, written by Prince, acknowledging receipt of £100, and suggestions as to further publicity in newspapers.

23 Copy of letter from E. Prince to J. Fairfax, 18 October 1842 concerning the shipment of seventy copies of parts 7 and 8 of *Birds of Australia* and other items.

24 Copy of letter from E. Prince to Rev. Ewing, 25 October 1842, concerning part 8 and other books being shipped.

25 Copy of letter from E. Prince to J. Fairfax, 26 December 1842, concerning the shipment of part 9 of *Birds of Australia* and the fact that total shipments to him to end of 1842 had been £2246.19 with only £222.6 received on account.

26 Copy of long letter from E. Prince to Rev. Ewing, 30 December 1842, concerning shipment of fourteen copies of part 9 and other books, a mention that part 6 of *Birds of Australia* was 'the first of the Birds drawn by Mr. Richter', receipt of a letter from Gilbert on his arrival at the Swan River area, and of Gould's having 'executed many very beautiful drawings [for *Odontophorinae*] in an entirely new style being a mingling of pencil and crayon the effect of which is astonishingly superior to anything he has yet done'.

27 Copy of letter from Gould to J. Fairfax, 13 May 1843, written by Prince, concerning receipt of £200, the shipment of fifty-two copies of part 10, and comments regarding the decrease in subscriptions.

28 Copy of letter from E. Prince to J. Fairfax of Sydney, 8 July 1843, concerning shipment of fifty copies of part 11 plus a complimentary copy for Mr Coxen, and a comment that Dr Bennett wishes to pay for his copies and not receive them free.

29 Copy of letter from E. Prince to Rev. Ewing, 10 July 1843, concerning the shipment of thirteen copies of parts 10 and 11 and some other volumes for Rev. Ewing, also the usual bookkeeping items.

30 Copy of letter from Gould to Mr Hagen of Adelaide, [date around August 1843, see letter 36], written by Prince, concerning Mr Hack of Adelaide, agent for Gould's works, now having transferred his business to Mr Hagen, Gould's concern over several accounts owed him, especially one of Capt. Sturt, the death of Mr Harvey, and the statement 'that my present work will require an outlay of nearly £15000'.

31 Copy of letter from E. Prince to Rev. Ewing, 29 December 1843, concerning part 9 of *Birds of Australia* being 'on board the unfortunate "Palmyra" and I presume was one of the packages thrown overboard', and concerning shipment of parts 12 and 13 with some other books for Ewing including three parts of the *Voyage of the Sulphur*.

32 Copy of letter from E. Prince to Rev. Ewing, 6 March 1843 [should be 1844], not written by Prince but signed by him, concerning the loss of part 9 on the *Palmyra*, directions to get the replacement copies of part 9 from Mr Fairfax of Sydney, the shipment of part 14 of *Birds of Australia* and two copies of parts 1 and 2 of *Macropodidae*, two copies of the first part of Mr [G. R.] Gray's *Genera of Birds*, a prospectus enclosed of the '*Mon. of Ortyginee* [sic] the first part of which will appear on the first of Sept.', and concern for Mr Gilbert's safety.

33 Copy of a letter from Gould to Mr Fairfax, 7 March [1844], written by a secretary, that was enclosed with the previous letter, concerning the transfer of copies of part 9 to Rev. Ewing, and also the statement 'I am quite in despair about my affairs in New South Wales'.

34 Copy of letter from E. Prince to Rev. Ewing, 1 August 1844, written and signed by Prince, concerning the shipment of part 15, with other books for Ewing and others, and further accounting concerning the loss on the *Palmyra*.

35 Copy of letter from E. Prince to Rev. Ewing, November [N.D.] 1844, written and signed by a secretary, concerning parts 16 and 17 of *Birds of Australia* being shipped, two copies of part 1 of *Odontophorinae*, two copies of parts 4, 5, 6 and 7 of Gray's *Genera* and fifteen other books, with a comment that 'Pt. 3 Icones is not yet ready' [it was never published as such], and that 'with the exception of the late Mrs. Goulds mother who has been very ill for the last months and who at her advanced time of life 77 cannot be expected to remain with us very long, all Mr. Goulds family are quite well'.

36 Copy of letter from John Gould to Mr Hagen of Adelaide, 11 December 1844, written and signed by a secretary, concerning a letter [number 30] written '16 months ago' sent on the *Augustus* to which no reply was received, where Gould begs the 'favor of an early answer'.

37 Copy of letter from John Gould to Dr Bennett of Sydney, 18 December 1844, written by a secretary and signed by Gould, concerning a request by Gould that Bennett take over the agency of Mr Fairfax of Sydney, the shipment of several parts of Gould's works, wishing to receive a letter from Mr Gilbert, and comments about Mr Strange collecting specimens for Gould or for sale and the specimens not being worth as much as Strange thought.

38 Copy of letter from John Gould to Mr Stokes of Western Australia, 28 January 1845, written and signed by a secretary, concerning Mr Gilbert having informed Gould that Stokes would 'receive and distribute to the Subscribers the few copies of my Birds of Australia and taken in your part of that country'.

39 Copy of letter from Edwin Prince to Rev. Ewing, 11 July 1845, written and signed by a secretary, concerning Ewing's account with Gould; a box sent to Gould from his 'old servant Benstead'; a list of the nine current subscribers in Van Diemen's Land ('Ewing, Gunn, Dunn, Archer, McLean, Bedford, Montague (or Museum), Dry & Breton'); concerning Ewing having been made a member of the Ray Society; the economic straits in Australia and an offer by Gould to let the subscribers have a 10 to 15 per cent discount on the current parts if they would continue the subscription; Gilbert's having joined the Leichhardt expedition from Moreton Bay to Port Essington; the publication of the first part of the *Mammals of Australia* on the first of May 'and as we are now in the height of the fashionable season there has been such a demand for it that I cannot get copies coloured fast enough'.

40 Copy of letter from John Gould to Dr Bennett, 12 July 1845, written by Prince and signed by Gould, concerning Gould's appreciation to Bennett for his contact with Gilbert, Strange and Fairfax, comment on the late Stephen Coxen, lengthy comment on the 'most perilous undertaking' by Leichhardt in his expedition to Port Essington, further comment on Strange having drawn another £40 on Gould and inadequate specimens to support this amount, that Fairfax was returning £900 worth of parts of Gould's *Birds of Australia*, but that Fairfax was still behind £900 in his payments to Gould and there had not been adequate accounting of the subscriptions, etc. etc.

41 Copy of letter from John Gould to Mr Fairfax, 26 September 1845, written by a secretary, concerning the 'vexation and loss' which has accompanied Gould's attempt to place subscriptions in the Sydney area, aggravated by inadequate records from Fairfax and the poor economy of the Colony.

42 Copy of an incomplete letter from Edwin Prince to Dr Bennett, 26 September 1845 written by a secretary, concerning shipment of parts 20 and 21 of *Birds of Australia*, and part 1 of *Mammals of Australia*, great concern over Gilbert's safety 'in daily expectation that our worst fears may be realized', and concluding with a list of the parts of Gould's works and other property that were in the possession of Gilbert, with the question of their present disposition.

43 Copy of letter from Edwin Prince to Rev. Ewing, 31 October 1845, written by a secretary, concerning the shipment of parts 20 and 21 with other books for Ewing, and also a gold watch Rev. Ewing had requested to be purchased by Gould.

44 Copy of letter from John Gould to Mr Fairfax, 16 December 1845, written and signed by Prince, acknowledging receipt of £50 on account, and requesting Fairfax to tranship 'the cases left in your charge only in the event of the sad report that has reached me of his [Gilbert] being murdered being confirmed'.

45 Copy of letter from John Gould to Dr Bennett, Sydney, written and signed by Prince, concerning receipt of £85.12.0 from sale of Gould's works; worst fears for Gilbert probably will be realized, he was 'a sacrifice to his love of science', and 'I have just learnt that the person whom he engaged to collect for me in W. Aust. has been barboursly [sic] murdered by the natives ... Johnson Drummond'. Gould then requests Bennett to collect any of Gilbert's property in the hands of Strange, Fairfax or Charles Coxen along with 'any letters or papers you may have that the same may be given to his family if necessary'.

46 Copy of letter from John Gould to Mr J. Stokes, 2 February 1846, written and signed by Prince, concerning thanks for the subscription business attended to by Stokes, the arrangement with Johnson Drummond was that 'the books sent for him were in payment for specimens sent or to be sent to me agreeably to an arrangement made with him by Mr. Gilbert', a balance was due Johnson so Gould suggested this arrangement be continued with Johnson's father, James Drummond, but 'if he will not, please to pay him out of any monies you may receive', regarding Gilbert 'none of them have since been heard of still I hope for the best', and finally Gould was agreeable to working out an exchange of books for specimens with Mr Lock Bossep [?].

47 Copy of letter from John Gould to Dr G. Bennett, 1 June 1846, written and signed by Prince, concerning 'a remittance of £59.6.6 on a/c of S. Coxen's Estate for which I am greatly obliged'; the statement that 'a report has just reached London that Leichardt [sic] Gilbert and party have arrived at Port Essington. I fear it is too good to be true but I do not yet despair of their safety and this it is that has prevented me from acceeding to Strange's solicitation to appoint him Gilbert's successor', concerning more specimens from Strange now to his credit, which could be paid by Fairfax 'who has still nearly £800 to a/c for', and parts 22 and 23 having been shipped, except for Sir G. Gipps copy which Gould retained since he was to arrive in London.

48 Copy of letter from John Gould to R. C. Gunn, Launceston, 11 December 1846, written by Prince, concerning a request for some specimens of small mouse-like opossums, bats, large quadrupeds, birds and eggs, also that since Rev. Ewing is now in London, Gould requested Gunn to take care of Gould's deliveries and collections in Van Diemen's Land, and that parts 22 to 25 were being shipped. A list of subscribers is appended, which includes Gunn, Bedford, McLean, Bicheno, Museum, Archer, Breton and Dry.

PART FOUR BIBLIOGRAPHY OF JOHN GOULD, HIS FAMILY AND ASSOCIATES

49 Copy of letter from John Gould to Dr Bennett, 15 December 1846, written by Prince, concerning Bennett's note on the Purple Waterhen, the shipment of parts 24 and 25 of *Birds of Australia* with other books, another request for any of Gilbert's property left with Fairfax or Strange, on the *Rattlesnake* having just sailed with MacGillivray as Naturalist, and concerning receiving a letter 'from Capt. Lynd with a loan of the animals Coll. by Dr. Leichardt [sic] belonging to your museum'.

[GOULD, JOHN.] Correspondence. 1840-69. 'John Gould and George French Angas/Autograph Correspondence relating/to Gould's "Birds of Australia",/6 Apr. 1840 - 18 Dec. 1869.' Collection: Mitchell Library, Sydney, item C204.

The first letter, dated 6 April 1840, is unrelated to the subsequent Gould-Angas correspondence. It is from John Gould to Sir Gordon Bremer concerning a request by Gould for assistance from Sir Gordon for John Gilbert's work in the Port Essington area.

The Angas correspondence begins with the date of 16 January 1860 with a letter from Edwin Prince to Angas in South Australia. The concluding letter is dated 15 December 1869 from Angas to Gould, with Angas back in England.

According to Whittell (1954) George French Angas was born at Newcastle-on-Tyne, England on 25 April 1822. After studying lithography and drawing, he went to Australia in 1843, returned to England, and again sailed for South Australia in 1849. In a letter to Gould in this Mitchell Library Collection dated 22 January 1863 Angas wrote that he was sailing for London 'on about the 15th of next month...with my wife and 4 girls'. Thus this Gould-Angas correspondence covers only three years of Angas's stay in Australia. Angas died in London on 8 October 1886.

There are eighteen letters in this Gould-Angas collection and one newspaper clipping. Thirteen of these letters are from Angas to Gould, three are rough drafts from Gould to Angas, one is a letter from Prince to Angas, and one is a letter from George Fife Angas, father of George French Angas, to his agents in London ordering Gould's *Birds of Australia*. The father eventually gave this set to his son, along with Gould's *Mammals of Australia*.

The newspaper clipping, which is undated, reports that 'some gentlemen [of Adelaide] have taken the matter in hand...[to obtain] the amount sufficient to purchase a copy [of Gould's *Birds of Australia*] for presentation' to the South Australian Institute. The 'work is published at £115'.

GOULD, JOHN. Correspondence. 1851. [Correspondence concerning exhibition of Gould's hummingbird collection at the Gardens of the Zoological Society of London.] Collection: The Zoological Society of London.

Included are five items of correspondence and a three-page tabulation of receipts for Gould's humming-bird exhibit in 1851 at the Gardens.

On 5 February 1851 Gould wrote to the Council of the Zoological Society 'Many scientific friends having strongly urged that during the approaching Exhibition the public should be allowed an opportunity of inspecting my fine collection of the Trochilidae or Humming Birds, I am induced to enquire if some arrangement may not be made for its exhibition in the Gardens of the Zoological Society?' Gould offered two proposals, namely, that the Society erect 'a house of about 60 feet by 30...to exhibit the collection for six pence each person on Mondays and one shilling on remaining week days; Fellows and two friends...free...the Society to be entitled to half the receipts', or Gould would erect the house himself, it would become the property of the Society at the end of six months, and the receipts would be his. Apparently, the Council accepted the latter proposal, with the admission charge going to

Gould (Scherren 1905, p. 106). He also requested permission 'to sell in the room a small popular Catalogue of the Collection' for six-pence each.

In a letter dated 5 November 1851 from Gould to D. W. Mitchell, Secretary of the Society, Gould requested that the building and collection be allowed to remain open for twelve months longer. 'The Collection has excited the highest interest... the number of visitors... have amounted to no less than 80,000.'

Having learned that the Council agreed unanimously to an extension of Gould's exhibit, but 'that the Commissioners of Her Majesty's Woods and Forests have determined that the building... shall be taken down... I beg to say I will re-erect the building at my own cost on any other site'. [According to Scherren the building was moved to the Middle Garden and housed Gould's collection until 1852.]

Two additional short notes are identical in text. One is dated 3 December 1851 unsigned, and the second is dated 4 December 1851 and is signed by Gould. Gould offered his collection of humming-birds at the disposal of the Society for twelve months for exhibition without charges to Fellows and visitors, with the Society assuming all the expenses.

The final item is a three-page accounting ledger labelled 'Humming Bird Account'. The dates covered are from 17 May 1851 to 8 November 1851 and show receipts of £1589.15.1. The largest attendance was in July 1851 when £463.17.0 was received.

GOULD, JOHN. Correspondence and statements of account. 1857-81. [Concerning subscriptions to the Melbourne Public Library.] Collection: La Trobe Collection, State Library of Victoria, Melbourne.

Through the kindness of the Trustees of the State Library of Victoria a copy was provided me of this set of thirty-two letters and statements of accounts, occupying sixty-two sheets. These items pertain to subscriptions by the Library for several of Gould's works, including *Trochilidae*, *Birds of Australia* with the *Supplement*, *Mammals of Australia*, *Ramphastidae* (new edition), *Odontophorinae*, *Trogonidae* (new edition), *Birds of Asia*, and *Birds of Great Britain*.

One item dated 18 February 1868 relates to a letter and statement of account to Melbourne University for part 4 of *Birds of Australia, Supplement*.

All of these subscriptions were shipped to Professor McCoy, Director of the National Museum of Melbourne, in the parts as issued. He then handled the distribution.

The thirty-two items which follow are as provided to me, apparently bound in a volume. The order is not completely chronological.

1 Copy of letter from John Gould to Augustus Tulk, Librarian of the Melbourne Public Library, 13 August 1862, written by Edwin Prince and signed by Gould, concerning a shipment to Professor McCoy of the works due the Library, that the *Trochilidae* are finished and ready for binding, that the first and second part of *Birds of Great Britain* will be published in a week or two and a request that they subscribe for this set. An account is enclosed for parts 21 to 25 of *Trochilidae* for £15.15.0 and for parts 13 and 14 of *Birds of Asia* for £6.6.0 to be payable through Messr. Drummonds of Charing Cross.

2 Letter from John Gould to Sir Redmond Barry, 22 May 1863 written by Prince and signed by Gould, concerning receipt of payment for the account rendered in the previous letter, and the hope that the Library would now subscribe for his *Birds of Great Britain*, two parts of which had been sent out for inspection through Professor McCoy.

PART FOUR BIBLIOGRAPHY OF JOHN GOULD, HIS FAMILY AND ASSOCIATES

3 Receipt dated 17 June 1859 signed by John Gould for £131.19.- from the Melbourne Public Library.

4 Copy of letter from Edward Namant [?] to His Honour Mr Justice Barry 30 June 1859 enclosing the receipt for £131.19.-.

5 Original letter of item 4.

6 Prospectus of two pages dated 1 August 1862. The first page announces 'The Birds of Great Britain' which 'will be published in Imperial Folio at the rate of two Parts a year, price Three Guineas each ... it is expected that the work will be completed in eight or nine years'. On the reverse twelve other works or editions are listed.

7 Letter from John Gould to Augustus Tulk, Librarian, 14 August 1857, written by Prince and signed by Gould, concerning part 8 of *Mammals of Australia* delivered to the Library by Dr Webber, it was not a donation but a continuation of the series, and further about parts 6 and 7 not being received yet with part 2 of *Birds of Australia, Supplement* all having been sent on the ship *Land O'Cakes*.

8 Letter from John Gould to A. Tulk, 19 July 1858, written by Prince and signed by Gould, concerning the Melbourne Public Library order for the monographs *Trogonidae, Trochilidae,* and *Birds of Asia,* and that part 9 of the *Mammals of Australia* for the Library had been shipped with a box of birds sent to the Public Museum.

9 Letter from John Gould to A. Tulk, 13 November 1858, written by Prince and signed by Gould, concerning shipment of parts 1 to 16 of *Trochilidae* (Professor McCoy had part 12 already), parts 1 to 10 of *Birds of Asia,* part of *Mammals of Australia,* and part 1 of a new edition of *Trogonidae* 'which will be completed in four parts'. An itemized statement of account for seven works, and the binding of two, dated 30 November 1858, for £131.19.- was enclosed. (See items 3 and 4 regarding payment.)

10 Letter from John Gould to A. Tulk, 16 December 1859, written by Prince and signed by Gould, concerning part 10 of the *Mammals of Australia* inadvertently not sent, and requesting notation of Gould's new address of 26 Charlotte Street.

11 Letter from John Gould to A. Tulk, 22 February 1861 written by Prince and signed by Gould, concerning the shipment of the parts of Gould's works for the year 1860 to Professor McCoy for the Library along with part 10 of *Mammals of Australia,* and with an itemized statement for the amount due for the years 1859 and 1860.

12 Letter from John Gould to A. Tulk, 13 August 1862, written by Prince and signed by Gould, concerning shipment of the parts due to the Library through Professor McCoy, that *Trochilidae* is finished and can now be bound which Gould offers to do for them if desired, that 'in the course of a week or two the first and second parts of my new work on the Birds of Great Britain will be ready for delivery' and would they allow Gould to send out the parts of this work? The statement for this transaction is item 14.

13 Letter from John Gould to Sir Redmond Barry, 19 September 1863, concerning answers to questions on the publication of *Birds of Great Britain* which the Library agreed to purchase, on the cost, number of years for completion, binding of a few parts at a time, etc.

14 Statement of account dated 13 August 1862 that obviously accompanied the letter of that date, item 12. Chronologically this statement appears here because it was returned to the Melbourne Public Library as requested and has dates of 27 April 1864 and 12 May 1864 written in. The total amount is £22.1.0 for parts 21 to 25 of *Trochilidae* (completing the work) for £15.15.0, and parts 13 and 14 of *Birds of Asia* for £6.6.0.

15 Letter from John Gould to Sir Redmond Barry, 23 February 1865, written by Prince and signed by Gould, concerning Gould's works of 1864 shipped to the Library via Professor McCoy. An additional note dated 2 May 1865 written by Henry Sheffield, Acting Librarian, requests that McCoy deliver the parcel of parts.

16 Letter from John Gould to the Librarian of the Melbourne Public Library, 26 May 1864, written by Prince and signed by Gould, requesting an answer to Gould's letter of 19 September 1863 (item 13) whether to send the *Birds of Great Britain* in parts or as the volumes when completed.

17 Statement of account dated 30 November 1864 for parts 1 to 6 of *Birds of Great Britain* for £18.18.0, parts 15 and 16 of *Birds of Asia* £6.6.0, and part 13 of *Mammals of Australia* for £3.3.0.

18 Letter from John Gould to Sir Redmond Barry, 24 August 1866, written by Prince and signed by Gould, concerning a request for remittance for the 1864 and 1865 parts previously sent to the Library. An itemized statement for £37.16.0 is appended.

19 Prospectus of two pages dated 1 September 1865. The first page announces the publication of 'A Handbook to the Birds of Australia'. 'This day is published, Vol. I. of the above work ... price £1.5s.; and on the 2nd of December will be issued the second and concluding volume, at the same price.' On the reverse is a note about 'The Birds of Great Britain', with a listing of ten other works.

20 Letter from John Gould to A. Tulk, 24 December 1866, written by Prince and signed by Gould, concerning receipt of the sum of £37.16.0, and the shipment of the 1866 parts.

21 Statement of account dated 7 December 1866 due from Melbourne Public Library for parts 9 and 10 of *Birds of Great Britain* for £6.6.0, and for part 18 of *Birds of Asia* for £3.3.0.

22 Letter from John Gould to A. Tulk, 17 February 1868, written by Prince and signed by Gould, concerning shipment of the 1867 parts including parts 11 and 12 of *Birds of Great Britain*, part 19 of *Birds of Asia* and part 4 of *Birds of Australia, Supplement*.

23 Statement of account dated 31 December 1867 for parts 9 to 12 of *Birds of Great Britain* for £12.12.0, for parts 18 and 19 of *Birds of Asia* for £6.6.0, and for part 4 of *Birds of Australia, Supplement* for £3.3.0.

24 Statement of account dated 25 November 1868 to the Melbourne Public Library for parts 13 and 14 of *Birds of Great Britain* for £6.6.0, and part 20 of *Birds of Asia* for £3.3.0.

25 Letter from John Gould to the Librarian of Melbourne University, 18 February 1868, written by Prince and signed by Gould, concerning the shipment of part 4 of *Birds of Australia, Supplement* for the University in the case to Professor McCoy. Enclosed is a statement for part 4 of the *Supplement* for £3.3.0.

26 Statement of account dated 1 September 1867 to the Melbourne Public Library (as all the following) for parts 15 and 16 of *Birds of Great Britain* for £6.6.0, for part 21 of *Birds of Asia* for £3.3.0, for part 5 of *Birds of Australia, Supplement* for £3.3.0 and for part 2 of the monograph *Trogonidae* for £3.3.0.

27 Statement of account dated 3 August 1870 for parts 17 and 18 of *Birds of Great Britain* for £6.6.0 and for part 22 of *Birds of Asia* for £3.3.0, along with the account for 1869 of £15.15.0.

28 Statement of account dated 14 October 1871 for parts 19 and 20 of *Birds of Great Britain* and part 23 of *Birds of Asia*.

PART FOUR BIBLIOGRAPHY OF JOHN GOULD, HIS FAMILY AND ASSOCIATES

29 Statement of account dated 13 August 1872 for parts 21 and 22 of *Birds of Great Britain* and part 24 of *Birds of Asia*.

30 Statement of account dated 8 June 1874 for parts 23, 24 and 25 of *Birds of Great Britain* and part 25 of *Birds of Asia*.

31 Statement of account dated 23 May 1876 for parts 26 and 27 of *Birds of Asia* for £6.6.0.

32 Letter from A. F. Tweedie to the Trustees, Public Library Museum and National Gallery, 17 March 1881 written 'on behalf of Mrs. Eliza Muskett, and Mr. Alex'r Forbes Tweedie, the Executors of the Will of the late Mr. John Gould, F.R.S.' stating that he died on the 3rd of February, 1881 and the sum of £31.10.0 appeared on Mr. Gould's books as owed by the Library. Under the date of 5 May 1881, in another hand, appears 'Correct, H. Sheffield'.

GOULD, JOHN. Correspondence. 1860-67. [Windsor Castle Receipts.] Collection: The Royal Library, Windsor Castle.

Three receipts and a letter were copied for me by the Librarian, Sir Robin Mackworth-Young. I acknowledge the gracious permission of Her Majesty Queen Elizabeth II to make use of this material from the Royal Archives.

The earliest receipt dated 21 February 1860 is accompanied by a letter from John Gould to H. F. Harrison, Esq. dated 24 February 1860. The receipt is as follows:

```
                                                    London Feb. 21, 1860
    Her Most Gracious Majesty
       The Queen
          To John Gould, F.R.S. etc.
             26 Charlotte St. Bedford Sq. W.
    1859
    July 4  To pts 7 to 18 of a Monograph
               of the Trochilidae
               or Humming Birds                          37:16.
            To pts 6 to 11 of the Birds
               of Asia                                   18:18:
                                                        £56:14.

    Received by Cheque on Messrs. Contts & Co.
       Feb 24; 1860
       John Gould [signed]
          Sent by Her Majesty's Command to H.M. the King of Portugal.
```

The receipt and the letter acknowledging the receipt are in the hand of E. C. Prince but both are signed by John Gould.

The two other receipts, also in the hand of Prince, are dated 30 June 1867 for two copies of part 19 of *Birds of Asia* ordered by the Queen, one copy for the King of Portugal, and the other for the 'Library of His late Royal Highness The Prince Consort'.

Gould signed both of the receipts under the date of 5 August 1867.

GOULD, JOHN. Correspondence. 1868-78. [Correspondence between John Gould and George N. Lawrence.] Collection: American Museum of Natural History, New York, N. Y.

Gould probably met Lawrence in 1857 when he visited New York City, but I have found no record of this meeting.

The first letter in this series of thirty-one letters (or drafts of letters) is dated 11 July 1869 from Gould to Lawrence. Twenty-two of the letters are from Gould to

Lawrence, five are from Lawrence to Gould, two are from Gould's secretary Amelia Prince in 1873 and 1874 to Lawrence, one letter from A. S. Bickmore to Gould, and one from Lawrence to J. Henry. The final letter in this series is dated 1 January 1878 from Gould to Lawrence.

The letters are concerned mainly with Lawrence acting as agent for Gould in America, and the exchange of specimens, particularly humming-birds and trogons. Individuals and institutions mentioned include the Academy of Natural Science of Philadelphia, Professor Baird, Dr Bryant, Buffalo Natural History Society, D. G. Elliot, J. Henry of the Smithsonian Institution, Professor Orton, Professor Owen, Mr Palmer, Peabody Institute, Osgood Salvin, P. L. Sclater, Smithsonian Institution. Professor Steere at Michigan, R. L. Stuart, Captain Taylor, Mr Turnbull, Dr Thomas B. Wilson, and Joseph Wolf.

GOULD, JOHN. Correspondence. *See* Sutton, J. 1929. Letters of John Gould to F. G. Waterhouse.

Twelve letters written by Gould are reproduced in their entirety.

GOULD, JOHN. Lithographic stones. [Several dates]. University of Kansas.

In the University of Kansas Spencer Library–Ralph Ellis Collection is a set of twelve lithographic stones. There are references to these stones in the Ellis Archives on pages 25, 108, 121, 176 and 270 as indexed by Sauer (1976 MS).

On page 108 is a note apparently for a display of a stone: 'The rock is Dusseldorf clay from the jurassic of north Germany — about 95,000,000 years'.

On page 121 is an itemized bill from Henry Sotheran, London, dated 3 September 1940 where there is a listing of '6 stones (additional) gratis'.

Five stones were for *Supplement to Trochilidae*, two for *Birds of Asia*, three for the *Birds of Great Britain*, one for *Birds of New Guinea* and the twelfth for Sharpe's *Birds of Paradise*.

The following information on each stone was deemed important:

1 *Rhamphomicron dorsale* (Simon's Thorn-bill)
- Published in: *Supplement to Trochilidae* pl. 43 J. Gould and W. Hart del et lith Walter, Imp.
- Measurements: 488 × 355 × 25 mm ($19\frac{3}{16}''$ × 14″ × 1″).
- Weight: Total: 14.38 kg (31 lb, 11 oz); stone alone estimated without box and plastic cover: 12.48 kg (27 lb, 8 oz).
- Notations: On left front in pencil: 'Suppl. Pt. 3 310 1882' or 'Suple V Pt 3.'
- Defaced and how: Yes. Perpendicular crossed scratched lines and J.W.S. scratched in lower left corner. J. W. Styche was a Director of Sotheran.
- Cracked: No.
- Other comments: In a wooden box or case open at the top.

2 *Metallura heterocerca (primolina)* (Guiana Coppertail)
- Published in: *Supplement to Trochilidae* pl. 45 W. Hart del et lith Mintern Bros Imp.
- Measurements: 559 × 340 × 40 mm (22″ × $13\frac{3}{8}''$ × $1\frac{9}{16}''$).
- Weight: Total: 21.3 kg (46 lb, 15 oz); stone alone estimated at 19.66 kg (43 lb, 5 oz).
- Notations: On right edge in pencil: '310 7/4 [?] /85'. On left edge in pencil: 'H. Ellitt'.
- Defaced and how: Yes. Perpendicular crossed lines and 'J.W.S.' in left lower corner.
- Cracked: No.
- Other comments: In a wooden box.

PART FOUR BIBLIOGRAPHY OF JOHN GOULD, HIS FAMILY AND ASSOCIATES

3 *Agyrtria bartletti* (Bartlett's Emerald)
Published in:	*Supplement to Trochilidae*, pl. 50 W. Hart del et lith Mintern Bros Imp.
Measurements:	464 × 362 × 38 mm (18¼″ × 14¼″ × 1½″).
Weight:	Total: 17.56 kg (38 lb, 11 oz); stone alone exact weight: 15.75 kg (34 lb, 11 oz).
Notations:	On left side in pencil '310'; on right side: 'H. Ellitt' [?]
Defaced and how:	Yes. X lines.
Cracked:	Yes, very thin line from upper edge to lower edge.
Other comments:	In a wooden box.

4 *Agyrtria fluviatilis* (Riverine Emerald)
Published in:	*Supplement to Trochilidae*, pl. 51 W. Hart del et lith Walter, Imp.
Measurements:	463 × 360 × 40 mm (18¼″ × 14 3/16″ × 1 9/16″).
Weight:	Total with box and plastic cover: 19.8 kg (43 lb, 1 oz); stone alone estimated: 17.62 kg (38 lb, 13 oz).
Notations:	On left side: '310'. On right side: 'H Ellitt' [?]
Defaced and how:	Yes. Four zig-zag lines.
Cracked:	Yes, long one in left upper corner.
Other comments:	In a wooden box.

5 *Agyrtria taczanowskii* (Taczanowski's Emerald)
Published in:	*Supplement to Trochilidae*, pl. 52 J. Gould & W. Hart del et lith Mintern Bros. Imp.
Measurements:	487 × 323 × 42 mm (19 3/16″ × 12 11/16″ × 1 5/8″).
Weight:	Total: 19 kg (42 lb) with box and cover; stone alone est.: 17 kg (37 lb, 8 oz).
Notations:	None.
Defaced and how:	Yes. Perpendicular crossed lines with 'J.W.S.'
Cracked:	No.
Other comments:	In a wooden box.

6 *Pitta coccinea* (Malaccan Pitta) (Figure 71)
Published in:	*Birds of Asia*, vol. V, pl. 68 J. Gould & W. Hart del et lith Walter, Imp.
Measurements:	489 × 328 × 35 mm (19¼″ × 12 7/8″ × 1 3/8″).
Weight:	Total: 17.76 kg (39 lb, 2 oz). Estimated stone alone weight – 15.89 kg (35 lb).
Notations:	None.
Defaced and how:	Yes. By X lines across stone, not straight lines.
Cracked:	Diagonally lower right and upper.
Other comments:	In a wooden case. Part of label torn off on edge at top.

7 *Pitta baudii* (Red-backed Pitta)
Published in:	*Birds of Asia*, vol. V, pl. 72 J. Gould & W. Hart del et lith Walter, Imp.
Measurements:	530 × 432 × 42 mm (20 7/8″ × 17″ × 1 5/8″).
Weight:	Total: 27.55 kg (60 lb, 11 oz); stone alone exactly 25.11 kg (55 lb, 5 oz).
Notations:	None.
Defaced and how:	Yes. Perpendicular crossed scratched lines and 'J.W.S.' in lower left corner.
Cracked:	No.
Other comments:	In a wooden case.

8 *Lanius minor* (Rose-breasted Shrike)
Published in:	*Birds of Great Britain*, vol. II, pl. 14 J. Gould & H. C. Richter del et lith Walter, Imp.
Measurements:	535 × 327 × 15 mm (21 1/16″ × 12¾″ × 9/16″).

Figure 71 Lithographic stone used for printing of *Pitta coccinea* for *Birds of Asia*, V: 68, Kansas-Ellis-Gould Collection.

Weight:	Total: 8.4 kg (18 lb, 8 oz); has only a small plywood board, without sides on back; also plastic cover. Stone alone estimated: 7.94 kg (17 lb, 8 oz). Ellis had this stone sliced thinner by Fairmount Monument Works (Sauer 1976 MS Ellis archives, p. 25).
Notations:	On left side front: 'G.B. Pt. 13/110 1883'.
Defaced and how:	Yes. Six crossed lines.
Cracked:	Yes, left upper corner, two parts glued together.
Other comments:	On wood backboard. How does one explain the 1883 date? Could it be the date of a second or additional printing run? (Also see stones 9 and 10).

PART FOUR   BIBLIOGRAPHY OF JOHN GOULD, HIS FAMILY AND ASSOCIATES

9 *Anthus arboreus* (Tree Pipit)
Published in:	*Birds of Great Britain,* vol. III, pl, 14, pt. IX, 1866 J. Gould & H. C. Richter del et lith Walter, Imp.
Measurements:	492 × 385 × 35 mm ($19\frac{3}{8}''$ × $15\frac{1}{4}''$ × $1\frac{3}{8}''$).
Weight:	Total: 20.97 kg (46 lb, 3 oz); stone alone estimated: 19.1 kg (42 lb, 1 oz).
Notations:	On left edge in pencil: 'G. B. pt. 9/110 1882'. In right upper corner in smaller letters in pencil: 'G. B. 196' (see stones 8 and 10).
Defaced and how:	Yes. With two scratches.
Cracked:	No.
Other comments:	Regarding the 'G. B. 196': Does 196 refer to plate number in List of Contents, or does it refer to an original or early printing run of 196, while the '110 1882' refers to an additional printing?

10 *Otocoris alpestris* (Shore-lark)
Published in:	*Birds of Great Britain,* vol. III, pl. 18 J. Gould & H. C. Richter del et lith Walter, Imp.
Measurements:	487 × 407 × 32 mm ($19\frac{3}{16}''$ × 16'' × $1\frac{1}{4}''$).
Weight:	Total: 20.2 kg (44 lb, 8 oz); stone alone est.; 18.5 kg (40 lb, 12 oz).
Notations:	On left edge in pencil: '109 printed Great B pt. 3 [?]'; also, smaller letters lower right corner 'G. B' at the corner of the stone where it is chipped. Possibly a number is missing. (See stones 8 and 9).
Defaced and how:	Yes. Four straight lines.
Cracked:	No.
Other comments:	In a wooden box. On top edge of the box is a torn label: 'Otocoris alpestris: ———8'

11 *Munia grandis* (Large rufous-and-black Finch)
Published in:	*Birds of New Guinea,* vol. IV, pl. 22 W. Hart del et lith Walter, Imp.
Measurements:	535 × 325 × 45 mm ($21\frac{1}{16}''$ × $12\frac{13}{16}''$ × $1\frac{3}{4}''$).
Weight:	Total with box and cover: 19.89 kg (43 lb, 13 oz); stone alone estimated: 17.85 kg (39 lb, 5 oz).
Notations:	In lower left on side of stone: '264/Guinea Pt. 14' [?]
Defaced and how:	Yes. Perpendicular straight crossed lines with 'J.W.S.' in lower left corner.
Cracked:	Yes, small area of left upper corner.
Other comments:	In a wooden box.

12 *Loboparadisae sericea* (Shield-billed Bower-bird)
Published in:	Sharpe's *Birds of Paradise,* vol. II, pl. 17 J. G. Keulemans & W. Hart del et lith Mintern Bros.
Measurements:	560 × 432 × 37 mm ($22\frac{1}{16}''$ × 17'' × $1\frac{7}{16}''$).
Weight:	Total: 27.3 kg (60 lb, 2 oz); stone alone estimated without box and plastic cover: 24.86 kg (54 lb, 12 oz).
Notations:	At top of stone in pencil 'Printed 350 March 10th 97 HEN' (the 'E' in centre is larger).
Defaced and how:	Yes. Perpendicular crossed scratched lines and 'J.W.S.' in lower left-hand corner.
Cracked:	No.
Other comments:	Whitish colour on stone in left upper corner. In a wooden case.
	Sharpe's 'Monograph of the *Paradiseidae*... and *Ptilonorhynchidae*' was published from 1891 to 1898.

GOULD, JOHN. Manuscript. [*c.* 1832] 'John Gould's manuscript for "The Birds of Europe" incorporating various letters and notes from other naturalists, vol. 1-4; 6-7.' Collection: Library of the British Museum (N.H.).

Volume 1 is labelled on the spine as *Raptores*; 2, 3 & 4 as *Insessores*; 6 as *Rasores* and *Grallatores*, and volume 7 as *Natatores* (Sauer 1976). Volume 5 is not in this set but is in the Library of the Academy of Natural Sciences of Philadelphia (Sauer 1978).

The six uniformly bound volumes measure approximately 250 x 220mm wide. Each pale-blue sheet has the scientific name of a bird at the top of the page. Under this are handwritten notes by Gould and his secretaries, plus notes from many other naturalists along with clippings from newspapers, magazines, and scientific journals relating to the specific bird.

Specifically noted in my cursory examination were notes by Marion Walker, John Gatcombe, J. H. Dunn, two drawings by Mrs Salvin, etc.

Birds of Europe was published from 1832 to 1837. It would undoubtedly be more correct to call this a Gould notebook or 'filing cabinet' for his *Birds of Europe* manuscript, rather than labelling it as the manuscript itself.

These six volumes and the one in Philadelphia should be collated and carefully compared with the published volumes.

GOULD, JOHN. Manuscript. [*c.* 1832] 'Europe/Insessores/Vol. V/Frillingillidae/Pars.' Collection: Number 71, Library of the Academy of Natural Sciences of Philadelphia.

The above title is on the spine of this volume. The pages of pale-blue paper measure 252 x 203mm wide and correspond in size and content with a set of six related volumes in the main library of the British Museum (N.H.). (See articles on these volumes (Sauer 1976 and 1978) and the preceding listing.)

There are several items of interest in this Volume V.

On page 1 is the note 'Handed to Dr. Roberts in 1952. Gift of R. M. DeSchauensee'. On page 2 is the notation, 'Gift Ja 13, 1953'.

Inside the cover of the book is a letterhead from Henry Sotheran Ltd of 2, 3, 4 and 5 Sackville Street, Piccadilly, London W1, not dated, which contains a typewritten note stating that this is 'John Gould's working notebook (binding labelled "Europe, Insessores, Vol. 5 - Fringillidae - Pars.") Containing Notes Sketches by him also Manuscript Contributions by Ornithologists who helped him in the compilation of his Works; includes some watercolour sketches of birds and details. Evidently used in the preparation of his work "The Birds of Great Britain" published 1862-73'.

Apparently this Volume V was purchased from Henry Sotheran and mistakenly they identified this as a working notebook for the *Birds of Great Britain* even though the binding was labelled 'Europe'. Where dates are included on notes or on drawings they are mainly of the 1850s and 1860s. One could assume that Gould had this volume, and the other six, bound as a notebook for his *Birds of Europe* published from 1832 to 1837, but that he continued to add to this 'filing cabinet' in later years, probably using some of the material also for his *Birds of Great Britain*.

It would be interesting to know how this one volume was separated from the other six volumes which are now in the British Museum (Natural History).

These seven volumes are similar in their arrangement, with each page being numbered in the right upper corner, and only one side of the sheet being used. On most pages the scientific name of a bird is written at the top of the sheet and occasionally a notation about 'Gould Plate 174', etc.

There are notes written in various hands, also seven sketches, apparently done by Gould himself and others by correspondents; one of these is a sketch of a bittern by John Gatcombe. Five letters are in a pocket at the back of the volume. These

PART FOUR BIBLIOGRAPHY OF JOHN GOULD, HIS FAMILY AND ASSOCIATES

letters are listed in the manuscript index of the Academy. Two letters are from Jno. W. Wheelwright to John Gould, dated 4 August 1862 and 21 February 1863, one from Osbert Salvin to Gould, 10 March 1856 [?], one from John Gatcombe to John Gould not dated, and one from John Gould to G. Ransome dated 24 January 1851.

On the final page, numbered 167, is an index to the Birds in the volume.

Then follows on the inside of the back cover an attached 'Index/to/G. R. Gray's/arrangement/of the/Trochilidae' which makes up an additional 43 pages in this volume. This is a manuscript index on paper that is watermarked 'C. Wilmot/1849'.

[GOULD, JOHN.] Manuscript. 1838-48. 'Gould papers, vol. 1.' Collection: Mitchell Library, Sydney, MS A172.

In 1938 Alec Chisholm visited England, and through an advertisement in the London *Times*, established contact with the descendants of John Gould. The great-grandsons of Gould, Drs Alan and Geoffrey Edelsten very magnanimously presented Chisholm with many items of great importance concerning Gould and John Gilbert.

In turn, Mr Chisholm eventually presented these documents to the Mitchell Library in Sydney.

Item A172 is volume one of these Gould papers, and in microfilm occupies frames 420 to 548 in CY Reel 461. An additional twenty-eight blank pages were not copied on microfilm. A photocopy of this item, and others of the series, was supplied to me by the Mitchell Library. The Library Council of New South Wales has kindly granted permission to me to list these documents and to publish quotations from them.

For purposes of organization, I will divide these documents in volume one into separate items:

1 The first printed page is a short introduction to this series as presented to the library by Mr Chisholm. Then follow two pages listing fifteen groups of documents which are in three volumes, plus a diary of John Gilbert when on Leichhardt's expedition of 1844 to 1845, and a letter from Leichhardt to 'my good companions'.

2 On four pages is copied Hindwood's 1938 article on 'Gouldiana' which appeared in the Gould Commemorative Issue of *Emu*, pages 235-8. In the *Emu* article Hindwood lists some of Chisholm's manuscript material, but the continuity does not conform to the items as bound in the volumes.

3 Letter from John Gould to 'My dear Sir', dated from 'Latt 14-21. N. Long. 22. 16. W/June 20th 1838', written by Gould but not signed. This six-page letter, and the following two items, apparently were drafts of letters intended for the Chairman of the Scientific Committee of the Zoological Society of London. Some of the information in these three items was published in the 1839 *Proc. Zool. Soc. London* 7: 139-45, from a letter by Gould, 10 May 1839, from Van Diemen's Land. This draft contains much more information on the early part of the voyage, than is in the published letter. The last date mentioned in this draft is 11 June. *See* Chronology at 20 June 1838 for entire letter draft.

4 Draft of letter from John Gould to 'My dear Sir', undated, written by Gould but not signed. Apparently this draft of four pages was written earlier than item 3, since the last date entered is 24 May.

5 A single sheet said to be 'from his journal, already copied in two previous letters'. This sheet contains entries from before 24 May and to 9 June 1838.

6 Letter from John Gould to his wife Eliza, 20 December 1838, written and signed by Gould. From 'Recherche Bay, *Thursday night*, 20 Decr. 38' this details some of

Gould's experiences on the voyage in a government ship with Lady Franklin when they attempted to reach Port Davey, Van Diemen's Land but were unsuccessful because of adverse winds. Gould did succeed in obtaining some bird specimens and he mentions them in a general way. This letter and the two following letters are published in their entirety under the respective dates in Part Three, Chronology.

7 Two letters from John Gould to his wife, Eliza, both from George Town, Van Diemen's Land, written and signed by Gould.

The first one is dated 8 January 1839. In it Gould relates that they had travelled overland from Hobart Town. 'I Collected a few birds on the way, among them a Splendid Eagle which I killed at Perth.' His plans were to leave soon for Flinders Island and then proceed to Kings Island which 'might take a month to accomplish the journey, that is if the winds prove contrary'. The statement 'I am happy to hear that Henry was drawing' is of interest, since there are bird egg drawings signed by him. The letter was addressed to 'Mrs. Gould/at Mr. Fishers/Davey Street/Hobart Town'.

The second letter from George Town is dated 'Sunday 20th Jany 1838 [1839]'. Gould reported they had 'arrived here yesterday from Flinders' since 'a fatal accident happened to one of the men who shot himself dead by uncautiously pulling the gun from the boat with the muzzle towards his chest, the cock of the gun caught the seat of the boat, and all was over with the poor fellow in half a minute... how great a shock I sustained... I almost hate the sight of a gun'. He gave up the idea of going to King Island, and was planning on returning overland to Hobart Town. This letter was addressed to Mrs Gould at Government House, Hobart Town.

8 A receipt for the purchase by Gould from Wm. Donald of shot, powder, percussion caps, etc. on 5, 7, and 23 January 1839.

9 Letter from John Gould to his wife, Eliza, 20 March 1839, written and signed by Gould from 'Dart Brook, River Hunter' relating to Gould's trip from Sydney through Maitland, comments on Charles and Stephen Coxen's extensive holdings at Dart Brook, on an expedition to the Liverpool Range, a drought for nine months in the area, etc. This letter is printed in Part Three, Chronology, under the date of 20 March 1839.

10 Letter from John Gould in London to Mr Lawton, George Town, 31 December 1840, written and signed by Gould, regarding the shipment of a 'Striped Hyaena or native Tiger', which Lawton said he had placed in salt, and which Gould was anxious to receive.

11 This consists of a series of 'Letters/to/Sydney/B. of A./Part 1'.

These twenty-three letters are written in Gould's worst hand, apparently meant as drafts for a secretary. They usually begin with 'I have at length the pleasure of sending to my various subscribers the first part of my new work on the Birds of Australia, etc.' A few are dated 31 December 1840 or 20 January 1841, but most are not dated.

The recipients of these letters, as deciphered by me, include Mr Aspinal, Ronald Gunn, Esq., Capt. Swanston, Mr Lewis, Sir G. Gipps, McArthur, McLeay, Dr O'Brian, and Alexander Walker Scot.

12 A draft of a letter from Gould undated, apparently from London after his return from Australia, to Mr Baker, gardener on an island near Newcastle. Gould reminds him of a request that his son obtain some eggs and nests for him, and states the prices he will pay for them, through Mr Walker Scot of Newcastle.

13 The title page of this item is 'Voyage from London/to/South Australia/1841/Longipennes/Descriptive Catalogue'. There are eleven pages in neat

PART FOUR BIBLIOGRAPHY OF JOHN GOULD, HIS FAMILY AND ASSOCIATES

handwriting listing fifty-five specimens of birds, mainly petrels, gulls, and albatross (Longipennes), collected on this voyage. The information on each specimen varies but includes date collected or seen, position of latitude and longitude, certain measurements, weight, gross description, sex, stomach contents, or whether entozoa were found or not. The first specimen was collected on 5 March 1841 at lat. 31°S., long. 21°18″W., which is off the west coast of Africa. The last date is 16 April 1841, but no location mentioned.

When this catalogue was listed by Hindwood (*Emu* 38: 237), the author or originator was unknown. Cleland (*Emu* 38: 536) correctly attributed it to George Grey, who, in 1841, travelled to Adelaide to become Governor of South Australia. Grey refers to this catalogue, one sent to Gould, and a copy sent to Mr Gray at the British Museum, in his letter to Gould from Adelaide, dated 21 November 1842 (*Emu* 38: 220).

14 On five photocopy pages is a very rough draft by Gould of his instructions to John Gilbert for his second, and fatal, collecting trip to Australia. Gilbert left England early in February 1842.

Another copy of these instructions, written in the careful hand of a secretary, was published by Longman (1922). The rough Gould copy contains two paragraphs not in Longman's article, and Longman's copy has extra paragraphs at the conclusion of the instructions.

15 'General Instructions for Mr. John Gilbert during his second journey to Australia...' In the hand of Edwin Prince, this is in the form of a letter, signed by John Gould, dated 21 January 1842 to Gilbert. On the fourth and final page is the statement written by Gilbert: 'Having carefully read over the preceding/details & propositions I agree thereunto/John Gilbert/Witness E. C. Prince/Signed Jan. 31, 1842'.

16 A receipt dated London 2 June 1842 'Received of Mr. Coxen by/payment of Mr. Gould the sum/of Ten pounds/£10.00/Mary Patterson/her X mark'.

17 A two-page letter from Gould, 29 April 1845, to I. R. Wheeler, Esq., written and signed by Gould, regarding the return of 'The letters of the immortal Linnaeus with many thanks for the loan of them'.

18 Letter of four pages from Gould, 13 March 1848, to an individual whose uncle is Robert Ransome, written and signed by Gould, concerning sending a hundred birds to him.

19 A small sheet written by Gould: 'for gen char [general characteristics?] that all I have records respecting the habits, manners and economy of the Australian Birds are written from observations made during the space of two seasons or ? months and etc etc etc.'

20 Six numbered sheets headed 'Introduction or Preface', apparently being draft notes for Gould's *An Introduction to the Birds of Australia*. It begins with 'When we consider the preponderance of sea over land in the Southern Hemisphere...'

21 A single sheet with a short pencilled notation by Gould headed 'Introduction'.

22 Draft of a letter from Gould, undated, to Lady Jardine, written and signed by Gould. Mainly concerned with the unexpected marriage of 'the great Philosopher and Mathematician Chester' who was between 60 and 70 marrying a 'fair young damsel of 25'.

23 The final item is a drawing of a Cassowary in three poses, with annotations. On the reverse is 'Supposed to be/Drawn by C. Gould/1880/In letter dated Jan. 14th.'.

[GOULD, JOHN.] Manuscript. 1838-61. 'Gould Papers, vol. 2/Letters from various persons,/1838-1861.' Collection: Mitchell Library, Sydney, MS A173.

Volume two of the Gould papers given Chisholm, occupy microfilm frames 549 to 776 in CY Reel 461.

For study purposes I have divided MS A173 into five folders.

Folder 1 Dated 11 May 1838. This is a four-page handwritten document titled 'Mr. John Gould/to Mr. William Gould & ors [others]/*Power of Attorney* to/carry on Business and act generally/in the affairs of Mr. Gould during/his absence from England'.

For this purpose Gould said 'I have determined to appoint my Uncle William Gould of Rochester in Kent [,] Robert Mitchell of Broad Street aforesaid gentlemen [,] William John Martin of Park Street Camden Town in the County of Middlesex Gentleman [,] and my Clerk Edwin Charles Prince and any two or more of them jointly and each of them severally to act for me as my Attornies and Attorney during my absence from England.' They were 'to manage carry on and conduct my business concerns as now carried on and conducted by me in Broad Street . . . and to sell and dispose or exchange all or any of my works on Ornithology and specimens of Natural History . . . to borrow for a temporary period any money from my Bankers Messieurs Drummond and Company . . . and if the said Robert Mitchell and Edwin Charles Prince or either of them shall so think fit for them or either of them to reside in my said dwelling house and to authorize and appoint retain and discharge and remove any Agent Clerk Workpeople Servants and others . . . also to make such payments and allowances to my Mother in Law Elizabeth Coxen for the maintenance of her the said Elizabeth Coxen and a female Servant in my said Dwelling house or elsewhere and also for the maintenance and education of my Children or for medical advice or for any purpose to promote their welfare and comfort . . . to make up adjust and settle all and every or any Accounts . . . and generally for me and in my name . . . to do execute and perform all and every other act and acts deeds and things which shall be necessary or expedient to be done . . . '

This document was dated 7 May 1838, signed by John Gould and witnessed by 'Joseph Baker/20 Broad Street'.

Folder 2 Here is a series of twenty-nine letters from Edwin C. Prince to John Gould, while the latter was on his Australian venture. These very important letters are briefly reported on by K. A. Hindwood (1938 *Emu* 38: 199-206). Prince's letters provide almost a day-by-day account of the activities of the London-based employees and family. Much is concerned with collecting money due on Gould's publications, paying bills which was a continuing struggle, and then follows an itemized account of receipts and expenditures for each month or so.

The steadiest and most substantial income received while Gould was away was that from the 'work room' or taxidermy business presided over by Joseph Baker. A great deal of information on Gould, the businessman, can be gleaned from these letters. Gould was indeed fortunate to have Edwin Prince acting so responsibly.

An index to the names of individuals mentioned in these letters would include most of the naturalists of England at that period. At some time these letters should be printed in toto.

The dates of the twenty-nine letters were from 31 May 1838 to 3 June 1840. This material is contained in sheets 5 to 120 in my photocopied collection.

Folder 3 A series of seventeen letters from John Gilbert to John Gould are in this section occupying sheets 121 to 194. The first letter is from Launceston, Van

PART FOUR   BIBLIOGRAPHY OF JOHN GOULD, HIS FAMILY AND ASSOCIATES

Diemen's Land, dated 30 January 1839 and the last from Perth, Western Australia, 15 December 1842, when Gilbert had returned to Australia the second time.

Four of these letters, along with the agreement between Gilbert and Gould concerning Gilbert's second trip to Australia, were printed in Chisholm's 1938 article (*Emu* 38: 186-99) in the Gould Commemorative Issue.

Whittell in his excellent series of three articles on Gilbert in Western Australia (beginning in *Emu* 41: 112) quotes from these letters quite completely, and adds many important notes on the species of birds collected, other collectors, and on the areas covered by Gilbert.

The place of origin and date of each one of these letters should be of interest in following Gilbert's travels.

1   Letter from Launceston, Van Diemen's Land, 30 January 1839.
2   From Perth, Swan River, 15 March 1839.
3   From Perth, Western Australia, 11 April 1839 (printed in Chisholm's 1938 article).
4   From Perth, Western Australia, 20 May 1839 (printed in Chisholm's article).
5   From Perth, 3 September 1839
6   From Perth, 4 September 1839.
7   From Perth, 28 October 1839.
8   From Perth, 14 January 1839 [1840]. This was addressed to 'J. Gould Esq. /Post Office/Sydney [crossed out] /Woollingong'. There is a Sydney postmark of February 25 [?] 1840.
9   From Sydney, 4 May 1840.
10 From Sydney, 15 May 1840.
11 From Sydney, 8 June 1840 (printed in Chisholm's article).
12 From Port Essington, 18 July 1840.
13 From Port Essington, 19 September 1840 (printed in Chisholm's article).
14 From H.M.S. *Pelorus*, Singapore, 29 April 1841.
15 Off Start [?] Point, 17 September 1841, returning to 'happy old England'.
16 From 'on board the Houghton Le Pherne/off the Nore Light/Saturday Feb. 5, 1842', on Gilbert's return trip to Australia.
17 From Perth, Western Australia, 15 December 1842, labelled as letter 'No. 13'.

Folder 4   Here is a series of six letters from Governor George Grey of Adelaide, South Australia, to John Gould from 13 August 1842 and ending 10 January 1843. These were published in their entirety by Chisholm in 1938 (*Emu* 38: 216-26). In addition, in this folder, is a two page 'List of Birds from Governor Grey's/collections sent to the British Museum' with the number of specimens adding up to 271 on the second page. These items occupy sheets 195 to 238 in my copy.

Folder 5   Two items complete this important MS. material. First is a letter from William Gilbert, father of John Gilbert, to John Gould, dated 21 August 1846, as follows:

Sir,
   I am extremely obliged to you for your kindness in forwarding to me as you received it, the news of the melancholy and untimely Death of my Son which I assure Sir has been a dreadful blow to me and my family, altho, I must say myself I was in a great measure prepared for it as I expected the same after hearing of the narrow Escapes of his first Expedition. I should have wrote to you Sir, before this, but the wrong Directions on the letters caused them to lay Dead in the post office for Several Days and as in your first letter you gave me an invitation to call on you at any time I might visit London. I will

make one any day and hour you please to let me know. I cannot come up without I know when I can see you Sir as I must lose a days work and travelling Expense. Anxiously waiting your answer Sir

I Remain Your humble Servant
William Gilbert
Acre Passage
Peascod Street
Windsor
Berks.

The final item is a letter from Charles Gould to his father, John Gould, dated 21 October 1861, from Hamilton, Tasmania. [This was published in 1939 by Chisholm (*Vic. Nat.* 56: 99-101).]

GOULD, JOHN. Manuscript. [*c.* 1838] 'Gould's Australian notebooks and mss.' Collection: Newton Library, Department of Zoology, Cambridge University.

A valuable set of nine folders containing 500 or so loose-leaf sheets. These were examined and photographed by me in September 1974 and McEvey (1968) also describes them.

The nine 'holders' are labelled with the orders of birds, such as *Insessores*, etc. An inside heavy paper folder has the Order and Generic names written on it.

Each enclosed page is devoted to a bird. Most typically the sheets have a lithographed coloured head of a bird at the top of the sheet. Then down the left side of the sheet is a column of a lithographed list as follows: 'Latin name, English Do, Colonial Do, Aboriginae Do, Synonyms, Locality inhabited, Permanent resident in . . . , Periodical visitant and at what season, Migratory or not, Occasional visitant, Haunts, Actions, Place of Breeding, Nest, No. and Colouring of Eggs, Differences in Colour of Sexes, Changes of Plumage, Young, Colouring of Bill,____eyes and feet, ____denuded parts.'

Some pages had only the printed list but a hand-drawn head or complete bird or two was added at the top of the sheet. Handwritten notes were placed on the lines adjacent to each item of the column. Some pages had many notes, others were blank but for the Latin name of the bird. Frequently extra sheets or notes were pinned to the pages or loosely inserted between the pages.

In my opinion, Gould had this 'notebook' made up before he left England for Australia. For illustrations he utilized the lithographic stones of the heads of the birds that appeared in his *Synopsis* issued from 1837 to 1838. In this way he could have at hand for reference and valuable notes the most complete 'book' on Australian birds. Further study, I believe, will bear out this premise.

What a treasure trove of Gouldiana!

[GOULD, JOHN.] Manuscript. [*c.* 1838] 'General List of Birds inhabiting/Australia and the adjacent/Islands.' Collection: Newton Library, Department of Zoology, University of Cambridge.

In September 1974 this bound notebook was examined and photographed by me. On the inside of the front cover at the top is 'Mr. Gould/20 Broad Street/Golden Square London' in Gould's hand.

The first page recto has the above title then space devoted to four birds with the names in Latin with synonyms, all written in a careful hand, mainly by Edwin Prince.

Unfortunately I did not have time to examine the notebook adequately. The librarian, Mr Ron Hughes, in response to my request for further information, states that this bound volume consists of sixty-eight recto pages with all the verso pages

PART FOUR BIBLIOGRAPHY OF JOHN GOULD, HIS FAMILY AND ASSOCIATES

blank except for page 57. This page has a rough sketch of two birds, obviously done by John Gould himself.

On many of the recto pages Gould has added notes, or other scientific names, or additional synonyms.

I believe that this volume, similarly to the nine folders of pages of Australian birds in this same Library, was also taken by Gould to Australia in 1838 to serve as a more handy notebook when on his exploration trips. This volume should be studied further.

[GOULD, JOHN.] Manuscript. [c. 1838] 'General List of Australian Birds.' Collection: Newton Library, Department of Zoology, University of Cambridge.

According to Mr Ron Hughes, librarian, who supplied me with photocopies, these are two unbound manuscript lists in a soft cover. Both items consist of twenty pages. The handwriting is not Gould's or Gilbert's. Some parts of the lists are written by Prince.

'Rough list No. 1' of twenty pages recto, has one verso used also. The Latin names of the birds are listed. On the nineteenth recto page there are notations by Gould on the page and on a sheet pinned on the page. They have to do with 'Rough Summary by Volume' of species in each volume of the *Birds of Australia* and some calculations of parts already done, and when the work will be completed.

'Rough list No. 2' also has twenty pages, recto only. The listing of species is practically identical with list No. 1 and in a careful secretarial hand. On page 8 are more amendments than usual and most of these are in John Gilbert's handwriting.

It is my opinion that these two lists of the known Australian birds were prepared before the Australian trip, and one was for Gould and one was for Gilbert.

[GOULD, JOHN?] Manuscript. [1840] 'Manuscript in Gould's Handwriting, containing Descriptions of 587 Birds, mostly one to a page. 4to. Letters "Boy's Birds." Loosely inserted is an older MS. on 12 ll. of paper, headed "Ornithological List" and containing Descriptions of 87 Birds'. Collection: Blacker-Wood Library McGill University.

These two manuscripts were listed as Item 1859 in Henry Sotheran's catalogue No. 81 of 1924 and were for sale for £7.7s.

'Boy's Birds' is listed under the Gould items by Casey Wood (1931) as 'an important and unique addition to the Blacker collection'.

Through the efforts of Eleanor MacLean (19 May 1978) five pages of 'Boy's Birds' were photocopied for me. All of these pages (1, 2, 149, 203 and 288) are written by the same hand, namely, Edwin Prince, in my opinion. Each sheet has a description of one bird species apparently from India. 22 January 1840 is a date [of collection?] on page 2, and 18 November 1840 is a date on page 288. The MS. was examined by me in September 1978.

The 'older MS' was also partially photocopied to include pages 1, 16, 17, and 22. The text contains descriptions of birds shot and observed in India. The writing is quite minute. On pages 16 and 17 the initials 'B.B.' occur three times. There are several dates of months and days throughout the descriptions. On the final page of the MS. (p. 22), however, are the dates of 'Nov'r 1842' and 'Nov'r 1843'.

In summary, it is my opinion that the 'Boy's Birds' MS. is a copy by Prince for Gould of notes by Captain W. J. E. Boys on birds of Asia. In a letter from Mrs Hugh Strickland to her father, Sir William Jardine, of 1 March 1848 (Roy. Scot. Mus.), there is a statement that Gould had purchased Captain Boys' set of birds and MS. notes. According to Salvin, (1882) Boys was 'an energetic collector of birds in Northern India'.

The second MS. is not truly an 'older' one. Could the initials 'B.B.' stand for Bill Boys? I wager that is true. Obviously, further study of both these manuscripts, and also of related Boys's material, is necessary to arrive at more than suppositions.

[GOULD, JOHN.] Manuscript. [*c.* 1842.] 'Catalogue of Skeletons/in Chip Boxes/purchased of Mr. Gould/in the Workroom upstairs No 1.' Collection: Royal College of Surgeons of England, item 275.h.5(34), by permission of the President and Council.

There are four sheets to this item. The first, not in Gould's hand, is the title-page as above. The second page is also a title page with 'Catalogue of Skeletons/in Chip Boxes./in the Top Work Room No. 1'. The third and fourth pages contain a listing in two columns of 'Number of Box' from 1 to 45, with second column containing each item in the box. The specimens include complete skeletons of birds and mammals, shells, tongue bone of a fish, bird skulls, organ of hearing of a whale, etc.

The collectors of some items are listed, and include Sir E. Hume, Mr South, Mr Bennett, Mr Clift, and Mr Owen.

[GOULD, JOHN.] Manuscript. [*c.* 1842.] 'List of Skeletons of quadrupeds and/Birds collected by Mr. Gould/during his Expedition into/Australia.' Collection: Royal College of Surgeons of England, item 275.h.5 (33), by permission of the President and Council.

Written in a flowing hand, not Gould's, is a seven-page list of specimens collected by Gould in Australia.

At the top of the list is 'Cranium of native of the Western Coast'. Next is a listing of '15 Entire Skeletons of 15 sp. of Mammalia', with the box number, common name and Latin name of each specimen.

Thirdly, is a list of '22 Crania of 19 species of Mammalia', each in a different box.

Fourth, is a list of 'Birds/Entire Skeletons', purported to be '141 Entire skeletons of 139 sp. of Birds'.

GOULD, JOHN. Manuscript. [1844]. 'On the sub-family *Odontophorinae*, or partridges of America. 8vo. p. 60. Original manuscript.' Collection: Blacker-Wood Library, McGill University.

Wood (1931) lists this under Gould material as 'John Gould's original manuscript in his own handwriting of a paper on the *Odontophorinae* which he read at the British Association meeting in 1844. It was never printed or published, only an abstract appearing in the British Association Report. Page 19 is unfortunately missing'.

Through correspondence with Eleanor MacLean, Librarian, and a visit to McGill in 1978, the following further information was ascertained.

This MS. is in a bound volume with a dark-green cover, and measures 9 x 7½ in. wide (228 x 190mm).

All the pages are written by a secretary, except for the final page 60. This is a sheet glued to a backing sheet. These 'concluding observations' are in Edwin Prince's handwriting. Throughout the MS. are a few notes in the hand of Gould.

The beginning of page one of the MS. is identical with the beginning of the 'Introduction' section in Gould's folio work on *Odontophorinae*.

For the missing page 19 of the MS., see the 'Introduction', page 20, on *Dendrortyx*.

[GOULD, JOHN.] Manuscript. [Several dates] 'Gould/Autograph/MSS and/Drawing on/Australian/Mammals/and Birds/c. 1848.' Collection: British Museum (N.H.).

PART FOUR BIBLIOGRAPHY OF JOHN GOULD, HIS FAMILY AND ASSOCIATES

This box was examined by me in 1974. On an attached slip of paper is this note: 'These manuscripts were discovered in the basement of the Bird Section [of the British Museum (Natural History)] and transferred to the Zoological Department Library in November 1960. F. C. Sawyer.' I was told it had been moved to the basement during World War II.

See Sawyer 1971 and Slater 1973 for very preliminary information on this box of notes, letters and drawings, at present uncatalogued, in the British Museum (N.H.). These relate primarily to Gould's and Gilbert's notes on and collections of Australian birds and mammals. There are probably 300 to 400 sheets of letters, drawings and list of specimens sent to Gould by Gilbert, Bynoe, MacGillivray, Krefft (with drawings by Angas), Strange and others.

GOULD, JOHN. Manuscript. 1878. John Gould's will. London, Somerset House.

I am indebted to Mrs Christine E. Jackson of Hertfordshire for sending me a copy of Gould's will. The will, dated 25 April 1878, is handwritten in small script on seven legal-sized pages. A codicil of half a page was dated 27 October 1880, approximately three months before Gould died.

Certain parts of the will, with insertion of a few punctuation marks, deserve quotation:

I John Gould of No. 26 Charlotte Street, Bedford Square in the County of Middlesex Esquire F.R.S. do hereby revoke all wills codicils and testamentary dispositions heretofore made by me and declare this to be my last will and testament. I bequeath to my daughter Eliza Muskett the wife of John Muskett of Diss in the County of Norfolk Esquire the complete set or copy of my Ornithological Works which I use as my own private Library copy, also my portrait recently painted in oils and all my Ornithological and other oil paintings absolutely, but my wish is that they shall all be kept in the family in the nature of heirlooms and my said daughter Eliza notwithstanding coverture shall have power by deed or will to give effect to my wish in this respect as she may be legally advised is practicable, and I make this bequest to my said daughter having regard to the fact that she has permanent accommodation for the same. I give to my son Charles Gould all of my fishing rods and fishing tackle of every description. I give to my daughters Louisa and Sarah in equal shares or to the survivor of them if one of them shall have died in my lifetime all my plates, plated articles, furniture, linen, glass, china, pictures, prints, musical instruments, books and other articles of household use or ornament, wines, liquors, and household stores and provisions which shall at the time of my death be in or about or belonging to or appropriated for my residence No. 26 Charlotte Street, but so nevertheless that this bequest shall not include any of my Ornithological Manuscripts, drawing books or specimens, which I intend so far as not specifically bequeathed to form part of my residuary estate. I bequeath to Miss Yates the Companion of my daughters and who has been in my establishment for upwards of forty years an annuity of fifty pounds during her life, And I direct that the said annuity shall be considered as accruing from day to day and shall be paid by equal half yearly payments the first payment thereof to be made at the expiration of six calendar months after my death. I give to my friend Miss Maryon Walker a legacy of fifty pounds. I give to the Widow and Daughters of my late confidential clerk and assistant Edwin Charles Prince the sum of one hundred pounds to be equally divided amongst them or the whole to go to the survivor of them at the time of my death. I Bequeath to my Artist H   C   Richter a legacy of one hundred pounds as a kind of remembrance for the purchase of a ring or any other article that he may prefer in memoriam, and to my artist   Hart who has a large family I give a legacy of two hundred pounds. I give to my man servant Molyneux a legacy of fifty pounds and to my female servant Sarah Bland a legacy of thirty pounds as an acknowledgement of their kind attention. And I give to my other two female servants, provided they shall have been in my service for one year or upwards at the time of my death, the sum of twenty pounds each. I devise and bequeath all my real and personal estate whatsoever or wheresoever of or to which I shall at my decease be seized possessed or entitled or overwhich I shall at my death have a general power of appointment (except what I otherwise dispose of by this my will or any codicil hereto) unto and to the use of the said John Muskett, my nephew John Cleve of the British Museum and Alexander Forbes Tweedie of No. 5 Lincoln Inn Fields their heirs executors and administrators respectively upon trust that they the said John Muskett, John Cleve and Alexander Forbes Tweedie or the survivors or survivor of them or the heirs executors or administrators respectively of such survivor shall in such manner and under such stipulations and upon such terms and conditions in all respects and at such period as they or he shall in their or his discretion think fit, sell, collect or otherwise convert into ready money according to the nature of the premises, all such parts of the same premises as shall not consist of ready money.

The income from 'one equal fifth part of the said trust premises' was to be paid to each of his daughters, Eliza Muskett, Louisa Gould, and Sarah Gould. The income from the final two-fifths was to be paid to Gould's son, Charles. If any of the children died without children, only £5000 was to be bequeathed to anyone else by them, and the balance remaining of the estate shares upon death was to revert back to the general estate for use by the remaining children. Any grandchild was to get the balance of the estate after the death of Gould's children.

Eliza Muskett and Alexander Forbes Tweedie were appointed Executors. Tweedie, as stated before was also a Trustee, and Gould also named him as his Solicitor. He wore three hats.

In the Codicil of 27 October 1880, Louisa Gould was also named a Trustee. Also 'I give to my Artist [blank space] Hart the sum of fifty pounds in addition to two hundred pounds given to him by my will, and I give to my man servant [blank space] Molyneux the sum of fifty pounds in addition to fifty pounds given to him by my will.'

The house at 26 Charlotte Street was to go to Louisa and Sarah. Gould's physician Dr Edward Meryon was also to receive one hundred pounds in addition to anything due him for medical services 'as an acknowledgement of my gratitude for his kind attention to me'.

The will and codicil was proved on 12 March 1881. Gould had died 3 February.

One cannot but help notice that Gould did not know the first names of Richter and Hart!

[GOULD, JOHN.] Newspaper. 1839. [Gould collection] Hobart Town, *Hobart Town Courier*, 24 May.

Whittell prints this article in his book (1954, 1:91) which states that by the above date Gould had collected 'about 800 specimens of birds, 70 quadrupeds (several of which are new), more than 100 specimens preserved whole in spirits, and the nests and eggs of about 70 species of birds, together with skeletons of all the principal forms'.

[GOULD, JOHN.] Newspaper. 1851. 'Mr. Gould's collection of Humming-birds at the Zoological Society's Gardens, Regent's Park.' *Illustrated London News*, 18: 479-80, 31 May.

Gould's 'plate glass cases, chiefly of octagonal form, are arranged throughout the room, in which the light is carefully modified with the view of giving effect to the jewel-like splendour which glitters in every direction'. A woodcut illustration of humming-birds with nest by Richter illustrates this long article.

[GOULD, JOHN.] Newspaper. 1852. 'Mr. Gould's collection of humming-birds in the Zoological Gardens, Regent's Park.' London, *Illustrated London News*, 12 June.

Christine Jackson (pers. corres. 6 June 1979) brought this article and the two accompanying illustrations to my attention. Since the article contains interesting or unique facts concerning Gould and his works before 1852, it is reprinted here:

This celebrated collection, which added so extensively to the attractions presented by the Zoological Society to their myriad visitors during last summer, has been recently re-opened on a new site, with the manifest improvement of no extra charge being made to the public for admission to it. The liberal spirit in which this change has been effected evinces a true desire, both on the part of the Society and of Mr. Gould, to meet the increasing taste of all classes for the cultivation of natural history, and therefore deserves our warmest commendation. The career of Mr. Gould has secured for him a place in the history of zoological literature which entitles him to a place in our gallery of portraits.

Mr. John Gould was born at Lyme, in Dorsetshire, on the 14th of September, 1804. At a very early age he evinced a strong desire for the study of nature; a desire which, we presume, may be termed innate, since no favourable opportunity presented itself by which it was called forth, neither had any inclination of the kind been manifested by any of the immediate members of his family. The interval between the

PART FOUR BIBLIOGRAPHY OF JOHN GOULD, HIS FAMILY AND ASSOCIATES

14th and 20th years of his age was spent under the fostering care of the late John Townsend Aiton, Esq., at the Royal gardens at Windsor, where a taste for botany and floriculture was added to his previous bent for zoology. Shortly after this he was induced to remove to London, as a field likely to afford a wider and more successful scope for his studies.

In the early part of 1830 a fine series of birds from the hill countries of India came into his possession. As this was the first collection of any extent which had been sent to this country from the great Himalayan range, most of the species comprised therein were new to science; and for this reason Mr. Gould was advised by the late Mr. Vigors to attempt the delineation and description of one hundred species under the title of "A Century of Birds from the Himalaya Mountains." This advice was followed, and the first part of a work, from the appearance of which dates the commencement of Mr. Gould's career as an ornithologist, was published on the 1st of January, 1831. The success of this work was far greater than could have been anticipated — so much so, indeed, as to induce Mr. Gould to commence another of a much more extensive and important character on the Birds of Europe, illustrated in 450 plates, forming five folio volumes. During the progress of this latter work the author's attention was directed to the South American group of Toucans, mainly from the circumstance of a living example being then in the possession of his friend Mr. Vigors. The study of this group induced him more than once to visit the principal museums of central Europe, in order to acquire the requisite information for a "monograph of the *Ramphastidae*." This being finished, he published a similar monograph of the *Trogonidae*. These various works having been completed, Mr. Gould, in the spring of 1838, left England for Australia, for the purpose of studying the natural productions of that country, of which previously so little had been made known; the result of his visit being the acquisition of a vast amount of most interesting information which has been duly laid before the public in "The Birds of Australia," in seven folio volumes, comprising figures and descriptions of upwards of 600 species, 300 of which were either new to science or very imperfectly known. Another result of his visit is the work on the Mammals of Australia, now in progress; and which, judging from the portion we have seen, bids fair to be a fit companion to the Birds, at present the most valuable and original of Mr. Gould's varied publications.

From the commencement of his career as an ornithologist, Mr. Gould's attention was especially directed to that lovely group, the *Trochilidae*, or humming-birds, but the late Mr. George Loddiges, of Hackney, being at that time engaged in adding to his fine collection, Mr. Gould surrendered the subject in favour of his friend, presuming that at some time or other he would be induced to favour the world with the result of his investigations: premature death having, however, prevented Mr. Loddiges from doing this, if, indeed, he ever contemplated it, Mr. Gould has been induced to resume his study of the subject, and to form the unrivalled collection now exhibiting to the public in the Gardens of the Zoological Society in the Regent's Park. The accompanying Engraving represents the interior of the elegant building erected for their reception, as it appeared in 1851: [Figure 47.] it has since been slightly altered, to afford greater facilities for the immense number of visitors who now have access to the collection. We give, also, the exterior of the building as it now appears. [Figure 49.] We must not omit to mention that Mr. Gould is now engaged in illustrating this very exquisite group in a style commensurate with their beauty; a recently-discovered process having enabled him to represent their metal-like brilliancy, as well as their rich and varied hues, in a manner which must be seen to be duly appreciated.

That the study of natural history, particularly of ornithology, is far greater than is usually admitted, must, we think, be evident when we find that the support conceded to Mr. Gould's publications has been such as to enable him to produce, by his own unaided efforts, a series of works, of which we have only enumerated a part, unrivalled for their extent and beauty. The list of the subscribers, with a copy of which we have been favoured, comprises indeed many of the crowned heads of Europe, the greater portion of her public libraries, and an immense number of the nobility and gentry.

The Portrait is copied by permission G. Ransome, Esq., from the Ipswich series of portraits of living scientific men.

The building in which the humming-birds are placed, and which forms the subject of our first Illustration, adjoins the Green Walk which leads from the Tunnel towards the Elephant's House, and that division of the Society's Garden in which the giraffes, the elands, the hippopotamus, and other African animals are crowded together, in such profusion as really astonishes when we take into consideration the difficulty of obtaining them in the first instance; and of preserving them in health within so limited an area in a climate which is so thoroughly unnatural to them.

The whole state of the establishment exhibits a continuance of the vigorous spirit of improvement which has made the Society celebrated throughout Europe; and rich as their collection was before, we are happy to observe that the additions which have been made to it since the close of last season are neither few nor unimportant. There is no period of the year at which the Gardens are more delightfully verdant, nor more abundant in colour, than at the present moment, when the hawthorns have not yet faded, and the rhododendrons are in the full luxuriance of bloom.

Of further interest is Item 507 in the University of Kansas Spencer Library–Ellis Collection. This is a 360 x 560mm-wide pencil sketch of the interior of the humming-bird house as it appeared in 1851 (Figure 49). The pencil sketch views the display from the same vantage point as in the picture from the *Illustrated London News* article (Figure 48).

The Kansas-Ellis Collection drawing appears to me to be by John Gould. Presumably this could have been Gould's sketch for the builder or architect.

[GOULD, JOHN.] Newspaper. 1854. 'The Crystal Palace, at Sydenham.' London, *Illustrated London News*, 7 January.

Mr B. Waterhouse Hawkins, who had earlier been employed by Gould as an artist and lithographer, was commissioned to construct 'gigantic restorations of the Extinct Inhabitants of the Ancient World', at the Crystal Palace in Sydenham:

On Saturday evening last (the last day of the year 1853) ... Mr. W. Hawkins ... invited a number of his scientific friends and supporters to dine with him in the body of one of his largest models, called the Iguanodon ... The incredible request [to dine] was written on the wing of a Pterodactyle, spread before a most graphic etching of the Iguanodon, with his socially loaded stomach ... Mr. Hawkins had one-and-twenty guests around him ... at the head of whom, most appropriately, and in the head of the gigantic animal, sat Professor Owen, supported by Professor E. Forbes; Mr. Prestwick [sic], the geologist; Mr. Gould, the celebrated ornithologist; and the Directors and officers of the Company.

The dinner, which was luxurious and elegantly served, being ended, the usual routine of loyal toasts were duly given and responded to ... after which this agreeable party of philosophers returned to London by rail, evidently well pleased with the modern hospitality of the Iguanodon. (Figure 53.)

[GOULD, JOHN.] Newspaper. 1857. [Steamship arrivals.] New York, *New York Evening Post*, 15 May.

The British Mail Steamer *Asia* had arrived in the morning of 15 May 1857. The boat had departed from Liverpool at 3 p.m. Saturday 2 May.

On page 2 of the newspaper is a note 'Passengers by the Asia ... Gould, Gould, Jr.'

[GOULD, JOHN.] Newspaper. 1857. [Lord Napier and Gould arrivals.] Washington, D.C., *Evening Star*.

From a search of this newspaper near the dates of Gould's visit to Washington, the following facts were uncovered:

Thursday 21 May 1857: 'Lord Napier, the new British Minister at Washington'. Gould met with Lord Napier in Washington, according to Charles Gould's letter of 11 June 1857. They possibly came over on the same vessel, the *Asia*, which arrived in New York City on 15 May.

Saturday 23 May 1857, page 4: 'Arrivals at the Hotels ... Willards' Hotel ... Mr. Gould, son, Eng.'

[GOULD, JOHN.] Newspaper. 1954. 'Australia's first big bird man liked Love Birds for breakfast'. Sydney, Australia, *Daily Mirror*, 11 May, p. 21.

A full-page dramatically written account of John Gould, 'an uncouth domineering ex-gardener's boy and protege of kings'. The article begins 'Late in 1848, the limited cultural, artistic and social circles in the embryo British colonies of Sydney and Hobart were thrown into a flutter by publication of the seventh and last volume of a monumental work costing the then enormous sum of £100. In delicate colours, the volumes depicted most of the birds of Australia, the first time they had ever been assembled on such a vast and luxurious scale.'

Other quotes: 'Gould made the world bird conscious and cashed in on it ... Working at the zoo convinced Gould that a fortune could be made by selling, stuffing, mounting, sketching and describing birds ... Coroneted carriages rolled up to his workrooms in Golden Square, London, bearing aristocrats with dead pets for mounting in life-like poses ... on the trip to Australia they took crates of preserving jars, barrels of preservatives, snares, guns, and letters of recommendations from high government circles ... he also ate birds ... Gould fell in love with the lovebirds, which

he was to introduce to the western world as pets... it did not prevent him, however, from shooting them for the pot... he declared their flesh "deliciously delicate" ... Gould found more than 300 new varieties of birds in Australia'. This is a good sample of the article. There are some errors in facts.

The article was sent to me by Alec Chisholm.

[GOULD, JOHN.] Obituary. 1881. *Nature* (London) 23: 364-5.

The writer, probably P. L. Sclater, speaks of Gould's 'qualities of a naturalist, an artist, and a man of business combined in one and the same person - the like of which will probably never be seen again'. Another statement was that 'no one ever heard him speak unkindly of any of his contempories [sic]'. This is quite a lengthy obituary and much of the information in it was also incorporated in Sharpe's biographical account of Gould in his *Analytical Index*.

[GOULD, JOHN.] Obituary. 1881. *Ibis* 5: 288-90.

Notable quotes from this obituary include the comment in regard to Gould's first work *Century of Birds* that 'In size this work rivalled the volumes of Levaillant, but far excelled them in the artistic excellence of the plates'. And later 'The plates of these [Gould's] works are of great merit; but, to our eye, there is always present in them too much of studied effect, which detracts from their scientific accuracy'. Finally 'With systematic ornithology Gould did not trouble himself much: he always used to say he was a follower of his first master, Vigors, and was content with his scheme of arrangement. But in discriminating minute specific differences between allied forms Gould had few equals; and though his judgment on such points sometimes carried him too far, it was seldom at fault'.

[GOULD, JOHN.] Obituary. 1881. *Proc. Linn. Soc. of London.* November 1880 to June 1882, pp. 17-18.

There are some sentences in this obituary that bear recording here because they indicate a close relationship of the author to Gould.

'[Gould's] early [bird] specimens obtained... when he was little more than fourteen years of age, show great skill in preparation.'

'Gould's drawings [for *Century of Birds*] afterwards were transferred to stone by his accomplished wife.'

'These various works were published by the author himself, owing to the fact that his first work went vainly begging for a publisher; afterwards, when advantageous terms were offered, he declined to treat. His enthusiasm for the objects of his care was such as to make him refuse to sell a copy of one of his books to a person who wanted to become a purchaser, but had thought fit to speak of him slightingly of the performance.'

The author of this obituary gives the date as 1827 when Gould came to London and 'received the appointment of Taxidermist to the Zoological Society's Museum'.

Gould 'was elected a Fellow of this [Linnean] Society on 15 January 1833.'

[GOULD, JOHN.] Obituary. 1881. *Zoologist*, third series, 5: 109-15.

This well-written obituary by J. E. Harting is a glowing one, but much of the material, especially the part on Gould's Australian journey, was taken from D. W. Mitchell's review of Gould's works that appeared in the 1841 *Westminster Review* (35: 271).

[GOULD, JOHN.] Obituary. 1882. *Proc. Roy. Soc. of London.* 33: 17-19.

P. L. Sclater was the author of this obituary.

Gould was elected a Fellow of the Royal Society of London in 1843. Sclater speaks of his Australian expedition as 'the most important and striking episode of Gould's career' and his publication of '41 folio volumes, illustrated by 2,999 plates, [as] a performance quite unrivalled in any other branch of literature'.

When this was written, in 1882, Gould's humming-bird collection had been purchased by the British Museum and consisted of upward of 5000 specimens.

[GOULD, JOHN.] Obituary. 1882. *Jahrb. der Erfindungen* (Leipzig) 18: 442.

This is short and consists mainly of a list of Gould's published works.

[GOULD, JOHN.] Original drawings. [*c*. 1846-84]. 'Gould's birds of Australia. Original watercolours [by] L.G.; W.H.B.G.; A.T.G.; L.P.G. 61 watercolour paintings, 1 lithograph (hand col.) in album 59 cm. x 48 cm. . . . Rex Nan Kivell collection, 5666.' Collection: National Library of Australia, Canberra.

The above entry is taken from the index card to this collection of drawings. Also on this card is the information that 'Some pages watermarked. Waterman's 1882, others 1876'.

Several of the drawings appear to be by John Gould himself.

Two different species of birds are often portrayed on one plate. Most species are from volume 7 of the *Birds of Australia*.

This set of drawings should be studied carefully and compared with the large collection at the University of Kansas Spencer Library.

[GOULD, JOHN.] Original drawings. 1862. 'Two framed water-colour paintings of Trout.' Collection: British Museum (N.H.).

'These were presented by Mrs. [Alan] Edelsten in 1967 and are attributed to J. Gould.'

1   Framed and measures 336 x 803 mm. The following is written in the right upper corner: 'Male *colne* trout (*Salmo faris*)/weight 12 lbs./killed in Mercers pool West Drayton/by J. Gould, F.R.S./15th May 1862/Length 27 inches,/girth 18¼".' The painting is in watercolour and pastel.

2   Framed and measures 407 x 886 mm. In right lower corners is: 'Male Thames trout *salmo faris*/killed at Cookham by J. Gould, F.R.S./26 April 1862/length 29 inches, girth 17 3/4".' This painting is also watercolour and pastel.

The drawings are not signed. Gould made a comment in a letter that Hart had drawn a fish for him.

GOULD, JOHN, AND OTHER ARTISTS. Original drawings. [No dates] [Original drawings] Collection: the author.

Beginning in 1947 I began purchasing original drawings that had been used for Gould's works. These came from the Henry Sotheran Ltd collection from Gould's estate. Two came from other sources.

Eighteen of the drawings are attributed to Mrs E. Gould, a few having notes in her hand. Eight are by John Gould, or they exhibit his pencilled corrections or additions. Other drawings are by H. C. Richter, W. Hart, Edward Lear, and Lieut. F. R. Stack.

Of special interest is a set of three items on the Great grey bower-bird illustrating the development of a lithograph. First is Gould's original rough sketch, then secondly the artist's (Richter's) finely finished rendition in watercolour, and finally the hand-coloured lithograph, which, incidentally, was never published. (Plate 29)

PART FOUR BIBLIOGRAPHY OF JOHN GOULD, HIS FAMILY AND ASSOCIATES

A table annotatating these items follows. This is the least information that should eventually be tabulated for *all* of 'Gould's' original drawings.

The drawings marked (K.U.) have been given to the University of Kansas Spencer Library by me.

Number	Name	Work	Provenance	Media	Signed	Attribution	Other
GS1 (K.U.)	Pomatorhinus temporalis Temporal pomatorhinus	*Birds of Australia* iv, pl. 20	Sotheran P.N. #22 item 2620; P.N. #9 item 1256	Pencil, ink & water-colour	No	E. Gould & J. Gould	Two birds on foliage. A 'J.G.' note about Pitta lithograph on reverse
GS2 (K.U.)	Turnix or Hemipode		Sotheran P.N. #22 item 2627	Pencil, ink & water-colour	'Mrs Gould's'	E. Gould	Two birds on ground
GS3 (K.U.)	Hemipodius velox Swift-flying Turnix or Hemipode	*Birds of Australia* v, pl. 87	Sotheran P.N. #9 item 799; P.N. #22 item 2621	Pencil, ink & water-colour	No	E. Gould	Two birds on ground
GS4 (K.U.)	Honey-eater or Thornbill		Sotheran P.N. #22 item 2624	Ink & water-colour; lower bird in pencil	Note in L.R.C. signed by E. Gould	E. Gould; lower bird by J. Gould	Two birds on green foliage
GS5	Melithreptus melano-cephalus Black-headed honey-eater	*Birds of Australia* iv, pl. 75	Sotheran P.N. #9 item 782*; P.N. #22 item 2611	Pencil, ink & water-colour	'V.D. Land' in L.L.C. in E. Gould's hand	E. Gould	Two birds in foliage
GS6	Pardalotus affinis Allied diamond-bird	*Birds of Australia* ii, pl. 39	Sotheran P.N. #9 item 775*; P.N. #22 item 2607	Pencil, ink & water-colour	No	E. Gould	Two birds in foliage
GS7	Cockatoo parrakeet	*Birds of Australia* v, pl. 45?	Sotheran P.N. #9 item 786; P.N. #22 item 2612	Pencil, ink & water-colour	'Mrs Gould'	E. Gould	Single bird on branch
GS8	Chlamydera nuchalis Great grey bower bird		Sotheran P.N. #22 item 2604	Pencil & water-colour rough sketch	No	J. Gould	Three birds on ground
GS9	Chlamydera nuchalis Great grey bower bird		Sotheran P.N. #22 item 2604	Pencil & water-colour finely finished	No	H.C. Richter	Three birds on ground

JOHN GOULD — THE BIRD MAN

Number	Name	Work	Provenance	Media	Signed	Attribution	Other
GS10	Chlamydera nuchalis Great grey bower bird		Sotheran P.N. #22 item 2604	Water-coloured lithograph	No		Three birds on ground; with pencilled notes
GS11 (K.U.)	Cinclosoma punctatum Spotted ground-thrush	*Birds of Australia* iv, pl. 4	Sotheran P.N. #2 item 2628	Pencil & water-colour; lower bird pencil only	No	E. Gould; lower bird by J. Gould	Two birds
GS12 (K.U.)	Bee-eater ? Merops		Sotheran P.N. #22 item 2625	Pencil, ink & water-colour	No	E. Gould	One bird
GS13 (K.U.)	Heteromyias cinereifrons	*Birds of New Guinea* ii, pl. 15	Sotheran letter of Feb. 25, 1954	Pencil & water-colour	'Drawing by W. Hart' in L.L.C.	W. Hart	Two birds in branch
GS14	Oriolus broderipi	*Birds of Asia* ii, pl. 73	Sotheran letter of Feb. 25, 1954	Pencil & water-colour	'Original drawing W.H.'	W. Hart	Bird on flower stem
GS15 (K.U.)	Ardea novae-hollandiae White-fronted heron	*Birds of Australia* vi, pl. 53	Sotheran P.N. #9 item 781; P.N. #22 item 2609	Pencil & water-colour	No	E. Gould & J. Gould	Bird on a limb
GS16	Australian goshawk	*Birds of Australia* i, pl. 17	Sotheran P.N. #9 item 780; P.N. #22 item 2608	Pencil, ink & water-colour	No	E. Gould	Two birds on a limb
GS17 (K.U.)	Meliphaga sericea White-cheeked honey-eater	*Birds of Australia* iv, pl. 25	Sotheran P.N. #9 item 782; P.N. #22 item 2610	Pencil & water-colour	No	E. Gould	Two birds in foliage
GS18 (K.U.)	Short-tailed petrel	*Birds of Australia* vii, pl. 56	Sotheran P.N. #9 item 791; P.N. #22 item 2616	Pencil & water-colour	'Mrs Gould'; also note in L.R.C. signed E. Gould	E. Gould	Bird on rock
GS19 (K.U.)	Great grey petrel	*Birds of Australia* vii, pl. 47	Sotheran P.N. #9 item 790; P.N. #22 item 2615	Pencil & water-colour	'Mrs Gould'; also in R.L.C. note signed E. Gould	E. Gould	Bird in water

PART FOUR BIBLIOGRAPHY OF JOHN GOULD, HIS FAMILY AND ASSOCIATES

Number	Name	Work	Provenance	Media	Signed	Attribution	Other
GS20 (K.U.)	Diving petrel	*Birds of Australia* vii, pl. 60	Sotheran P.N. #9 item 789 P.N. #22 item 2614	Pencil & water-colour	No	E. Gould	Two birds in water
GS21	Daption capensis Cape petrel	*Birds of Australia* vii, pl. 53	Sotheran P.N. #9 item 788; P.N. #22 item 2613	Pencil & water-colour	'Mrs Gould'	E. Gould	Two birds in water, one wounded
GS22	Procellaria conspicillata Spectacled petrel	*Birds of Australia* vii, pl. 46	Sotheran P.N. #9 item 794; P.N. #22 item 2617	Pencil & water-colour	'Mrs Gould'	E. Gould	Two birds in water, one wounded
GS23 (K.U.)	Flying fish		Sotheran P.N. #22 item 2629	Pencil & water-colour	No	E. Gould	Fish in air
GS24	Trogon		Sotheran P.N. #9 item 1924; letter of Feb. 25, 1954	Water-colour	Edward Lear	Edward Lear	On reverse is a detailed rough drawing in ink and pencil of another trogon
GS25	Trogon		Sotheran P.N. #9 item 1923; P.N. #22 item 2596	Pencil & water-colour	No	Edward Lear	One bird in watercolour and another in pencil
GS26 (K.U.)	Chlamydera maculata Spotted bower-bird	*Birds of Australia* vi, pl. 8	Sotheran P.N. #22 item 2619	Pencil & water-colour	No	J. Gould	In note in ink on my P.N. #22 is written 'preliminary sketch in colour, *not by Mrs Gould.*' On the accompanying folder is 'The Spotted Bower-Bird ...the original sketch for same by John Gould.'
GS27	St. Helena pheasant		Sotheran P.N. #9 item 2115; P.N. #22 item 2639	Ink	'F.R. Stack for J. Gould'	F.R. Stack	Note in hand of Lieut. Stack; male and female

JOHN GOULD — THE BIRD MAN

Number	Name	Work	Provenance	Media	Signed	Attribution	Other
GS28	St. Helena pheasant, No. 2		Sotheran P.N. #9 item 2115; P.N. #22 item 2639	Ink	'F.R. Stack for J. Gould'	F.R. Stack	Note in hand of Lieut. Stack; male with female on nest surrounded by seven young
GS29	Harpactes kasumba Kasumba trogon	*Birds of Asia* i, pl. 74; Mono. Trogons, Ed. 2, pl. 37	Sotheran letter of Nov. 15, 1967	Ink & watercolour	No	W. Hart	Finely finished male and female
GS30	Eurylaimus javanicus Javan eurylaime	*Birds of Asia* i, pl. 57	Sotheran letter of Nov. 15, 1967	Watercolour	No	W. Hart	Two males and a female in foliage
GS31	Cotyle riparia Sand piper	*Birds of Great Britain*, ii, pl. 7	Sotheran letter of Nov. 15, 1967	Pencil & watercolour	No	J. Gould (?)	Single bird with many others in distance flying around a sand cliff; published plate is *not* reversed
GS32	Rhipidura leucothorax	*Birds of New Guinea* ii, pl. 27	Ralph Ellis gave this to Dr Portia Bell Hume; gift to author on 11 Dec. 1975	Pencil, ink & watercolour	No	W. Hart	Two birds on foliage
GS33	Androdon aequatorialis	*Supplement to Trochilidae* pl. 1	Purchase & trade Aug. 1975 from Taylor Clark	Pencil, ink & watercolour	No	W. Hart & J. Gould	Three birds on a palm with a fourth sketched roughly in pencil by J. Gould; also note in hand of J. Gould

[GOULD, JOHN.] Original drawing. [No date.] 'Birds Eggs/Gould'. Collection: Mitchell Library: Sydney.

The catalogue entry for this item (M.L. Ref.: A3927) is as follows:

  Drawings of Australian birds' eggs by J. H. Gould, presumably Dr. John Henry Gould, son of John Gould. Watercolours. 72 plates. 30 cm. x 24 ½ cm.
(Drawings are on sheets of drawing paper mounted on the pages of the volume. One of the plates is signed 'J. H. Gould', another 'J. H. Gould del', others are initialled 'J. H. G.' and three have the initials followed by 'del'. There is some variation in the handwriting. These drawings are probably those referred to by J. Gould in letter to E. P. Ramsay, 23 Nov. 1867. The vol. came from a great grandson of Gould, Dr. G. Edelsten. It is described in detail in the *Emu* Nov. 1952, and in the *Victorian Naturalist* April 1955).

PART FOUR BIBLIOGRAPHY OF JOHN GOULD, HIS FAMILY AND ASSOCIATES

In the Chisholm article (1952 *Emu* 52: 301-5) he states that there are seventy-two plates, with eighty-seven species represented, and 180 eggs figured. The initials of 'J. H. G.' on many of the plates were felt to be those of Gould's son, John Henry Gould. *See* Gould, John Henry; Gould, Louisa, etc. [Egg drawings.]

I have not examined this item.

[GOULD, JOHN.] Portrait. 1851. George Ransome - T. H. Maguire. Ipswich Museum. [A series of portraits of scientific men of that institution.]

Ransome, in September 1851, published a series of portraits, including one of John Gould. (Figure 44)

The title on this frequently reproduced portrait is as follows: 'John Gould, F.R.S., F.L.S., etc./Honorary Member of the Ipswich Museum/M. & N. Hanhart, Lith. Printers.'

In the lower right corner of the portrait is: 'T. H. Maguire/1849'. Thus this portrait is usually listed as being by Maguire.

The portrait measures 290 mm vertical x 240 mm wide, and is pasted on a larger sheet measuring 540 x 380 mm.

Gould was 45 years of age in 1849.

[GOULD, JOHN.] Portrait. 1857. J. E. McClees, Philadelphia.

In the Library of the Academy of Sciences of Philadelphia, in Collection 457, are three states of a portrait of John Gould.

The first item is a 90 x 55 mm-wide photograph of Gould (Figure 58).

Secondly, there is a 280 x 220 mm-wide watercolour portrait of Gould, quite similar to the smaller photograph. On the reverse of this portrait is written 'To the Philadelphia/Academy of Sciences/with compliments of/Jas. E. McClees./John Gould/ Author of Monograph of the Trochilidae, etc.' Below this is a printed label as follows: 'J. E. McClees'/Philadelphia/Photograph Establishment,/No. 626 Chestnut St., below Seventh, ... ' (Figure 59).

The third finely finished [watercolour?] portrait, different but quite similar to portraits one and two, appears as the frontispiece in volume three of Wm. Baily's unpublished 1855-58 'Illustrations of the Trochilidae'. It measures 255 x 210 mm wide.

The question is, did Gould sit for the first small photograph when in Philadelphia in June 1857, or did he supply McClees with the photograph as one taken previously?

[GOULD, JOHN.] Portrait. 1875. Messrs. Maull & Fox, London.

In March 1875 Messrs. Maull & Fox took a photograph of Gould, aged 71 (Figure 63).

GOULD, JOHN. Prospectus. 1831. *The Birds of Europe.*

Dated 24 October 1831, this prospectus was bound in the folder with Numbers 15-20 of Gould's *Century of Birds* in the Kansas–Ellis Collection copy F-163. It is illustrated in Figure 4 in Part Two under *Birds of Europe.*

GOULD, JOHN. Prospectus. 1836. *The Birds of Europe.*

Dated April 1836, Gould states that 'fifteen Parts are already published'. This single sheet is bound in the Kansas–Ellis copy (Aves E62) of Part 1 of the *Synopsis.*

GOULD, JOHN. Prospectus. 1844. 'A Monograph/of/the Ortyginae/or/Partridges of America'.

This sheet is dated 4 April 1844 and measures 220 x 143 mm wide.'... It is proposed to publish this Monograph in three Parts, each Part to contain Ten Coloured Plates, representing as many species of the natural size, accompanied with appropriate scenery, descriptive letter-press, etc. The size will be imperial folio, and the price of each Part to subscribers £2.10s.... The first Part will appear on the 1st of September, and the succeeding Parts at intervals of three months.'

This item is in the University of Kansas Spencer Library Ellis Collection (MS P463B:1). It is illustrated in the section on Gould's works under this publication.

GOULD, JOHN. Prospectus. 1844. 'A Monograph/of/The Odontophorinae,/or/Partridges of America ...'

This 'November 1st 1844' prospectus in the Ellis Collection at the University of Kansas is of a large size, measuring 443 x 280 mm wide. It is item F122.

'It is proposed to publish the Monograph in three Parts, each Part to contain Ten Coloured Plates ... The price of each Part to Subscribers to be £2.10s.'

GOULD, JOHN. Prospectus and List of Subscribers. 1852. Prospectus/and/list of subscribers/to/Mr. Gould's works.

This is the earliest prospectus with a list of subscribers that I have found. It is dated 1 May 1852.

Three pages are devoted to the first section which is a listing or prospectus of his works. The listing is quite valuable because it contains much information on the completed works, such as the number of parts, number of plates per part, cost, and a *raison d'être*. Some incompleted works are also listed, and in the remarks by Gould relating to these works we see proposals that may or may not have been carried to completion. For instance, he contemplated three parts for his monograph *Macropodidae* and only two were published. Also for these incomplete works there is the mention of the number of parts published to date. Thus for *Mammals of Australia* Gould provides us with the information that only 'three Parts have been published'.

The second section is a list of 'names of the subscribers, patrons, and possessors of the various works'. This list occupies twenty-nine pages with 736 possessors of Gould's works. The individual or institution is listed in a left-hand column, and twelve additional columns extending over the two facing pages are used for each work published to date. An asterisk appears in the column under the appropriate work if purchased.

Needless to say, the listing of names with address is very helpful, particularly for one trying to decipher Gould's handwriting in his correspondence.

This copy is in the library of the Academy of Natural Sciences of Philadelphia.

See Sauer (1980) for a booklet on this edition and the following four editions of Prospectus and List of Subscribers.

GOULD, JOHN. Prospectus and List of Subscribers. 1856. Prospectus/of the works/on ornithology,/and on/the mammals of Australia,/by/John Gould, F.R.S., etc.

There is no title-page but the above appears on page one. The date of issue is 1 January 1856. The format is essentially the same as for the 1852 copy. This copy is in the Yale University Library.

GOULD, JOHN. Prospectus and List of Subscribers. 1866. A/prospectus/of, and/list of subscribers/to,/the works on ornithology, etc./by/John Gould, F.R.S.

PART FOUR BIBLIOGRAPHY OF JOHN GOULD, HIS FAMILY AND ASSOCIATES

The prospectus is dated 1 January 1866 in three places. The names are listed of 1010 'subscribers, to, or possessors of, the various works'.

The 1856 Prospectus and List of Subscribers is updated by this 1866 copy. The first section on works consists of four pages. The second section of subscribers or possessors of Gould's work has the individuals numbered, for a total of 1010. Data are included for only ten of Gould's works. A new format for this section appears with more compact columns so that the list of subscribers and the works possessed are on one page only, not on two facing pages as earlier.

Many statistics and conclusions can be wrought from this compilation, if one is interested. Palmer (1895 p. 71) for instance, had access to a prospectus of 1 January 1866 and he counted the number of Monarchs on the list (12), the Royal Highnesses, Dukes and Duchesses, etc. Palmer also added up the number of subscribers for each work, an exercise of greater value, and then, by further computation, concluded that by 1866 Gould had a subscription list amounting to £143,000.

This copy is in the Library of Congress, Washington, D.C. (Z8362 G691).

GOULD, JOHN. Prospectus and List of Subscribers. 1868. A/prospectus/of/the works on ornithology, etc./by/John Gould, F.R.S./(with a list of subscribers to, or possessors of,/the works.)

This prospectus of works and list of subscribers is similar to the one published in 1866. It was published 1 January 1868. Obviously it is updated, both with information on the works completed or in contemplation, and with the listing of persons or institutions.

The first section on works makes up four pages. Much of what was in the 1866 copy is identical.

The second section of subscribers or possessors of Gould's works has the individuals or institutions numbered. Data are included for only ten of his works. In these twenty pages, 1050 subscribers are listed.

The Honourable East India Company was marked as having forty copies of Gould's *Birds of Asia*. I wonder what happened to all of those copies.

This 1868 prospectus is in the library of the Academy of Natural Sciences of Philadelphia.

GOULD, JOHN. Prospectus and List of Subscribers. 1870. [No title-page, but it is a Prospectus of Gould's works and a list of names of subscribers to or possessors of his various works.]

This copy is dated 1 January 1870 and is similar to the 1866 and 1868 copies I have examined, but, of course, the 1870 copy is updated.

On twenty-one pages are listed 1062 names of individuals or institutions possessing Gould's ten works.

The National Library of Australia, Canberra has this prospectus.

GOULD, JOHN. Prospectus. 1874. 'List/of works/on ornithology etc./by/John Gould, F.R.S., etc.' London, published by the author.

Dated 1 January 1874, this is a single sheet printed on both sides. It measures 10¼ x 7⅞ inches wide (258 x 202 mm). Author's collection.

Twelve works are listed with prices, even if they were 'all sold'.

[GOULD, JOHN.] Review(s). In addition to being mentioned under the appropriate work, a few of the earlier reviews of Gould's works are indexed more completely in the following section.

[GOULD, JOHN.] Review. 1832. 'Gould, John, A.L.S.: The Birds of Europe... Part I. 20 plates; imperial folio. 2 l. 10s. plain; 3 l. 3s. coloured.' *Mag. Nat. Hist.* 5:535-6.

Appearing in the July 1832 issue of this journal, the review is quite complimentary: 'The artist and those associated in the undertaking appear equally qualified to depict the various inhabitants of the air, the trees, the land, the marsh, or the water; and much of that which constitutes real excellence will be found in this work, which we have examined with pleasure and recommend with sincerity.'

The reviews are signed 'J.D.', who apparently is J. Denson, co-editor with J. C. Loudon.

The species illustrated are not listed, but nine or so are mentioned by their common names.

[GOULD, JOHN.] Review. 1833. 'Gould, John, F.L.S.: The Birds of Europe... In Parts, imperial folio, each containing 20 plates...' *Mag. Nat. Hist.* 6: 135-6.

This concerns parts 2 and 3 'and [they] bear evident proof of increasing excellence as the numbers proceed... The figures, beautiful and varied, are drawn with great truth and knowledge of the subjects, the colouring is natural.' At the conclusion, in a critical vein, the reviewer (J.O.W.) writes 'There is, occasionally, though rarely, a light appearance of stiffness in the drawing of some of the figures, and the tone of colouring, in one or two instances, is too uniform.'

Seventeen or so species of birds are referred to by their common names.

[GOULD, JOHN.] Review. 1833. 'Transactions of the Zoological Society of London. Vol. 1. Part 1.' *Mag. Nat. Hist.* 6: 501-4.

There were ten essays in this first part, and Gould's article was the tenth. The title was 'On a new genus in the family of *Corvidae* (*Dendrocitta*)'. Gould's article is reviewed in a paragraph.

[GOULD, JOHN.] Review. 1837. 'A Synopsis of the Birds of Australasia [sic]. By John Gould, F.L.S., etc.... London, 1837'. *Mag. Nat. Hist.* 1 (New series): 51-2.

'We have hastily looked over the plates of Mr. Gould's new work, which is announced for publication in the 2d of this month [January 1837]... [they] will not lessen the author's high reputation as an artist and practical ornithologist.'

An abstract then follows consisting of five paragraphs of the prospectus. This is identical with the prospectus found bound in the Ellis Aves E62 copy in the original parts.

[GOULD, JOHN.] Review. 1837. 'A Synopsis of the Birds of Australia and the adjacent islands. By John Gould, F.L.S., etc. Part II.' *Mag. Nat. Hist.* 1 (New Series): 270.

'Nearly half the birds in this and the preceding part of Mr. Gould's "Synopsis" are new to science... it must constitute one of the most important ornithological works that have appeared in this country... for the plates are executed with such minute fidelity, and the characters exhibited by the head alone are, in most instances, so well marked, as to render a figure of the entire bird unnecessary.'

This journal was issued in May 1837. No species are listed.

[GOULD, JOHN.] Review. 1837. 'A Monograph of the Family Ramphastidae. By J. Gould, F.L.S. Three parts, folio. 1833-36.' London. *Magazine of Zoology and Botany* 1: 187-92.

This review of Gould's first work on the Toucans, obviously written by Sir William Jardine, is quite detailed. Comparisons are made between Gould's descriptions and

PART FOUR BIBLIOGRAPHY OF JOHN GOULD, HIS FAMILY AND ASSOCIATES

Wagler's 1827 descriptions of twelve species of *Ramphastos* and fourteen species of *Pteroglossi*.

Unfortunately for systematists, the reviewer does not state which species were issued in which of the three parts.

At the beginning of this review, a short history is given of Gould's earlier publications, but with no definitive dates of appearance of these works. Covered are his *Century of Birds*, *Birds of Europe* ('the first seventeen numbers of the "ouvrage de luxe" have appeared'), and *Trogonidae* which 'has reached its second number'.

[GOULD, JOHN.] Review. 1837. 'A Synopsis of the Birds of Australia and the adjacent Islands. By John Gould, F.L.S., etc. Part I. Royal 8vo. 1837.' *Magazine of Zoology and Botany* 1: 571-2.

By way of introduction, the reviewer, undoubtedly Jardine or Selby, quotes two long paragraphs from Gould's prospectus to this work. 'The work will be published in parts, each of which will contain eighteen plates, with letter-press descriptions. The price of the work is £1.5s. coloured; 15s. uncoloured; and the letter-press may be had separately for 5s.'

Several species are particularly noted by the reviewer.

[GOULD, JOHN.] Review. 1837. 'Birds of Europe. By John Gould, F.L.S. London, Folio.' *Magazine of Zoology and Botany* 1: 572-3.

An interim review, or really an announcement, that 'Parts 20 and 21 of this fine work are nearly completed, and will appear together. These numbers were expected to have contained all the European birds, but it has been found that another (Part 22) will still be necessary'. Some rare birds will be figured.

[GOULD, JOHN.] Review. 1838. 'A Synopsis of the Birds of Australia and the adjacent Islands. By John Gould, F.L.S. Part II Royal 8vo. 1837.' *Magazine of Zoology and Botany* 2: 266-7.

'The second number of this peculiarly managed work has just been forwarded to us. It equals its predecessor in the beauty of its finishing, and we have illustrations of the characters of forty species.'

[GOULD, JOHN.] Review. 1838. 'The Birds of Australia and the adjacent Islands. By John Gould, F.L.S. Part I. Folio 1837.'

'Icones Avium, or Figures and Descriptions of new and interesting Birds from various parts of the Globe. By John Gould, F.L.S. Forming a Supplement to his former works. Part I. Folio. 1837.' *Magazine of Zoology and Botany* 2: 357-8.

'The two works ... have been sent to us by their indefatigable author.'

'The Birds of Australia ...' contains figures and descriptions of ten species, which are listed.

Regarding the *Icones Avium*, the reviewer elaborates on the descriptions and naming of several species, including *Ortyx plumifera* 'procured in California by the late David Douglas'.

[GOULD, JOHN.] Review. 1838. 'The Birds of Australia and the adjacent islands. By John Gould, F.L.S. Part II. Folio. London 1838.' *Ann. Nat. Hist.* 1: 223-4.

First, the author [Jardine?] alludes with enthusiasm to Gould's contemplated trip to Australia and gives some suggestions for collections. Then he comments on the ten birds figured in Part II.

This article appeared in Number III of volume one, dated May 1838.

[GOULD, JOHN.] Review. 1839. 'Icones Avium... By John Gould, F.L.S. Folio. August 1838. Part II. Monograph of the Caprimulgidae.' *Ann. Mag. Nat. Hist.* 2: 222-3.

Eight species are figured and described with no less than five new generic names.

[Gould was asked to prepare and write this monograph at the 1837 meeting of the British Association. It was to be presented at the 1838 meeting at Newcastle. In the meantime Gould had sailed for Australia so he asked Jardine to lay the Part before the meeting. Apparently Jardine did not do this for Gould. *See* British Association Reports.]

[GOULD, JOHN.] Review. 1841. 'The Birds of Australia. By John Gould, F.L.S., etc. Part First. Oblong folio. Published by the author, London, December 1840.' *Ann. Mag. Nat. Hist.* 6: 471-2.

The first part of *Birds of Australia* was issued with seventeen plates at a cost of three pounds, three shillings. Each subsequent part was to be issued at three-month intervals.

The author of this review, undoubtedly Sir William Jardine, also stated that Gould made collections in Australia of specimens other than birds and mammals. 'To Mr. Brown has been sent the collections of plants, the Reverend Mr. Hope has the insects, and to Professor Owen has been entrusted all the preparations fitted for dissection; even Mr. Denny has not been neglected.'

[In a letter to Jardine, 13 February 1841 (Roy. Scot. Mus.), Gould made a comment relevant to this review: 'Why do you or Taylor call my work oblong folio which happens to be just the reverse of its form.']

[GOULD, JOHN.] Review. *See* Mitchell, D. W. 1841 for reviews of six of Gould's early works.

[GOULD, JOHN.] Review. 1842. 'Gould's Birds of Australia. Parts I to VI. Folio. 1841-1842.' *Ann. Mag. Nat. Hist.* 9: 337-9.

The following quotations summarize this review: 'Great as is the excellence of Mr. Gould's former publications, there can be no doubt that the present work exceeds them all, both in an artistic and in a scientific point of view. Additional practice in designing and additional opportunities of studying animated nature have greatly improved his pictorial powers, while his recent excursions in the wilds of Australia have supplied him with a mass of novel and original information of the highest value to the ornithologist.'

'Mr. Gould's designs show a remarkable freedom from mannerism. Whether he wishes to represent the torpor of the drowsy Podargus, the dignified repose of the eagle, the pert Malurus, the restless parakeet, or the lean and anxious wader, he is equally successful in his efforts. No attitude of action or of repose which is consistent with natural habits comes amiss to him, and in this respect he preserves a happy medium between the stiff formality of Temminck's 'Planches Coloriees,' and the occasional extravagances of Audubon.'

[GOULD, JOHN.] Review. 1843. 'The Birds of Australia by J. Gould, F.L.S., etc. Parts VIII, IX. Oblong folio.' 1842. *Ann. Mag. Nat. Hist.* 11: 58.

The comment is made by the anonymous author that 'we have illustrations of many new genera, (perhaps too many.)'

[GOULD, JOHN.] Review. 1869. 'Gould's Birds of Great Britain parts xiii and xiv'. *Ibis* 5 (new series): 108-9.

PART FOUR BIBLIOGRAPHY OF JOHN GOULD, HIS FAMILY AND ASSOCIATES

This review, undoubtedly by the Editor Alfred Newton, is critical in several minor points. The reviewer feels that it is a mistake to call a bird a British species if it is only a stray. Then apparently Gould overlooked some breeding records, one which had appeared in *Ibis*.

Finally, the reviewer requested that the non-binomial 'names' of Brisson not be used.

[GOULD, JOHN.] Review. 1889. 'Gould's "Birds of New Guinea".' *Auk* 6: 269.

A 'magnificent contribution to ornithology' states J.A.A. [J. A. Allen]. Only twelve parts were issued before Gould's death in 1881, but R. Bowdler Sharpe prepared the thirteen remaining parts, the last part issued in December 1888.

[GOULD, JOHN.] Review. 1889. 'Gould's Supplement to the Trochilidae.' *Auk* 6: 334.

The reviewer, J.A.A., [J. A. Allen] states 'that the magnificent monograph of the most beautiful and interesting family of birds, illustrated with a delicacy and gorgeousness well befitting the gems of bird life it portrays, is not only a monument to the author, but one of the most attractive ornithological works yet produced'.

GOULD, JOHN. Specimens. [Various dates.] Bird Specimens, Yale University Peabody Museum of Natural History.

As an incentive for a museum to purchase his publications, Gould would often offer a set of duplicate skins.

As noted in the Gould correspondence with James D. Dana and Edward C. Herrick of Yale University, Gould offered fifty skins to them. Yale in turn purchased directly from Gould at least one set of folio volumes, namely, *Trochilidae*.

In an attempt to trace these bird specimens, I sent a letter to the Peabody Museum. The response from Fred Sibley was more than anticipated.

First, he was able to find the original catalogue containing around fifty specimens with the notation 'Pres. by John Gould'. Then he supplied me with an update on the scientific names of the birds on that list. Finally, he was able to identify twenty-three specimens currently in the Yale collection as the original Gould material. It is possible other Gould specimens may be uncovered.

Mr Sibley's detective work is greatly appreciated and proved productive.

GOULD, JOHN. Stamp. 1976. Famous Australian Stamps.

Included with John Gould (1804-1881) in this series of four commemorative stamps from Australia are Sir Thomas Laby (1880-1946), Sir Baldwin Spencer (1860-1929) and Professor Griffith Taylor (1880-1963).

GOULD, JOHN. 1861. *An introduction to the Trochilidae*. London, Taylor and Francis.

This is copy C447a in the University of Kansas–Ellis Collection. Ellis wrote on the inside front cover that this was 'Gould's own proof copy'. Mengel agreed and added 'compared with Gould MS on hand and evidently true'.

There are a few notations throughout the book, some in Gould's hand (pp. iii, 14, 62) but the majority are by E. Prince (pp. 1, 63, 99). Some sections of the book have been removed by scissors.

This volume is in an original black binding. According to a note by Ellis, it was purchased from 'H. Sotheran £0/18/9 Mar. 11, 37'.

GOULD, JOHN. 1861. *An introduction to the Trochilidae*. London, Taylor and Francis.

This copy in the University of Kansas–Ellis Collection (C447b) has many annotations and comments written in a shaky hand, that is not Gould's in my opinion. The author especially comments on grammar (pp. 3, 4), and on genera, types, and species names (pp. 40-41). Some comments are clever, namely 'Type specimens, Type species, Type forms, Type genera, type fiddles and fiddlesticks'.

In an original red binding. Ellis noted that this copy was purchased from 'H. Sotheran £0/18/9 Mar. 11, 37'.

GOULD, JOHN. [1958]. *I nostri ucelli*. Milano, Görlich.

The Library of Congress copy is bound, but I would assume that this originally was a folder of quarto-size pages.

The fourteen plates by Gould of 'Our birds' are tipped-in on heavy sheets. The only printing on the plates is the name of the bird. There is no text.

GOULD, JOHN. [1966]. *Schöne alte Vogelbilder. John Gould und die grobe Zeit der Vogelmaler geschildert von Richard Gerlach*. [Klogenfurt], Kaiser.

Sixteen coloured plates from several of Gould's works illustrate this volume.

GOULD, JOHN. 1972-75. *The birds of Australia*; with the *Supplement*; and the *Handbook*; a facsimile set, 9 vols. Melbourne, Lansdowne Press.

This facsimile edition was limited to 1000 sets, with the plates of the eight imperial-folio volumes produced by chromolithography. Those responsible deserve plaudits on the printing and binding. Chromolithography from photographs of these large plates is never too satisfactory, but Lansdowne Press has done an admirable job.

Most facsimile reprints unfortunately are just that. No additional or updated information on the work or the author is appended. Fortunately, in the one volume facsimile of the *Handbook* which originally was bound in two volumes when published by Gould in 1865, there are two worthwhile prefatory articles.

The first is 'A note on the Handbook' by the Australian authority on John Gould, A. R. McEvey. In six pages there is quite a bit of factual updating by McEvey, covering the period from Gould's first *Synopsis* begun in 1837, through five additional works by Gould on Australia, and concluding with a mention of the 1926-67 'Handbook/RAOU check list' with supplements.

Regarding the Gould *Handbook*, McEvey has this to say: 'In addition to the extraordinary amount of good observation it records, and to its plentiful snippets of ornithological and, indeed, Australian history, the *Handbook* offers pleasing samples of natural history prose of the nineteenth century which, verbose or not, reflect the dignity that the age expected of its science.'

The second prefatory article is 'John Gould: a biographical note' by John Currey. This is essentially an abstract of R. B. Sharpe's biographical memoir in his 1893 *Analytical Index*. The date of Gould's death is in error — he died 3 February 1881, not 3 September.

GOULD, JOHN. 1979. *A synopsis of the Birds of Australia and the adjacent islands* by John Gould, F.L.S., etc. . . . Facsimile. Melbourne, Queensberry Hill Press.

A fine example of the book-publishing art. The colour plate reproductions are of excellent quality. Limited to 387 copies.

Following the title-page is a three-page 'Note on Synopsis' by Allan McEvey. A 'Publishers Note' by Peter Marsh is primarily concerned with some copies of the work having the plate on *Acanthorhynchus* incorrectly captioned as *Canthorhynchus*. Then follows an undated original two-page 'Prospectus', and a two-page list of 181 'Subscribers' to the 1979 facsimile edition.

Plate 35 Drawing in watercolour, signed W. Hart, of *Lophorhina minor*, for *Birds of New Guinea*, I: 18 and Sharpe's *Birds of Paradise*, II: 16, Kansas-Ellis-Gould MS 973.

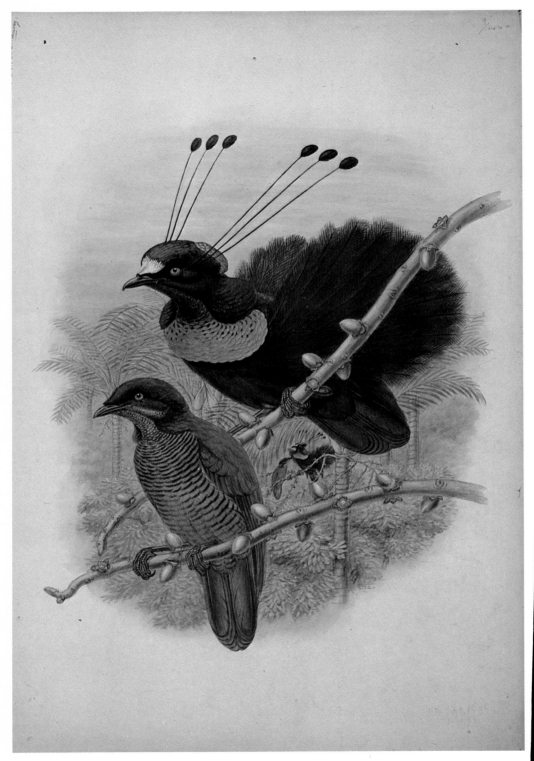

Plate 36 Drawing in watercolour, signed W. Hart del, of *Parotia sexpennis*, for *Birds of New Guinea*, I: 26 and Sharpe's *Birds of Paradise*, II: 12, but not similar in design to either of the two lithographed plates, Kansas-Ellis-Gould MS 974.

PART FOUR BIBLIOGRAPHY OF JOHN GOULD, HIS FAMILY AND ASSOCIATES

A separate folder with six extra colour plates accompanied my copy.

GOULD, JOHN. 1979. *The birds of Australia, and the adjacent islands.* Facsimile. Melbourne, Lansdowne Editions.

A fine four-star bird book with beautiful and accurate reproduction of the twenty colour plates with text. Limited to 500 copies.

Allan McEvey, in five appendices, elaborates in depth on this Gould publication of 1837-38. He utilizes many Gould-Jardine letters for relevant information.

In a table of nomenclature McEvey supplies current names for Gould's twenty species, with other valuable comparisons.

GOULD, JOHN. 1981. *A monograph of the Macropodidae or family of Kangaroos.* Facsimile. Melbourne, Lansdowne Editions.

A full-size facsimile of the two parts Gould published on the Kangaroos in 1841-42. Lansdowne Editions sent me an unbound copy to examine for inclusion in this work.

The two covers of the parts as issued are the only prefatory material. The thirty colour plates are exquisitely reproduced. The paper is of excellent quality, with a sheet separating each plate from the text page. A four-star publication.

The Gould facsimile part is beautiful enough, but the most valuable addition to this volume is the material appearing in the five appendices at the back of the publication. Here Joan M. Dixon, Curator of Mammals at the National Museum of Victoria, has done what I wish could be done for every one of Gould's works. (In April 1979 I wrote Ms. Dixon and suggested that this kind of study on the Kangaroos would be very worthwhile.) In these appendices Dixon covers the known history of Gould's work on the Kangaroos, an essay about John Gilbert, Gould's principal collector, taxonomic notes on the Kangaroos, a table comparing the nomenclature in this work with that in volume two of Gould's *Mammals of Australia* and with current names, and, finally, recent information on the species figured in this volume. She has done a very scholarly job on these subjects.

GOULD, JOHN HENRY; GOULD, LOUISA, AND OTHER ARTISTS. Drawings. Several dates including [1850], 1860 and 1862. [Egg drawings and lithographs.] Collection: Author.

A collection of egg drawings in a box with 'Oological Drawings/Louisa Gould' on the spine.

Enclosed is a sheet of paper with the following text: 'These egg plates were done by Louise (second daughter of John Gould) unmarried. These plates have been in the family since John Gould died & were passed down to my husband, the late Dr. Alan Edelsten, by his mother who was the only grandchild of John Gould. Some of the plates may have been drafted by John Gould initially. [Signed] Grace E-D Edelsten 11-9-77.'

This collection was purchased from a British dealer. It consists of fourteen original watercolour drawings of eggs, twenty hand-coloured lithographs, and 45 uncoloured lithographs, for a total of seventy-nine items.

In Lord Derby's Old Knowsley Library is a slim volume entitled 'Gould's Drawings of Eggs'. This contains nineteen egg drawings, twelve of which are signed in pencil 'Henry Gould del 1850'. The other seven drawings are the originals for seven lithographs in my collection. They are unsigned, and have the title written by a different hand from the other twelve. Presumably these seven could have been drawn by Mrs Gould, John Gould, or by another artist. (I acknowledge the assistance of the present Lord Derby, and his librarian, Diana Kay, in providing me photocopies of these original egg drawings.)

A third set of egg drawings is in the Mitchell Library, Sydney. According to Suzanne Mourot, librarian, this volume is entered as 'Drawings of Australian birds' eggs by J. H. Gould, presumably Dr. John Henry Gould, son of John Gould. Watercolours. 72 plates. 30 cm. x 24 ½ cm.' This volume was presented to the library by Alec Chisholm, who in turn had received it as a gift from Dr Geoffrey Edelsten of England. *See* Chisholm 1952 *Emu* 52: 301-05. It would be valuable to compare further these three collections of egg drawings and lithographs.

The seventy-nine items in my collection are delineated here. All the drawings and lithographs are unbound, and, unless otherwise noted, the sheets measure approximately 11 x 7½ inches wide (280 x 190 mm).

Number	Species as on sheet	Description	Media	Signed	Other
GCS 0-1	Grus Australasianus	Two large eggs	Drawing-Watercolour	L.G. May 2/? in Rt. Lower Corner	
GCS 0-2		One small egg	Drawing-Watercolour	No	Watermark (partial): J. Whatman
GCS 0-3	Moorhen (Gallinula chloropus)	One egg	Drawing-Watercolour	L.G. June 3rd 1860 in R.L.C.	
GCS 0-4	Moorhen (Gallinula chloropus)	One egg, different from 0-3	Drawing-Watercolour	L.G. June 5th/1862 in R.L.C.	
GCS 0-5	Thrush (Turdus musicus)	Two eggs	Drawing-Watercolour	May 28th/ L. Gould 1862 in R.L.C.	In L.L.C. 'from nature.' On verso are drawings of two cubes
GCS 0-6	Susura inquala [?]	Two eggs in different manner	Drawing-Watercolour	J.H. Gould faintly in L.L.C.	Watermark: J. Whatman/ 1850
GCS 0-7		Two eggs	Drawing-Watercolour	No	
GCS 0-8	Tropidarhynchus corniculatus	Two eggs, both similar to those in 0-6	Drawing-Watercolour	J.H.G. in L.L.C.	
GCS 0-9		Three eggs but only one finished	Drawing-Watercolour	No	
GCS 0-10	Kingfisher's Eggs	Two eggs	Drawing-Watercolour	L.G. June 10th 1/6 [?] in R.L.C.	
GCS 0-11		One egg	Drawing-Watercolour	No	
GCS 0-12		Two eggs, only one finished	Drawing-Watercolour	No	Watermark: J. Whatman/1852
GCS 0-13	Malurus longicaudatus [?]	Two eggs	Drawing-Watercolour	J.H. Gould in L.L.C.	
GCS 0-14		One large egg	Drawing-Watercolour	E.G. in L.L.C.	

PART FOUR BIBLIOGRAPHY OF JOHN GOULD, HIS FAMILY AND ASSOCIATES

Number	Species as on sheet	Description	Media	Signed	Other
GCS 0-15		Three eggs	Hand-coloured lithograph	No	
GCS 0-16 [Same as 0-15]					
GCS 0-17	Corvus Coronoides	Three eggs	Hand-coloured lithograph	No	Original is Number 11 in Lord Derby's Collection, unsigned. Reversed from lithograph.
GCS 0-18 [same as 0-17]					
GCS 0-19	Otis Australasianus	Two large eggs	Hand-coloured lithograph	No	Original is No. 12 in Lord Derby's Coll., unsigned. Not reversed.
GCS 0-20	Phaeton phoenicuris	Two large eggs	Hand-coloured lithograph	No	Original is No. 14 in Lord Derby's Coll., unsigned. Reversed.
GCS 0-21	Pandion leucocephalus	Two large eggs	Hand-coloured lithograph	No	
GCS 0-22 [Same as 0-21]					
GCS 0-23	1. Pomitorhinus superciliosus 2. _____ temporalis	Four eggs	Hand-coloured lithograph	No	Original is No. 10 in Lord Derby's Coll., unsigned. Not reversed.
GCS 0-24 [Same as 0-23]					
GCS 0-25	1. Ieracidea Occidentalis 2. _____ Berigora	Four eggs	Hand-coloured lithograph	No	Original is No. 8 in Lord Derby's Coll. unsigned. Not reversed.
GCS 0-26 [Same as 0-25]					
GCS 0-27	Esacus magnirostris	Two eggs	Hand-coloured lithograph	No	Original is No. 13 in Lord Derby's Coll., unsigned. Reversed
GCS 0-28 [Same as 0-27]					
GCS 0-29	Corcorax leucopterus	Two eggs	Hand-coloured lithograph	No	Original is No. 9 in Lord Derby's Coll., unsigned. Reversed.

JOHN GOULD — THE BIRD MAN

Number	Species as on sheet	Description	Media	Signed	Other
GCS 0-30 [Same as 0-29]					
GCS 0-31 Falco melanogenys		Two eggs	Hand-coloured lithograph	No	
GCS 0-32 [Same as 0-31]					
GCS 0-33 Thalasseus poliocerca		Six eggs	Hand-coloured lithograph	No	
GCS 0-34 [Same as 0-33]					
GCS 0-35		Two large eggs	Uncoloured lithograph	No	
GCS 0-36 [same as 0-35]					
GCS 0-37		Two large eggs	Uncoloured lithograph	No	
GCS 0-38 [Same as 0-37]					
GCS 0-39		Two medium eggs	Uncoloured lithograph	No	
GCS 0-40 [Same as 0-39]					
GCS 0-41		Large sheet, 22 × 15 inches wide, with two folds, forming four plates, numbered 41a, b, c & d	Uncoloured lithograph	No	
GCS 0-42 [Same as 0-41b]		Five eggs	Uncoloured lithograph	No	
GCS 0-43 [Same as 0-41c]		Two eggs	Uncoloured lithograph	No	
GCS 0-44 [Same as 0-41d]		Four eggs	Uncoloured lithograph	No	
GCS 0-45 [Same as 0-41a]		Three eggs	Uncoloured lithograph	No	
GCS 0-46		Large sheet as 0-41, numbered 46a, b, c & d. Two eggs on each plate.	Uncoloured lithograph	No	
GCS 0-47		Five small eggs	Uncoloured lithograph	No	
GCS 0-48 [Same as 0-47]					
GCS 0-49		Six eggs	Uncoloured lithograph	No	

PART FOUR · BIBLIOGRAPHY OF JOHN GOULD, HIS FAMILY AND ASSOCIATES

Number	Species as on sheet	Description	Media	Signed	Other
GCS 0-50 [Same as 0-49]					
GCS 0-51		Two eggs	Uncoloured lithograph	No	
GCS 0-52 [Same as 0-51]					
GCS 0-53		Two small eggs	Uncoloured lithograph	No	
GCS 0-54 [Same as 0-53]					
GCS 0-55		Two eggs	Uncoloured lithograph	No	
GCS 0-56 [Same as 0-55]					
GCS 0-57		Four eggs	Uncoloured lithograph	No	
GCS 0-58 [Same as 0-57]					
GCS 0-59		Three eggs	Uncoloured lithograph	No	
GCS 0-60 [Same as 0-59]					
GCS 0-61		Four eggs	Uncoloured lithograph	No	
GCS 0-62 [Same as 0-61]					
GCS 0-63		Two large eggs	Uncoloured lithograph	No	
GCS 0-64 [Same as 0-63]					
GCS 0-65		Large sheet, 11 × 15 inches wide, with one fold, forming two plates, numbered a & b. 0-65a has two eggs, but are identical. 0-65b has two sets of eggs which are identical.	Uncoloured lithograph	No	
GCS 0-66		Large sheet, as 0-65, but with only one egg on 0-66a, & two eggs on 0-66b.	Uncoloured lithograph	No	
GCS 0-67 [Same as 0-66]					
GCS 0-68 [Same as 0-66]					
GCS 0-69 [Same as 0-66]					

Number	Species as on sheet	Description	Media	Signed	Other
GCS 0-70 [Same as 0-66]					
GCS 0-71	1. Cinclosoma castanotus. 2. ———— punctatum.	Five eggs	Uncoloured lithograph	No	
GCS 0-72	Milvus isurus	Two eggs	Uncoloured lithograph	No	
GCS 0-73 [Same as 0-72]					
GCS 0-74		Two large eggs	Uncoloured lithograph	No	
GCS 0-75 [Same as 0-74]					
GCS 0-76		One large egg	Uncoloured lithograph	No	
GCS 0-77 [Same as 0-76]					
GCS 0-78		Five eggs	Uncoloured lithograph	No	
GCS 0-79 [Same as 0-78]					

[GOULD, JOHN HENRY.] 1854. *Letters from an Assistant-Surgeon in the Honourable East India Company's Service, to his father.* London, privately printed.

'The object of putting these Letters into type is to answer the following inquiry from many kind friends: — "Have you heard from your Son lately; how is he getting on, and what is he doing?" '

This pamphlet of thirty-two pages was kindly loaned to me for copying by Dr Geoffrey Edelsten.

There are twelve letters from Gould's son, Dr John Henry Gould, while the latter was in the Red Sea and India area. Letter No. 1 is dated 8 February 1854 from the Red Sea, and No. 12 dated 7 November 1854 from Camp, Kurrachee. These are interesting, chatty, and informative letters concerning the areas in which Dr Gould was stationed, the people, the natural history, and some accounts of his Government duties.

Dr Gould died on 4 October 1855 in Bombay, apparently of 'fever'.

GOULD, JOHN HENRY. *See* [GOULD, JOHN.] Original drawings. [No date] 'Birds Eggs /Gould.' Collection: Mitchell Library, Sydney.

GRAVES, CHARLES. 1964. *Leather Armchairs, the book of London Clubs.* New York, Coward-McCann.

The Athenaeum was founded in 1824. Gould was elected under Rule II in 1854 according to correspondence in October 1974 with the Secretary, G. L. E. Lindow. Rule II 'provides for the annual introduction of a certain number of persons of distinguished eminence in science, literature or the arts'.

In Graves's book are two illustrations of the club and a four-page account of the club.

GRAY, GEORGE ROBERT. 1841. *A list of the genera of birds.* 2nd ed. London, Taylor. Included is an unbound 'Appendix' of 16 pages dated 1842.

Based on 'the inflexible law of priority', Gray attempts a careful listing of the correct names for the genera of birds with their synonyms. He accepts the generic names beginning with the first 1735 edition of Linnaeus' 'Systema naturae'.

There are many names attributed to Gould.

[GRAY, GEORGE ROBERT.] Manuscript. [After 1849] 'Index/to/G. R. Gray's/arrangement/of the/Trochilidae.' Collection: Acad. Nat. Sciences, Phila.

Phillips and Phillips (1963) list this item under Collection 71 as being 'in the hand of Gould'. This is not in the handwriting of John Gould, nor is it the handwriting of Edwin Prince, Gould's secretary.

Coues in his *Ornithological Bibliography*, part three, refers to a separate 'Handlist' of humming-birds by Gray, but I have not found any further reference to such a listing.

In this forty-three-page manuscript booklet three columns are entered. The first column contains the species name of the humming-bird, the second column the generic name, and the third entry is a number from 1 to 101, probably referring to pages. The entries are alphabetical.

Thus the first entry is as follows: Abeillei Mellisuga 37

and the last entry: Yarrelli Calothorax 7

The paper is watermarked 'C. Wilmot/1849'.

This booklet is attached to the inside back cover of Gould, John. Manuscript. [No date] 'Europe/Insessores/V.'

GRAY, GEORGE ROBERT. 1863. *Catalogue of British birds in the collection of the British Museum.* London, Trustees.

In this 248-page book is a list of all the species of birds recorded by 1863 in Great Britain and Ireland. In regard to the specimens in the British Museum at that time, the source is listed, and quite a few were from Gould's collection. If a drawing of the bird appeared in Gould's *Birds of Europe* or another folio publication, this is also listed. Gould's *Birds of Great Britain* had not been published by 1863.

GRAY, GEORGE ROBERT. 1869-71. *Hand-list of genera and species of birds, distinguishing those contained in the British Museum.* 3 vols. London, Trustees British Museum.

Any species of bird listed that is figured in Gould's works is so noted.

GRAY, JOHN EDWARD, Editor. 1831 to 1844. *The Zoological Miscellany.* London, Treuttel, Wurtz and Co. (A facsimile reprint was published in 1971 by the Society for the Study of Amphibians and Reptiles, Athens, Ohio.)

These notes were published in three parts on irregular dates and in an irregular manner, apparently as a way for Gray to expedite his own papers and those of a few of his close colleagues.

Of Gouldiana interest, Gray published some notes on snakes and batrachians collected by Gilbert at Port Essington.

In the issue of March 1842 on pages 43 and 45, are described four snakes of the family *Boidae*, one named *Nardoa Gilbertii* [sic] by Gray.

In the April 1842 issue, sixteen reptiles and batrachians are listed in the British Museum collection as obtained by Gilbert as Gould's collector from Western Australia and Port Essington. One of these species is named by Gray as *Lophognathus Gilberti*. In the May 1842 issue, Gray describes two water-snakes as collected by Gilbert at Port Essington.

GREY, GEORGE. 1841. *Journals of two expeditions of discovery in north-west and western Australia during the years 1837, 38 and 39* . . . 2 vols. London, T. & W. Boone. (Australiana Facsimile Editions No. 8, Adelaide Libraries Board of South Australia, 1964.)

The beginning of volume one is concerned with quite a detailed account of the voyage of Grey on the *Beagle* to Santa Cruz, and Cape of Good Hope, en route to the Swan River area of Western Australia. Interestingly the voyage duplicates the one Gould and his party are to take in 1838, but with more accurate sightings by Grey of both birds and fauna, and latitude and longitude.

In volume 2 in the Appendix on pages 397-414 is a paper on 'Contributions towards the geographical distribution of the Mammalia of Australia . . . ' by J. E. Gray dated 10 July 1841. Gould and his collections are referred to many times.

Appendix D, pages 415-21 is a list by Gould 'of the Birds of the Western Coast'.

Further references are made in Appendix E to Gould's collections of reptiles and amphibia in a paper by John Edward Gray.

GÜNTHER, ALBERT. 1885. *A Guide to the Gould collection of Hummingbirds.* London, British Museum (Natural History).

This is a twenty-two-page pamphlet. After a short introduction, which contains several paragraphs on Gould's mounted humming-bird collection, the species of humming-birds exhibited in sixty-two cases at the British Museum (Natural History) are each given a short paragraph.

The only illustration is a map of the Americas showing the distribution of humming-birds.

In the Ellis-Kansas collection this guide with the date 1887 is bound together with similar guides to reptiles, fishes, mollusca, etc.

GÜNTHER, ALBERT C. L. G. 1900. 'The unpublished correspondence of William Swainson with contemporary naturalists (1806-40).' *Proc. Linn. Soc. London,* from Nov. 1899 to June 1900, pp. 14-61.

Here is an account of Swainson's life with forthright comments on 934 letters written to him by 236 correspondents from the years 1806 to 1840.

The remarks about the Audubon, Rafinesque, and Prince Bonaparte letters are the most detailed. 'Audubon's letters are rather disappointing.'

Gould wrote three letters in this collection, dated Dec. 12, 1830, April 30, 1834 and Jan. 21, 1837. They have to do with exchanging publications and Gould sending Swainson some birds on loan or for sale. [I have photocopies of these letters.]

GÜNTHER, ALBERT E. 1975. *A century of zoology at the British Museum through the lives of two keepers 1815-1914.* London, Dawsons.

An interesting fact-filled volume on the natural scientists of an important century.

Part I is concerned with John Edward Gray (1800-75), and Part II with Albert Gunther (1830-1914).

PART FOUR BIBLIOGRAPHY OF JOHN GOULD, HIS FAMILY AND ASSOCIATES

The Gould references are five. Gunther's first impressions of John Gould at the meetings of the Zoological Society in 1857 and 1858 (p. 267) are worth repeating here:

John Gould (1804-1881) was bon-ami with everyone who could be of any use to him. Nothing could provoke him to any show of temper. It was most amusing to hear and see him holding forth in the most nonchalant way about the characters of a finch or pigeon, the type of a new species or even genus, as if it were one of the grandest contributions to science. Then Gray would commence to chaff him, and throw cold water on the whole subject, Gould winding up the comedy by telling the meeting that our chairman knew nothing about the matter. Gould had never received more than a very common school education. He was very fond of bird skins but he hardly ever read a book, and ornithology afforded him a fine income. But there was no one who contributed so much to ornithology, becoming one of our most popular branches of zoology. Like some of our American friends, he was quite innocent of classics (for nomenclature) and when he came to the Museum to have a chop with me I had to accompany him to the bird room to coin a classical name for a new genus. Many of the Gouldian names came from my mint.

GURNEY, JOHN HENRY [Senior]. 1864. *A descriptive catalogue of the Raptorial birds in the Norfolk and Norwich Museum.* London, Van Voorst.

Part one of a proposed series, this part was the only one published. Specimens from Gould are acknowledged on pages 23, 28 and 29, while references to Gould's illustrations are on pages 53, 69 and 78.

GURNEY, JOHN HENRY [Senior]. 1884. *A list of the diurnal birds of prey.* London, Van Voorst.

Contains many references to Gould's works and illustrations as they pertain to birds of prey.

GURNEY, J. H. [Junior]. 1894. *Catalogue of the birds of prey.* London, Porter.

In the prefatory comments on the first three pages, the naturalists are listed who corresponded with or sent specimens to J. H. Gurney, Senior. Gould's name is included.

HACHISUKA, MASAUJI. 1931-35. *The birds of the Philippine Islands.*   H

According to Tate (1979) some of the colour plates for this book were taken from Gould's *Birds of Asia*.

HACHISUKA, MASAUJI. 1953. *The Dodo and kindred birds.* London, H. F. and G. Witherby.

There are two minor references to Gould on pages 78 and 89.

HALL, ROBERT. 1900. *The insectivorous birds of Victoria.* Melbourne, privately printed.

Half of the fifty-five figures are reduced in black and white from Gould's *Birds of Australia* according to Zimmer 1926.

A popular account of a number of birds of Victoria arranged according to their food habits. Not examined by me.

HALL, ROBERT. 1906. *A key to the birds of Australia.* 2nd ed. Melbourne, Walker, May and Co.

The many illustrations are black-and-white photographs from Gould's *Birds of Australia* and one from his *Synopsis*.

Species of birds with *Gouldi* include numbers 20, 59, 118, 153, 303, 418, 721 and 726.

The first edition of this work was in 1899. There was only one illustration; it showed the structure of a bird.

HALLOCK, MARY. 1978. 'John Gould: History honors him as "The bird man".' Griggsville, Illinois, *Purple Martin News* (May 29) 13: 18-19.

In a hypothetical and interesting way Mrs Hallock traces the steps one would follow to publish a book on natural history in the nineteenth century. The steps proceed from selection of a subject, procuring of specimens, study and research on the subject necessitating membership in prestigious organizations, obtaining an artist, if not yourself, finding an engraver or lithographer, obtaining subscribers, issuance of a prospectus, giving a copy to royalty as a form of sponsorship and advertisement, publishing the work in parts with plates and text, delivering the parts by horse-drawn coach or rail [or by person], collecting the fee, and finally getting new subscribers to replace those lost over the years of issue of the parts.

Then follows a short résumé of Gould's early life and a discussion of his first work, *A Century of birds hitherto unfigured from the Himalaya Mountains*.

Mrs Hallock mentions that Gould was president of the Zoological Society but, to my knowledge, he only attained the honour of vice-president.

Part 2 of this series of Gould articles appeared in the *Purple Martin News* of 26 June 1978, 13: 16-17. This is concerned with *Birds of Europe*, and Gould's monographs *Ramphastidae* and *Trogonidae*.

The third article by Mrs Hallock appeared on 31 July 1978, 13: 14-15 and has to do with the several publications on the birds of Australia.

The fourth article of this series appeared in the *Purple Martin News* of 28 August 1978, 13: 18-19. The works mentioned include the *Mammals of Australia* and his monograph *Trochilidae*.

The fifth article appeared on 25 September 1978, 13: 21. *Birds of Great Britain*, *Birds of Asia*, and *Birds of New Guinea* are considered.

The sixth and final account of this *Purple Martin News* series on Gould appeared on 30 October 1978, 13: 20-1. This was concerned with Edwin C. Prince, Gould's secretary and business manager. The final paragraphs are on 'Gould as a person', and repeats the comments made by R. B. Sharpe and by Joseph Wolf.

HARRIS, HARRY. Manuscript. 1921. 'Biographical index to the Ibis (first ten series) 1859-1918.' Kansas City, Mo., not published.

The copy in hand is from the University of Kansas Department of Ornithology Library and is the second carbon copy. Harris wrote on the title page that the 'original and first carbon copy given to Dr. T. S. Palmer who is to furnish additional biographical data (honors, academic degrees, etc.) and who will stand as senior author. This is the second carbon copy. K[ansas] C[ity] 1921.'

This is a typescript copy. Gould has eleven lines devoted to a listing of reviews of his works, obituary, collections and auction prices current. Since this listing has not been published, I will do so here. Mrs Harry Harris kindly gave permission for me to publish this compilation of her late husband.

A complete listing with notes of *Ibis* references to Gould and his works is listed under "Journal. The Ibis." Harris missed only a few reviews.

Gould, John
1859, 99-101, 319-320, 454-455 (revs.); 1862, 392 (ed.n.); 1863, 102-103 (rev.); 1864, 116 (rev.); 1865, 98-99, 220 (revs.), 239 (ed.n.), 526 (rev.); 1866, 111-113, 409-410 (rev.); 1867, 371-372 (rev.); 1868, 216-218,

PART FOUR BIBLIOGRAPHY OF JOHN GOULD, HIS FAMILY AND ASSOCIATES

472 (revs.); 1869, 108-109 (rev.); 1870, 118-121 (rev.); 1871, 437-438 (rev.); 1872, 188, 428-429 (revs.); 1873, 453 (rev.); 1874, 172-173 (ed.n.), 450 (ed.n.), 452-453 (rev.); 1875, 394 (ed.n.); 1877, 377-378 (revs.);1878, 109-110 (rev.); 1879, 94-95, 356-357 (revs.); 1880, 134-135, 474-475 (revs.); 1881, 160 (rev.), 288-290 (obituary), 497-498 (collections); 1882, 167-168, 461, 601-602 (revs.); 1883, 214, 379, 568 (revs.); 1884, 104, 208-209, 341-342 (revs.); 1885, 227-228, 316-317 (revs.); 1886, 194-195 (rev.); 1887, 108 (rev.); 1888, 271-272 (rev.); 1889, 247 (rev.); 1893, 461 ("Index"); 1899, 164-165 (auction prices current of monographs)
Born, Lyme, Dorsetshire, England, 1804.
Died, London, England, February, 1881.

HAWKER, JAMES C. 1899. *Early experiences in South Australia*. Adelaide, E. S. Wigg & Son. 1975 facsimile. Adelaide, Libraries Board of S.A.

Hawker writes of a visit from Gould as follows (p. 45):

> While located at the Onkaparinga I had a very pleasant experience. Mr. John Gould, F.R.S., F.L.S., F.Z.S., etc., etc., etc., who had lately arrived from Van Diemen's Land, was visiting Australia for the purpose of making a collection of the birds, having been previously employed similarly over Europe. Letters of introduction to all colonial Governors, naval and military officers, etc., were furnished him by the Secretary of State when he left England, in which these gentlemen were requested to give Mr. Gould every assistance in their power towards the accomplishment of his object. The Government sent instructions to Mr. Maclaren to let me accompany Mr. Gould in his researches for specimens in this district, and for several days I had the pleasure of his society and instruction in taxidermy, in which he was a thorough expert. Some distance down the river there was a little cliff, and this used to be frequented by two fine eagles — or, as they are now called, eaglehawks. Mr. Gould was most anxious to get one or both for specimens, but the birds were too wary for a gun. I had a long American pea rifle, very true for 150 yards, so I took this down one day, and while the birds were sitting on the cliff I ventured a long shot at one. Some feathers flew out, and the bird reeled, apparently badly wounded, to Mr. Gould's delight, who started at the run to make sure of him with his gun, but, alas, his delight was premature, the bullet must have only grazed it, and both birds took a quick departure and gave us no other chance of a shot. After ten years hard work, and in some cases very dangerous, Mr. Gould published in 1848 his magnificent work of "The Birds of Australia," in seven volumes, dedicated by permission to Her Majesty Queen Victoria — a work unrivalled for its comprehensive information and exquisite coloured illustrations. Mr. Gould's collection consisted of 1,800 specimens of 600 species of birds, and the eggs of more than 300. This valuable collection cost him over £2,000, and he offered it to the Trustees of the British Museum for half that amount, but they actually declined the offer, and this unique collection was immediately purchased for Dr. T. B. Wilson, of Philadelphia, and was sent to his residence in North America — an incalculable loss to scientists in England.

HEMSLEY, W. BOTTING. 1906. *A new and complete index to the Botanical Magazine from its commencement in 1787 to the end of 1904 . . .* London, Lovell Reeve & Co.

A very complete history of the *Botanical Magazine* precedes the Index section.

There is obviously nothing on John Gould but there are quite a few notations about William Aiton, Thomas Drummond and his brother James Drummond, and a John Gould Veitch! on page XLVI.

Gould did make use of Curtis's *Botanical Magazine* for the flower or plant backgrounds of his plates, and acknowledged this aid.

HENEY, HELEN M. E. 1961. *In a dark glass.* Sydney, Angus & Robertson.

An interesting book on Paul Edmund Strzelecki. He arrived in Australia in April 1839 and stayed four years. His main interest was in geology, but his book in 1845 covered many subjects relating to Van Diemen's Land and New South Wales.

His friendship with Sir John and Lady Franklin along with his natural history interests brought him in contact with the Goulds. The Goulds are mentioned on pages 44, 112, 117, and 212.

HERRICK, FRANCIS HOBART. 1917. *Audubon the naturalist.* 2 vols. New York, D. Appleton and Co.

Two important sets of references to Gould appear in the second volume of Herrick's book on Audubon.

Under 'Sidelights on Contemporaries,' on pages 120 to 124, Herrick quotes from the letter of 13 June 1839 from Audubon to Havell where Audubon states 'I have ...no 5th vol. of Biog. for Mr. Gould; let the Gentlemen [Bonaparte and Gould] purchase or procure what they want where they can'.

Bonaparte's comparative reference to Audubon's and Gould's works is quoted, stating that Gould's *Birds of Europe* 'is the most beautiful work that has ever appeared' (Bonaparte 1838). This comment so greatly peeved an American reviewer, Charles Winterfield, that he claimed that Gould copied some of Audubon's drawings. Herrick concludes this paragraph by stating 'Suffice it to say that there is little or no substantial basis for such odious charges'.

Finally, in this section on page 123, Herrick details an anecdote from Bowdler Sharpe regarding Gould's dining at a country estate. Gould left the dinner table to look at a bird, interested a nobleman in the habits of the bird, and sold him a set of the *Birds of Europe*.

On pages 223 to 224 a letter from William Yarrell to Audubon dated 4 March 1841 is quoted by Herrick. Yarrell relates about Gould's return from 'an absence of two years and 8 months' on his Australian venture, and that the two parts of Gould's *Birds of Australia* is 'improved in execution' as compared with his *Birds of Europe*.

HINDWOOD, K. A. 1938. 'John Gould in Australia'. *Emu* 38: 95-118.

This is the second-longest article in the Gould Commemorative Issue. It contains a wealth of material, dates, anecdotes, and comments about Gould's exploration of Australia. Included are three maps showing principal localities visited by Gould, and also two portraits of Gould, the Maguire 'engraving' at the age of 45, and the Maull and Fox photograph of him at the age of 71.

At the end of the article is a list of literature consulted that is quite complete through 1938.

HINDWOOD, K. A. 1938. 'Mrs. John Gould'. *Emu* 38: 131-8.

Following some introductory comments about Mrs Gould's artistry and devotion to John and her family, Hindwood publishes, with a few comments, the complete portion of Mrs Gould's diary written while she was in Australia, covering the period from 21 August to 30 September 1839. [This diary was listed for sale by Sotheran in No. 9 'Piccadilly Notes' on page 97, item 1438 for £10/10, and was purchased by the Mitchell Library of Sydney in 1938.]

In this incomplete diary, Mrs Gould mentions several excursions she took with her husband, one to Mosquito Island near Newcastle, into the bush near Maitland, then to her brother Stephen's homestead at Yarrundi. [*See also* Enwright (1939)]

HINDWOOD, K. A. 1938. 'The letters of Edwin C. Prince to John Gould in Australia'. *Emu* 38: 199-206.

Mr Prince was Gould's 'Assistant' for some forty years. His position was that of secretary and business manager, and he obviously served Gould well.

Hindwood quotes several extracts from the thirty letters written by Prince to Gould when the latter was in Australia. These letters provide a considerable amount of information on the operation of Gould's London business with comments about his scientific friends and relatives. [A typescript of these letters is being completed.]

PART FOUR BIBLIOGRAPHY OF JOHN GOULD, HIS FAMILY AND ASSOCIATES

HINDWOOD, K. A. 1938. 'Some Gouldian letters.' *Emu* 38: 206-12.

This pertains to twelve letters from Gould to Edward Pierson Ramsay of Dobroyde, Sydney dated from 26 October 1865 to 29 January 1869. Ramsay apparently sent Gould quite a few eggs and skins of Australian birds.

HINDWOOD, K. A. 1938. 'Gouldiana'. *Emu* 38: 235-8.

Hindwood lists thirteen items of additional miscellaneous manuscript obtained by Chisholm in 1938 from Gould descendants in England.
  [See also Cleland 1939 and Somerville 1939 in regard to Item 6.]
  [A typescript of this material is being prepared. The original collection is in the Mitchell Library, Sydney.]

[HODGSON, BRYAN HOUGHTON.] Manuscript. 1837. [Manuscript concerning a proposition with John Gould for publication of a work on the birds of Nepal.] Collection: Zoology library, British Museum (N.H.).

This item consists of four parts. The first part is labelled 'Mr. Gould's Proposition' and is five handwritten pages on legal size paper dated 6 March 1837. It is a copy from the original.
  Gould was replying to B. H. Hodgson's father who had relayed a communication from his son concerning Gould publishing with young Hodgson a work on the birds of 'Nepaul'. Gould replied that he could publish such a work, as a continuation of his *Century of Birds*, with certain considerations. Gould estimated the cost would be 'from 9 to £10,000' for an imperial-folio-size publication, issued 'in Parts, each containing 20 Plates... the price to be £3.10 each part ... not [to] exceed 20 in all ... published every 3 months'.
  Hodgson had made the statement that he had 350 subscribers in India, so Gould included as a condition that Hodgson would take one hundred copies at 20 per cent off the fixed price.
  Some plates from *Century of Birds* were to be copied for inclusion in this new work. Gould said he was aware of Hodgson's collection of birds, 'at least those forwarded to the Zoological Society (which are of course kept sacred) and on my own part I could add considerably to the number of species'. Gould mentioned his own expanded collection since the completion of *Century of Birds*, and specimens available to him in the collections at 'Fort Pitt, Chatham, the United Service Museum', and others.
  Hodgson was to supply 'any manuscripts or other matters connected with the subject'.
  The second part is a point-by-point commentary by Hodgson's father on Gould's proposition. To some points of Gould's proposal, he had 'a serious objection' such as Gould only wanting to publish a work on the birds of Nepal, and not to include the mammals also as Hodgson wished. There were several comments about 'Gould & Co.'. These mainly were relating to Gould wanting to call this a continuation of *Century of Birds*, and also that Gould appeared to be in no great hurry to get this work going. Mr Hodgson construed this as a delaying tactic so that Gould might come upon some new species from this area of Nepal through his own sources, thereby negating Hodgson's novelties.
  The third part was a letter from father Hodgson to Gould dated 10 March 1837 voicing some of these objections to the propositions.

The fourth part is eight similar handwritten pages headed 'My own remarks on it'. This statement is signed by B. H. Hodgson and dated at Nepal, July 1837. Seventeen points are raised.

Hodgson begins by stating 'I am not likely ever to assent to that proposition, as it now stands, but, eventually and failing other arrangements, I may, on certain conditions, place the whole of my Birds at Mr. Gould's disposal, to be published by him, on his own account'.

Further he stated that 'Mr. Gould's scale of Illustration seems to me very needlessly large and proportionately expensive... [they] are indeed nobly efficient but... no text should be mixed [with the illustrations]... The text attached to Gould's Century and to his Birds of Europe is miserably trivial... In Audubon the flowers and trees obscure the Birds, and double the cost. So they would in such a work as Gould proposes'.

The final point is Hodgson's concern over the failure of the Zoological Society of London, for the past five years since his collection had been in their hands, to have published 'the novelties on its own responsibility or on mine... That Society has evaded the task, and has, at the same time, suffered a publisher of its own, and one aiming directly to occupy my field, to have access to my Stores, contrary to my known wish and to its own pledge. Mr. Gould meanwhile intimates *other* sources of supply... not shown to be such, and of date prior to 1832, or the date of my deposit in London'.

[This is a familiar type of squabble that has developed between some scientific personages. Gould's *Century of Birds* was completed early in 1832, probably March. Also Gould acknowledged Hodgson as the collector of the bird portrayed in plate 1 of the *Century of Birds*. Further than this, I cannot comment.]

The courtesy of the Trustees of the British Museum (Natural History) is acknowledged for quotations from this manuscript.

HOPE, REV. F. W. 1842. 'Observations on the *Coleoptera* of Port Essington, in Australia'. *Ann. Mag. Nat. Hist.* 9: 423-30.

This long article on over fifty specimens of insects is based on material collected for Gould, probably by Gilbert. A reference to 'Mr. Gould's Collection' is made on page 425.

Hope named several insects after Gould.

HUDSON, WILLIAM HENRY. 1892. *The naturalist in La Plata*. London.

Hudson writes quite a long soliloquy on Gould's lack of artistry in portraying humming-birds. He felt that a proper portrayal of a humming-bird in flight was to show the wings as 'a mist encircling the body'.

HYMAN, SUSAN. 1980. *Edward Lear's birds*. London, Weidenfeld and Nicholson.

Hyman is another author who has sided against John Gould, the 'shrewd entrepreneur and ruthless businessman'. Also Philip Hofer, in the Introduction, views Gould's occasional use of Lear's drawings with lack of acknowledgement as 'wicked'. But, in the text, one after another author or artist is cited as not giving proper credit when Lear's illustrations were used; so it was a common practice at that time. Still, that is no excuse for Gould's occasional remissness.

All in all this is an excellent large-sized book about Edward Lear, with beautiful illustrations, mainly from Gould's works, but some are of original Lear drawings. Unfortunately the quotations from Lear's letters are not referenced.

PART FOUR BIBLIOGRAPHY OF JOHN GOULD, HIS FAMILY AND ASSOCIATES

IREDALE, TOM. 1924. 'Lhotsky's lament'. *Australian Zoologist.* 3: 223-6.

Dr John Lhotsky published the New South Wales *Literary, Political and Commercial Advertiser.* In No. II, in early 1835, there is a note entitled 'Australian Geography. Private Interior Discovery'. The note continues to relate that:

Mr. C. Coxen, who arrived in this Colony with orders from the Zoological Garden, London, has finished his first trip to the interior. He started on the 26th of December last from the Hunter, and penetrated on the banks of the Nammoi [sic], so far as 100 miles beyond the last station (one of Sir John Jamison's), situated on this River ... As far as collecting is concerned Mr. C. was very successful, having discovered as much as twenty-six very rare species of birds, amongst which twenty at least, are entirely new to science. The greater part are of the parrot and pigeon tribe, the former ones of a very splendid plumage.

IREDALE, TOM. 1937. 'The last letters of John MacGillivray'. *Australian Zoologist* 9: 40-63.

MacGillivray was collector for Gould and many others. These letters cover a period from 9 July 1865 to 1 March 1867. He died on 6 June 1867 from a combination of heart disease and asthma, aged 43.

The letters have great interest for those early Australian years of natural-history collectors. He worked with James F. Wilcox in the Grafton area north of Sydney. There are only a few references to Gould.

IREDALE, TOM. 1938. 'John Gould: The Bird Man'. *Emu* 38: 90-5.

This is an excellent short general summary about Gould and his works. The two introductory paragraphs merit quoting:

One hundred years ago an Englishman landed in Australia upon a remarkable quest — the investigation of our bird-life for the publication of a work upon the subject. Probably his most sanguine anticipations did not foresee the results actually secured. It was a venturesome undertaking, as only fifty years previously — a very short time in the history of a country — Phillip had made the first settlement at Port Jackson. No white settlers had paved the way; it had been absolutely virgin land, and at first sight not too attractive, either. Yet in that short space of time hardy pioneers had launched out in every State, Tasmania, Western Australia, Northern Australia, Moreton Bay, South Australia, and Melbourne coming into being. Still the settlements were of very slight extent, and one at Port Essington of dubious stability, whilst the immense interior was absolutely unknown, the phantasy of a huge inland sea being still implanted in many minds, and savage blacks still seriously to be feared. Indeed, Richard Cunningham had lost his life amongst the blacks only a little time before Gould's arrival, whilst even at Port Jackson there had been trouble from this cause.

To adventure into such a country, with his whole fortune at stake, must have seemed blameworthy to his friends, but Gould probably weighed up all the chances very deliberately before he engaged upon this mission. He came provided with letters of introduction to all the Governors of the States, and his organizing ability had been proved by his previous work. Everything he had undertaken had been run to a steady and quick time-table, and that requires great foresight and a calculating brain. Upon his arrival here everything moved like clockwork, with no delays or hesitancy, and his accomplishments in the short time he was in our land read almost miraculously when we remember the disabilities of travel a century ago. When his results only are criticized the data appear more remarkable still, as his collections were almost unbelievable in extent. When Gould arrived here Australia was scarcely more than a name and that not of the greatest esteem. When his work on The Birds of Australia was completed, ten years later, the name was so familiar that it only needed the discovery of gold to place it in the forefront of the younger nations, a place worthily held ever since. It would be just to claim that Gould was the first and greatest ambassador with which Australia has been blest.

Further comments by Iredale are also most flattering and illuminating.

IREDALE, TOM. 1945. 'Jules Verreaux'. *Australian Zoologist* 11: 71-2.

Apparently in the 1840s Jules Verreaux, of the famous Verreaux family of bird collectors, visited Tasmania and New South Wales for the Paris Museum.

Gould's Australian birds were shipped to Maison Verreaux for preparation for shipment to Philadelphia when they were purchased by Dr T. B. Wilson.

IREDALE, TOM. 1950. *Birds of Paradise and Bower Birds.* Melbourne, Georgian House.

Illustrated by Lillian Medland (Mrs Iredale), this is as complete a treatise on these complicated birds as was possible in 1950.

Gould and Sharpe are referred to in many accounts. Since there is no index, the following pages have references to Gould: 16, 21, 22, 80, 111, 158-61, 173, 174, 175, 177, 187, 201, 209, 219-21, 225, 235, 237, 238.

IREDALE, TOM. 1951. 'The Humming Bird'. *Australian Zoologist.* 11: 314-15.

This short article is concerned with a periodical entitled *The Humming Bird* published from 1891 to 1895 by Adolphe Boucard of London.

In the article Iredale makes references to Gould's 'hobby' of the study of Humming Birds and his 1851 exhibition of them in London.

IREDALE, TOM. 1951. 'Again Gould'. *Australian Zoologist* 11: 316-17.

This short but interesting article relates to comments on a set of pattern plates now preserved at the Commonwealth National Library at Canberra. A pattern or 'key' plate is a hand-coloured lithograph that is used as the accurate colour-pattern for the colourists.

This series includes most of the pattern plates for *The Birds of Australia, Supplement* and for *Mammals of Australia.* Iredale does not give the number of plates present in the collection.

A few of the pencilled comments written on these plates by Gould or by his artists are quoted in the article.

IREDALE, TOM. 1951. 'Audubon in Australia'. *Australian Zoologist* 11: 318-21.

Gould gave Audubon two parts each of his *Trogonidae* and *Icones avium*, and also the two rare cancelled parts of *Birds of Australia and adjacent islands*, before Gould left for his Australian adventure.

Part 2 of the monograph *Trogonidae* has written at the top of it in Gould's handwriting 'J. J. Audubon, March 21st. 1836, London'.

Mr Len Audubon, a great-grandson of J. J. Audubon, was born in Australia and had these books in his possession.

The article by Iredale is concluded with a complete collation of 'A Monograph of the Trogonidae', 1835 to 1838.

IREDALE, TOM AND HALL, A. F. B. 1927. 'A monograph of the Australian Loricates'. Appendix C. Biography and Bibliography. *Australian Zoologist* 4: 339-59.

This is a list of naturalists from Australia who collected molluscs. Many also collected for Gould.

IRWIN, RAYMOND. 1951. *British bird books: an index to British ornithology.* London, Grafton.

Gould's applicable works are listed on pages 12 and 253.

J JACKSON, CHRISTINE. 1975. *Bird illustrations.* London. Witherby.

As Mrs Jackson states, 'This book tells the story of the use of lithography for bird book illustration'. In the first chapter is a review of the development of lithography for bird illustration with notes on early printing firms, mainly in England.

Chapters are devoted to William Swainson, Edward Lear, John Gould, and Josef Wolf among others, concluding with ones on Thorburn and Lodge. The Gould chapter has quite a complete account of his life and works, interestingly written.

JACKSON, CHRISTINE E. 1978. *Wood engravings of birds.* London, Witherby.

The woodcut lives in this small volume. Mrs Jackson has selected for illustrations eighty-five fine examples of this form of engraving.

Chapters are devoted to Thomas Bewick, T. C. Eyton, William Yarrell, William Dickes and Benjamin Fawcett but many other artists and engravers appear in the text. It is interesting to contemplate that so many of these persons were contemporaries of John Gould. Gould is referred to in four places.

Another fact-filled but most interesting book by Mrs Jackson.

JACKSON, CHRISTINE E. 1978. 'H. C. Richter—John Gould's unknown bird artist'. *J. Soc. Biblphy Nat. Hist.* 9: 10-14.

Henry Constantine Richter was the full name of the artist who began working for Gould in 1841, after the death of Mrs Gould.

Richter was born 1821 at Brompton, son of Henry James Richter and Charlotte Sophia. The father was a noted historical painter, who died in 1857.

H. C. Richter was around twenty years old when he began working for Gould. In addition to his drawings and transfers to lithographic stones for Gould, Richter drew a few birds for George Robert Gray, and Richard Owen. Mrs Jackson estimates that Richter was responsible for over 1600 of Gould's '2999 bird designs'.

Richter never married and died at the age of 81 on 16 March 1902.

JACKSON, HOLBROOK. 1951. *The complete nonsense of Edward Lear.* New York, Dover.

Jackson has written a short but interesting biography of Lear with comments on his nonsense drawings and verse. Gould is mentioned.

JACKSON, SIDNEY WM. 1908 'A trip to the Upper Hunter River District, N.S.W.' *Emu* 8: 11-23.

Jackson retraces the steps of Gould's visit to the Upper Hunter in 1839 and 1840 with stops at Segehoe, Dartbrook, Yarrundi, Aberdeen, and, further up the Hunter, Belltrees and Ellerston.

JARDINE, SIR WILLIAM. 1858. *Memoirs of Hugh Edwin Strickland.* London, Van Voorst.

This is a very valuable volume which, in addition to the account of Strickland's short life, contains contemporary accounts of English and European scientists and ornithologists. Strickland visited Europe on his honeymoon and wrote back about many of the museums there.

In 1843 Strickland suggested in a letter to his future father-in-law, Sir William Jardine, that a group of ten individuals should, at £10 each sponsor a collector for a year to go to Honduras, Central America. 'It would be well to consult with Gould on the subject, as he is well acquainted with the practical part of these arrangements, and would perhaps become one of the subscribers.' It was decided to contact Mr Dyson to be the collector, and arrangements were almost completed, when unfortunately they broke off.

Jardine reprinted an article by Strickland from the 1844 Report of the British Association. On page 274 there are glowing comments about Gould 'as a man of genius and of enterprise', and the statement that 'the many new and interesting details of natural history which [Gould's] work contains indicate powers of observation and of description which will place the name of Gould in the same rank with those of Levaillant, Azara, Bewick, Wilson, and Audubon'.

Further in this reprinted article (pp. 289-92) Strickland elaborates at length on the methods of ornithological illustration.

The references to Gould are on pages 260-1, 273, 284, 285, and 291.

JARDINE, WILLIAM AND SELBY, PRIDEAUX JOHN. 1825-43. *Illustrations of ornithology.* 4 vols. Edinburgh, Lizars.

According to Engelmann (1846) the title-page of this work was published 'with the co-operation of J. E. Bicheno, J. G. Children, Hardwicke, Horsfield, R. Jameson, Stamf. Raffles, N. A. Vigors, and John Gould'. The Kansas-Ellis copy (E-45), with many original drawings interleaved, does not have Gould's name on the title page. Zimmer (1926) also does not have Gould's name listed in this context, but in his commentary he lists Gould as one of the artists for the hand-coloured plates, along with Jardine, Selby, E. Lear, Thompson, James Stewart, A. F. Rolfe and R. Mitford.

The Kansas-Ellis copy is unique. In the first three volumes there is an uncoloured plate following each coloured plate. Also there are sixteen or so original drawings for the plates. The artists are Sir William Jardine and P. J. Selby, with two by Mrs Gould in volume two, plates 109 and 110. Ralph Ellis wrote extensive notes on these four volumes on the inside of the back of volume one. Ellis lists this as the 'author's copy', meaning Jardine's copy.

JENKINS, ALAN C. 1978. *The naturalists.* New York, Mayflower.

When the beautiful and colourful bird figured on the front wrapper of a book is by John Gould, but is labelled as being by Audubon, one is concerned over the reliability of the book's contents. However, it is a good book.

Jenkins begins with the primitive hunter and ends with the use of the camera and the helicopter in the quest for nature facts.

The chapters on Carolus Linnaeus, Charles Waterton, David Douglas, and Philip Henry Gosse reveal especial interest on the part of the author in these naturalists.

Audubon and Gould are the subjects in the chapter on 'Monuments to Nature'. The 1849 Maguire portrait of Gould is illustrated. Comparisons between Audubon and Gould are inevitable and interesting.

The indexed references on Gould are correct. On page 134 is the same colour-plate as on the wrapper. The Red-billed Blue Magpie, of 'south-east Asia habitat', is not by Audubon, but by Gould.

JENYNS, LEONARD (afterward Blomefield). 1885. *Reminiscences of William Yarrell.* Bath, privately printed.

A twelve-page booklet. Yarrell 'was hospitable and fond of giving small dinner parties; entertaining a select few, from time to time, in a quiet way... I have met at his table formerly, Thos. Bell, Edward Turner Bennett, Bicheno, Gould the ornithologist, John Edward Gray, Marshall Hall, and William Thompson of Belfast'.

Yarrell 'was always fond of the theatre, and he would often pick up and get quite by heart some humorous song of the day... I remember one evening, when he and I went together to a large evening party at Gould's, the ornithologist, he suddenly,

PART FOUR BIBLIOGRAPHY OF JOHN GOULD, HIS FAMILY AND ASSOCIATES

at the request of a friend standing by, struck out in a stentorian voice — "I'm *afloat*! I'm *afloat*!" — the first words of an amusing sailor's song, which immediately brought to a stand the whole party engaged in conversation at the time'.

JOOP, GERHARD. 1968. *Paradiesvögel.* [Braunschweig], G. Westermann.

A three-page note on John Gould and his works precedes twelve colourful plates of birds of paradise by Gould and Hart.

JOURNAL. *Annals and Magazine of Natural History.* 1837 to date. London.

Several changes in names occurred in the early years of the existence of this important journal.

It was first published in Edinburgh in 1837 as the *Magazine of Zoology and Botany*, with Sir William Jardine as chief editor, according to Wood (1931). In 1838 it merged with Sir W. J. Hooker's *Companion to the Botanical Magazine*, and appeared under the name of *Annals of Natural History*. When the office of publication was transferred to London in 1841 its full title became *The Annals and Magazine of Natural History including Zoology, Botany and Geology*. Jardine ceased being the principal editor in 1857. Since that date it has been under the editorship of a distinguished array of British naturalists.

In the early pages of this periodical are reviews of Gould's folio works. His first article for this journal was in the 1837 *Mag. of Zool. & Bot.* 1: 62-4 'On the Genus *Paradoxornis*'.

Many of the articles in this publication that were on newly discovered birds or mammals, were reports of meetings of the Zoological Society of London, and they would be also published, usually at a later date than in the 'Annals', in the Zoological Society 'Transactions' or 'Proceedings'.

Salvadori (1881) and Sharpe's *Analytical Index* lists Gould's published articles rather completely.

As reported in the 'Annals', Gould was on many occasions in the 1850s 'in the chair' presiding over the Zoological Society meetings.

The last article by Gould in the 'Annals' appeared in the 5th series, 1880, volume 5, page 488, on two new humming-birds from Bolivia.

There was no obituary or other notice of Gould's death in this magazine.

JOURNAL. *The Auk.* 1884 to date. The American Ornithologists Union.

Under the editorship of J. A. Allen, this important journal began as a quarterly in 1884.

Only two direct references to Gould are found in the 'Auk Index'. Both are reviews by Allen of his works or rather their continuation by Sharpe, namely on Gould's *Birds of New Guinea* (1889 6: 269) and his *Supplement to Trochilidae* (1889 6: 334).

JOURNAL. *Canadian Journal of Industry, Science and Art.* 1857. [Note concerning Gould in Toronto] 2: 328.

In further pursuit of information on Gould's trip to America in 1857, Mrs Rachel Grover of the Thomas Fisher Rare Book Library, University of Toronto, found a reference in the James Baillie Collection that led her to this journal.

Dr James Bovell of Toronto wrote this note in the Canadian Journal of September: 'During the present summer we were visited by Mr. John Gould, the distinguished Naturalist, whose chief object in his tour through Canada was for the purpose of studying the habits and manners of the species of *Trochilus* frequenting this portion of the North American Continent. Shortly after his return to England, at a meeting of the London Zoological Society, Mr. Gould detailed some of the results of his observations. He arrived in Canada just before the period of the migration of these beautiful little birds from Mexico to the north.' The note continues with some observations by Gould on flight and feeding activities of humming-birds. Gould was 'successful in keeping one alive in a gauze bag attached to his breast button for three days, during which it readily fed from a small bottle filled with syrup of brown sugar and water.'

JOURNAL. *Contributions to ornithology*, edited by Sir William Jardine. 1848-52. London, Highley.

One set of this periodical in the Ellis Collection at the University of Kansas Spencer Library is in two volumes. Volume 1 contains the text for the five years and volume 2 contains the plates.

The test in volume 1 has irregular pagination, and missing title-pages. Gould, in his letters to Sir William now in the collection of the British Museum (Natural History), made several references to the sloppy way in which the 'Contributions' were published. Gould complained about his inability to refer to the articles published in the 'Contributions' because of the lack of title-pages and incompleteness of the sets sent to him. Ralph Ellis made many annotations on this copy with reference to the irregularities.

Here follows the Gould references in this journal. Some of the scientific notes by Gould are not listed in Sharpe's *Analytical Index* to Gould's works.

In the 1848 section there is a note on page 7 about an arsenical soap Gould used for preserving bird skins in Australia.

In the 1849 issue on page 27 is a reference to a letter from Linnaeus in the possession of Gould and three others he had borrowed. The four letters are printed – in Latin – with the dates of 20 January 1772, 7 August 1772, 2 January 1774 and 3 July 1774.

On pages 51 to 59 are notes by Gould and John Gilbert on *Artamus sordidus*, wood swallow.

On page 66 is a note on a new humming-bird from Mr Jameson to be figured by Mr Gould, labelled *Oreotrochilus Jamesonii* Jardine.

Dr T. T. Kaup on page 96 expressed 'warmest thanks to my friend... Gould ...'.

On page 137 are two short notes by Gould on *Boshas bimaculata* and on *Motacilla boarula*. Concerning the latter, he said he found two nests of this bird when on a trout excursion in June 1848 to Chemies, 40 kilometres north-west of London.

In the 1850 issue on page 2 is a reference to 'Gould's figure' of *T. temminckii*.

A minor note is on page 41.

On page 86 it is mentioned that Gould read papers at the 20th meeting of the British Association for the Advancement of Science at Edinburgh, 30 July 1850. [However, in the published Report of this meeting there is no record of this paper being read.]

On pages 92 to 105 is the first formal article by Gould entitled 'A brief account of the researches in natural history of John M'Gillivray, Esq., the naturalist attached to H.M. Surveying Ship the Rattlesnake...'. In this article Gould has abstracted parts from letters written by MacGillivray to him dated 12 May 1847, 6 February 1848, 19 May 1849, and 29 February 1850, with many notes of birds of the Atlantic and Indian Oceans, Australia, and New Guinea. Apparently this is the paper Gould read at the British Association meeting in July 1850.

In the 1851 issue on page 2 under 'Ornithology in 1850' is a short review of British

PART FOUR BIBLIOGRAPHY OF JOHN GOULD, HIS FAMILY AND ASSOCIATES

ornithology including mention of Audubon's death. Gould's publication of *Odontophorinae* is mentioned, also the first part of his *Trochilidae*, and the first two parts of *Birds of Asia*.

On pages 10 to 13 is an informal account of Gould's description of *Balaeniceps rex* procured by Mr Mansfield Parkins.

On page 79 is a note by Gould on *Trochilus verticeps*.

On page 111 is a short reference to Gould.

On pages 115 to 118 is an article by Gould on *Scytalopus* dated 7 August 1851.

On pages 137 to 138 is an article 'On a new species of *Musophaga*', named by Gould *Musophaga rossae*. In the article is a reference to a drawing of the pheasant by Lt. J. H. Stack, which is probably in error for Lt. F. R. Stack. [These two ink drawings of the St. Helena pheasant are in my collection.]

On pages 139 to 140 is an article 'Descriptions of three new species of Hummingbirds' dated 1 October 1851.

In the 1852 issue on pages 1 and 2 are notices of the publication of part 1 of the supplement of *Birds of Australia*, part 3 of *Birds of Asia*, and part 2 of *Trochilidae*. Mention is also made of Gould's exhibition of 'Hummingbirds' and the 'elegant arrangements', with 'accompaniments of wax flowers, the modelling of which has now been brought to so great perfection'.

On page 39 is a Gould drawing of an egg.

On pages 91 to 92 is an article by Dr Baron J. W. Van Muller entitled '*Balaeniceps Rex*, Gould' with a short description of the bird observed by him on the White Nile.

On pages 93, 109, and 139 are short references to Gould.

On pages 135 to 137 is an article 'Descriptions of three new species of humming birds' by Gould, dated 22 October 1852.

In volume 2 of this set of Jardine's 'Contributions' there are 101 plates, most of which are coloured. The plates accompany the text. The artists are W. J. (William Jardine), H.J. (?), and mainly C.D.M.S. (Catherine D. M. Strickland), Jardine's daughter, who was married to Hugh Strickland. [In the Royal Scottish Museum, Edinburgh, are several letters from Mrs Strickland to her father relating to these drawings. I have photocopies of three of these letters.]

A second set of this periodical in the Ellis Collection is bound in two matched volumes for the years 1848, 1849, and 1850. The text and coloured plates are contiguous. A third volume, in a different binding, contains the years 1851 and 1852, but the plates are uncoloured.

JOURNAL. *The Emu.* 1938. Gould Commemorative Issue. Vol. 38, pt. 2, pp. 89-244, Oct. 1st, 1938. Melbourne, Royal Australasian Ornithologists Union.

To commemorate the centenary of Gould's visit to Australia in 1838, this special issue of *The Emu* was published (Figure 72). This was the most valuable contribution to Gouldiana published since Sharpe's *Analytical Index* in 1893.

The authors and their articles are as follows:

Anonymous. — Gould League of Bird Lovers
Bryant, C. E. (also Editor). — Gould Miscellanea and Some Anecdotes
Campbell, A. G. — John Gould amongst Tasmanian Birds
Cayley, N. W. — John Gould as an Illustrator
Chisholm, A. H. — John Gilbert and Some Letters to Gould
    Some Letters from George Grey to John Gould
Conigrave, C. P. — The 'Gilbert Country' in Western Australia

Figure 72 Cover and first page of 'Gould Commemorative Issue' of *The Emu*, October 1938.

Dickison, D. J. — A Resume of Gould's Major Works
Hindwood, K. A. — John Gould in Australia
  Mrs. John Gould
  The Letters of Edwin C. Prince to John Gould in Australia
  Some Gouldian Letters
  Gouldiana
Iredale, Tom. — John Gould: The Bird Man
  Gould as a Systematist (with G. M. Mathews)
Mack, George. — John Gould's Correspondence with Sir Frederick McCoy
Mathews, Gregory M. — Gould as a Systematist (with Tom Iredale)
  John Gould: An Appreciation
Stone, Witmer. — The Gouldian Collection in Philadelphia
Whitley, Gilbert P. — John Gould's Associates
Whittell, Major H. M. — Gould's W.A. Birds: With Notes on His Collectors

Each of the above articles is discussed separately in this bibliography under the name of the author or first author.

The following is an index to all the individuals mentioned in this commemorative issue of *Emu*. It does not include the authors of the articles listed above, or the individuals on pages 116 to 118, since for the most part, they are listed in this bibliography under their own names as authors.

A

Adelaide, Queen, 141
Aiton, J. T., 203
Albert, Prince, 141
Allan, W., 159
Allis, —, 201
Allport, M., 161, 208
Anderson, Major, 159
Andrews, F. W., 159
Angas, G. F., 161
Argyle, Duke of, 122
Audubon J. J., 126, 171, 228
Austin, Robert, 184

PART FOUR BIBLIOGRAPHY OF JOHN GOULD, HIS FAMILY AND ASSOCIATES

B

Bachman, Dr, 126
Baily, W. L., 128, 233
Baird, S. F., 122, 231
Baker (gardener), 236
Baker (taxidermist), 202
Bankier, Dr R. A., 153
Bauer, —, 157
Bayfield, —, 125, 169, 170, 201
Becker, L., 158
Bennett, Dr George, 100, 135, 147, 148, 151, 161, 194, 207, 208, 209, 211
Bennett Mrs, 135
Benstead, James, *see* James
Bentinck, Lord Wm., 194
Blanchard, Laman, 133
Blyth, 129
Bonaparte, Prince C. L., 200, 202
Booth, Lt., 155
Bourcier, M. J., 128
Bowler, —, 201
Bremer, Sir Gordon, 197
Brete, Capt., 177, 185
Breton, H. W., 155, 177, 185
Brockman, Mrs Robert, 145
Brown, —, 136
Brown, George, 131
Brown, Robert, 157
Bunsome, Robert, *see* Ransome, R.
Burges, —, *see* Burgess
Burgess, L., 159, 181, 185
Bynoe, B., 153, 182, 185

C

Caley, 123
Calvert, —, 150
Campbell, Archibald, 124
Cassin, John, 231
Chambers, Capt., 154
Charlesworth, —, 202
Chester, —, 238
Cockerell, —, 126
Cohans, —, 136, 137
Colthard, — (binder), 201, 202
Coxen, (?) 149, 151, 237
Coxen, Charles, 98, 100, 107, 137, 142-3, 187, 194, 206
Coxen, Mrs Elizabeth (E. Gould's mother), 186
Coxen, Stephen, 98, 100, 124, 126 (?), 137, 143

Craufiurd, Mrs E., 162

D

D'Albertis, —, 131
Dale, Mr, 158
Dark, —, 159
Darwin, C., 153, 200
Delafresnaye, Baron, 201
Derby, Earl of, 121, 154, 200
Dickinson, —, 209
Dieffenbach, Dr, 149
Donald, Wm., 236
Donaldson, —, 194
Douglas, David, 128
Dring, J. E., 153
Drumond, James, 179
Drummond, Johnston, 151, 152, 179, 189

E

Edelsten, —, 186
Edelsten, Dr Alan, 186, 188, 217
Edelsten, Dr Geoffrey, 186, 188, 217
Elliott, Mr (fisherman), 203
Elliott, Henry, 152
Elsey, J. R., 126, 156
Elwes, R., 162
Emery, J. B., 153, 182
Erskine, Mr., 136
Ewing, T. J., 162, 211
Eyre, J. E. [sic], 120, 159

F

Fairfax, John, 147, 195, 211
Friend, Lt., 155
Franklin, Lady Jane, 96, 152
Franklin, Sir John, 96, 152
Franklin (sister of Lady Jane), 152
Fuertes, L. A., 171-2

G

Galbraith, A. L., 162
Gardner, J., 162
Gawler, Col. G., 99, 152, 180
German collector, 163
Gilbert, John, 96, 101, 109, 119, 124, 125, 127, 142-52, 156, 179, 181, 182, 183, 185, 186-99, 203, 206, 227, 233-5, 237
Gilbert, William, 187
Gipps, Gov., 135
Goldie, 131
Gould, Charles, 105, 133, 164, 200, 213, 238

Gould, Miss Eliza, 133, 200
Gould, Mrs Elizabeth, 92, 96, 100, 101,
　105, 121, 123, 124, *131-8*, 142, 166, 167,
　168, 169, 186, 215, 217, 218
Gould, Franklin, 96, 101, 130, 133, 135,
　204, 206, 213, 217
Gould, John (on every page)
Gould, John Henry, 96, 101, 133, 164
Gould League, 240-4
Gould, Miss Louisa, 133, 200
Gould, Miss Sai, 133
Gourlie, Wm., 166
Governor of Western Australia, 193
Gray, 120 (*see* Sir George Grey)
Gray, Mr (Sir George Grey's secretary),
　217, 218
Gray, George Robert, 200, 204, 217
Gray, John Edward, 155, 221
Grey, Sir George, Lt. and Gov., 120, 127,
　152, 177, 180, 185, 186, 193, 216-26
Gregory, F. T., 156, 184, 185
Griffith, Edward, 201
Guillemard, Dr, 131
Gunn, Ronald, 105, 162
Gyde, —, 201

H

Hack, Mr., 218
Hart, W., 122, 128, 130, 131, 133, 168, 170
Harvey, J. B., 159
Hawker, J. C., 160
Herring, — (binder), 202
Hills, —, 159
Hooker, Wm., 153, 179
Hullmandel, C., 201
Hutt, John, 159
Hutt, William, 221

I

Ince, J. M. R., 155
Iredale, T., 116

J

Jackson, Mr, 136
Jackson, S. W., 211
James (James Benstead), 135, 137, 217,
　220, 221, 223, 237
Jardine, Lady, 238
Jardine, Sir Wm., 126, 168, 200, 201, 238
Jemmy, 100

Jerdon, —, 129
Jerrold, Blanchard, 133
Joseph (taxidermist), 202, 206
Jukes, J. B. 155

K

Kermode, —, 159
Knight, —, 225
Krefft, J. L. G., 143

L

Lambert, A. B., 174
Latham, John, 174
Lawrence, G. N., 122
Lawton, —, 236
Lear, Edward, 94, 121, 123, 132, 168
Legge, W. V., 139
Leichhardt, F. W. L., 109, 119, 143, 144,
　147, 148-51, 187, 211
Leigh, — (lawyer), 203
Leycaster, A. A., 159
Linnaeus, C., 174, 238
Loddiges, George, 128
Lonsdale, Mr, 161
Lowe, Lt., 114
Lushington, Lt., 216

M

MacGillivray, John, 126, 154, 155, 161,
　174, 211
Macleay, —, 158
Marriott, Rev., 162
Martin, —, 201, 202
Martin (lawyer), 202, 205
Martin, William Charles, 201, 202, 204
Maxmilian [sic], Prince, 122
McArthur, —, 158
McCormick, R., 154
McCoy, Frederick, 158, 212-16, 227
Milligan, Alex W., 234
Mitchell, Mr, 203
Moon, Mrs Eliza Gould, 133
Muirhead, Dr, 159
Murphy, John, 149
Musignano, Prince of, (*see* Bonaparte), 202

N

Natterer, Johann, 122
Natty, 100, 108, 159
Neill, J. "Patrick", 160

PART FOUR BIBLIOGRAPHY OF JOHN GOULD, HIS FAMILY AND ASSOCIATES

Newton, Alfred, 200, 228
Nurse (Mrs Gould's), 217, 218, 219

O

Old Sportsman, 163
Owen, Prof. R., 98, 121, 149, 200, 221, 222

P

Palmer, T. S., 175
Patterson, Mary, 237
Pennant, Thomas, 200, 238
Pouncey (paper co.), 201
Preiss, Johann A. L., 190, 191
Prince, E. C., 94, 101, 103, 120, 125, 135, 142, 187, 196, 199-206, 228

R

Raffles, Sir S., 91
Rainbird, John, 208
Ramsay, E. P., 108, 162, 206-12
Ransome, Robert, 238
Rawnsley, Henry C., 211
Rayner, F., 154
Reeves, Thomas, 128
Richardson, John, 154, 160
Richter, H. C., 105, 121, 122, 124, 125, 128, 129, 138, 168, 169
Robertson (bookseller), 207
Robertson, Dr, 154
Roe, J. S., 182
Roper, J., 150
Ross, James, 154

S

Salvadori, Count, 130
Salvin, O., 122
Schomburgk, Richard, 163
Sclater, P. L., 122, 208
Scott, A. W., 163
Selby, P. J., 201
Severtzoff, Dr, 230
Sharland, M. S. R., 116
Sharpe, R. B., 112, 119, 122, 128, 131, 132, 170, 199, 200, 226, 229, 231
Shore, J. C., 120
Sibbald, Dr, 153
Simkinson, Mrs, 152
Skinner, G. U., 128
Snooke, Mr, 220

Sotheran, H., Ltd, 119, 128, 207, 226
Stanley, Capt. Owen, 154, 194
Stokes, J. L., 153, 177, 182, 211
Strange, F., 147, 148, 151, 152, 159-60, 211
Strickland, H., 165, 200
Strzelecki, Paul Edmund de, 120
Stuart, John McDouall, 158
Sturm, —, 121
Sturt, Charles, 97, 99, 156
Swainson, Wm., 204
Symers, Capt., 194

T

Taylor's (printers), 201
Temminck, C., 121, 200
Theakes, Capt., 194
Thorburn, A., 172
Throsby, C., 159

V

Verreaux Brothers, 232
Victoria, Queen, 124, 141
Vidler, E. A., 161
Vigors, N., 120, 132, 133, 172

W

Wall, Thomas, 157
Wall, W. S., 157
Wallace, A. R., 122, 215
Waller, Eli, 161, 211
Wards, Mr, 162
Warszewiez, M., 128
Waterhouse, F. G., 158, 227
Watson, Mary, 96, 101, 135
Wheeler, I. R., 238
White, Rev. Gilbert, 238
White, S., 163
White, S. A., 163
Wickham, J. C., 153, 177
Wilcox, James F., 126, 161, 211
William IV, King, 120, 141
Wilmot, Henry, 221
Wilson, Edward, 152, 231
Wilson, H. M., 184
Wilson, Dr T. B., 152, 231
Wolf, Joseph, 125, 128, 129, 170, 227-30

Y

Yarrell, William, 201, 204

JOURNAL. *The Ibis.* 1859 to date. London.

This prestigious journal was first published in 1859. As the organ of the newly formed British Ornithologists Union it had as its initial editor Philip Lutley Sclater.

In the Preface to volume one is the statement that:

for some years past a few gentlemen attached to the study of Ornithology, most of them more or less intimately connected with the University of Cambridge, had been in the habit of meeting together, once a year, or oftener ... to talk over ... this branch of Natural History ... In the autumn of 1857 the gathering of naturalists was greater than it had hitherto been and it appeared ... there was a strong feeling that it would be advisable to establish a Magazine devoted solely to Ornithology ... In November 1858, the annual assemblage took place at Cambridge and ... it was determined by those present that a quarterly Magazine of General Ornithology should be established ... and that the subscribers should form an "Ornithological Union", their number at present not to exceed twenty.

I have detailed this prefatorial statement by way of introduction, and also possibly to explain why John Gould was not a member. The fact undoubtedly, however, is that Alfred Newton, the force at Cambridge, blackballed him. Gould was definitely the most famous ornithologist in England at that time, and it was an obvious slap at him never to be invited to join the Union, even when their membership enlarged. Gould never remarked on this Union in any correspondence seen by me. He did send communications on birds to the editor which were published, and all of his works from 1859 on were reviewed, usually very favourably, by the various editors.

For completeness it will be necessary, but painful, to go through the first seven series of this journal for references to Gould, to his papers, to any of his correspondence, and for reviews of the parts of his works as they appeared. Harris (1921) did this for the reviews of Gould's works and a few other items but, unfortunately, all Gould references are not listed. There are many references to Gould in articles by others in these volumes, but an index to these will have to be a pastime of my retirement days, if such occur.

First series

The first series of six volumes appeared from 1859 to 1864 and was edited by P. L. Sclater.

Volume 1, 1859, January. Pages 99-101. Reviews of 'Mr. Gould's magnificent series of illustrated works claims our first attention'.

Trochilidae, part 15, 1 May 1858 with list of fifteen species illustrated; part 16, 1 September 1858 with list of fifteen species illustrated.

Birds of Asia, part 10, no date, with list of sixteen species figured.

'Trogons, first part of a new edition' issued in 1858 is mentioned.

April. Pages 206-7, a letter dated 3 February 1859 from Gould to the editor concerning 'a splendid female Goshawk *(Astur palumbarius)*' recently shot in Suffolk on 24 January. 'I was shooting at Somerleyton at the time.'

July. Pages 319-20, review of *Birds of Asia*, part 11, no date, with a list of sixteen species.

October. Pages 454-5, review of *Trochilidae*, parts 17 and 18, no date and no list; *Birds of Australia, Supplement*, part 3, no date, list of fifteen species.

Pages 473-4, a letter from Gould on 'obtaining from Siam a specimen of the splendid Pheasant named by Mr. Blyth *Diardigallus fasciolatus*'.

Volume 2, 1860, July. Page 309, a note through Professor S. F. Baird concerning the discovery by Mr Xantus of a new humming-bird of the genus *Amazilia* and the fact that 'Mr. Gould has just received specimens of this bird, and of three other *Trochilidae* recently described by Mr. G. N. Lawrence of New York'.

PART FOUR BIBLIOGRAPHY OF JOHN GOULD, HIS FAMILY AND ASSOCIATES

October. Pages 423-5, a note extracted from the 'Proceedings of the Academy of Natural Sciences of Philadelphia' for 20 March 1860 is concerned with Dr T. B. Wilson's bird collection presented to the Academy. Mr Gould's Australian collection consists of 2000 specimens 'which are the types of his great work, "The Birds of Australia", and embracing all the species then known, except five only'.

Volume 3, 1861, January. Page 112, letter from Gould to the editor on the occurrence of the Chiff-chaff *(Phylloscopus rufus)* in Norway where Gould visited in the summer of 1856 with Joseph Wolf.

Volume 4, 1862, January. Pages 72-4, review of *Trochilidae* on its completion and mention of *Introduction to Trochilidae*, both completed in 1861. Some very flattering comments are made about Gould's courage in undertaking such a publication.

October. Page 392, an announcement that 'Mr. Gould is engaged in preparing for publication the two first numbers of the ... "Birds of Great Britain" '.

Volume 5, 1863, January. Pages 102-3, review of Gould's *Birds of Great Britain*, parts 1 and 2, dated 1 October 1862, no list. Another glowing tribute to Gould's works past and present. 'Each part contains fifteen plates.'

Volume 6, 1864, January. Page 116, review of *Birds of Great Britain*, parts 3 and 4, no dates, but with listing of fifteen species for each part.

Page 130, letter from Gould of 30 November 1863 concerning a communication from Edward H. Rodd on *Muscicopa parva* obtained in Britain.

Second series

The second series of *Ibis* was published from 1867 to 1870, edited by Alfred Newton.

Volume 1 (2nd series) 1865, January. Pages 98-9, review of *Birds of Great Britain*, parts 5 and 6, no dates but fifteen illustrations in each are listed.

Pages 114-6, letter from Gould dated December 1864 on a Rock-Pipit, possibly *Anthus spinoletta*.

April. Page 220, review of *Birds of Asia*, part 16, no date, with list of sixteen species.

Page 239, announcement of the forthcoming publication of Gould's *Handbook*.

October. Page 526, review of *Birds of Great Britain*, parts 7 and 8, no date, with list of fifteen species for each part.

Volume 2 (2nd series), 1866, January. Pages 111-13, review of *Handbook* with some critical comments on genus-formation.

October. Page 409-10, review of *Birds of Asia*, part 17, April 1865 date, with sixteen species listed.

Volume 3 (2nd series), 1867, April. Pages 237-8, review of *Birds of Great Britain*, parts 9 and 10, dated 1 August and 1 September 1866 respectively, with list of fifteen illustrations for each part.

July. Pages 371-2, review of Birds of Asia, part 18, published 1 April 1866 with sixteen species listed.

Volume 4 (2nd series), 1868, January. Pages 117-8, review of Sylvester Diggles's *The Ornithology of Australia*, parts 1 to 10, no date but mentioned as monthly publications, six plates in each part. Diggles's work is mentioned for interest.

April. Pages 216-8 contain reviews of three Gould works.

Birds of Great Britain, part 11, 1 August 1867 with list of fifteen illustrations; part 12, 1 September 1867 with list of fifteen illustrations.

Birds of Asia, part 19, 1 May 1867, with list of sixteen species.

Birds of Australia, supplement, part 4, 1 December 1867, with list of sixteen species.

Page 250, letter from Gould dated 8 February 1868 concerning *Psaltria concinna* of Gould or *Aegithaliscus anophrys* of Swinhoe.

July. Page 348, review of Diggles's work on Australia, parts 11 to 15, no date.

October. Page 472, review of *Birds of Asia*, part 20, 1 April 1868, with list of sixteen species.

Volume 5 (2nd series), 1869, January. Page 108-9, review of *Birds of Great Britain*, part 13, 1 August 1868, and part 14, 1 September 1868, both with fifteen illustrations.

Pages 127-8, letter from Gould to the editor dated 30 December 1868 on *Emberiza (Euspiza) melanocephala* and *Turdus atrigularis* being found in England.

Volume 6 (2nd series), 1870, January. Pages 118-21, review of several works by one of 'the "untiring divinities" of the ornithological Olympus'.

'Trogons, second edition', part 2, 1 March 1869 with thirteen species listed as illustrated.

Birds of Asia, part 21, 1 April 1869, with fifteen species listed.

Birds of Australia, Supplement, part 5, 1 August 1869 with fourteen species listed.

Birds of Great Britain, part 15, 1 August 1869, and part 16, 1 September 1869, both with fifteen species listed. Newton apologizes for an error he wrote in the 1869 review on pages 108-09.

Page 135, short review of Diggles's Australian work, parts 16-20.

Third series

Series 3 from the years 1871 through 1876 were edited by Osbert Salvin. The membership in 1871 was fifty-five regular members, two extraordinary members, and still Gould is not on the list??

Volume 1 (3rd series), 1871, October. Pages 437-8, reviews of two works.

Birds of Asia, part 22, 1 March 1870, with list of sixteen species.

Birds of Great Britain, parts 17 and 18, 1870, with the thirty species in one list.

Four of Gould's papers are also reviewed on page 438.

Volume 2 (3rd series), 1872, October. Pages 428-9, reviews of two works and six papers.

Birds of Asia, part 23, 1871, with list of fifteen species.

Birds of Great Britain, part 19, August 1871, and part 20, September 1871 with twenty-nine species listed together.

Volume 3 (3rd series), 1873, supplement. Pages 453-4 include reviews of two works and five papers.

Birds of Asia, part 24, 1872 with fifteen species listed.

Birds of Great Britain, part 21, August 1872 and part 22, September 1872, with twenty-six species listed.

Volume 4 (3rd series), 1874, April. Pages 172-3 contain a short statement that Gould's 'great work on the Birds of Great Britain is now complete . . . no bird-book, it is whispered, has ever had such a financial success'. A minor Gould reference is also on page 178.

October. Page 450 has mention of Gould's *Introduction to the Birds of Great Britain*, 1873.

Birds of Asia review on pages 452-3, part 26, 1 August 1874, no list. Page 454 has a sentence on Gould's 'excessive multiplication of species'.

Volume 5 (3rd series), 1875. July. Page 394 a sentence that another part of *Birds of Asia* has been completed, probably part 27.

PART FOUR BIBLIOGRAPHY OF JOHN GOULD, HIS FAMILY AND ASSOCIATES

Volume 6 (3rd series), 1876. July. Page 363, mentioned as 'preparing for issue' of *Birds of New Guinea*, part 3, May 1876, with thirteen species listed. This note appears in an article by P. L. Sclater.

Fourth series

Series 4 of *Ibis* appeared from 1877 to 1882 and was edited by Osbert Salvin and P. L. Sclater.

Volume 1 (4th series), 1877. July. Page 377, review of *Birds of New Guinea*, part 4, 1877, with list of thirteen species.

Pages 377-8, review of *Birds of Asia*, part 29, 1877, with list of thirteen species. These two are somewhat critical reviews.

Volume 2 (4th series), 1878. January. Page 109, review of *Birds of New Guinea*, part 5, 1877, with list of thirteen species.

Pages 109-10, *Birds of Asia*, part 30, 1877, with list of thirteen species.

Volume 3 (4th series), 1879. January. Page 94-5, review of *Birds of New Guinea*, part 6, 1878, with thirteen species; part 7, 1878, with thirteen species, and part 8, 1878 with twelve species. 'In spite of severe ill health, Mr. Gould continues his great work on the birds of New Guinea.'

July. Page 356, review of *Birds of New Guinea*, part 9, 1879, with twelve species listed.

Volume 4 (4th series), 1880. January. Page 134-5 has one Gould article reviewed and two works.

Birds of New Guinea, part 10, 1879, with twelve species listed.

Birds of Asia part 31, 1879, with thirteen species.

July. Page 379 contains the announcement that 'Mr. Gould, we understand, has in contemplation the issue of a supplementary volume to his great work on the Trochilidae... '

Page 384 has a list of ornithological works in course of publication:

Birds of Asia, part 31, July 1879.

Birds of New Guinea, part 10, September 1879.

October. Pages 474-5 have reviews of one article and two works:

Supplement to Trochilidae part 1, 1 August 1880, with list of twelve species. 'Undeterred by serious illness.'

Birds of New Guinea, part 11, February 1880, with list of thirteen species.

Volume 5 (4th series), 1881. January. Page 160, review of *Birds of Asia*, part 32, 1880 with list of twelve species.

April. Page 288-90 is an obituary of Gould.

July. Page 497-8 has a paragraph on 'Mr. Gould's works and collections'. It is quoted in full here:

The entire stock of the late Mr. Gould's illustrated ornithological works, as also of the 'Mammals of Australia,' with all the copyright and other interests involved in them, has been purchased from the executors by Messrs. Henry Sotheran & Co. of Piccadilly, who, we believe, are taking steps to complete the unfinished portions. Mr. Gould's famous collection of Humming-birds, together with his extensive collection of unmounted birdskins has, as already stated in our last issue, been purchased by the Trustees of the British Museum for £3000; and the Humming-birds are already exhibited in the bird-galleries of the Museum. At present the specimens are not labelled; but we are informed that the names will shortly be placed upon them and a 'Guide' to the collection issued.

Page 498 has an article entitled 'Value of ornithological works' as recently realized from the sale of Gould's library.

Volume 6 (4th series), 1882. January. Page 167-8, review of *Supplement to Trochilidae*, part 2, 1881, with ten species listed. A statement is made that Henry Sotheran will continue the work with help of Salvin & Sharpe.

Page 168, *Birds of New Guinea*, part 12, 1881, with list of twelve species.

July. Page 461, *Birds of Asia*, part 33, 1882, with list of thirteen species.

October. Page 601-02, *Birds of New Guinea*, part 13, 1882, with list of thirteen species.

Fifth series

Series 5 appeared from 1883 to 1888 and was edited by P. L. Sclater and Howard Saunders.

Volume 1 (5th series), 1883. April. Page 214, review of *Supplement to Trochilidae*, part 3, 1883, with fourteen species listed and illustrated, with sixteen additional species listed only.

July. Page 379, *Birds of Asia*, part 34, 1883, with list of thirteen species.

October. Page 568, *Birds of New Guinea*, part 14, 1883, with thirteen species listed.

Volume 2 (5th series), 1884. In the January issue on pages 49 to 60 is a reprint of R. B. Sharpe's 'Introduction to Gould's "Birds of Asia"' that appeared in the concluding number of Gould's work.

Interestingly, on the following pages (60 to 66) is a notice of the 'Inauguration of the American Ornithologists' Union'.

Page 104, *Birds of Asia*, part 35, 1883, is reviewed and seven species are listed as figured. This final part 'also contains the title-pages and lists of plates for the seven volumes in which it is proposed that the work shall be bound, and the preface and introduction by Mr. Sharpe...which we have reprinted...in this journal'.

April. Pages 208-9, *Birds of New Guinea*, part 15, 1883, with thirteen species listed as illustrated.

July. Page 341-2, *Birds of New Guinea*, part 16, 1884, with thirteen species illustrated.

Volume 3 (5th series), 1885. April. Page 227-8, *Birds of New Guinea*, parts 17 and 18, 1884. Both parts have thirteen species illustrated.

July. Page 316, *Birds of New Guinea*, part 19, 1885, with thirteen species illustrated.

Page 316-17, *Supplement to Trochilidae*, part 4, 1885, has twelve species listed as illustrated.

Volume 4 (5th series), 1886. On page 97 is a short review of 'Waterhouse on the Dates of Publication of Gould's Works'. It is stated that the biographical sketch included is 'reprinted, with slight alterations, from "Nature"'.

April. Page 194-5, *Birds of New Guinea*, part 20, 1885, lists thirteen species illustrated.

Volume 5 (5th series), 1887. January. Page 108, *Birds of New Guinea*, parts 21 and 22, 1886, are reviewed with thirteen species figured in each part.

July. Page 352-3, *Supplement to Trochilidae*, part 5, 1887, with twelve species illustrated, and twenty-nine other species described in the text. This is the final part to be published. The title-page, preface, and contents are also included. 'The total number of Humming-birds treated of in the Supplement is 122, of which 60 are figured. In the original work 360 species were illustrated. This would make the total number of species about 482.'

Volume 6 (5th series), 1888. April. Page 271-2, *Birds of New Guinea*, parts 23 and 24, 1887 and 1888 respectively, with thirteen species illustrated in each part.

A short obituary of Edward Lear appears on page 286.

The first review of *The Audubon Magazine* appears on page 481.

PART FOUR BIBLIOGRAPHY OF JOHN GOULD, HIS FAMILY AND ASSOCIATES

Sixth series

Series 6 appeared from 1889 to 1894 and was edited by P. L. Sclater.
   Volume 1 (6th series), 1889. April. Page 247, *Birds of New Guinea*, part 25, 1888, with ten species illustrated. This is the final part, the total to be bound as five volumes with 320 plates. 'It seems rather a pity to stop the work when, as we are told in the preface, only some 300 species have been figured out of a total of 1030 belonging to the Papuan Avifauna; but we suppose the publishers had no choice but to adhere to the promise made to the subscribers to finish the book in twenty-five parts. Of these only twelve were issued during Gould's lifetime, the remaining thirteen having been prepared by Mr. R. B. Sharpe.'
   Volume 5 (6th series). 1893, on pages 461 to 462 is a review of Sharpe's *Analytical Index*.

Seventh series

In Volume 5 (7th series) 1899, on pages 164 to 165 (January) is a note on 'Sales of Gould's Bird-books'. Forty-four volumes, with eight of these in parts, from the library of the late Edmond Coulthurst, sold for £430.
   A careful comparison with Harris' list will reveal that he missed only a few of the reviews of Gould's works in *Ibis*.
   A complete run of *Ibis* is in the Linda Hall Library, Kansas City, Missouri.

JOURNAL. *The Ibis. See* Harris, Harry. 1921. Manuscript: Biographical index to *The Ibis* (First ten series) 1859-1918.

JOURNAL. *The Ibis. See* Sclater, P. L., and Evans, A. H. 1909. *The Ibis* jubilee supplement.

JOURNAL. Linnean Society. *See* Linnean Society Journal. Index. 1896.

JOURNAL. *Linnean Society Transactions. See* Linnean Society Transactions. Index. 1867.

JOURNAL. *Magazine of Natural History and Journal of Zoology, Botany, Mineralogy, Geology and Meteorology.* 1828-1829 to 1840. London.
According to Sharpe's *Analytical Index*, Gould published his second article, in this journal (1832) 5: 309-11, 'A notice of the Reed-Warbler (*Curruca arundinacea*, Briss.)'.
   Series 1 of this magazine was published from 1828-1829 to 1836, and series 2 from 1837 to 1840. Series 1 was conducted by J. C. Loudon, and series 2 by Edward Charlesworth. The University of Kansas Watson Library-Ellis Collection has a complete set.
   Wood (1931) states that it was a forerunner of the *Annals and Magazine of Natural History*.

JOURNAL. *The Zoological Journal.* 1824-35. London, W. Phillips.
This journal was published from 1824 to 1835 and appeared in five volumes of twenty numbers. I have all the volumes except Volume I.
   On the first page of volume II under 'Contents/No. V April 1825' is an Introductory Address by the Chairman of the Zoological Club, Rev. William Kirby, on the foundation of the club, of which this journal was the organ. The club was founded on 29 November 1823, as a section of the Linnean Society.

On page 133 of volume V is the announcement of the special meeting of 30 November 1829 where, after six years of meetings, it was voted to discontinue the Zoological Club. Volume V was published in 1835 and part xx contains the final articles dating from 1832 to 1834.

The only Gould reference in the *Zoological Journal* is his article on page 102 of volume V entitled 'On the occurrence of a new British Warbler' presented before the Zoological Club of the Linnean Society at the meeting of 24 November 1829 as noted on page 132. This is John Gould's first scientific article.

Every volume has a few coloured engravings of birds but, except for plate 1 of volume IV being drawn by Sir W. Jardine, the artists are not acknowledged. Supplemental plates were also issued in five separate parts.

There are some interesting notes on the early formation of the Zoological Society of London. In volume II, page 284 is recorded the Prospectus for the establishment of a 'New Zoological Institution' with a menagerie and a museum. This prospectus was dated 1 March 1825. An organizational meeting of the Friends on June 22 [1825] is recorded where Sir Stamford Raffles was appointed chairman.

In volume III, page 143 is the next note on the Zoological Society and 'the first meeting for the formation of the Society took place at the House of the Horticultural Society on the 29th of April, 1826, when Sir T. Stamford Raffles was unanimously elected President of the Society'.

A second meeting 'took place on the 7th of March, 1827 for the purpose of electing a President in the place of Sir T. Stamford Raffles [who had died in 1826], when the Marquis of Lansdown was unanimously chosen to that office'. The Council and Officers were listed.

Volume III also contains a long article on Sir Thomas Stamford Raffles by E. W. Brayley, Jr. and an engraving of his bust appears at the end of this volume.

JOURNAL. *The Zoological Magazine.* 1833. London, Whittaker, Treacher & Co.

A short-lived journal that was published monthly from 1 January 1833, through 1 June 1833 for only six issues. The Kansas-Ellis Collection copy is complete.

According to *Dict. Nat. Biog.* 'Owen started the "Zoological Magazine", [in January 1833] which, however, he ceased to edit and sold in July.' Owen was Professor and later Sir Richard Owen, a brilliant and productive British anatomist and naturalist.

The first article in this journal is on 'The Giraffe', which was eventually stuffed by Gould. [*See* Owen (1833) and L. S. Lambourne (1965).]

JOURNAL. *Zoological Miscellany. See* Gray, John Edward, Editor. 1831 to 1844. *The Zoological Miscellany.*

JOURNAL. *Zoological Record.* 1864 to date. London.

Alfred Newton was the author of the Aves section for the first volume and through 1869. R. B. Sharpe and H. E. Dresser were authors for 1870, Sharpe alone from 1871 to 1873, Sharpe and James Murie for 1874, Osbert Salvin for 1875 and 1876, Howard Saunders from 1877 through 1881, and R. B. Sharpe again in 1882.

The reviews in this 'Record' are all quite perfunctory except for the one on Gould's *Handbook* which appeared in the 1865 volume. Any information on the various parts of Gould's works that is noteworthy is so mentioned by me in the appropriate section on 'Works'.

See also Waterhouse, C.O. 1902. *Index Zoologicus.*

PART FOUR BIBLIOGRAPHY OF JOHN GOULD, HIS FAMILY AND ASSOCIATES

JOURNAL. *The Zoological Society of London, proceedings of the committee of science and correspondence of* 1830-32. London.

This is the earliest printed record of the Zoological Society meetings. In 1833 the 'Proceedings of the Zoological Society of London' appeared in place of these earlier 'Proceedings'.

This 1830-32 journal was published in two parts.

Part one, published from 1830 to 1831, consisted of pages 1 to 182. There are no Gould articles in this part but there are records of presentations by N. A. Vigors on birds from the Himalaya Mountains. These birds were portrayed by John and Elizabeth Gould in Gould's first folio publication *A Century of Birds*. Vigors wrote the letterpress.

Vigors delivered his papers on the Himalayan birds on the following dates. (Zimmer's (1926) records for these dates are in error.)

On pages 7 to 9, at the meeting of 23 November 1830. At this meeting some plates from Goulds *Century of Birds* were exhibited by Vigors.

On pages 22 to 23, at the meeting of 11 January 1831.

On page 35, at the meeting of 8 February 1831.

On pages 41 to 44, at the meeting of 22 February 1831.

On pages 54 to 55, at the meeting of 22 March 1831.

On pages 170 to 176, at the meeting of 27 December 1831. In this paper Vigors refers to the fact that 'the monthly publication' of the parts of Gould's *Century of Birds* had been in progress.

Part two appeared in 1832, and consisted of 215 pages. In this part are four articles by Gould as follows:

On pages 129 to 130, 'On a new species of Wagtail (*Motacilla* L.) (*M. Flava*, Ray and *M. neglecta*)'.

On page 130, 'Report on the second specimen of *Cypselus alpinus*, Ill. in England'.

On pages 139 to 140, 'On a new species of woodpecker (*Picus*, L.) from California (*Picus imperialis*, Mas.)'.

On page 189, 'On a collection of birds from the Orkneys'.

The Linda Hall Library has the two parts of this journal. The first part only is also in the Kansas Spencer Library bound in a volume with miscellaneous pamphlets from the O'Hegarty Collection, item C595.

JOURNAL. *The Zoological Society of London, proceedings of.* 1833 to date.

The following historical information is from pages 2 and 3 of the Index 1860-1870.

Two parts in 1830-31 and 1832 appeared under the name of 'Proceedings of the committee of science and correspondence of the Zoological Society of London'.

The first series of the regular annual 'Proceedings' began with part I, 1833 and concluded with part XV, 1847, with an Index to the fifteen volumes, plus the earlier two volumes.

The second series continued with part XVI in 1848 through XVIII in 1860 with an Index to these thirteen volumes.

From then on, the annual volumes were not numbered. An Index covering 1860-70 was also published.

The great majority of Gould's scientific publications appeared in these 'Proceedings'.

In the Index of 1830-47 there are 135 articles listed beginning with 'On a new species of Wagtail (*Motacilla*, L.) in the 1832 part, page 129. The great majority of these articles are but two or three paragraphs in length usually beginning with a statement such as 'Specimens were exhibited of the adult male ... and at the request of the Chairman, Mr. Gould made some observations'.

In the Index of 1848-70 there are eighty-eight Gould contributions listed.

The Index for the next decade (1871-80) shows that age had taken its toll. Gould published only six articles in these 'Proceedings' the final one in 1875, six years before his death.

Thus, a total of 229 contributions by Gould were printed in these 'Proceedings'. Many of the articles related the description of new species or genera.

JOURNAL. *The Zoological Society of London, transactions of.* 1835 to date.

The first ten volumes, issued irregularly from 1835 to 1879, cover the period of Gould's life. An Index for the first ten volumes was published in 1881.

Gould published only four articles in the 'Transactions'.

Volume I published in 1835 has 'On a new genus in the family of Corvidae', read 14 May 1833, on pages 87-90. A species *Dendrocitta leucogastra* was illustrated by J. and E. Gould as plate 12.

'Description of a new species of the genus *Eurylaimus* of Dr. Horsfield', is the second article communicated 10 December 1833, on pages 175-8. This new species, *Eurylaimus lunatus*, was illustrated by Gould as plate 25.

J. and E. Gould also provided the illustration for an article by W. Yarrell on *Apteryx australis* as plate 10.

Lear provided plates 3, 4, 21, 24, 37 and probably 42 in this first volume of the 'Transactions'.

Short references to Gould in addition to his two articles are on pages iii, iv, 76, 331, and 403.

Volume II published in 1841 has J. and E. Gould illustrations (plates 1, 2, 3 and 4) for an article on *Coturnix* (quails) by Sykes.

A reference to Gould is on page 206.

Lear contributed plates 6, 16, and 17.

Volume III published in 1849 has the third Gould article 'On a new species of genus *Apteryx* (*A. owenii*)', communicated 8 June 1847, pages 379-80, illustrated with plate 57 by H. C. Richter.

References to Gould are on pages 69, 372, 373 (2) and 376.

Volume IV published in 1862 contains the fourth article by Gould, 'Remarks on *Notornis Mantelli*, pages 73-4, illustrated by J. Wolf as plate 25. There are additional Wolf plates in this volume for three articles by Sclater, namely plates 58, 59, 60, 61, 62, and 63, on hawks and owls. Also plate 64 on *Balaeniceps rex* is for the article by W. K. Parker. Finally, plate 77 by Wolf is of an eagle. 'J.W.' [Wolf?] drew in ink [?] figures on pages 347, 355, 356, 357 and 358.

References to Gould are on pages iii, 13, 70-2, 356, 360, 362.

Volume V published in 1866 has nothing on Gould but has some uncoloured Wolf drawings on apes.

Volume VI published in 1869 has a reference to Gould on page 376 about the Dodo, and two plates by Wolf on an otter and a hawk.

Volume VII published in 1872 has two plates by Wolf of mammals.

Volume VIII published in 1874 has many of Gould's scientific names in an article by Arthur on birds of the Islands of Celebes.

Volume IX published in 1877 has references to Gould on pages 461-76, 484, 485, 488, and 494. Wolf contributed plates 95 to 99 on Rhinoceroses.

Volume X published in 1879 is void of Gould or Wolf references or plates.

PART FOUR BIBLIOGRAPHY OF JOHN GOULD, HIS FAMILY AND ASSOCIATES

JUKES, J. BEETE. 1871. *Letters and extracts from the addresses and occasional writings of J. Beete Jukes.* London. p. 220.

Whittell (1954, I: 98) quotes that Jukes wrote from Sydney on 4 March 1844 that 'I go with him [Dr. George Bennett] and Gilbert (Gould's agent and collector) to look at some gigantic birds' bones from New Zealand this evening'. Book not examined by me.

KNIGHT, DAVID. 1977. *Zoological illustration.* Folkestone, Kent, Wm. Dawson.   K

An interesting and perspicacious essay 'towards a history of printed zoological pictures'. Chapters cover the purposes of zoological illustrations ('to show what the animal is like' which 'is more complicated than it appears'), a chapter on techniques of zoological illustration, and then three chapters concerned with a discussion of zoological works up to the early 1900s.

Rather complete listings of bibliographical material and illustrated works follow each chapter. The Gould references are listed in the index, except for the addition of page 113. Generally however, the index is quite incomplete.

LACK, DAVID. 1947. *Darwin's Finches.* Cambridge, University Press.   L

This is a quite comprehensive descriptive study of the finches on the Galapagos Islands followed by Lack's interpretation of their descriptive points in relationship to evolutional theories.

There are four colour plates portraying seven species of the finches as drawn by Gould. Since 'The Zoology of the Voyage of H.M.S. "Beagle" ' was published in 1841, the finished drawings were undoubtedly done by Mrs Gould. [Some of these original drawings are in Lord Derby's collection.]

Gould's role in naming the finches is discussed by Lack, and to me, the passage of time serves to demonstrate again how outstanding Gould was as a taxonomist.

LAMBOURNE, L. S. 1965. 'A giraffe for George IV'. *Country Life,* Dec. 2, 1500-1502.

This is an interesting, informative and entertaining article written by the husband of one of Gould's descendants. When the giraffe died in 1829 it was stuffed for the King by John Gould.

LAMBOURNE, MAUREEN. 1964. 'Birds of a feather: Edward Lear and Elizabeth Gould'. *Country Life,* 25 June, 1656-1657.

The great-great grand-daughter of John Gould has written this short article prompted by the discovery of two unpublished letters by Lear. These letters were written to John Gould from Lord Derby's Knowsley Hall upon the death of Mrs Gould. In this article are portraits of Mrs Gould and Lear, and also five bird illustrations.

LAMBOURNE, MAUREEN. 1967. 'A Victorian illustrator of wild life'. *Country Life,* 20 April, 908-910.

The great-great grand-daughter of John Gould writes on Joseph Wolf. Palmer's book on Wolf is her main source of information.

The Calkin portrait of Wolf is reproduced and also six watercolour drawings by Wolf of birds and mammals in the possession of the Zoological Society of London and the Gould family descendants.

LAMBOURNE, MAUREEN. 1969. 'John Gould and his illustrators'. *Birds* (RSPB Magazine) 2: 168-71.

Mrs Lambourne is a fourth-generation descendant of John Gould and has illustrated this article with five photographs of original watercolours, one a portrait of Gould by Miss M. Walker, and four of original bird drawings in the family collections by Richter, John Gould, Wolf and Hart. Mrs Gould and Lear are represented by photographs of lithographs.

While the majority of the textual material in this article is from Sharpe's *Analytical Index*, Mrs Lambourne does quote from writings of a daughter of Gould's concerning her mother, and a description of Gould's studio in their home.

LEACH, J. A. (revised and edited by Charles Barrett). 1944. *An Australian bird book*. 8th ed. Melbourne, Whitcombe and Tombs.

The black-and-white plate of the noisy Scrub-Bird, facing page 164 is from Gould's *Birds of Australia*.

LEICHHARDT, DR LUDWIG. 1847. *Journal of an overland expedition in Australia from Moreton Bay to Port Essington... 1844-1845.* London, T. & W. Boone. (Australian Facsimile Editions No. 16, Adelaide, Libraries Board of South Australia, 1964.)

A notable and valuable personal account of this famous expedition. Much has been written about it especially by Aurousseau (1967) and Chisholm (1955).

Parts in this volume relating to John Gilbert bear reprinting.

From pages xv and xvi:

My means, however, had since my arrival [in Brisbane] been so much increased, that I was after much reluctance prevailed upon to make one change, —to increase my party; and the following persons were added to the expedition: — Mr. Pemberton Hodgson, a resident of the district; Mr. Gilbert; Caleb, an American negro; and "Charley," an aboriginal native of the Bathurst tribe. Mr. Hodgson was so desirous of accompanying me that, in consideration of former obligations, I could not refuse him, and, as he was fond of Botanical pursuits, I thought he might be useful. Of Mr. Gilbert I knew nothing; he was in the service of Mr. Gould, the talented Zoologist who has added so much to our knowledge of the Fauna of Australia, and expressed himself so anxious for an opportunity of making important observations as to the limits of the habitat of the Eastern Coast Birds, and also where those of the North Coast commence; as well as of discovering forms new to Science during the progress of the journey, that, from a desire to render all the service in my power to Natural History, I found myself obliged to yield to his solicitations, although for some time I was opposed to his wish. These gentlemen equipped themselves, and added four horses and two bullocks to those already provided.

From pages 307 to 311:

At the end of our stage, [on June 28, 1845] we came to a chain of shallow lagoons, which were slightly connected by a hollow. Many of them were dry; and fearing that, if we proceeded much farther, we should not find water, I encamped on one of them, containing a shallow pool; it was surrounded by a narrow belt of small tea trees, with stiff broad lanceolate leaves. As the water occupied only the lower part of this basin, I deposited our luggage in the upper part. Mr. Roper and Mr. Calvert made their tent within the belt of trees, with its opening towards the packs; whilst Mr. Gilbert and Murphy constructed theirs amongst the little trees, with its entrance from the camp. Mr. Phillips's was, as usual, far from the others and at the opposite side of the water. Our fire place was made outside of the trees, on the banks. Brown had shot six Leptotarsis Eytoni, (whistling ducks) and four teals, which gave us a good dinner; during which, the principal topic of conversation was our probable distance from the sea coast, as it was here that we first found broken sea shells, of the genus Cytherea. After dinner, Messrs. Roper and Calvert retired to their tent, and Mr. Gilbert, John, and Brown, were platting palm leaves to make a hat, and I stood musing near their fire place, looking at their work, and occasionally joining in their conversation. Mr. Gilbert was congratulating himself upon having succeeded in learning to plat; and, when he had nearly completed a yard, he retired with John to their tent. This was about 7 o'clock; and I stretched myself upon the ground as usual, at a little distance from the fire, and fell into a dose, from which I was suddenly roused by a loud noise, and a call for help from Calvert and Roper. Natives had suddenly attacked us. They had doubtless watched our movements during the afternoon, and marked the position of the different tents; and, as soon as it was dark, sneaked upon us, and threw a shower of spears at the tents of Calvert, Roper, and Gilbert, and a few at that of Phillips, and also one or two towards the fire. Charley and Brown called for caps, which I hastened to find, and, as soon as they were provided,

they discharged their guns into the crowd of the natives, who instantly fled, leaving Roper and Calvert pierced with several spears, and severely beaten by their waddies. Several of these spears were barbed, and could not be extracted without difficulty. I had to force one through the arm of Roper, to break off the barb; and to cut another out of the groin of Mr. Calvert. John Murphy had succeeded in getting out of the tent, and concealing himself behind a tree, whence he fired at the natives, and severely wounded one of them, before Brown had discharged his gun. Not seeing Mr. Gilbert, I asked for him, when Charley told me that our unfortunate companion was no more! He had come out of his tent with his gun, shot, and powder, and handed them to him, when he instantly dropped down dead. Upon receiving this afflicting intelligence, I hastened to the spot, and found Charley's account too true. He was lying on the ground at a little distance from our fire, and, upon examining him, I soon found, to my sorrow, that every sign of life had disappeared. The body was, however, still warm, and I opened the veins of both arms, as well as the temporal artery, but in vain; the stream of life had stopped, and he was numbered with the dead.

As soon as we recovered from the panic into which we were thrown by this fatal event, every precaution was taken to prevent another surprise; we watched through the night, and extinguished our fires to conceal our individual position from the natives.

A strong wind blew from the southward, which made the night air distressingly cold; it seemed as if the wind blew through our bodies. Under all the circumstances that had happened, we passed an anxious night, in a state of most painful suspense as to the fate of our still surviving companions. Mr. Roper had received two or three spear wounds in the scalp of his head; one spear had passed through his left arm, another into his cheek below the jugal bone, and penetrated the orbit, and injured the optic nerve, and another in his loins, besides a heavy blow on the shoulder. Mr. Calvert had received several severe blows from a waddi; one on the nose which had crushed the nasal bones; one on the elbow, and another on the back of his hand; besides which, a barbed spear had entered his groin; and another into his knee. As may be readily imagined both suffered great pain, and were scarcely able to move. The spear that terminated poor Gilbert's existence, had entered the chest, between the clavicle and the neck; but made so small a wound, that, for some time, I was unable to detect it. From the direction of the wound, he had probably received the spear when stooping to leave his tent.

The dawning of the next morning, the 29th, was gladly welcomed, and I proceeded to examine and dress the wounds of my companions, more carefully than I had been able to do in the darkness of the night.

Very early in the morning we heard the cooees of the natives, who seemed wailing, as if one of their number was either killed or severely wounded: for we found stains of blood on their tracks. They disappeared, however, very soon, for, on reconnoitring about the place, I saw nothing of them. I interred the body of our ill-fated companion in the afternoon, and read the funeral service of the English Church over him. A large fire was afterwards made over the grave, to prevent the natives from detecting and disinterring the body. Our cattle and horses fortunately had not been molested.

LEWIN, JOHN WILLIAM. 1828. *Natural history of the birds of New South Wales* ... 3rd ed. London, H. G. Bohn.

On the title-page is stated 'Contributed by Mr. Gould, Mr. Eyton, and other scientific gentlemen' (Engelmann 1846).

LEWIN, JOHN WILLIAM. 1838. *A natural history of the birds of New South Wales.* London, Bonn.

'New and improved edition,/to which is added/a list of the synonymes of each species,/incorporating the labours of/T. [sic] Gould, Esq., N. A. Vigors, Esq., T. Horsfield, M.D., and W. Swainson, Esq.' This is taken from the title-page in the University of Kansas Spencer Library-Ellis Collection volume, E-23,1838.

According to a note by the publisher, Gould also loaned some specimens of birds to insure accurate colouring of the plates.

There are twenty-six hand-coloured engraved plates in the volume.

LEWIN, WILLIAM. 1838. *The birds of Great Britain*, new and improved edition. London, Johnson.

The title-page states 'incorporating the labours of T. Gould, N. A. Vigors, T. Horsfield and W. Swainson' (Engelmann 1846). Not examined.

LEWIS, FRANK. 1974. *A dictionary of British bird painters.* Leigh-on-Sea, England, F. Lewis.

It is amazing, but probably understandable, why John Gould, Mrs Gould, Hart and Richter are left out of this book. The probable reason Lewis didn't include them is because he knew of no signed drawings by them, but they do exist.

Lear, Wolf, and W. Hawkins are included. An uncoloured plate by Lear of a Blue and Yellow Macaw is portrayed.

LINNEAN SOCIETY CATALOGUE. 1925. *Catalogue of the printed books and pamphlets in the library of the Linnean Society of London.* London, Longmans, Green.

In 1925 the library had the five-volume set of Gould's *Birds of Europe*, five other single volume monographs, two of the octavo *Introductions*, plus *Synopsis* and *Handbook*.

LINNEAN SOCIETY JOURNAL. Index. 1896. *General index to the first twenty volumes of the Journal (zoology)... 1838 to 1890, of the Linnean Society.* London, Longmans, Green.

Four items on Gould are listed: 'Obituary, Proc. (1880-1882) 13, 17; *Balaeniceps Rex*, Gould, exhib., Proc. II 109; exhib. lizard from Australia, Proc. i 81; *Menura Alberti*, Gould, a new Lyre-bird from Richmond River, exhib., Proc. ii. 67'.

LINNEAN SOCIETY TRANSACTIONS. Index. 1867. *General index to the transactions of the Linnean Society of London.* Vols. I to XXV. London, Longmans, Green.

Gould did not publish any articles in the *Transactions*.

LLOYD'S REGISTER. 1838. London.

Through the courtesy of Carolyn Ritger of the Mariners Museum, Newport News, Virginia the following data are supplied on the ships used by the Goulds in their Australian venture. This information on each ship is for the date 1838, but since some of these ships were used by the Goulds in 1839 and 1840, the data for those dates would be different in Lloyd's Register.

Ships	Masters	Tons	Build Where	When	Owners	Port belonging to	Destined Voyage	No. Yrs. first assigned	Classification Character for Hull & Stores
Black Joke	Bg J. Miller	219	Havan		J. Miller	London	Lon. India	—	AEl
Katherine Stewart Forbes	Bk I.B. Fell	457	Nrthflt	1818	Chapman	London	Lon. Austra	—	AEl
Kinnear	Bk C. Millard	369	Yrmth	1834	Ellice & C.	London	Lon.	10	Al
Mary Ann	Bk	290	Sndrld	1834	G. Sparks	Sndrlnd	Sld. Ptrsbrg	8	Al
Parsee	Bk M'Kellar	349	Dmbtn	1831	J. Gordon	London	Lon. Hob. T.	8	Al
Potentate	Bk M'Gilchst	302	Grnck	1838	M'Cunn	Gren'ck	Cly. Jamai	12	Al
Susanna Ann	Sr Gaynor	79	Bridpt	1814	Street	London	Lon. NSWls	—	AEl

LONGMAN, HEBER. 1922. 'John Gould's notes for John Gilbert'. *Memoirs Queensland Museum*, 7: 291-4.

This report supplies a verbatim text of Gould's manuscript instructions to Gilbert presumably for his second collecting trip to Australia. It is not dated.

The majority of the instructions for collecting has to do with the procuring of kangaroos and other mammals in New South Wales and Western Australia. Gould also wanted specimens of fish, shells, plants, sponges, corallines, reptiles, and insects, plus, of course, birds, nests, and their eggs.

LORD, CLIVE E. 1933. 'A review of Tasmanian ornithology'. *Emu* 32: 211-20.

In 1642 the Dutch Tasman first described the Australian island later to be named Tasmania, and noted 'a large number of gulls, wild ducks and geese'. The 'geese' were Black Swans, according to Lord.

When Gould visited Tasmania, or Van Diemen's Land as it was then known, in 1838 and 1839 'a very definite advance was made as regards the knowledge of the ornithology of the island'. Lord then mainly recounts Lady Franklin's writings from her diary and her letters, particularly concerning Gould's voyage with her to Recherche Bay. Lord makes a few comments on the birds found by Gould on this trip (*See also* Mackaness 1947).

LYSAGHT, A. M. 1975. *The book of birds.* London, Phaidon.

The sub-title is 'Five Centuries of bird illustration'. There is quite an emphasis on bird illustration before 1800 with over half the illustrations devoted to this early period.

The reproduction of the forty colour plates is excellent; they were printed in the Netherlands.

The plates No. 94, 98, 103, 110, 129 and 135 are beautiful Gould plates. Plate 84 is by Lear, and several plates are by Joseph Wolf. The Wolf watercolour of *Balaeniceps rex* (Shoebill), which was described by Gould, is spectacular in colour and composition.

MACGILLIVRAY, JOHN. 1852. *Narrative of the voyage of H.M.S. Rattlesnake, commanded by the late Captain Owen Stanley, R.N., F.R.S., etc. during the years 1846-1850.* 2 vols. London, T. & W. Boone.

Item No. IV in the Appendix of Volume two is a 'Catalogue of the birds of the north-east coast of Australia and Torres-Strait'. MacGillivray states 'With the kind assistance in determining the species of Mr. Gould, who has elsewhere published similar lists (in the works of Strzelecki and Eyre, and Introduction to the Birds of Australia) of the birds of other parts of Australia, the annexed Catalogue has been made out'. This section covers pages 355 to 359.

MACGILLIVRAY, WILLIAM. 1901. *A memorial tribute to William MacGillivray.* Edinburgh, private circulation.

William MacGillivray (1796-1852) was the author of 'History of British Birds' (1837-52) and also contributed 'scientific details' for Audubon's *Ornithological Biographies.*

A journal of his visit in 1833 to the museums of Glasgow, Liverpool, Dublin, Bristol and London is very interesting. His visit to the Museum of the Zoological Society on Bruton Street is informative since he describes both floors with their contents (p. 107).

On Saturday, 21 September 1833 he attempted to meet John Gould, the 'celebrated preparer of objects of natural history', on Broad Street, Golden Square, but he was out. He adds that 'some of his performances, however, I had seen in the Zoological Society's collection, and they are highly creditable'. For MacGillivray that is quite a compliment as he was highly critical of almost all the museum collections seen.

MACGILLIVRAY, W. D. K. 1932. 'The Flock Pigeon'. *Emu* 31: 169-74.

MacGillivray writes on renewing his acquaintance with the Flock Pigeon (*Histriophaps histrionica*) that Gould had met with in immense numbers in December 1839 near the Nundewar Range on the Namoi.

MACK, GEORGE. 1938. 'John Gould's Correspondence with Sir Frederick McCoy'. *Emu* 38: 212-16.

Sir Frederick McCoy was the first Director of the National Museum, Melbourne. Soon after assuming this responsibility, he wrote Gould commissioning him to purchase 'an entire set of generic types of birds'. Gould accepted this arduous task, determining every specimen as labelled, and from January 1858 until 1870, for twelve years, Gould sent thousands of birds to the Australian museum.

The exact amount of correspondence between Gould and McCoy is not mentioned. [The National Museum of Victoria has kindly supplied me with photocopies of these thirty letters, from which I have made typescripts.]

MACKANESS, GEORGE. 1947. *Some private correspondence of Sir John and Lady Jane Franklin*. Sydney, privately printed.

This material is published in two parts, part I containing twenty-six letters and diary extracts over a period from 6 January 1837 to 13 April 1841, and part II containing twenty-four letters from 21 April 1841 to 19 May 1845. The great majority of the letters were written by Lady Jane.

The Goulds stayed in Hobart, Van Diemen's Land (now Tasmania) on the first part of their trip to Australia. Soon after their arrival, they moved into Government House, the home of Sir John and Lady Franklin.

Since these two parts are not indexed, the following pages have references to the Goulds: part I, pages 35 (a portrait of John Gould), 38, 39 to 53 (on an excursion to Recherche Bay, December 1838), 55, 59, 62, 64, 68, 76, 80, 86, 97, 108, 121, 123, 124; and part II page 20. Quotations from these letters and diary appear in the Chronology section of December 1838.

MADSEN, CECIL. 1974. *World treasury of birds in colour*. New York, Galahad.

It is really pathetic that the fine printing and extensive work in collecting the coloured figures for this book should be wasted on a worthless text. Here is a quarto-sized volume with one to five coloured bird figures on every page. The illustrations are not in perfect colour shades, especially the Flamingo, but they are finely printed. The author states once on the title page that these are 'engravings by Audubon, Gould and several artists' but, nowhere else are the artists mentioned or even associated with their drawings! The text is childlike.

Gould and his artists are responsible for over one-third of the 215 or so illustrations.

MAHONEY, ELIZA SARAH. 1928. 'The first settlers at Gawler'. *Proceedings Royal Geographical Society of Australasia: South Australian Branch* 28: 53-82.

In 1898, Mrs Mahoney, née Reids, wrote a notebook reminiscing about her migration to South Australia some sixty years earlier in January 1839. Their homestead of 'Clonlea' was at Gawler, north of Adelaide.

Gould visited their homestead in 1839 with Evelyn Sturt, brother of Captain Sturt, the famous Australian explorer. [She remembered the date as 'Christmas Day' 1839, but correctly it had to be the end of June or the early part of July 1839.] The account in Mrs Mahoney's notebook follows:

> On this Christmas Day, [error, see above] 1839, an orderly arrived with dispatches for the Governor, whom he was told to meet at our house. Next morning Colonel Gawler arrived with his A.D.C., Mr. Gill. (Some years afterwards he was drowned going to New Zealand. Like many more of the small and only traders of that day, the ship was never heard of after leaving port.) They had missed the track the night before, so had to sleep in the bush with but one blanket each. They had come from the Murray.

PART FOUR BIBLIOGRAPHY OF JOHN GOULD, HIS FAMILY AND ASSOCIATES

Evening brought the rest of the large party, and all dined with us — Captain and Mrs. Sturt, Miss Gawler (about my age), and several gentlemen. They were all riding, the cart following with the tents, which were pitched near. Captain Sturt was the great explorer who had discovered the Murray River, and had been some years in New South Wales before coming to Adelaide, where he was Colonial Secretary. His brother Evelyn came, I remember, one morning soon after on his way to Adelaide, to send back provisions to his party, whom he had left quite out of flour. Gould, the ornithologist, was with him, and as they had been some days without anything to eat, except what they shot, they enjoyed a good breakfast. Gould gave me a lesson in skinning and stuffing; or rather preserving. His wonderful book shows the number of new birds (his wife painted them) he found in Australia, and when with us he was much pleased with a white kangaroo-skin he preserved, the first I had ever seen.

MAIDEN, J. H. 1907. 'A century of botanical endeavor in South Australia'. *Australasian Association for the Advancement of Science*, 11th meeting, Adelaide, pp. 158-99.

Following an historical account of early botanical investigation in South Australia, Maiden gives short biographical accounts of many botanical collectors. Those associated with Gould include Charles Sturt, John McDouall Stuart, Edward John Eyre, George French Angas, James Backhouse, Richard Schomburgk, and F. G. Waterhouse.

Since the Northern Territory was at that time a political division of South Australia, Maiden then discussed that area and Ludwig Leichhardt and J. Armstrong among others.

MAIDEN, J. H. 1908. 'Records of Australian botanists'. *Proc. Royal Society of New South Wales* 42: 60-132.

Contains many names of botanical collectors in Australia with a short account of their life. Included are James Backhouse, Dr George Bennett, Philip Parker King, Friedrich W. L. Leichhardt, Frederick Strange (with a portrait), Charles Stuart, etc. Many of those named also collected for Gould.

MAIDEN, J. H. 1909. 'Records of Western Australian botanists'. *J. West Aust. Nat. Hist. Soc.* 6: 5-27.

As a continuing effort 'to publish my notes on the botanists of other Australian States', Maiden performs an admirable service in this article for the Western Australia group. Many of the botanical collectors also collected birds, some for Gould.

Biographical notices of the following are relevant: James Backhouse, Benjamin Bynoe (with a listing of the ships upon which Dr Bynoe served as Surgeon, including three stints on the *Beagle*), James Drummond, Edward John Eyre, John Gilbert (a full page devoted to him), George Grey, Ludwig Preiss, and J. S. Roe.

MANDER-JONES, PHYLLIS. 1972. *Manuscripts in the British Isles relating to Australia, New Zealand and the Pacific.* Honolulu, University Press of Hawaii.

A valuable compendium of manuscript material very adequately indexed. [A great 'thank you' to Ms. Mander-Jones.]

Gould MS. is included in this listing from the British Museum (Natural History); Linnean Society; Royal College of Surgeons of England; Zoological Society of London; Newton Library, Cambridge; Scott Polar Research Institute, and the Earl of Derby, Knowsley Hall.

MANNERING, EVA. 1955. *Mr. Gould's Tropical Birds.* New York, Crown Publishers.

This folio-size thin volume contains twenty-four beautifully reproduced colour plates from John Gould's works. The four-page text on Gould's life and his publications is a condensation of the information in Sharpe's *Analytical Index*. Next follow very short descriptions of the birds portrayed 'taken from the original text of the Gould folios'.

MANNERING, EVA. 1964. *Gould's Tropical Birds*. London, Ariel Press.

Twenty beautifully reproduced colour plates from Gould's works dominate this small quarto book that has eight pages of text. This text is primarily a repetition of the information that appears in Sharpe's *Analytical Index*.

This edition is a smaller-sized version of Mannering's 1955 book, but with fewer plates.

MANSON-BAHR, PHILIP. 1959. 'Famous British ornithologists'. *Ibis* 101: 53-64.

Included are biographical notes on Alfred Newton, Abel Chapman, Henry E. Dresser, R. Bowdler Sharpe, William Eagle Clarke, and E. G. B. Meade-Waldo.

MANSUETTI, ROMEO. 1953. 'The Goulds – olympian bird artists'. *Nature* 46: 314-17, 333.

Mr Mansuetti has written a long and thoughtful article on Gould, his art, and his artists. An original pencil sketch by John Gould in Dr Herbert Friedmann's office at the U.S. National Museum is described. He stated that 'the striking elements [of this sketch] are the deft pencil strokes, accurate anatomical details, and a realistic pose, suggesting a Louis Agassiz Fuertes' sketch'.

MARSHALL, A. J. 1954. *Bowerbirds*. Oxford, Clarendon Press.

Marshall supports with gonadal studies that, in agreement with 'Gould and Darwin, the bowers, display, grounds, and the activities that take place near them are primarily concerned with sexual reproduction'.

He further states that 'the poetic family name "bower-bird" was probably coined by the nineteenth-century English naturalist, John Gould. On his return from Australia, Gould announced to the Zoological Society of London, in 1840, that bowerbirds had the extraordinary habit of building what colonists called "runs". Of bower-bird's runs John Gould said, "These runs are perfectly anomalous in the architecture of birds ... [they] consist of a collection of pieces of stick or grass, formed into a bower ... One of them (that of the *Chlamydera*) might be called an avenue ... They are used by the males to attract the females.' Fifty-three years later, Alfred Newton observed, 'this statement, marvelous as it seemed, has been proved by many subsequent observers to be strictly true'.

'Gould's information caught the imagination of the Victorian public, soon many were speaking affectionately of the "bower-bird".'

Additional references to Gould, John Gilbert, Charles Coxen, and Sharpe are listed in the index.

Marshall's classification of these birds differs from and is simpler than that of Iredale.

MARTIN, W. C. L. 1852. *A general history of Humming-birds, or the Trochilidae: with especial reference to the collection of J. Gould, F.R.S. etc. now exhibiting in the gardens of the Zoological Society of London*. London, H. G. Bohn.

'How crowded are the gardens of the Zoological Society of London, and there, how attractive is Mr. Gould's magnificent Cabinet of Humming-Birds!' Further references to Gould are on almost every page of this octavo book.

On page 127 is the statement that 'It is but proper to observe, that our descriptions are all taken from specimens in the cabinet of Mr. Gould, without access to which we could scarcely have ventured upon the present undertaking'.

Fifteen hand-coloured illustrations, including one frontispiece, are inserted throughout the volume. There is no mention of the artist.

An 1861 edition was also published, which appears to be identical.

MASON, ALEXANDRA. 1974. 'The adventure of the California box cars, the Nebraska farmhouse, the Finnish wheelbarrow, and the coal shed in Alabama'. *Kansas Alumni* 72: 22-5.

An informal article on the interesting events of book-acquiring for the University of Kansas libraries. Anecdotal accounts of the Ellis and Fitzpatrick acquisitions are reported.

MATHEWS, G. M. 1910-27. *The birds of Australia.* 12 vols. London, Witherby.

A monumental work with 600 colour plates. Five supplements were issued from 1920 to 1925, the last two containing a bibliography of the birds of Australia.

Needless to say, much is on John Gould and the birds discovered and studied by him.

MATHEWS, G. M. 1912. 'Dates of issue of Lear's Illustrations Psittacidae'. *Austral avian record* 1: 23-4.

From a work in parts, Mathews gives the dates of issues of each of the twelve parts, and the species in each part. Three to four plates were published in each part for the total of forty plates. Parts I and II were dated 1 November 1830, parts III to VIII were issued from 1 January 1831 to 1 October 1831. Parts IX to XII were not dated on the wrapper but the title page issued with part XII is dated 1832.

[From the John Gould and Edward Lear correspondence to Sir William Jardine there are several references to the fact that Gould purchased from Lear his remaining stock and publications rights. These letters are in the Royal Scottish Museum dated January [before the 16th] 1834 (Lear to Jardine); 16 January 1834 (Gould to Jardine), and 23 January 1834 (Lear to Jardine).

In the second letter of 16 January 1834 Gould wrote 'Mr. Lear's Parrots stopped at the 12 nos. I have purchased from him the whole of his Stock, so that if you are not complete as far as published I can make them so; I have some idea of finishing them myself'.]

MATHEWS, G. M. 1915. 'Diggles' ornithology of Australia, and other works'. *Austral avian record* 2: 137-44.

From the *Zoological Record* of 1866 to 1870 Mathews found four notes on the various parts of the 'Companion to Gould's handbook'. The species in each of the twenty-one parts is listed with the year of the notice in the *Zoological Record.*

MATHEWS, G. M. 1917. 'Silvester Diggles, ornithologist' (with portrait). *Austral avian record* 3: 98-108.

Diggles (1817-80) of Brisbane, Queensland was a naturalist, artist and musician. Mathews describes some of Diggles's drawings of birds and other natural history objects found in a sketch-book.

Mathews also gives a complete collation of Diggles's *Ornithology of Australia* from the twenty-one parts as issued. This is a correction and elaboration of Mathews's 1915 article on this work.

MATHEWS, G. M. 1919. 'Samuel Albert White' (with portrait). *Austral avian record* 3: 162-6.

An account of the son of Samuel White. The father provided Gould with specimens and data. Mathews states that the father 'practically sacrificed his life in his pursuit of ornithology and whose talents were never recognized by Gould'. 'Never' is a strong word, and erroneous in this instance.

MATHEWS, G. M. 1920. 'New species of Australian birds since the time of John Gould'. *Emu* 20: 96-8.

Mathews states that 'From Gould's time to the end of 1899, 22 species had been added to the Australian list'. He then gives a breakdown of 'purely Australian species', 'visitors or sub-species', birds occurring less than three times, etc. A quite confusing listing in reference to the number 22, but ornithologists may understand it.

The final paragraph is clear. Of 668 species of Australian birds, Gould described 188, Latham 81, Gmelin 35, Linné 34, Temminck 30, Vieillot 28, etc. Of these top describers, Gould was the only one to visit Australia.

MATHEWS, G. M. 1920. 'Dates of ornithological works'. *Austral avian record* 4: 1-27.

This is an amplification and revision of a list of exact dates of publication of the ornithological works dealing with Australian birds that appeared 'as Appendix B. in Part V. of Volume VII. of my Birds of Australia'. Many early journals have been researched and reported.

The dates of publication of the separate parts of the works of many authors have also been ferreted out. For Gould, Mathews gives the dates of publication of the parts of Gould's Australian works, including *Synopsis, Birds of Australia and adjacent islands, Birds of Australia* and the *Supplement, Handbook*, and *Birds of New Guinea.*

Many pearls are hidden in this paper.

MATHEWS, G. M. 1920-25. *The Birds of Australia, Supplement.* 1 vol., 5 parts. London, H. F. & G. Witherby.

The Library of Congress has these five supplements to Mathews's major work bound in one large volume. The original covers are bound in also.

The first three supplements have to do with a 'Check-list of the Birds of Australia'. Supplement No. 1 or Part 1 of the check-list is dated 16 February 1920, supplement 2 or part 2, 26 July 1923, and supplement 3 or part 3, 8 September 1924. This section closes with an index.

The 'Bibliography of the Birds of Australia' is in two parts. Supplement No. 4 which is Part 1 of the Bibliography was published 6 April 1925, and supplement 5 or part 2 of the Bibliography was published 22 June 1925 according to the covers of the supplements.

To the bibliographer these last two supplements by Mathews are quite important. His 'Introduction' contains 'life sketches of the two friends who have helped me for so many years', namely Charles Davies Sherborn and Charles Wallace Richmond.

Then follows an extensive bibliography of persons and journals related to Australian ornithology. Mathews's personal comments add interest to this valuable listing.

Whittell (1954) performed a similar task but had the advantage of Mathews's earlier work. Whittell covers far more individuals.

The pages devoted to Gould are thirteen. Mathews lists the new species of Australian birds named in Gould's works, as does Whittell.

MATHEWS, G. M. 1921. 'Fort Pitt, Chatham, bird collection'. *Austral avian record* 4: 147-8.

Mathews came across a 'Catalogue of the Collection of Mammals and Birds at Fort Pitt, Chatham' published in 1838 by Edward Burton. Gould had described in 1837 two Australian birds from this collection, namely *Anthochaera lunulata* and *Eopsaltria griseogularis*.

MATHEWS, G. M. 1923. 'Gould's "North-west coast of Australia" '. *Emu* 23: 76-7.

Quoting some passages from Gould's *Handbook*, Mathews adds support to his contention that when Gould stated that a specimen was from the 'north-west coast of Australia' he at least sometimes included Port Essington in this locale. [For those not familiar with Australia, the Port Essington area is truly north-central Australia.]

MATHEWS, G. M. 1925. *Bibliography of the birds of Australia.* London, Witherby.

Here is an extensive listing of the Australian birds described and named by Gould. Mathews's listing and comments must be compared with Whittell's similar but more complete bibliography of Australian ornithology (1954). Mathews limited this bibliography to a listing of 'publications in which new scientific names of Australian birds were proposed' (Whittell 1954, IX).

[MATHEWS, G. M.] Review. 1927. 'Mathews' "The Birds of Australia" '. *Auk* 44: 435-42.

The best way to summarize this in-depth review by Witmer Stone of Mathews's twelve-volume work is to give some quotations:

> We have always had great respect for his [Mathews's] sincerity of purpose and for the great service that he has done to ornithology in bringing to light so many overlooked publications and in settling so many problems depending on dates of publication. We have endorsed his researches in nomenclature and have agreed to most of his findings. His main effort has been to arouse Australian ornithologists to a realization that Gould had not said the last word on the birds of the Antipodes and that they could not sit back and ignore the rest of the world and all early authors, but must move along with the rest of us and bring their nomenclature up to date. The old school conservatives would have none of this and bitterly opposed Mr. Mathews' ideas but with the passing of the years the younger men have in a great measure come into agreement with him.
> In the matter of subspecies Mr. Mathews was an extremist and proposed them at a rate unequalled we believe by any other writer.

Stone concludes with a 'collation' of each volume, with dates of publication of each part, pages, plates, and some notes.

MATHEWS, G. M. 1927-30. *Systema avium Australasianarum. A systematic list of the birds of the Australasian region.* 2 vols. London, British Ornithologists Union, Taylor and Francis.

A valuable but apparently controversial nomenclature of Australian birds with synonyms.

MATHEWS, G. M. 1930. 'Some interesting links with John Gould'. *Emu* 30: 141-3.

'Many years ago there came into my possession, through a daughter of John Gould, some interesting bird plates sent from Australia by Eli Waller, of Brisbane ... Kendall Broadbent was collecting birds at the time and sending the material to Waller ... Presumably Waller painted the birds.' Four of these drawings were sent to Gould for naming in 1874. Interestingly, E. P. Ramsay named these birds first.

Mathews then proceeds to list sixteen more 'original drawings' for Gould's plates that are in his possession, the majority being Birds of Paradise.

[This article raises two questions. First, where are these original 'Gould' drawings now, and secondly, if Mathews had contact with Gould's daughter in the 'many years

ago' before 1930, how or why did he lose this contact, so that Chisholm in 1938 had to re-discover the descendants in England? *See* Mathews, G. M. 1966.]

MATHEWS, G. M. 1938. 'John Gould: an appreciation'. *Emu* 38: 239-40.

Mathews writes in this Gould Commemorative Issue many statements of praise such as ' "The Birds of Australia" is an inimitable classic'; 'it was the genius of Gould himself that carried the work through', he 'was a very excellent all-round ornithologist', '... intense personality ...'; '... intense energy'; and also that his wife 'was a victim, absolutely unintentional, of overwork'.

MATHEWS, G. M. 1942. *Birds and books.* Canberra, Verity Hewitt.

This interesting autobiographical account of the author of another great *The Birds of Australia* contains brief mention of Gould on pages 23, 39, 44, 48, and 52.

Mathews also knew and worked with R. Bowdler Sharpe, as did Gould. Another similar circumstance between Gould and Mathews is that while the British Museum declined to purchase Gould's unique collection of Australian birds and they ended up in America, Australia also declined to purchase Mathews's collection of Australian birds, and they eventually also migrated to America.

MATHEWS, G. M. 1946. *A working list of Australian birds, including the Australian Quadrant and New Zealand.* Sydney, privately printed.

'The present List only records sub-species which can be accepted by workers using the small imperfect collections available at present in Australian Museums. The many sub-species not listed, are not to be rejected, but are sub-species that require the comparison of better material.' Mathews states that his 1931 'A list of the birds of Australasia' is a more complete list with every synonym included and indexed.

The earliest reference for any species is given in full, so the list is replete with references to Gould. As an example of Mathews's meticulous research, not only is the year of publication given for the earliest reference, but also the month and day, if known. Thus, for species first described in Gould's *Birds of Australia*, Mathews gives the part in which it appeared and the date.

MATHEWS, G. M., collection. 1966. *Checklist to the Mathews ornithological collection.* Canberra, National Library of Australia.

Gregory Mathews in 1939 donated his library of works relating to Australian ornithology to the National Library of Australia.

This checklist is in three parts. Section one is concerned with Printed Works. There are twenty items related to John Gould, some being imperfect sets of works. Essentially everything is related to Australia and the adjacent areas. One Gould letter is mentioned.

Section two is on Manuscripts. The Gould material consists of photocopies or typescript copies of correspondence, and reprints of articles on Gould and Gilbert.

Section three is Pictorial Material. Here is listed a collection of eighty-three pattern plates to Gould's *Birds of Australia, Supplement*, 179 pattern plates to Gould's *Mammals of Australia*, and '17 water colours of birds, 3 lithographs bound in 1 vol. (also includes a Wolf water colour)'. The artists for these seventeen watercolours are not stated.

MATHEWS, GREGORY M. AND IREDALE, TOM. 1922. 'Thomas Watling, artist'. *Austral avian record.* 5: 22-32.

PART FOUR BIBLIOGRAPHY OF JOHN GOULD, HIS FAMILY AND ASSOCIATES

There are numerous references to John Latham, G. R. Gray, Gould, R. B. Sharpe and Hugh Strickland in this article. Mathews attempts to clear up some questions relating to the authorship of a series of drawings, originally known as the Lambert drawings, now in the British Museum (Natural History).

MATHEWS, GREGORY M. AND IREDALE, TOM. 1938. 'Gould as a systematist'. *Emu* 38: 172-5.

A very complimentary article about Gould's ability as a systematist. They state 'Gould's judgment (regarding generic names) was generally correct, as no one was better qualified to judge differences, his training, first as a "bird-stuffer," then as an illustrator, and afterwards as a bird observer and describer, enabling him to provide unbiased opinions . . . Gould was never afraid of changing his mind when the occasion demanded, but he must have facts, and thus his work as a systematist claims praise from every serious investigator after a century's trial'.

MCDONALD, N. J. 1952. 'John Gould: The Bird Man'. *Frontiers*, 16: 140.

This is a three-page article of general information about Gould with special emphasis on the purchase of Gould's Australian bird collection for the Academy of Natural Sciences of Philadelphia. *Frontiers* is published by the Academy. Mention is made of Gould's presence at the Academy meeting in Philadelphia on 19 May 1857 and of Gould's visit with W. L. Baily who showed Gould his first live humming-bird in the Bartram Gardens.

A photograph of T. H. Maguire's engraving of Gould at the age of 45 is printed in the article, along with a photograph of Gould's type specimen of the Pygmy Goose from the Academy's collection.

MCEVEY, ALLAN. 1967. 'John Gould's ability in drawing birds'. *Art Bull. of Victoria*, 1967-68, pp. 13-24, Melbourne.

In 1955 McEvey visited the National Library at Canberra and the Mitchell Library at Sydney to examine the Gouldiana in these libraries. In 1966 he visited the Balfour and Newton Library at Cambridge University. As an ornithologist and an art historian, McEvey in this article discusses the Gould material at these three institutions with regard to an attempt to arrive at an answer to the question of 'Gould's own artistic ability and contribution' to his works.

McEvey especially comments on some sketches in the Cambridge University collection which he attributes to John Gould. His conclusion is that Gould 'had at his best, a genuine ability for drawing birds and considerable aptitude for catching the essential character of a species together with the ornithological rightness so important in bird illustration'.

MCEVEY, ALLAN. 1968. 'Collections of John Gould manuscripts and drawings.' *La Trobe Library Journal* 1: 17-31.

McEvey has compiled and commented on the world collections of manuscripts and drawings relating to John Gould. Twelve collections in Australia are mentioned, the Mitchell Library at Sydney having the largest and most significant holdings. Seven collections are listed in England including the valuable family material. The McGill University collection in Montreal, and four collections in the United States complete the compilation. The Ellis collection at the University of Kansas Spencer Library is the most extensive in existence.

McEVEY, ALLAN. 1973. *John Gould's Contribution to British Art.* Sydney, Sydney University Press.

This forty-six-page booklet is the first detailed evaluation of the artistry of John Gould and his artist associates, including his wife, Richter, Hart, Lear and Wolf.

McEvey has the fortunate combination of being both a distinguished naturalist and also an art historian. With these two backgrounds he very diligently attempts to solve some of the riddles associated with the question of who drew what for Gould's works.

The booklet is divided into two parts. Part one is concerned with Gould's role as an artist. Here McEvey discusses the evidence that Gould made many of the rough sketches of the birds upon which his other artists elaborated. His conclusion was that Gould's two main delineators, Richter and Hart, were 'faithful servants but failed to achieve a distinction' that exceeded that 'conceived by Gould's eye and sometimes performed by his hand'. 'Gould therefore played a continuing and significant role. He was no doubt incapable of producing the technical refinement of Richter's or Hart's finished drawings but this was, in part, technique rather than art. Without Gould there was, broadly speaking, neither ornithological sketch nor drawing force; his native talent though limited was real.' McEvey continues that 'After full tribute to the devoted talents of Richter and Hart has been paid and it must always be recognition of high order and after a distinctive place in her own right has been granted to gentle Elizabeth Gould, the plates, in their grand conception, their genesis, their spirit and their achievement, cannot in these senses fairly be attributed to anyone but Gould.'

Wolf, on the other hand, was an artist in his own right, and his original drawings demonstrate his qualities of outstanding draughtmanship, his handling of plumage and his ability to infuse life into his subjects. The contributions by Lear, according to McEvey, remain to be clarified.

The second part of the booklet has to do with 'overseas collections examined' and here McEvey also comments in a detailed manner on the artistic styles of Gould and the other artists. Eleven collections are mentioned. Then follow forty-one plates of various stages of artistry to illustrate McEvey's discussions and conclusions. Plate 34 is reproduced in colour on the cover of the booklet.

McEVEY, ALLAN 1975. *Victorian ornithology.* Victoria, National Museum of Victoria.

A large folded chart and a smaller eight-page pamphlet comprise this chronology of Victorian (Australia) ornithology.

As McEvey states, Gould did not visit the Port Phillip District of New South Wales, which was in 1851 to become the separate state of Victoria. But Gould's contribution to Australian ornithology was so significant, that in McEvey's chart, the years 1838 to 1850 are called the 'Gould Period'.

This is a very valuable chronological compilation of the discoveries by explorers and naturalists of the Victorian area birds. The years covered are from 1770 with Cook's first ornithological reports, down to the modern era in 1970 of the Royal Australasian Ornithologists' Union Period. The societies, publications and field work of many groups and naturalists are listed to form a wealth of information on Victorian ornithology.

McEvey was assisted in this project by L. Reid.

PART FOUR BIBLIOGRAPHY OF JOHN GOULD, HIS FAMILY AND ASSOCIATES

McEvey, Allan. 1976. 'John Gould – some unanswered questions'. *Books and Libraries at the University of Kansas* 13 (part 2): 5-8.

It is gratifying that McEvey of Australia and Chittenden of Kansas both agreed to my suggestion to prepare articles for this Gould issue of the 'Books and Libraries'.

McEvey, being both an art historian and an ornithologist, brings to the study of Gould's work a high index of inquisitiveness which is evident in this paper. He raises and elaborates on many unanswered questions for the Gould scholar. His special interest to date has been attempting to answer the question of the precise role that Gould himself played in the production of his prints from the rough sketch to the final lithographic plate. He concedes that this is not solved completely but a theory exists, namely that Gould's 'guiding and inspiring role throughout the complex productions of the plates was dominant enough for the plates to be fairly attributed, in the broad sense, to him'.

Meinertzhagen, R. 1959. 'Nineteenth century recollections'. *Ibis* 101: 46-52.

An interesting chatty article with anecdotes about Darwin, Herbert Spencer, Thorburn, Wolf, Lord Lilford, Sharpe, Abel, Hartert, Dresser, Seebohm, and the Rothschilds.

Meinertzhagen and his brother Dan said that Gould's *Birds of Great Britain* became their 'Bible and we never tired of poring [sic] over its pages'.

Mengel, Robert M. 1957. *A catalogue of an exhibition of landmarks in the development of ornithology.* Lawrence, Univ. of Kansas Libraries.

This thirty-three-page booklet contains a short history of the development of ornithology as illustrated by works in the Ellis Collection at the University of Kansas. Mengel begins with comment on 'renaissance ornithology' and concludes with the 'new systematic'.

A paragraph is included on John Gould, 'a creditable ornithologist' whose 'output doubtless [was] unequalled by another ornithological contribution'. Mengel, in consultation, said this was a printer's error for '... by *any other* ornithological contribu*tor*.'

Mengel, Robert M. 1962. 'News from the Ellis collection'. *Books and libraries at the University of Kansas*, 2 (number 3): 1-4.

A note on the expanding ornithological collection at the University of Kansas with a gift by Mrs Marcia Brady Tucker of some of the library of Jonathan Dwight. Mengel also comments on Ralph Ellis of California who acquired the large collection of Gould original drawings now residing at Kansas.

Mengel, Robert M. 1966. 'Bibliography and the ornithologist'. From Buckman, T. R. *Bibliography and natural history.* Lawrence, University of Kansas Libraries, pp. 121-30.

Philosophical eruditions by a frustrated bibliographer. Mengel had assumed the enormous task of cataloguing the Ellis ornithological collection at the University of Kansas. This time-consuming and almost never-ending job he felt could be justified by his statement 'a science uninspected by its own devotees and insensitive of its past is a science not fully comprehended'.

The Ellis collection contains the largest number of 'Gould' drawings in existence.

Mengel, Robert M. 1972. *A catalogue of the Ellis Collection of ornithological books in the University of Kansas Libraries*, Vol. 1, A & B. Lawrence, University of Kansas Publications.

'This catalogue is a blend of ornithology, ornithological bibliography, and bibliography in that order, because the author is an ornithologist by training and inclination, an ornithological bibliographer by necessity, and so much of a bibliographer as he may claim to be quite by accident', so states Mengel in the Preface.

Necessarily a biographical note on Ralph N. Ellis, Jr. appears. [Ralph's life story would be a best-selling novel if the reader were interested in the irresponsible antics of a self-centred individual who had cyclic tender sores in his mouth, and had the knack, and the money, for collecting bird books. (*See* Sauer. 1976. Ellis archives.)]

Ellis's greatest ornithological acquisition was from Henry Sotheran, Ltd of London, when in 1937 he acquired over 1200 original sketches and drawings used for John Gould's folio works.

There are references to Gould on pages 7, 84, and 164; and to a Mr Prince on page 80.

Whether this University of Kansas catalogue will ever be completed remains to be seen, but this volume of 259 pages covers only A & B of the Ellis collection. Another monumental venture.

MEYER. 1876. 'Convers'. *Lexicon* 8: 6.

Salvadori (1881) gives this reference on Gould. Not found.

MEYER DE SCHAUENSEE, RODOLPHE. 1957. 'On some avian types, principally Gould's, in the collection of the Academy'. *Proc. Acad. Nat. Sciences of Philadelphia* 109: 123-246.

The major paper on the Gould collection of Australian birds sold to the Philadelphia Academy in 1847.

The scientific and historical facts of this valuable acquisition make such a fascinating story that I am printing the entire introduction to this article. The kind permission of the Academy is acknowledged.

The first paper devoted to the types of birds contained in the ornithological collections of the Academy was written by Dr. Witmer Stone in 1899 (Proc. Acad. Nat. Sci., 51, pp. 5-92). This account listed all of the types with the exception of those of Gould from Australia.

In 1912, with the assistance of Gregory Mathews, Stone undertook the difficult task of trying to select what he thought to be suitable types from Gould's material. This was done by selecting birds which agreed more or less with Gould's measurements and descriptions, but no reference was made to any pertinent data from Gould's writing which might have helped to establish the validity of the specimens as actual holotypes or cotypes, Stone calling the birds types, rather than lectotypes which they actually were.

Since then considerable controversy has existed over these types, many persons often refusing to acknowledge the validity of any, due mainly, I imagine, to the fact that the claim of some was rather shaky.

In this paper I have carefully studied each of the types of Gould's species listed by Stone and have endeavoured to extract from this list the types which I feel have a claim, which can be substantiated, to the status of holotype or cotype.

Before going into the history of our Gould Australian Collection and the facts pertinent into the types contained therein I shall list the other kinds of types which will be found in this paper.

1.—Types of birds described by Academy staff members in other than Academy publications.

2.—Types of birds described by other than Academy staff members in non-Academy publications, based on Academy specimens.

3.—Types of birds described before 1900 and discovered in the Academy's collection since Dr. Stone's paper was written.

4.—Types on which additional information is pertinent.

Many of the latter are found in the Rivoli Collection, one of the most important contributions of Dr. Thomas B. Wilson to the Academy. It might be of interest here to give a short summary of its history.

The Rivoli collection was formed by Victor Masséna, Duc de Rivoli, and Prince d'Essling, the son of Napoleon's "Enfant chéri de la Victoire", Marshall André Masséna who distinguished himself in the battle of Rivoli, near Turin in 1796 and at the battle of Essling, near Vienna, in 1809.

The main part of the collection, consisting of 12,500 specimens, arrived at the Academy in 1846.

The actual purchase of the collection was made for Dr. Wilson by Dr. John Edward Gray, Director of the British Museum. Edward Wilson, acting for his brother, told Dr. Gray that he was desirous of

PART FOUR BIBLIOGRAPHY OF JOHN GOULD, HIS FAMILY AND ASSOCIATES

purchasing a good ornithological collection for the Academy and Dr. Gray advised him to purchase the Rivoli Collection, one of the greatest private collections in Europe, which was then for sale. Many years later Dr. Gray wrote an amusing account of his negotiations in Paris with Masséna and his agents:—"On my arrival in Paris, I put up at Meurice's and at once sent a messenger with a note to the Prince Masséna saying that I was willing to purchase the collection of birds at the rate of four francs per specimen, and that I was prepared to pay for it in ready money. While sitting at dinner at the *table d'hôte*, an aide-de-camp came in, all green and gold with a cocked hat and a large white feather, to inquire for me, with a message from the Prince to inquire what I intended with ready money, and, when I explained, to inquire if I was ready to pay the sum that evening. I said no, that I had only just arrived in Paris ... but I would be ready to pay as soon as the bank opened the next morning. He said the bank opened early, and would I come to the Prince at seven o'clock? to which I assented ... I kept my appointment; the Prince met me, declared the collection agreed with the catalogue, on which I gave his highness a check on Messrs. Green; and he gave me a receipt and handed me the keys to the cases, and I sealed them up, the affair being settled in a few minutes ...

Mr. Wilson was much pleased with the purchase, and afterwards purchased ... the specimens of the parrots which were not contained in the Catalogue." (Ann. Mag. Nat. Hist., 1869, *3*, p. 318.)

In recognition of Dr. Gray's assistance Dr. Wilson presented a number of specimens from the Rivoli Collection to the British Museum.

THE GOULD COLLECTION

One of the most notable possessions of the Academy is John Gould's collection of Australian birds. On this collection he based his great work "The Birds of Australia" commenced in 1840 and completed in 1848. The collection first offered by Gould to the British Museum, was purchased for Thomas B. Wilson, by his brother Edward Wilson, between May and July 1847. The bird skins were then shipped to Verreaux Frères (Jules and Edouard) in Paris to be mounted, and upon completion of this task, the collection was shipped to Philadelphia where it arrived by June of 1849. In Paris the Verreaux's unfortunately removed Gould's labels, but copied them on the bottom of the stands an abbreviated transcription of the original data. Thus localities appear as "V.D. Land" (Van Diemen's Land), "W. Austr.", "N.S. Wales"; only a few have more complete data, and all bear, in addition, the legend "Type de Gould, Bds. Austr." regardless of whether the species was described by Gould or not. In all, 1848 specimens were shipped accompanied by a manuscript catalogue in the handwriting of Jules Verreaux. It was understood at the time of purchase that the collection contained Gould's original material, or types of many of the species described by him. Reporting on the collection at the time the collection was purchased, Cassin wrote as follows:

"When I inform the Society [Academy of Natural Sciences] that this collection [of Gould] contains specimens of all the known Australian birds, except five species ... its peculiar value will be immediately understood. I may be excused for remarking, however, that Mr. Gould's collection acquires additional interest from the consideration that it contains the original specimens from which many of the numerous species described by him were first characterized ..." (Proc. Acad. Nat. Sci. Phila., 1847, p. 347).

Gould did not designate individual birds as types (holotypes) as is usually done today, but based his description on one specimen, a pair of birds (cotypes), or a series (cotypes), often not even saying of what his material consisted. Stone, assisted by Gregory Mathews, wrote a paper on our Gould collection, selecting a specimen which agreed best with Gould's description as the type (lectotype) of Gould's description, saying that this selection was "arbitrary and final" (Austral Av. Rec., 1913, no. 6, 7, pp. 129-180).

In his "List of the Specimens of Birds, in the Collection of the British Museum" (pt. 1, pp. V, VI and VIII, 1844) Dr. Gray wrote:

"Thus among the collections which have been received as presents, those presented by ...

"*Capt. William Chambers, R.N.*, as the types of the species from North West Australia, described in Mr. Gould's 'Birds of Australia'.

"*Capt. George Grey*, as the types of species described by Mr. Gould in his 'Birds of Australia' ...

"Amongst the specimens ... procured by purchase ...

Mr. Gould's collection, as the originals of the species described by that author in the 'Proceedings of the Zoological Society' and figured in the 'Birds of Australia'."

In 1848 Dr. G. R. Gray again wrote in the introduction to his "List of the Specimens of Birds in the Collection of the British Museum (2nd. Ed.)" the Museum owned birds presented by:

"*Capt. William Chambers, R.N.*, as the types of the species from North West Australia described in Mr. Gould's 'Birds of Australia' ...

"*The Earl of Derby, Lieut.-Col. Sir Thomas Mitchell, His Excell. Capt. George Grey, Capt. Sturt and Mr. Jukes*, as the types of species described by Mr. Gould in his 'Birds of Australia' ...

"Amongst the specimens or collections which have been procured by purchase or exchange, the following may be especially indicated: viz., those from ...

"*Mr. Gould's Collection*, as the originals of the species described by that author in the 'Proceedings of the Zoological Society,' and figured in the 'Birds of Australia' ". (This statement was probably based on the assumption that the British Museum was about to buy Gould's collection.)

This led Mathews, in 1921, to make the rather sweeping statement that "many if not a large majority of types of the species of birds described by Gould from Australia previous to 1847 are in the British

JOHN GOULD — THE BIRD MAN

Museum and not in Philadelphia ... This explains the action of the Trustees of the British Museum in refusing to purchase the Gouldian Bird collection". (Bds. Austr., 9, p. 107.)

Gould, however, carefully stated if Sturt, Chambers or any one else had lent him a bird to describe, and the specimens credited by Gould as belonging to other collections than his own were not claimed by Stone for the Academy. In 1925 Mathews, somewhat modified his first statement by saying of birds "named by Gould in the early 1840's were missing" from the British Museum (Birds Austr., Bibliogr., p. 58). Unfortunately Major H. M. Whittell quotes Mathews first opinion (Lit. Austr. Birds, 1954, p. 688).

In the first edition of "List of Specimens, etc." (1844, pt. 1, p. VIII) Gray wrote that among the specimens purchased by the British Museum were:

"*Mr. Audubon's Collection*, as the types of the species figured in his large work on the 'Birds of America'."

I should think that this statement ought to have warned Mathews that Dr. Gray used the term "type" in a rather looser sense that it is used today for he could not possibly have believed that the types of the birds figured by Audubon were all in the British Museum.

The following two letters to Edward Wilson, the brother of Thomas B. Wilson which have never been published before are highly important, and shed considerable light on the subject. They are both preserved in the archives of the library of this Academy. The first is from Gould himself which is given in full. The second is a letter from Sir William Jardine (1800-1874) in which he pleads with Edward Wilson to relinquish his purchase of the Gould collection so that it may go to the British Museum. He hardly would have done this had the British Museum already acquired Gould's birds as Mathews supposed. Gould's letter follows:

"London, 20 Broad Street
Golden Square, May 8th., 1847

"Dear Sir,

You know that we have more than once spoken about the destination of my collection of Australian Birds and you are aware that it was my wish that they should be deposited in the national museum. This, however, I feel will not be the case. I have offered them to the Trustees who now have declined the purchase; in consequence, I believe, of the Government not being willing to sanction any heavy outlays for such purposes at this moment. Having done what I consider my duty by offering them to the Trustees, I am now in a position to dispose of them elsewhere and I give you the preference. The enclosed is a copy of the letter I addressed to the Trustees which will serve for your guidance. Since I saw you I have determined upon publishing a work on the Oology of Australia, consequently the eggs, unless you wish it, need not be taken. At a word then the collection of Australian Birds is offered to you without the Eggs for £800.—or with the Eggs for £1,000.—. I am sure you do not know or can have taken into consideration the value of this collection. Pray, recollect it comprises the complete Ornithology so far as discovered of one entire quarter of the Globe. The specimens are in the finest condition and their value is certainly highly enhanced by their sexes having been ascertained by dissection and will ever be of interest as being the originals of my work. You have already purchased a magnificent collection in Paris [the Rivoli collection]. Will you not make it more celebrated by adding this thereto. If you refuse the collection, I then brake it up under the following scale of price, giving the British Museum the first choice as they purchased so largely of the mammalia:

	£
For the first hundred specimens	250.—
second	150.—
third	100.—
fourth	75.—
fifth	50.—
sixth	40.—
seventh	30.—
eighth	28.—
ninth	26.—
tenth	24.—
eleventh	22.—
twelfth	21.—
thirteenth	19.—
fourteenth	17.—
fifteenth	15.—
sixteenth	13.—
seventeenth	10.—
eighteenth	10.—

20 males or young males will be added to the first hundred making a hundred and twenty specimens and ten to each of the next four hundreds. I have at least six persons wanting parts of this collection, consequently I shall have no difficulty in selling it at the full price. Will you then, my Dear Sir, favour

PART FOUR BIBLIOGRAPHY OF JOHN GOULD, HIS FAMILY AND ASSOCIATES

me with a reply to this letter as early as possible. Should the collection be taken entire, I shall feel it a duty and pleasure to give the owners, whoever they may be, the first offer of any novelties I may subsequently acquire. So as to keep it complete, I have men actively employed in Australia and we may be certain there are many fine things yet to come and which will offer materials for a supplement to my work.

Awaiting your reply,

I remain my Dear Sir,
Yours most truly,
John Gould

to E. Wilson, Esq."

The enclosed copy of the letter addressed to the Trustees of the British Museum is missing. Wilson, I may add here, purchased Gould's Egg collection as well as the birds.

Sir William Jardine's letter to Edward Wilson, written just two months after Gould's is quoted below. The last part of this letter deals with an endeavour on his part to sell his own collection of British birds to Edward Wilson for the sum of £500.[1] Not being pertinent to the present subject it has been omitted:

"Jardine Hall by Lockerbie
9th July, 1847.

"Dear Sir,

From the very short time I had the pleasure of seeing you in London and from my being hurriedly called to Scotland I had no opportunity of learning your views as to Mr. Gould's birds, which we (ornithologists) all so much regret may soon leave Europe. I know that you agree with me in thinking that such a collection as that you have purchased as the types of an extensive and important work should not leave this country, but should have been taken by some public museum where they could at all times be accessible for study or reference. Our Government and the trustees of the British Museum did wrong in refusing Mr. Gould's very liberal offer, at the same time I think Mr. Gould scarcely made his conditions sufficiently public.

We cannot however go back to these circumstances, the plain question to be asked of you is whether any inducement could be held out, which would prevail on you to relinquish your bargain with Mr. Gould and allow his collections still to be purchased by this country, or, if there are duplicates that a series of named specimens authenticated by Mr. G. with his figures should be selected from the collection. I know that it would be a great stretch of liberality in you to give up such a collection. At the same time, under the circumstances you will forgive such an application being made to you. I will state at once on what terms if any, you would be willing to renounce your arrangement with Mr. Gould . . .

Believe me,
Sincerely yours,
Wm. Jardine

to J. Wilson, Esq.,
London —

"Mr. Gould is aware that I have written to you and on what subjects."

"To J. Wilson, Esq." is of course a mistake, as the letter is written to Edward Wilson.

It is evident that both Gould and Jardine had no hesitation in stating that a great number of Gould's original specimens were in the collection purchased by Mr. Wilson.

That this was generally regarded as a fact in England is again demonstrated by the statement made in 1856 by Dr. P. L. Sclater (1829-1913) the first editor of the Ibis, and for many years Secretary of the Zoological Society of London. Writing of the Academy's bird collection he said: "The general Collection formed by Prince Masséna . . . and the types of the species described by Gould . . . were the first and largest of Dr. Wilson's contributions" (Proc. Zool. Soc. London, 1857, p. 1).

Finally I think it must be acknowledged that Dr. R. Bowdler Sharpe, (1847-1909) Keeper of the Bird Collection in the British Museum and author of many volumes of the British Museum's "Catalogue of Birds" must have known what he was talking about when in 1893 he wrote the following regarding the Gould Collection: "It is not for the present writer, who of all men, has felt most keenly the absence of the Gould collection, with its hundreds of types, from the series of bird skins in the British Museum, to criticize the action of the officers of the British Museum at this distant date, but there can be no doubt that, scientifically, the loss of this historical series was nothing less than a national disaster . . . Gould told me in later years that he had never intended that the collection should leave the country and he regretted ever afterwards that, in a moment of chagrin at the unexpected refusal of his offer to the nation, he accepted Dr. Wilson's offer and allowed his treasure to go to America. It is pleasing to remember how, after Gould's death, the Trustees of the British Museum promptly secured his remaining collections of birds and eggs on the recommendation of Dr. Günther" (An Analyt. Index to the Works of John Gould, pp. XVIII, XIX).

---

[1] He did not buy it.

JOHN GOULD — THE BIRD MAN

It need hardly be pointed out that if a majority of Gould's types had been in the British Museum, Sharpe would not have considered it "a national disaster" that Dr. Wilson bought Gould's collection. It is apparent from the above statements that the best and most qualified authorities in England regarded it as a fact from the time that Wilson purchased the collection, that the types were included. This certainly is true for at least 64 years!

Stone listed the location of Gould's types of Australian birds as follows:

Academy of Natural Sciences	314
British Museum	71
Zoological Society of London	5
King's College (lost)	4
United Services Museum (lost)	4
Lord Derby's Collection	2
Fort Pitt, Chatham (lost)	2
South Australia Institute	1
Lost	17
Total	420

Of the 314 lectotypes listed by Stone I have recognized 158 as valid holotypes or cotypes. Of course among the rest are many of Gould's original specimens but today it is impossible to prove which are and which are not the ones on which the descriptions were based.

Almost all of these 158 species were described in the Proceedings of the Zoological Society of London. However, due to delays in the publication of that journal,[2] parts of the Birds of Australia, containing plates and descriptions of the same birds were published before the "Proceedings" appeared and must be considered the original reference. Dr. Stone apparently was not aware of this. As has been seen, the collection of Gould's birds bought by Dr. Wilson was stated by both Gould and Cassin to contain all the birds used to illustrate Gould's "Birds of Australia" and it is justifiable to assume that the birds used as models for the illustrations were those described.

In the cases where birds were described in the Proceedings of the Zoological Society only a few months before the appearance of the plate depicting the corresponding bird in the "Birds of Australia," our specimens of these species have also been considered to have a legitimate claim to typeship, on the assumption that the birds used to illustrate Gould's book were those he described.

In some cases Gould stated definitely that he had one specimen or one pair of birds up to the time he finished his "Birds of Australia" and in these cases, of course, it is simple to state definitely that our specimens are the type or types.

THE ACADEMY'S TYPES

Dr. Stone's two papers listed a total of 665 types as being in the collection of the Academy. Of these, 314 were lectotypes of species described by Gould from Australia. Since Dr. Stone's two papers were published the number of types in the Academy's collections have almost doubled, the total now stands at 1130.

The following list gives the number of species described by each author and it is of interest because it shows the wide use that the collection has been put to, not only by American ornithologists but also by those in Europe. Among the names listed below will be found many famous ones, such as Audubon, Cassin, Gould, Sclater, Temminck, Bonaparte, Lesson, Lafresnaye to mention but a few, and to whom modern ornithology owes so much.

The type collection is of great interest for it goes back to 1811, when Alexander Wilson described the Mississippi Kite, and it contains birds secured on such early and famous journeys as John K. Townsend's across the continent with pony train to the Columbia River in 1834; the "Voyage au Pôle Sud" by the Astrolabe and the Zélée under the command of Dumont-Durville from 1837 to 1840; the United States Exploring Expedition around the World under Captain Wilkes from 1838 to 1842; Gould's and Gilbert's Explorations in Australia beginning in 1838 and culminating in the latter's death during the Leichhardt Expedition in 1845; Audubon and Harris Expedition to the "Kickapoo Country" on the Upper Missouri River in 1843; the "Pacific Railroad Survey" of 1853-1856; Commander Perry's Naval Expedition to Japan in 1854, and Paul du Chaillu's pioneer work in Gaboon from 1855 to 1859.

Audubon, J. J.	10	Bryant, H.	1
Ashby, E.	1	Bourjot-St. Hilaire, A.	1
Baird, S. F.	13	Cabot, S. Jr.	1
Bonaparte, C. L.	6	Carriker, M. A., Jr.	129
Bond, J.	19	Carriker, M. A. & Meyer de Schauensee, R.	5
Bond, J. & Meyer de Schauensee, R.	41	Cassin, J.	168
Bowen, W. W.	15	Coues, E.	1

---

[2] Exact dates of publication are given by P. L. Sclater in the Proceedings of the Zoological Society's July issue of 1893.

PART FOUR BIBLIOGRAPHY OF JOHN GOULD, HIS FAMILY AND ASSOCIATES

Deignan, H. G.	4	Mayr, E.	1
Delacour, J.	1	Mayr, E. & Meyer de Schauensee, R.	5
Eyton, T. C.	2	McCall, G. A.	4
Finsch, O.	2	Meyer de Schauensee, R.	101
Friedmann, H. & Bowen, W. W.	2	Meyer de Schauensee, R. & Ripley, S. D.	15
Gambel, Wm.	11	Nuttall, T.	3
Gentry, A. F.	1	Oberholser, H. C.	1
Geoffrey-St. Hilaire, I.	2	Ogden, J. A.	3
Gervais, F. L. R.	1	d'Orbigny, A. D.	2
Gould, J. (Australia)	327	Peale, T. R.	14
Gould, J. (not Australia)	9	Prevost, F.	1
Gray, G. R.	1	Reichenbach, H. G. L.	1
Gurney, J. H.	1	Rhoads, S. N.	1
Harper, F.	1	Ridgway, R.	3
Harris, E.	1	Ripley, S. D.	1
Heermann, A. L.	2	Rothschild, L. W.	1
Heine, F.	1	Schafer, E.	4
Henry, T. C.	2	Sclater, P. L.	6
Hombron, J. B. & Jacquinot, H.	1	Scott, W. E. D.	1
Hoopes, B. A.	1	Sharpe, R. B.	5
Hoy, P. R.	3	Smith, A.	31
Huber, W.	3	Stone, W.	15
Jardine, W.	1	Strickland, H. E.	11
Jones, W. H.	1	Temminck, C. J.	2
Kelso, L. & Kelso, E. H.	1	Todd, W. E. C.	2
Knip, P. & Prevost, F.	1	Townsend, J. K.	9
Krider, J.	1	Verreaux, J.	1
de Lafresnaye, F.	23	Verreaux, J. & E.	19
de Lafresnaye, F. & d'Orbigny, A.	27	Verreaux, J. & Des Murs. M. A. F. O.	1
Lawrence, G. N.	2	Wilson, A.	2
Layard, E. H. & Layard, E. L. C.	1	Zimmer, J. T.	4
Lesson, R. P.	5	Total	1130
Masséna, V. & Souancé, C.	7		

SCOPE OF THIS PAPER

The present paper deals with 270 types. Of these, 158 are considered to be Gould holotypes or cotypes from Australia; 46 are types of species discovered in the collection of the Academy, many of them in the Rivoli Collection, after Dr. Stone's paper was published in 1899; 14 are species described before 1900 but acquired by the Academy after that date; 21 are types of species, based on Academy specimens described by non-Academy personnel in non-Academy publications; 8 are species described by Academy staff members in outside publications; 21 are types on which additional information is deemed necessary; 2 are shown to be artifacts.

In the preparation of this paper I have examined each of the 1130 types in the Academy's collection and made notes on each one. These notes are deposited in the Academy's Library and are available to any one interested in types not included in this paper. As these types have been listed by Dr. Stone, and as a paper covering them all is a very lengthy one, amounting to about 450 typewritten pages, it has seemed best to confine this study to the 270 which are enumerated in the following pages, as they are the ones to which new information is attached or else have not been listed previously.

With regard to the Gould types not mentioned in this paper, but listed by Stone, many of course are Gould's actual types, but as this cannot be proved they can of course be still regarded as lectotypes by those who attach value to such a designation.

Obviously, in the listing and description of the types that follow these are many valuable references to Gould's publications. Meyer de Schauensee's detailed comments complement these references and must be read carefully by anyone interested.

[MILLAIS, SIR JOHN EVERETT.] 1967. Millais: an exhibition arranged by the Walker Art Gallery Liverpool and the Royal Academy of Arts London.

Through the interest of L. E. James Helyar, Curator in Graphics, Spencer Library, University of Kansas, the following association item was noted in a catalogue of Millais paintings.

The Millais painting 'The Ruling Passion', later called 'The Ornithologist', now in the Glasgow Art Gallery and Museum, portrays an invalid on a couch surrounded

by and examining a collection of bird skins. His family looks on. The quotation in the catalogue is that 'The idea for the picture came from a visit made to John Gould (1804-81), an eminent ornithologist, then an invalid. It was started in the early spring of 1885 and finished in time for the Academy.' The principal figure portrayed, however, was not Gould but another Royal Academician, T. Oldham Barlow.

MILLIGAN, A. W. 1904. 'Notes on a trip to the Wongan Hills, Western Australia, with a description of a new *Ptilotis.' Emu* 3: 217-26; 4: 2-11.

Milligan and a group of ornithologists went 'to view scenes which Gilbert saw and to tread [on] what may be truly termed "hallowed ground" '.

See also the letters from James Drummond to Hooker (Drummond 1844).

MITCHELL, DAVID WILLIAM. 1841. 'Gould's birds'. *Westminster Review.* 35: 271-303.

This very long and detailed article is a review of Gould's *Century of Birds, Birds of Europe,* the monographs *Ramphastidae* and *Trogonidae,* parts I and II of *Icones Avium,* and part I of *Birds of Australia.*

The author begins with a few philosophical thoughts on natural history in the past, on the establishment of the Zoological Society of London, and then proceeds with a detailed review of the above works, which is especially detailed on Gould's Australian trip. Mitchell included excerpts from the letter Gould sent to the Zoological Society from Australia, which was published in full in the *Proceedings of the Zoological Society* of 8 October 1839.

The whole article is most flattering and complimentary of Gould and the artistry of his wife.

MITCHELL, P. CHALMERS. 1929. *Centenary history of the Zoological Society of London.* London, Zoological Society.

The Zoological Society of London apparently had its origin in 1824 with an organizational committee formed at that time. The first headquarters were at 33 Bruton Street from 1826 to 1836. From 1836 to 1841 it was at 28 Leicester Square; from 1841 to 1843 in a warehouse near Golden Square; from 1843 to 1883 at No. 11 Hanover Square; from 1883 to 1910 at No. 3 Hanover Square; and finally after 1910 the new office and library were opened at the Zoological Gardens.

Sir Stamford Raffles was the first President and the co-founder of the Society with Sir Humphry Davy. Raffles died unexpectedly in 1826. The Marquis of Lansdowne was his successor from 1827 to 1831; Lord Stanley, afterward Earl of Derby, from 1831 to 1851; the Prince Consort from 1851 to 1862; Sir George Clerk from 1862 to 1867; Viscount Walden (afterwards Marquess of Tweeddale) from 1868 to 1878; and Professor Flower was president from 1879 to 1899. These dates cover the Gould years.

Nicholas Aylward Vigors was the first secretary of the Zoological Society until 1832. Then followed Edward T. Bennett, 1833 to 1836; William Yarrell, 1836 to 1838; Rev. John Barlow, 1838 to 1840; and William Ogilby, 1840 to 1847. The Society was in its most depressed period during the tenure of the last two secretaries. From then on the secretary was a paid officer and David William Mitchell served from 1847 to 1859; Philip Lutley Sclater from 1859 to 1902, and after four months with an interim secretary, Chalmers Mitchell assumed the duty of secretary in 1903.

Gould was appointed Curator and Preserver of the Zoological Society in 1827 (p. 97) when the offices and museum were at 33 Bruton Street.

PART FOUR BIBLIOGRAPHY OF JOHN GOULD, HIS FAMILY AND ASSOCIATES

He remained in the service of the Society until 1837, during which period he not only got the museum in order and catalogued it, but produced his "Birds of Europe", in full folio volumes, a "Monograph of the Toucans", and two parts of another on the Trogans. He then resigned his post to go to Australia ... He was made a Corresponding Member of the Society, and soon after his return, in 1840, became a full Fellow, in later years serving repeatedly on the Council. He died in 1881, and if the Zoological Society had no other claims, its existence would be justified from the fact that it gave Gould his opportunity.

G. R. Waterhouse, father of F. H. Waterhouse, succeeded Gould as the Curator of the Museum in 1836 and was succeeded by Louis Fraser. After 1845 the museum was in charge of caretakers only. Gould had, with J. O. Westwood, appraised the museum collection of birds mainly, and also mammals, reptiles, fishes and insects (p. 100) at £10,965 in 1841. With increasing and burdening acquisitions, and with the museum income very small, it was decided to transfer the museum collections to the British Museum (Natural History) in 1857, then at Bloomsbury. Gould was on the Council of the Zoological Society at the time (p. 104) and 'made an offer, which was accepted, to buy the birds' skins not otherwise disposed of for the sum of £40'.

The Zoological Society's gardens with the zoo were becoming increasingly popular and lucrative. 'In 1851 Gould lent to the Society his collection of mounted humming birds, and these were exhibited in a special building near the Lion House'. According to Mitchell, they 'attracted over 75,000 visitors and without doubt added to the popularity of the Gardens in the year of the Great Exhibition' (p. 102). The building was taken down and re-erected in what is now known as the middle Garden. It became the Parrot House and with modification has remained as such (p. 136). [In the library of the Zoological Society are several letters and proposals from Gould regarding this humming-bird exhibition. Photocopies of these items are in my collection. *See* Gould, John. Correspondence. 1851.]

The portrait of Gould facing page 208 of the volume is the one from the Ipswich Museum.

MONTREAL NATURAL HISTORY SOCIETY. Minutes of Meeting. 1857. Manuscript. Collection: Blacker-Wood Library of Zoology, McGill University.

At my suggestion Librarian Eleanor MacLean reviewed the minutes of the Montreal Natural History Society for June 1857 when Gould was visiting in Montreal. Included in the minutes is a report to the Society dated 29 June 1857 written by William Stewart M. D'Urban, Sub-Curator of the Society's Museum as follows:

At the beginning of this month the celebrated Ornithologist Mr. Gould F.R.S. author of those magnificent works "the Birds of Europe, of Asia, of Australia, Monograph of the Humming Birds Etc. Etc. paid Montreal a flying visit of 3 or 4 days. He crossed the Atlantic expressly to see the Humming Birds of this country and to obtain Specimens.

He brought his Son, an Icthyologist [sic] and Concologist [sic] with him and they both visited the Museum. Mr. Charles Gould kindly promised to speak to his friends in London and at the British Museum to request them to exchange with us and I trust they will do so. On behalf of the Society I gave such assistance as I could in obtaining specimens of such Fish etc. as he wanted.

Hoping that this Report will meet with your approval.

MOON, ELIZA GOULD MUSKETT. Manuscript. 1885. 'Some pages of reminiscences of her early life.' Collection: Maureen Lambourne.

These reminiscences are part of a 7-by-9-inch book (178 x 203 mm) now in the possession of Mrs Maureen Lambourne, great-great-grand-daughter of John Gould. A photocopy was very kindly supplied to me by Mrs Lambourne. In her letter to me of 1 December 1978, she wrote that these pages were started on 31 March 1885 (3

pages) and resumed on 9 April 1885 (31 pages). Mrs Moon wrote them for the 'amusement of my little Helen when older, knowing what interest there is to a child, in the past of her belongings. Unfortunately up to 1880 I have little else than memory to help me and that a bad one'.

Mrs Lambourne wrote that 'the book was later used by her daughter (Mrs. Helen Edelsten) in the 1930's as a scrapbook for oddments, about 200 pages of items such as newspaper cuttings of the 1930's, descriptions of family events, some inserted sketches by Eliza, Sai, and Louisa Gould of views abroad, and flower studies, dated between 1896 and 1912'.

These thirty-four pages of reminiscences are most interesting and provide considerable information on the John Gould family life, all of the children, their summer excursions, their childhood pleasures, their relatives, and father John Gould's daily schedule. There is a charming section on the famous Broad Street water pump (*See* Snow 1936 and Barton 1959):

One of our great amusements was watching the grooming of [father's horse] Georgie in the yard below from our high nursery window at the back of the house. Another was to watch the commers and goers to the pump in the Street from the room called the 2nd floor, which was school room of a morning, when we grew older, & general living room for us children. There was always *something* going on at the pump, the water was considered very good then & people sent jugs to be filled from all round ... The water cress women would pump on their cresses, & then sit down on the curb, & shake each bunch & re-arrange them in their baskets, & all day long children were playing round the pump, making drinking cups of their hands, or caps, sending the handle high up & riding down on it & all sorts of tricks.

The pleasant memories go on.

An important section by Mrs Moon relates a lot about her father, John Gould's, daily routine, as follows:

We saw very little of him all day, not even at meal times, as we had early breakfast & tea in the school room, or 2nd floor as it was sometimes called, & dined in a front room downstairs next to the kitchen to save the servants trouble, whilst the dining room proper, was considered especially his room. The drawing room when I first remember it was very pretty with creamy walls & gold & rather pale blue curtains & chair covers & a warm coloured all overish patterned carpet, but as fathers collection of humming birds grew larger & he mounted fresh cases for want of other room they were collected there until it became almost too full to move about. The housemaid not allowed in it with broom or duster except on rare occasions, so in time it looked anything but pretty & of course was not used as a drawing room. I don't know one was needed as we were children & father having no wife did not have ladies to see him, except it was to see the birds & he generally dined at his club while we were little. We always saw him in the morning between nine & ten when he went down to breakfast, & all made a rush at him for a kiss nearly pulling his head off, he used to say, & again when he came up to change his coat for late dinner. Other times he was in his office at the back of the house busy with his books & birds in all of which Mr. Prince was his help & Mr. Richter & afterwards Mr. Hart alone did the lithographing or drawing on stone from father's sketches and coloring from the same. Of course the greater part of the coloring & all the printing was done away from the house & by others but all the copy plates were done at home and everything was overlooked by father or Mr. Prince & the latter had to make fair copies for the printers of the Fathers writing no light task sometimes as his hand was none of the clearest to make out.

I certainly can vouch for that last statement. A valuable record that should be printed in its entirety.

MOORE, T. J. 1891. 'History of the living collections at Knowsley'. *Transactions Biological Society of Liverpool* 5: 1-18.

An anecdotal account of the hundred-acre (40-hectare) aviary and menagerie built by Lord Derby at Knowsley Hall. The thirteenth Earl of Derby succeeded to the title and estates of his father in 1834. From then on he devoted his life to the study of birds and animals. He maintained a fine museum at Knowsley and also collected and bred 'such beasts as would be likely to be ornamental or useful if successfully naturalized in this country'. He spared neither time, labour, nor expense. He improved

the estate buildings and grounds and 'commenced and carried on to completion a high and well built stone wall round the whole of the part measuring 10 or 12 miles in length with numerous tastefully built stone lodges'. An important structure was a 'new aviary' of large extent. [In 1974 when we visited Knowsley, the stone wall, lodges and main Hall were an impressive sight.]

Moore comments (p. 4) on the stuffing of the Giraffe of King George IV by Gould and says 'this brought Gould to light'.

Lord Derby died in 1851 and the Derby Museum was removed to Liverpool. The aviary and menagerie were dispersed, some specimens going to Windsor for Queen Victoria and a few others to the Zoological Society Gardens.

MOOREHEAD, ALAN. 1963. *Cooper's Creek*. London, Hamish Hamilton.

Concerning the Sturt expedition into central Australia in 1845, Stuart's expedition of 1861, and the Burke and Wills expedition of 1860 to 1861.

Both Sturt and Stuart collected for Gould.

MOOREHEAD, ALAN. 1969. *Darwin and the Beagle*. New York, Harper and Row.

A fascinating and lavishly illustrated book on Darwin's life, from his youth, his college days, his trip around the world on the *Beagle*, concluding with the great debate of 1860 at Oxford between the clergy and the scientists on the origin of the species.

Eight Gould illustrations are reproduced, five coloured and three uncoloured, of the birds found during the *Beagle's* journey.

On page 111 is this quote from Darwin about the *Rhea darwinii*: 'Mr. Gould in describing this new species, has done me the honour of calling it after my name'.

MORRIS, FRANCIS ORPEN. 1851-57. *A history of British birds*. 6 vols. London, Groombridge.

Many of the hand-coloured wood engravings of the rarer birds for this work were drawn 'after Gould' by Benjamin Fawcett.

MOYAL, ANN M. 1976. *Scientists in nineteenth century Australia: A documentary history.* Melbourne, Cassell Australia.

Not seen, but this was reviewed in the *J. Soc. Biblphy. Nat. Hist.* 8: 93, 1976. Gilbert and Gould are discussed.

MUELLER, FERDINAND. 1858. 'An historical review of the explorations of Australia'. *Trans. Phil. Institute Victoria* 2: 148-68.

Mention is made of John Gilbert's participation in Leichhardt's expedition.

MULLENS, W. H. AND SWANN, H. KIRKE. 1917. *A bibliography of British ornithology from the earliest times to the end of 1912.* London, Macmillan.

This monumental compilation was published in six parts from 20 June 1916 to 29 June 1917. A supplement entitled 'A Chronological list of British birds' was published in 1923.

The works of Gould relating to Britain are discussed and listed on pages 240-2 along with a biography of him.

There are also trivial references to Gould or his works on pages 298, 658, 663, and on pages 28 and 30 of the supplement.

MUMBY, FRANK ARTHUR. 1931. *Publishing and bookselling.* New York, Bowker.

An historical account of Henry Sotheran, Ltd is on pages 289-90. It is said that this firm was interested in natural history books, 'their most important venture being in connexion with the ornithological and other books of John Gould, which were taken over in 1881'.

MUSGRAVE, ANTHONY. 1932. *Bibliography of Australian entomology 1775-1930.* Sydney, Royal Zoological Soc. of N.S.W.

John Gilbert (-1845) is noted on page 120: 'He obtained a number of insects at Port Essington which, being sent to Gould and then being distributed by him, are apt to be attributed to Gould... *Stigmatium gilberti* A. White, 1849, from Port Essington from Mr. Gould's collection recalls him'.

John Gould (1804-1881) appears on page 128: 'In addition to birds, he collected insects... *Allecula gouldi* Hope, 1843; *Cisseis gouldi* (Hope, 1846 *Ethon*); *Dineutes gouldi* Hope, 1842; *Sericesthis gouldi* Hope, 1842 recall this collector.

N NEUFFER, CLAUDE HENRY. 1960. *The Christopher Happoldt Journal.* Charleston, S. C., Charleston Museum.

At the age of 14, Christopher Happoldt was chosen to accompany Dr John Bachman on a trip to Europe. Young Happoldt kept a journal of this tour from June to December 1838.

The journal of this trip is published here and provides considerable information on Happoldt and Dr Bachman, and also Happoldt's ideas and thoughts on the England and Europe of 1838. They visited the museum of the Zoological Society of London, the British Museum and the Zoological Gardens.

NEWTON, ALFRED. 1869. 'The Strickland Collection in the University of Cambridge'. *Ibis.* 5 (new series): 320-4.

Of contemporary interest concerning the construction of appropriate cabinets for Strickland's collection, the quality of the skins, and the large quantity of 5802 specimens referable to 3031 species, with quite a few types.

NEWTON, ALFRED. 1896. *A dictionary of birds.* London, A. & C. Black.

This volume is divided into four sections all dated 1896 in my copy. A March 1893 date ends a three-page 'note' at the beginning of part 1.

Newton states that this Dictionary had 'as its foundation a series of articles contributed to the ninth [1875-89] edition of the "Encyclopaedia Britannica"'. An alphabetical arrangement is used for the definition of the ornithological listings.

An 'Introduction' of 124 pages is bound at the front of my copy of this Dictionary, but was printed as a section of part 4. This introduction is also very similar to the section on Ornithology in the ninth edition of the *Encyclopaedia Britannica.* Some paragraphs are identical. The section about Gould is more up-to-date in this Dictionary.

With reference to Gould, Newton writes that the year 1832 'saw the beginning of the marvellous series of works by which the name of John Gould is likely to be always remembered'. A listing of his folio publications follows and then the following comment by Newton:

PART FOUR BIBLIOGRAPHY OF JOHN GOULD, HIS FAMILY AND ASSOCIATES

The earlier of these works were illustrated by Mrs. Gould, and the figures in them are fairly good; but those in the later, except when (as he occasionally did) he secured the services of Mr. Wolf, are not so much to be commended. There is, it is true, a smoothness and finish about them not often seen elsewhere; but, as though to avoid the exaggerations of Audubon, Gould usually adopted the tamest of attitudes in which to represent his subjects, whereby expression as well as vivacity is wanting. Moreover, both in drawing and in colouring there is frequently much that is untrue to nature, so that it has not uncommonly happened for them to fail in the chief object of all zoological plates, that of affording sure means of recognizing specimens on comparison. In estimating the letterpress, which was avowedly held to be of secondary importance to the plates, we must bear in mind that, to ensure the success of his works, it had to be written to suit a very peculiarly composed body of subscribers. Nevertheless a scientific character was so adroitly assumed that scientific men—some of them even ornithologists—have thence been led to believe the text had a scientific value, and that of a high class. However, it must also be remembered that, throughout the whole of his career, Gould consulted the convenience of working ornithologists by almost invariably refraining from including in his folio works the technical description of any new species without first publishing it in some journal of comparatively easy access.

NISSEN, CLAUS. 1936. *Schöne Vogelbücher.* Wein, Reichner.

This 96-page paper-bound booklet contains a page of text about John Gould and sixteen black-and-white full-page plates from his ornithological works. All of Gould's publications are listed in the second part of this booklet.

NISSEN, CLAUS. 1953. *Die illustrierten vogelbücher.* Stuttgart, Hiersemann.

Divided into several sections, Nissen gives an historical perspective of illustrated bird books, a bibliographic section with a list of works and references to these works (even with a reference to my first 1948 paper), a list with index to artists (including 'Hart, Will.'), a section on families of birds covered by the various authors, a section on the countries covered by the bird books, and finally an author or editor index.

Plate 14 is from Gould's 1833-35 monograph *Ramphastidae*, uncoloured.

Book dealer catalogues refer to this bibliography as Nissen IVB.

NISSEN, CLAUS. 1966-76. *Die zoologische buchillustration.* Stuttgart, Hiersemann.

Ornithological works are not included in this compendium.

On page 173 item 1661 is listed as Gould's 'Mammals of Australia', 13 parts, 182 illustrations, 1845-1863; item 1662 is 'Macropodidae', 2 parts, 30 illustrations, 1841-1842.

In book-dealer catalogues this bibliography is referred to as Nissen ZBI.

NOAKES, VIVIEN. 1969. *Edward Lear.* Boston, Mifflin.

A very readable and quite well-researched book on Lear. Mrs Noakes had access to the Lear letters and diaries in the Houghton Library at Harvard and other manuscript material in England.

The Gould references are mainly in Chapters 2, 3 and 4 with others on pages 265, 322, 323, 337, 343, and 344.

Elizabeth Gould is in error given the name of Edith.

An illustration by Lear of Tengmalm's Owl for Gould's *Birds of Europe* is on page 39.

NORELLI, MARTINA R. 1975. *American wildlife painting.* New York, Watson-Guptill.

This beautifully illustrated book with sensible text covers the works of Catesby, Wilson, Audubon, Martin Johnson Heade, Abbott Handerson Thayer and Fuertes.

In the section on Heade she states that 'Heade knew Gould and his work' and possibly this caused Heade to abandon his idea of publishing a contemplated work on Brazilian Humming-birds.

O [OWEN, RICHARD.] 1833. 'The giraffe'. *Zoological Magazine* 1: 1-14.

According to L. S. Lambourne (1965), Richard Owen wrote this article for his *Zoological Magazine* which ceased publication after only six monthly issues.

A young female giraffe arrived in England 11 August 1827, as a gift from the Pasha of Egypt for King George IV. It grew rapidly but did not prosper, apparently having been injured or mishandled during its long journey from Egypt. It is stated that 'it died the following year'.

Mr Davis, the animal painter, executed several portraits of the living giraffe for the King, and apparently Sir Everard Home dissected it. [There is no mention of Gould stuffing it, but he did.] 'The skin, however, and skeleton, both beautifully prepared, are preserved in the museum of the Zoological Society.'

[See also the 1965 article by L.S. Lambourne, and the copy of the 1830 bill from Gould to the King for stuffing the giraffe (The Royal Library, Windsor Castle).]

OWEN, RICHARD. 1879. *Memoirs on the extinct wingless birds of New Zealand.* 2 vols. London, John Van Voorst.

The frontispiece of volume one (text only) is an uncoloured triple-folded plate of *Notornis mantelli* Owen, as drawn and lithographed by Gould and Richter for *Birds of Australia, Supplement.*

OWEN, REV. RICHARD. 1894. *The life of Richard Owen by his grandson.* 2 vols. London, John Murray.

Some of the most personal of notes in reference to Gould are in this work; one wishes there were more of this type. The diary and journal were written by Mrs Owen:

March 27 [1841]. – Lord Northampton's evening party. Richard [Owen] very tired, and thought he would not go, but about eight Dr. Buckland looked in, bag and all, and said "Oh, you better come." So after some dinner R. felt better, and they started off. He was glad afterwards he went, for Prince Albert was there, and Mr. Gould brought his singing New South Wales parrots.

November 15th [1841]. — Went to see Gould's birds — not to be imagined till seen. The great dragon lizard now set up excellently. Strange that the Chinese should have the idea of a creature so much like it. After dinner this evening Mrs. Darwin, Mr. Gould, and his brother came here for some music. [Whose brother, Owen's or Gould's??]

October 24 [1850]. — Note from Mr. Gould to ask us to step round and see the skin of a notornis which has been sent him.

In volume 2, page 140

In September 1863, Owen, accompanied by John Gould, was the guest of the Duke of Northumberland at Alnwick Castle. Afterwards they went to Lord Tankerville's, where, among other things, they inspected the Chillingham cattle. 'I arrived here [at Alnwick] to dinner on Wednesday' he writes to his wife (September 5), 'and found Professor Tyndall and Lord and Lady Tankerville, etc. On Thursday there was a grand flower and fruit show in the grounds; since then have arrived, among others, Sir R. Murchison, Captain Grant of Nile celebrity, Sir William Armstrong, and a dark native of Ceylon in gorgeous costume.'

From page 249:

In his declining years it was one of Owen's favourite amusements to observe the habits of the birds which frequented his garden. The notes which he made upon his feathered visitors were, as he writes in his diary, 'communicated to my friend Robinson's weekly paper "The Garden" in successive numbers.' A few extracts from these 'Notes on Birds in my Garden,' which were published in 1883, throw light on Owen's interests and occupations. The number of birds which the Professor noted in his garden at Sheen Lodge is surprising: —

'I have entered in my garden book,' he writes, 'the name of every kind of bird which I have noted there, distinguishing the permanent dwellers from the occasional residents, and the latter according to the periods of their temporary sojourn, whether to breed or to feed — in other words, the summer and winter visitors.

'The list, however, would have been incomplete without the aid of my lamented friend, John Gould. It was ever with him a favourite summer afternoon's holiday, after a ramble in the park, to pass an hour

PART FOUR BIBLIOGRAPHY OF JOHN GOULD, HIS FAMILY AND ASSOCIATES

in the garden. On one of these occasions, in early June, we rested on a seat overshadowed by a weeping ash, but allowing a view of the lawn. Happening to show him my ornithological list at that date, Gould said, "You have got more birds in the garden than I see here, I expect." Now he possessed in a remarkable degree the faculty of imitating the various notes of all our vocal species. He bade us sit still and be silent; then began. After emitting a particular "motivo" for a few minutes, he would quietly point to a little bird which had flown from an adjoining bush upon the lawn, and was there hopping inquisitively to and fro, gradually nearing the locality of its specific song. We could then recognise the species to which Gould gave the name. This attraction and its result was repeated; and we enjoyed the same instructive amusement in subsequent summer vacations, to which I am indebted for additions that would otherwise probably have escaped my observation.'

Professor Owen died if 1892.

PALMER, A. H. 1895. *The life of Joseph Wolf, animal painter.* London and New York, Longmans, Green & Company.

Palmer has written an enjoyable, readable and quite accurate account of the life of Joseph Wolf.

Wolf moved to London from Germany in 1848. Gould had previously commissioned him to do a small watercolour drawing entitled 'Partridge Dusting' before he came to England (p. 50). Later, in 1849, a second commission from Gould entitled 'Woodcocks seeking Shelter' was hung at the Royal Academy 'on the line' (p. 80).

Gould met Wolf shortly after he came to England and they became friends, even travelling together to Norway in 1856 to collect birds.

Palmer relates quite a few anecdotes about Gould and they are not too flattering. Palmer feels that Wolf was a much greater artist than Gould and his artists [and he was], and that Gould was too business-like and rough for Wolf's sensitiveness and homeliness (p. 93).

Since Palmer's book is not indexed I submit an index of the persons, institutions and journals associated with Gould:

Anderson, C. J., 119–23
Ansdell, Richard, 86
Argyll, Duke of, 64, 65
Audubon, J. J., 284
Bartlett, Abraham D., 56
Darwin, C., 63, *192–8*
Derby, Lord, *82–7*
Dresser, H. E., 57, 106, 115, 118, 224, 254, 257, 271, *283*
Elliot, Daniel Giraud, 112, 179, 198, 230, 243
Ford, G. H., 63
German Athenaeum, 174
Giraffe stuffed, 70
Gould, 38, 50, *69–80*, 88, 100, 117, 233, 260, 275, *295–8*, *322*, *326*
Gray, George Robert, 50, 53
Gray, J. E., 68, 82
Grosvenor, Earl, 88 (later Duke of Westminster)
Gurney, J. H. Sr., 99
Hawkins, Waterhouse, 83, 96
Hulmandel & Walton, 55
Ibis, The, 105–6, 306–8
Institute of Painters in Water Colours, 216
Keulemans, J. G., 106, 112
Knox, A. E., 116

Landseer, Sir Edwin, 64, 68, 80, 199
Lear, E., 83, 284
Lorne, Lord, 224
Mitchell, David William, 50, 52, 53, 61, 111, 284
Newton, Prof., 68, 90, 105, 111, *283*
Owen, Prof., 62, 188
Richter, H. C., 96
Rossetti, W. M., 67
Royal Academy, 125
Russell, William, 64
Science, Quarterly Journal of, 103
Sclater, P. L., 103, 109, 111
Severtzoff, Dr., 74–5
Sharpe, R. B., 188
Thorburn, A., 179, 286
Voorst Van, 116
Westminster, Duke of, 88 (*see* Grosvenor)
Whymper Engravers, 201–4
Whymper, Charles, 286
Woolmer, Thomas, 63, 67
Yarrell, William, 56, 96
Zoolog. Gardens, 183
Zool. Soc. of London, *94–105*, 109–10, 245
Zool. Soc. of London Proceedings, 293–304
Zool. Soc. of London Transactions, 305

PARKER, SHANE A. 1980. 'Samuel White's ornithological explorations in northern South Australia in 1863'. *South Australian Ornithologist* 28:113-19.

A well-researched article on Samuel White (1835-80), and his attempt in 1863, at the age of 28, to cross the Australian continent from south to north.

Parker was provided copies of four letters from White to John Gould. These letters, plus Gould's published references concerning the receipt of birds and eggs from White which he had collected on this expedition, provide some dates and places for White's journey.

PARKIN, THOMAS. 1911. 'The Great Auk. A Record of sales of birds and eggs by public auction in Great Britain, 1806-1910.' Hasting's and East Sussex Naturalist 1, part 6, p. 1.

This is a list of sales at the Great Auction Rooms, 38 King Street, Covent Garden, London.

On page 7 is Sale #5, on July 11th, 1865 as follows: 'Egg V — Lot 141. Another differently marked. Purchased by Mr. John Gould, for Mr. G. Dawson Rowley, of Brighton, for £33-0-0.

'Now (1911) in the possession of Mr. G. Fydell Rowley, being one of six eggs inherited by him from his father, Mr. G. Dawson Rowley.'

PEATTIE, DONALD CULROSS. 1939. *A gathering of birds.* New York, Dodd, Mead.

Gould, as an author, was not highly esteemed enough by Peattie to be included in this anthology of famous bird writings. On page 7, however, Gould is mentioned in the biography of W. H. Hudson to the effect that Hudson 'was inclined to quarrel with ornithologists like John Gould and Alfred Newton, and to dispute that they knew birds at all in the sense that he knew them. His taste was rather for bird lovers of the stamp of Grey of Fallodon, his friend'.

PESCOTT, EDWARD E. 1938. 'The Botany of the "Birds of Australia".' *Vic. Nat.* 55: 103-4.

The author states that 'many, if not all, of the botanical studies in the great work were done by Mrs. Gould'. Mention is made that a series of original botanical drawings of Australian plants, flowers and foliage, painted by Mrs Gould, are now (1938) on sale in London. [This collection of botanical drawings was listed in the 1934 Henry Sotheran's No. 9 'Piccadilly Notes' Catalogue as item 800 on page 83: 'Botanical Drawings by Mrs. Gould: 74 Original Drawings (mostly in Water-colour), of Australian Plants, Flowers or Foliage, specially made for incorporation in the backgrounds of the plates to the Birds of Australia; many being endorsed V[an] D[iemen's] Land; N[ew] S[outh] W[ales], or Yarrundi, some are dated 1838 or '39, 2 are signed J. Gould, and E. Gould, £105'.

The same collection was again listed for sale in their 1937 No. 22 'Piccadilly Notes' as item 2622 on p. 197 for the same price.]

PHILLIPS, V. T. AND PHILLIPS, M. E. 1963. *Guide to the manuscript collections in the Academy of Natural Sciences of Philadelphia.* Philadelphia, Acad. Nat. Sciences of Philadelphia.

This is another of those monumental tomes that take a religious devotion to complete. A few persons have persevered with cataloguing manuscripts so that others have only to consult an index.

PART FOUR BIBLIOGRAPHY OF JOHN GOULD, HIS FAMILY AND ASSOCIATES

The Academy Collection #71 contains '16 items' related to John Gould. There are five letters, a pencil sketch of a Yellow-billed Cuckoo done by Gould in 1857 when he was in North America, a humming-bird watercolour study, and a workbook. [An article on the pencil sketch of the Cuckoo was written by Sauer in 1976.]

The workbook is volume 5 of a series of seven volumes used by Gould as a 'filing cabinet' for preparation of his *Birds of Europe*. [The other six volumes are in the library of the British Museum (Natural History). Sauer (1976) is an article on the British Museum volumes, and Sauer (1978) is an article on the Philadelphia Academy volume.]

There are other Gould-related items in this Guide, namely, Collection 11 on William Lloyd Baily with a portrait of Gould; Collection 53 on Jules Verreaux who wrote a catalogue listing of Gould's Australian birds sold to Dr T. B. Wilson for the Philadelphia Academy in 1847; Collection 155 on William Lloyd Baily containing three letters from Gould; Collection 169 on Thomas Bellerby Wilson; Collection 190 on Rodolphe Meyer de Schauensee which is the typescript of his article 'On some avian types, principally Gould's in the collection of the Academy' published in 1957; Collection 297 on the Academy bird department; Collection 364 on John Torrey which contains a letter from Gould to E. Wilson offering his Australian bird collection for sale; Collection 502 on minutes of the Academy meetings with a note that Gould was present at the meeting of 19 May 1857; and Collection 936 on Edward Wilson.

POLLARD, JACK (ed.). 1967. *Birds of paradox, birdlife in Australia and New Zealand.* Melbourne, Lansdowne.

Pollard has collected narratives from various authors concerning Australian and New Zealand birds.

Four articles are exerpted from Gould's *Birds of Australia*. Beginning on page 7 is one on the Black Swan, on page 36 on the Bower-birds, on page 79 on the Channel-bill and on page 94 on the Great Skua.

Gilbert's report on the Mallee Hen begins on page 10, and Charles Coxen's story on the Regent's Bird's bower begins on page 29.

POVEY, DOROTHY. 1954. 'A naturalist's books'. *North Western Naturalist* 2 (new series): 7-8.

As the Librarian at Knowsley Hall, Miss Povey wrote to reassure those interested that with the 'reorganisation of the library [in 1954]... Lord Derby is not disposing of his collection of natural history books, manuscripts and drawings, except for a few duplicates'.

Then follows a short account of the history of this library, and especially the important role played by Edward Smith, 13th Earl of Derby, in acquiring unique natural history items.

As Miss Povey says many country houses and libraries have the large folio volumes by Audubon, Gould, Redoute, Lear, Buffon, and others, but 'what gives the Knowsley collection its special character is the number of albums of original drawings, some being studies later reproduced, some unpublished'. These include works by Edward Lear, Waterhouse Hawkins, Joseph Wolf and John Gould. [For my comments on this Gould MS. material see Sclater 1896.]

PUTTICK AND SIMPSON. 1881. *Catalogue of the ornithological library of the late John Gould Esq. F R S... which will be sold by auction by Messrs. Puttick and Simpson.* London.

> # CATALOGUE
> OF THE
> # ORNITHOLOGICAL LIBRARY
> OF THE LATE
> ## JOHN GOULD ESQ. F.R.S.
> (*Author of* "THE BIRDS OF EUROPE" *etc.*)
>
> COMPRISING A VALUABLE
>
> ## COLLECTION OF STANDARD AUTHORITIES
> ENGLISH AND FOREIGN
> ON THE
> ## NATURAL HISTORY OF BIRDS ETC.
> INCLUDING THE FOLLOWING:—
>
> IN OCTAVO. Audubon's Ornithological Biography—Bewick's Birds—Bree's Birds of Europe—Curtis' Botanical Magazine 3rd series—Harvey's Phycologia Britannica—Heuglin's Ornithologie Nordoest Afrikas—Hewitson's British Oology—Hume and Marshall's Game Birds of India—The Ibis complete to Oct. 1880—Jardine's Contributions to Ornithology—Lefevre's Birds' Eggs—Lesson, Histoire des Oiseaux—Magazine of Natural History complete to 1881—Morris' British Birds—Ray Society 37 vols.—Hume's Stray Feathers—Swainson's Zoological Illustrations—Wilson and Bonaparte's American Ornithology—Yarrell's Birds and Fishes—PUBLICATIONS of the LINNEAN, GEOGRAPHICAL, GEOLOGICAL, ZOOLOGICAL, ROYAL and and other LEARNED SOCIETIES, etc. etc.
>
> IN QUARTO. Buller's Birds of New Zealand—Edwards' History of Birds—Jerdon's Indian Ornithology—Latham's History of Birds—Legge's Birds of Ceylon—Marshall's Capitonidæ—Mulsant and Verreaux, Histoire des Oiseaux-Mouches Palæontographical Society—Rowley's Ornithological Miscellany—Sharpe's Birds of Europe and Monograph of Alcinidæ etc.
>
> IN FOLIO. Angas' South Australia—Audebert, Histoire des Colibris—Buonaparte, Fauna Italica—Catesby's Carolina LARGE PAPER—Desmarest Histoire des Tangaras—Elliott's Monograph of Tetraoninæ—Gould's Himalayan Birds—Gray's Genera of Birds—Knip, Les Pigeons—Le Vaillant Histoire des Oiseaux de l'Amerique, Oiseaux de Paradis, Oiseaux d'Afrique etc. LARGE PAPER—Malherbe, Monographie des Picidés—Siebold, Fauna Japonica—Wolf's Zoological Sketches
>
> **BEAUTIFUL DRAWINGS OF HUMMING BIRDS (114)**
> ETC. ETC.
>
> 𝔚hich will be 𝔖old by 𝔄uction,
> BY MESSRS.
> ## PUTTICK AND SIMPSON,
> AUCTIONEERS OF LITERARY PROPERTY AND WORKS OF ART,
> AT THEIR GALLERY,
> No. 47, LEICESTER SQUARE, W.C.
> ON WEDNESDAY, MAY 4TH, 1881,
> AT TEN MINUTES PAST ONE O'CLOCK PRECISELY.
>
> MAY BE VIEWED TWO DAYS BEFORE THE SALE.

Figure 73 First page of Puttick and Simpson's 1881 sale catalogue of Gould's library.

This pamphlet consists of two preliminary pages of a summary of the sale items, date of sale 'Wednesday, May 4th, 1881', and a statement of 'Conditions of Sale', and then eighteen pages listing items for sale from Gould's library (Figure 73).

Included under 'Octavo et infra' is Audubon's *Ornithological Biography* 5 vols.; Curtis's *Botanical Magazine Third Series*, 20 vols.; Darwin's *Descent of Man*, 2 vols.; separate Gould papers; *The Ibis* complete to October 1880; Zoological Society Transactions complete to 1880, etc. – a total of 193 items.

Under 'Quarto' is listed G. Edwards's seven-volume work of 1751; Gould's 4 parts

PART FOUR BIBLIOGRAPHY OF JOHN GOULD, HIS FAMILY AND ASSOCIATES

of *Ramphastidae* published in Germany; works by Bonaparte, Baron Cuvier, John Latham, Prof. Owen, Thos. Pennant, R. B. Sharpe & Swainson.

Under 'Folio' sized items are works by Bonaparte, Catesby (2 vols., 1754); D. G. Elliott's Grouse monograph; 'Gleanings from the Menagerie & Aviary at Knowsley Hall', 1846; Lear's Parrots; works by Levaillant, Alf. Malherbe, 4 volumes, 1861-62, etc.

Item 293 is listed 'Ornithological Collection. A quantity of manuscript notes on birds, with numerous figurings of the heads of different species arranged according to orders in 9 portfolios. An interesting and valuable collection'. [This undoubtedly is the set of nine portfolios in the Newton Library at Cambridge University. See Gould, J. Manuscript. [*c.* 1838]. "Gould's Australian notebooks and mss."]

The total number of items for sale is 301.

QUARITCH, BERNARD (Book Dealer). 1881. *Catalogue of works on natural history, etc.* London, Quaritch. Q

This November 1881 catalogue contains a listing of 'The Ornithological Library of John Gould Esq. FRS, FZS, etc.' on pages 320 to 329. There is also a list of books from the libraries of Sir William Jardine, and T. C. Eyton.

Some of Gould's personal books are listed for sale on pages 348 to 359 also, and a few are scattered throughout this 471-page catalogue.

Most of the items have an identical listing in Puttick and Simpson's catalogue of Gould's library sold at auction on 4 May 1881, so Quaritch apparently purchased a large portion of this original sale.

A comprehensive index covers the 4542 items listed for sale.

QUEEN VICTORIA. Manuscript. 1851. Queen Victoria's Journal. Collection: The Royal Library, Windsor Castle, through the courtesy of Her Gracious Majesty Queen Elizabeth II.

The entry on 10 June 1851 was copied by the Librarian, Sir Robin Mackworth-Young, and is as follows:

After breakfast we drove with our 3 girls, Alexandrine, & the 2 Ernests to the Zoological Gardens, where we saw the lions, tigers & leopards, which are very fine, — also a collection (in a room specially arranged for the purpose) of Gould's stuffed Humming Birds. It is the most beautiful & complete collection ever seen, & it is impossible to imagine anything so lovely as these little Humming Birds, their variety, & the extraordinary brilliancy of their colours.

RAY SOCIETY. 1845. *Reports on the progress of zoology and botany, 1841, 1842.* Edinburgh, Ray Society. R

Formed for the publication of 'works on Natural History... which would otherwise be inaccessible', this is the first volume presented to the members of the newly formed Ray Society.

The society was founded in 1844. A 'Report of the First Annual Meeting' on 2 October 1844 with the laws of the society and a list of over 400 members including John Gould, forms the final item in this volume.

Three papers constitute the core of this work, all being translated to English, having originated in 'foreign countries... on the continent'.

The first is a paper by Charles Lucian Bonaparte read in 1841 in Florence, Italy entitled 'Observation on the state of zoology in Europe', translated by H. E. Strickland. The opening sentence on the Great Britain section is concerned with Gould's *Birds*

of Australia, stating 'that distinguished naturalist entrusted to me the first number of his work, that I might present it to this meeting'. Then follows a review of zoology for Great Britain and the rest of Europe in 1841. Bonaparte's paper is rich in names of persons, museums, collections and specimens.

The second paper is 'Reports on the progress of zoology for the year 1842' translated from the German by W. B. Macdonald. The sections on mammals and birds, both by Andr. Wagner of Munich, have many references to Gould's works, as well as a wealth of information about other contemporary naturalists and their publications.

The third paper is on botany in 1841 by Dr H. F. Link.

A somewhat comparable report on the state of ornithology for 1844 was authored by Strickland and appears in the 1844 Report of the British Association.

READE, BRIAN. 1949. *Edward Lear's Parrots.* London, Duckworth.

A short book of thirty-two pages, it contains an excellent essay on Edward Lear, with twelve coloured illustrations of Lear's parrots.

Lear never spoke too kindly about Gould, and Reade emulates Lear. He ends one paragraph by stating that 'In fairness to Gould it should be added that Lear was predisposed to be hostile to business men'.

READE, BRIAN. 1978. *An essay on Edward Lear's illustrations of the family of Psittacidae or Parrots.* London, Prion and New York, Johnson Reprint.

Through the courtesy of Eleanor MacLean at McGill University, I was given a reprint of this essay. It appeared as the introduction to a 1978 facsimile reproduction of Lear's work on Parrots.

The Preface is by Gavin Bridson and contains the dates of publication and the contents of the twelve parts of Lear's *Parrots*.

The essay by Reade is almost word-for-word the same as when it appeared in his 1949 *Edward Lear's Parrots*.

An additional plum is the inclusion of eight figures, on two pages, of original drawings of parrots by Edward Lear from the Blacker-Wood Collection at McGill University. Five figures are in colour.

[Very interestingly, I had the opportunity of examining this collection of original Lear drawings in 1979. Many of the thirty or so Lear watercolours in the McGill collection have annotations on them by John Gould. It soon became obvious on making comparisons that some were drawn for Gould's *Birds of Europe*. In this reprint, figures 7 and 8 which are in colour, are heavily annotated by Gould.]

REEVE, LOVELL. (ed.). 1863-67. *Portraits of men of eminence in literature, science, and art, with biographical memoirs.* The photographs from life, by Ernest Edwards, B.A., ed. by Lovell Reeve... vol. 6. London, L. Reeve & Co.

According to 'The National Union Catalog Pre-1956 imprints', six volumes were published. Edward Walford edited volumes 3 to 6.

Gould is seated at a desk in the portrait.

The biographical memoir of Gould occupies six pages. There are a few statements not noted in other biographical accounts. As an infant he moved with his parents to the neighbourhood of Guildford, Surrey and:

it was on the wild commons and heaths of that district that at the early age of five to six years he ran about in search of flowers and insects. Soon he commenced to make collections of them, and at the age of bird-nesting to ramble in search of eggs, which he strung and hung around the cottage window. In

PART FOUR BIBLIOGRAPHY OF JOHN GOULD, HIS FAMILY AND ASSOCIATES

1818 John Gould's father obtained an appointment as foreman in the Royal Gardens of Windsor under Mr. J. T. Aiton, and having removed his family thither, young Gould now commenced the active business of life in the royal gardens. "I've gathered many a bunch of dandelions", we once heard the ornithologist remark, "for Queen Charlotte's German salads."

[Interestingly, later accounts state that Gould collected the dandelions for Queen Charlotte's tea.]

There are some further comments regarding humming-birds that are of interest. The question is asked, then answered, 'How was it possible to illustrate a monograph of the humming birds in the style which our ornithologist had made his own, from a contemplation of their stuffed skins!' Gould answered by stating 'when inquirers of a strong will really set themselves to attain a definite object they generally attain it' and in his case he eventually travelled to America to observe the humming-bird in its native woods. [But that was in 1857, when in fact the first part of his folio work on these birds was published in 1849.]

The memoir concludes with the statement that Gould was elected a Fellow of the Royal Society in 1843.

REVIEWS. *See* Harris, Harry. Manuscript. 1921. 'Biographical index to the Ibis (First ten series) 1859-1918.'

RICHARDSON, DR JOHN. 1842. 'Contributions on the ichthyology of Australia'. *Ann. Mag. Nat. Hist.* 9:15-31; 120-31; 207-18; 384-93.

This is a series of articles on a collection of dried skins of one side of thirty-seven fish procured by Mr Gilbert at Port Essington for Gould. Richardson states that they were 'in excellent condition'. 'Mr. Gould destines his collection for the British Museum.'

Serranus Gilberti (p. 19) was named after Gilbert.

RIDE, W. D. L. 1970. *A guide to the native mammals of Australia.* Melbourne, Oxford University Press.

In the Introduction to this book is a reference about Gould's exploration in Australia 'the then newly occupied island continent'. Gould spoke of 'Australia — a part of the world's surface still in maiden dress... those charms will not long survive the intrusion of the stockholder, the farmer, and the miner...'.

There are other references to Gould's 'pioneer work' on Australian mammals.

RIDGWAY, ROBERT. 1890. 'The humming birds'. Report of National Museum, 1890, 253-383.

This article is a comprehensive summary of information on humming-birds, containing sections on history, origin of names, geographical distribution, migrations, habits, abundance, flight, intelligence, nests and eggs, voice, food, anatomy, variations and colours. Ridgway then gives detailed descriptions of the seventeen or so species found in the United States.

Obviously there are many references to John Gould's works and there are pages of direct quotations. Of the forty-six black-and-white plates, twenty-two are 'after Gould'.

RIDGWAY, ROBERT. 1911. *Birds of North and Middle America.* Part V. Washington, Gov't Printing Office.

This part of Ridgway's monumental work covers the humming-birds and trogons and therefore has a multitude of references to Gould. The humming-bird section alone encompasses 386 pages. Ridgway (p. 303) states that 'the vernacular names are mostly taken, with or without modification, from Gould's "Monograph of the Trochilidae" ... that magnificent work'.

RIPLEY, S. DILLON. 1952. 'The Coe ornithological collection'. *Yale University Library Gazette* 27: 66-70.

This is a short article on the material in the William Robertson Coe ornithological collection presented to Yale University in 1952.

A complete set of John Gould's publications is stated to be in the collection. More important is 'an unpublished series of watercolour drawings, plates, proofs, and pencil sketches by John Gould for a proposed monograph on the Cassowaries of the Australian region'. [Also see Sotheran's 'Piccadilly Notes' No. 22, page 195, where this collection is listed in detail.]

RIPLEY, S. DILLON AND SCRIBNER, L. L. 1961. *Ornithological books in the Yale University Library*. New Haven, Yale University Press.

There is no duplication in this catalogue of biographical information concerning ornithological volumes that appears in Wood, Zimmer or Nissen concerning the number or the dates of the parts of any work.

In addition to the folio and other works by Gould, the Yale library Coe Collection has 'a collection of original watercolours and sketches, hand coloured plates with some uncoloured duplicates and proofs from "The Birds of Australia" and "The Birds of New Guinea"...'. All of these drawings and plates are of the Cassowaries. [Sotheran, 'Piccadilly Notes' No. 22 lists this collection in detail on page 195.]

Several Gould letters are also in the Coe collection.

Gould is referred to on pages 21 under Barrett, 59 under Chisholm, 71 Darwin, 78 Diggles, 91 Eyre, 91 Eyton, 108 Gilbert, 112-14 Gould, 118 Grey, 133 Hinds, 137 Houghton, 163 Lack, 179 MacGillivray, 183 Mannering, 186 Martin, 261-3 Sharpe (a MS. is listed), 267 Sitwell, 279 Strzelecki, 280 Sulphur, and 301 Vigors.

All the sections pertaining to Gould's published works in this bibliography are reprinted here (Figure 74). The courtesy of the Yale University Library in allowing this reproduction is gratefully acknowledged.

RONSIL, RENÉ. 1948-1949. *Encyclopédie ornithologique*, VIII & IX. *Bibliographié ornithologique Française*. 2 vols. Paris, Lechevalier.

Item 1250 lists eight articles by Gould published in France in *l'Institut*, and one obituary from Nouvelle III, 1881, pages 375-6. [This obituary reference appears to be incorrect. The 1881 journal I examined at the Linda Hall Library did not have these pages, nor could I find the obituary elsewhere in this journal.]

An additional article by Gould and Léopold de Beauffort is listed as item 3330 in the Supplement.

ROYAL COMMISSION. 1851-52. *Official descriptive and illustrated catalogue (of the great exhibition of the works of industry of all nations, 1851)*. 4 vols. London, Spicer Bros.

These volumes were contemplated as an illustrated catalogue with detailed notes on the exhibitors at the 1851 great exhibition. Three volumes were thought to be necessary

Figure 74 Pages concerning Gould's works from Ripley and Scribner 1961 *Ornithological books in the Yale University Library*.

PART FOUR BIBLIOGRAPHY OF JOHN GOULD, HIS FAMILY AND ASSOCIATES

title "Illustrations of the birds of Jamaica."
— A naturalist's sojourn in Jamaica. By Philip Henry Gosse ... Assisted by Richard Hill ... London, Longman, Brown, Green, and Longmans, 1851. 1 p.l., [v]-xxiv, 508 p. front., VII col. pl. 19 cm.
Many notes on birds throughout.
— Popular British ornithology, containing a familiar and technical description of the birds of the British Isles ... 2d ed. London, Reeve and Co., 1853. viii, 320 p. col. front., XIX col. pl. 17 cm.
— Letters from Alabama, (U.S.) chiefly relating to natural history ... London, Morgan and Chase, 1859. xii, 306 p. illus. 18 cm.
Includes accounts of birds.
— The romance of natural history ... London, J. Nisbet and Co., 1860. 2 v. fronts., plates. 20½ cm.
Yale also has copies of the Boston, 1861 and 1864 editions of the 1st series and London, 1875 edition of the second series.
— Land and sea ... London, J. Nisbet & Co., 1865. ix p., 1 l., 425 p. illus. 17 cm.
Many references to birds throughout.
Cover title: Sea and land.
BIOGRAPHY
GOSSE, Sir EDMUND WILLIAM, 1849-1928. The life of Philip Henry Gosse ... London, K. Paul, Trench, Trübner & Co., Ltd., 1890. viii p., 1 l., 387 p. 22½ cm.
Imperfect: the frontispiece is wanting.

GOSSE, PHILIP HENRY GEORGE, 1879- ... Notes on the natural history of the Aconcagua valleys ... 1899. 338-376 p. (In Fitzgerald, Edward Arthur. The highest Andes. A record of the first ascent of Aconcagua and Tupungato in Argentina and the exploration of the surrounding valleys ... London, Methuen & Co., 1899)
Birds, p. 342-352.
No. 43 of a special edition of 60 copies.
— Memoirs of a camp-follower ... London [etc.] Longmans, Green and Co., 1934. xiv, 299, [1] p. front. (port.) 22½ cm.
"Adventures and impressions of a doctor in the great war. Much about beasts and birds." — Pref. p. xi, xiv.

GOTHENBURG, SWEDEN. MUSEUM. ZOOLOGISKA AVDELNINGEN. Medd. Göteborg. 1- 1913-

GOULD, AUGUSTUS ADDISON, 1805-1866, ed. The naturalist's library: containing scientific and popular descriptions of man, quadrupeds, birds, fishes, reptiles, and insects, compiled from the works of Cuvier, Griffith ... and other writers on natural history. Arranged according to the classification of Stark ... New York, C. Wells [185-?] xxi, [23]-880 p. illus. 23½ cm.
Birds, p. [406]-684.

GOULD, Mrs. ELIZABETH (COXEN) 1804-1841. BIOGRAPHY: Chisholm, Alexander Hugh, 1890- The story of Elizabeth Gould ... Melbourne, Hawthorne Press, 1944. 4 p.l., 74 p., 1 l. front. (port.) 22 cm.
One of 350 copies printed.

GOULD, JOHN, 1804-1881. A century of birds from the Himalaya Mountains ... London, 1832. 6 p.l., 72 l. 80 col. pl. 56 cm.
Letterpress by N. A. Vigors. This work appeared in two forms, one with the backgrounds of the plates colored and the other uncolored. The Yale copy has the backgrounds uncolored.
— A monograph of the Ramphastidæ, or family of toucans ... London, Published by the author, 1834. 3 p.l., [14] p., 34 l., [4] p. 34 pl. (33 col.) 57 cm.
Yale also has copies of the "Supplement to the 1st edition," London, 1855, and the rev. ed., London, 1854, as well as the German edition, Nürnberg [1841]-47, the title of which is from an original wrapper which is bound in.
— The birds of Europe ... London, Printed by R. and J.E. Taylor. Pub. by the author, 1837. 5 v. 449 col. pl. 56 cm.
— The birds of Australia and the adjacent islands ... London, Printed by R. and J.E. Taylor. Pub. by the author, 1837-38. 2 pts. in 1 v. 20 col. pl. 57 cm.
Cover title.
— Icones avium; or, Figures and descriptions of new and interesting species of birds from various parts of the globe ... Forming a supplement to his previous works ... London, Printed by R. and J.E. Taylor, Pub. by the author, 1837-38. 2 pts. in 1 v. 18 col. pl. 56 cm.
Cover title.
— A synopsis of the birds of Australia, and the adjacent islands ... London, Pub. by the author, 1837-38. 2 p.l., 75 l., 8 p. 73 col. pl. 29 cm.
"Descriptions of new species of Australian birds," 8 p. at end.
Yale has two copies; one is bound and the other in four parts as issued. The second copy lacks the 2 preliminary leaves (including the title page).
— A monograph of the Trogonidæ, or family of trogans ... London, Printed by R. and J.E. Taylor. Pub. by the author, 1838. 3 p.l., vii p., [38] l. 36 col. pl. (1 fold.) 56 cm.
Yale also has a copy of the 2d ed., 1875, with 47 colored plates.

327

────── The birds of Australia ... London, Printed by R. and J.E. Taylor. Pub. by the author, 1840-48. 7 v. illus., 600 col. pl. (2 fold.) 56 cm.

Yale has two copies; one is bound and the second copy is in the original 36 parts as issued. This set contains 1 extra plate each of Euphema splendida and Falco hypoleucus, intended to replace earlier plates. Yale also has a copy of the "Supplement," London, 1869, which is bound.

────── A collection of original water colors and sketches, hand colored plates with some uncolored duplicates and proofs from "The birds of Australia" and "The birds of New Guinea" enclosed in a cover of Part 4 of "The birds of New Guinea." 1 v. in portfolio.

Included are two pencil sketches of close-up of head detail of Casuarius australis; a pencil drawing of the female of C. Beccari and a water color sketch of the head of the male with MS note in Gould's hand: "Sent from the Gardens Dec. 9, 1876 ... A large bird exceeding like Casuarius galeatus on the cheeks and neck, except on the back which is orange, red, dark blue, all these parts covered with black hairs: helmet dark horn colour round at the apex like the galeatus not thin as in Cas. Australis: legs horn colour, short, wattles on the neck small owing in this specimen I am sure to immaturity, each lappet free, hanging down beside each other 2 inches in length, these appendages light (almost white) like that seen in my drawing of Cas. galeatus but more developed. In conclusion I might say that I see insufficient differences in the example now before me to separate this bird from Cas. galeatus. In the plumage of the present specimen a little of the brown immaturity remains on the hinder part. The colouring of the rest of the feathers raven black, eyes hazel brown. The above specimen is added to the British Museum collection."; A pencil drawing of the heads of a pair of birds and a water color of one of the heads of C. Bennetti; a water color of a pair of heads of C. bicarunculatus and a pencil sketch giving details of plumage; a water color of C. Picticollis and 2 rough sketches (1 in color); and a water color and pencil sketch of detail of head of C. Westermanni.

────── ... Birds ... with a notice of their habits and ranges, by Charles Darwin ... and with an anatomical appendix by T.C. Eyton ... illustrated by numerous coloured engravings. 1841. 4 p.l., ii, [3]-156 p., 4 l. 50 col. pl. (In Darwin, Charles Robert, ed. The zoology of the voyage of H.M.S. Beagle, under the command of Captain Fitzroy ... during the years 1832 to 1836. Published with the approval of the Lords Commissioners of Her Majesty's Treasury ... London, Smith, Elder and Co., 1839-42. Pt. 3)

────── [A list of birds of the western coast of Australia] 1841. 415-421 p. (In Grey, Sir George. Journals of two expeditions of discovery in north-west and western Australia, during the years 1837, 38, and 39, under the authority of Her Majesty's government ... London, T. and W. Boone, 1841. v. 2)

────── Birds [of the islands of the Pacific] 1844. [37]-44 p. 19-34 col. pl. (In Hinds, Richard Brinsley, ed. The zoology of the voyage of H.M.S. Sulphur, under the command of Captain Sir Edward Belcher ... during the years 1836-42 ... London, Smith, Elder and Co., 1844. v.1)

Yale also has a copy of the plates, uncolored, and in portfolio, with dates 1843-46.

────── [Birds of New South Wales and Van Diemen's land] 1845. 317-329 p. (In Strzelecki, Sir Paul Edmund de. Physical description of New South Wales and Van Diemen's land. Accompanied by a geological map, sections, and diagrams, and figures of the organic remains ... London, Longman, Brown, Green, and Longmans, 1845)

────── List of birds known to inhabit southern Australia ... 1845. 440-448 p. (In Eyre, Edward John. Journals of expeditions of discovery into central Australia, and overland from Adelaide to King George's Sound, in the years 1840-1: sent by the colonists of South Australia, with the sanction and support of the government: including an account of the manners and customs of the aborigines and the state of their relations with Europeans ... London, T. and W. Boone, 1845. v.1)

For discussion of ornithology, cf. Whittell. A bibliography of Australian ornithology. Part 2, p. 233.

────── An introduction to The birds of Australia ... London, Printed for the author, by R. and J.E. Taylor, 1848. viii, 134 p. illus. 23 cm.

────── A monograph of the Trochilidæ, or family of humming-birds ... London, Printed by Taylor and Francis, pub. by the author, 1849-61. 5 v. 360 col. pl. 56 cm.

Yale also has a copy of the "Supplement; completed after the author's death by Bowdler Sharpe." London, Sotheran and Co., 1887.

Yale also has another set of the original parts as issued; this set was ordered when the work was published. There is a letter to Edward C. Herrick, the acting

PART FOUR BIBLIOGRAPHY OF JOHN GOULD, HIS FAMILY AND ASSOCIATES

librarian, from John Gould acknowledging Yale's order and another to Professor J.D. Dana dated November, 1857, stating that along with the volumes he was sending 50 specimens from tropical America and Australia from his own collection; these specimens are in the Yale Peabody Museum.

Yale also has various other letters from John Gould written during 1857 and 1858.

——— A monograph of the Odontophorinæ, or partridges of America... London, Pub. by the author, 1850. 4 p.l., [11]-23 p., 33 l. 32 col. pl. $56 1/2$ cm.

——— The birds of Asia... London, Printed by Taylor and Francis, pub. by the author, 1850-83. 7 v. 530 col. pl. $56 1/2$ cm.

Completed after the author's death by R. Bowdler Sharpe.

In v.1, pl. 21 and 25 and accompanying text are interchanged.

——— Catalogue of the birds of the northeast coast of Australia and Torres Strait... 1852. See Macgillivray, John, 1822-1867. Narrative of the voyage of H.M.S. Rattlesnake... 1852.

——— Prospectus and list of subscribers to Mr. Gould's works on ornithology, &c. [London, Pub. by the author, 1856] 4, 29 p. tables. 27 cm.

Cover title.

——— An introduction to the Trochilidæ, or family of humming-birds... London, Printed by Taylor and Francis, 1861. 4 p.l., iv, 216 p. $22 1/2$ cm.

Advertising matter, p.[213]-216.

Yale has two copies; one is the author's presentation copy to Edward Newman and the other to G.N. Lawrence with Lawrence's MS notes.

——— Handbook to The birds of Australia... London, Pub. by the author, 1865. 2 v. $25 1/2$ cm.

See also Diggles, S. Companion to Gould's Handbook. 1877.

——— The birds of Great Britain... London, Printed by Taylor and Francis. Pub. by the author, 1873. 5 v. 367 col. pl. 57 cm.

20 plates were republished by Phyllis Barclay-Smith in "Garden birds" in 1952.

——— An introduction to the birds of Great Britain... London, Printed for the author by Taylor and Francis, 1873. 2 p.l., [iii]-iv p., 1 l., 135, 14 p. $22 1/2$ cm.

——— The birds of New Guinea and the adjacent Papuan Islands, including many new species recently discovered in Australia... Completed after the author's death by R. Bowdler Sharpe... London, H. Sotheran & Co., 1875-88. 5 v. 320 col. pl. 56 cm.

——— Monograph of the Pittidæ... London, Pub. by the author, 1880-81. 2 pts. in 1 v. 10 col. pl. 57 cm.

Cover title. The original cover of pt.1 is mounted on cover and the one for pt.2 is bound in.

——— Tropical birds, from plates by John Gould. With an introduction and notes on the plates by Sacheverell Sitwell. London [etc.] B.T. Batsford [1948] [4], 12 p. illus., 16 col. pl. 24 cm. (Batsford colour books)

"Handlist of Gould's works," 4th preliminary page.

——— Mr. Gould's Tropical birds; comprising twenty four plates selected from John Gould's folios, together with descriptions of the birds taken from his original text, edited and introduced by Eva Mannering. New York, Crown Publishers [1955] xvi p. illus., col. plates. 41 cm.

BIOGRAPHY

BARRETT, CHARLES LESLIE, 1879-
The bird man, a sketch of the life of John Gould... Auckland [N.Z., etc.] Whitcombe & Tombs, Limited [1938] 4 p.l., 7-51 p. incl. front. (port.) illus. plates, port., facsims. $18 1/2$ cm.

INDEX

SHARPE, RICHARD BOWDLER, 1847-1909. An analytical index to the works of the late John Gould... With a biographical memoir and portrait. London, H. Sotheran & Co., 1893. 3 p.l., [v]-xlviii, 375 p. front. (port.) $29 1/2$ cm.

"Bibliography," p. [xxvii]-xlviii.

GOVAN, Mrs. ADA CLAPHAM. Wings at my window... illustrated by Dorothy Bayley. New York, Macmillan Company, 1940. xiv p., 2 l., 198 p. illus. 21 cm.

GRAAH, WILHELM AUGUST, 1793-1863. Undersögelses-reise till östkysten af Grönland. Efter kongelig befaling, udfört i aarene 1828-31... Kiöbenhavn, Trykt hos J.D. Qvist, 1832. xvii, [1], 216 p. VIII col. pl. (incl. front.) fold. map. $25 1/2$ x $21 1/2$ cm.

"Fugle," p. 193-195.

Yale also has a copy of the English edition, London, 1837.

GRAESER, KURT, 1850- Der Zug der Vögel. Eine entwicklungsgeschichtliche Studie... Zweite, verm. Aufl. Berlin, H. Walther, 1905. 167 p. illus. 24 cm.

GRAHAM, EDWARD HARRISON, 1902-
The land and wildlife... New York, Oxford Univ. Press, 1947. xiii, 232 p. 32 pl. on 16 l. 22 cm.

Many references to birds throughout.
"Bibliography," p. 211-224.

GRAHAM, HENRY DAVENPORT, 1825-1872. The birds of Iona & Mull...

originally but by 1852 a fourth volume was added as a supplement. Due to the tremendous detail in the four volumes, the works were not published until the completion of the exhibition.

There is an index in volume I and there are several Goulds mentioned. The reference to 'Gould, J. XXVIII. (Fine Art Court), 247' refers to volume II, page 836 and exhibitor 247. The following is noted:

Gould, J., 20 Broad Street, Golden Square – Inventor.
A new mode of representing the luminous and metallic colouring of the *Trochilidae*, or humming birds.
The effect is produced by a combination of transparent oil and varnish colours over pure leaf gold laid upon paper, prepared for the purpose.

Thus John Gould had an exhibit at this great 1851 exhibition on how he produced the metallic colouring on his humming-bird prints.

One wonders if any of the other Goulds mentioned in the index are related to John Gould.

ROYAL SOCIETY OF LONDON. 1867-1925. *Catalogue of scientific papers (1800-1900)*. 19 vols. London, C. J. Clay, 1867-1902; Cambridge University Press, 1914-25.

A monumental compilation. In volume II are listed John Gould's papers from number 1 of 1829, 'On the occurrence of a new British Warbler', through paper number 190 of 1863. Paper number 39, however, encompasses twenty-seven articles published in the *Zool. Soc. Proc.* from 1836 through 1850 entitled 'On Australian Birds, with characters of the new species'.

Two articles by Charles Gould are also listed in this volume, and one article by John Henry Gould.

Volume VII has John Gould's articles listed from number 191 of 1864 through number 229 of 1873.

In volume X are listed the papers by Gould numbering from 230 in 1874 through 238 in 1880, the final one 'On two new hummingbirds from Bolivia'.

ROYAL SOCIETY OF LONDON. 1883. *Catalogue of the scientific books in the library of the Royal Society.* London, Royal Society.

Thirteen folio works of Gould are listed, plus his *Introduction to the Birds of Australia* and his *Handbook*.

RUTGERS, ABRAM. 1966. *Birds of Europe.* London, Methuen.

The 160 colour plates are taken from Gould's *Birds of Europe* and *Birds of Great Britain* and are printed moderately well by a Holland firm. The text by Rutgers accompanying each plate is new 'taking into account findings since Gould's day'.

This is the first of the five Rutger's series of volumes. Some or all of these sets may have been divided and published in two volumes.

RUTGERS, ABRAM. 1967. *Birds of Australia, illustrations by John Gould.* 2 vols. London, Methuen.

Volume 1 contains 161 pages with 80 coloured plates and volume 2 contains pages 162 to 321 with another 80 coloured plates. The plates are from Gould's work on Australia and are reproduced moderately well.

Another copy of this work is complete in a single volume.

This was the second set of this Rutgers series.

PART FOUR BIBLIOGRAPHY OF JOHN GOULD, HIS FAMILY AND ASSOCIATES

RUTGERS, ABRAM. 1969. *Birds of Asia.* New York, Taplinger.

This volume contains 160 colour plates from Gould's *Birds of Asia* and is the third of this Rutgers series of volumes. Apparently a London edition was also printed.

RUTGERS, ABRAM. 1971. *Birds of New Guinea.* New York, St. Martin's Press.

This is the fourth set of 160 Gould colour plates with the new text by Rutgers.

RUTGERS, ABRAM. 1972. *Birds of South America.* London, Methuen.

The 160 colour plates for this Rutgers volume were selected from Gould's monographs *Trochilidae, Ramphastidae, Trogonidae* and *Odontophorinae.* According to the publisher this is the fifth and final series of these colour-plate books.

RUTGERS, A. AND NORRIS, K. A. 1970-77. *Encyclopaedia of aviculture.* 3 vols. London, Blandford.

These are beautifully illustrated volumes, with quite a comprehensive coverage of the bird families depicted.

Volume one has fourteen colour plates by van den Broecke and Slijper. Of the 209 black-and-white photographs, 90 per cent of these are of Gould's plates. No acknowledgement of this fact is mentioned in volume one, but it is cursorily mentioned in volume two.

In volume two, the majority of the pages and illustrations are devoted to the parrot family. Again almost all of the 120 black-and-white photographs are of Gould's plates. There are 27 colour-plates by van den Broecke and Slijper.

The section on the Budgerigar refers to the fact that on Gould's return to England from Australia in 1840 he brought with him the first living examples ever to have been seen in Europe. 'He introduced them by their aboriginal name of Betcherrygah which later became corrupted to Budgerigar', now further shortened to Budgie.

A short biographical note on Gould appears in volume one on page 34, with only a few other scattered references.

The third volume was not available for examination.

RUTHERFORD, H. W. 1908. *Catalogue of the library of Charles Darwin now in the Botany School, Cambridge.* Cambridge, University Press.

Gould's works from Darwin's library now at Cambridge include two copies of 'Birds. Zoology of... H. M. S. Beagle, part 3', the two-volume 'Handbook to the Birds of Australia', 'An Introduction to the Birds of Australia', 'An Introduction to the Trochilidae...' and 'An Introduction to the Birds of Great Britain'.

SALVADORI, CONTE TOMMASO. 1881. 'Della vita e della opere dell' ornitologo Inglese John Gould.' *Atti della R. Accademia Scienze di Torino* 16: 789-8.

S

The opening sentence states that Gould was elected a Corresponding Member of the Royal Academy in 1841.

Salvadori visited Gould in London in the autumn of 1877 and 'admired the stupendous collection of Humming Birds, which was then placed in a gallery expressly built in the house which he inhabited. The illustrations for the Monograph [on the

Humming-birds] were prepared from sketches by Gould, and were lithographed by Richter.'

This memoir of Gould's life and works is nicely done, and the listing that concludes this article of 302 'opere minori' is very valuable. Only a few of Gould's scientific papers were missed. Sharpe (1893) acknowledged Salvadori's compilation and copied this list of papers without changes.

SALVIN, OSBERT. 1882. *A catalogue of the collection of birds formed by the late Hugh Edwin Strickland, M.A.* Cambridge, Cambridge University Press.

The Strickland collection of 3117 species of birds with 6006 skins was presented to the University of Cambridge by Mrs Strickland in 1867.

A valuable part of this volume is a listing under the 'origins of the specimen' which contains the names of many world-wide bird collectors, some little known, and quite a few bird skin dealers.

In this list Salvin states that fifty-four skins came from John Gould.

SAUER, GORDON C. 1948. 'Bird art and artists: John Gould'. *American Antiques Journal* 3: 6-9.

This was my first article on John Gould, relating especially toward information on his prints as sold by antique dealers. Sharpe's *Analytical Index* and the Sotheran 'Piccadilly Notes' provided the majority of my material. It was illustrated by seven photographs of lithographs and a series of three illustrations of my series on the Australian Bower-bird showing Gould's rough pencil and watercolour sketch, the amplification by Richter, probably, and then thirdly, the finished hand-coloured lithograph.

SAUER, GORDON C. 1956. 'Gouldiana'. *Books and Libraries at the University of Kansas* 1: 2-3.

This is a cursory listing of the extensive manuscript and other Gould material in the Ellis collection at the University of Kansas. Ralph Ellis of California purchased this collection in 1937-38 mainly from Henry Sotheran, Ltd, London, who in turn had purchased Gould's estate in 1881 upon his death.

SAUER, GORDON C. 1976. 'Summary and index of material in the file boxes of the Ellis Archives in the Kenneth Spencer Library, University of Kansas, Lawrence.' Typescript.

When the University of Kansas obtained the vast collection of ornithological books, journals, pamphlets, manuscripts, and original art accumulated by Ralph Ellis of California, it also acquired fifty-seven large file-boxes of Ralph's personal and ornithological correspondence, bird collecting notes and family records.

Because these boxes contained much material relevant to Ralph's purchases of the Gould estate collection of over 1200 original drawings of birds and mammals, plus an almost complete set of his folio works, I decided to inspect, catalogue and, finally, index these files. Over a two-year period I made several three-day trips to Lawrence. The information gained was worthwhile and fascinating.

I doubt if any individual ever kept more complete records of his life than Ralph Ellis, Jr. When it is known that he lived to be only 37, had an illness that recurred approximately every 21 days (which incapacitated him for three to four days at a time), and was in a mental sanitarium for almost three years, then the true magnitude

PART FOUR BIBLIOGRAPHY OF JOHN GOULD. HIS FAMILY AND ASSOCIATES

of his book-collecting accomplishments becomes more obvious. He wisely accumulated a unique book collection of 21 688 volumes, mainly ornithological, over 40 000 scientific journals and separates, around 3000 original drawings, including the largest collection of John Gould original drawings in existence, a mammal collection of 2782 skins and skeletons, a large bird-skin collection, and then he meticulously chronicled all of this, and much more, in fifty-seven large file-boxes.

Obviously he had a lot of help. He had his family's money, servants, tutors, secretaries, book-dealers, physicians, and two wives. But the dominant feeling as one reads all of his voluminous letters, notes, and comments is that he himself existed under the most adverse circumstances. First and foremost was his recurring illness of a cyclic neutropenia that caused deep sores in his mouth so that he could not eat for several days. This obviously contributed to his periods of mental derangement. He spent much of his life occupied with this illness, and devoted many hours and days with physicians and in hospitals seeking a cure, or even relief, from this severe incapacitating problem. Medical treatments included frequent injections with liver, vitamins, hormones and even experimental drugs. He became very introspective, suspicious and demanding. In later years he compounded his medical and emotional problems by drinking excessively and becoming even more violent. Alcohol and possibly pneumonia, coupled with depression over a pending second divorce, eventually caused his demise in 1945.

The other dominant theme in his voluminous files is his obsession with money, or the relative lack of it. This problem surfaced after his father's death when Ralph was 22. Then begins a cancerous and consuming demand for the money necessary to have expensive cars, to have life-memberships in over two hundred scientific societies, and to support a relentless drive to accumulate rare bird-books and prints. The object of his personal wrath was his mother. Page after page of the files are devoted to letters to his mother asking for money, questioning her sale of the family's numerous properties, criticizing her handling of her own money (which was quite appreciable), and always accusing her of not giving him enough money for his very 'good investments' in his books. When he did come into a $103,000 trust fund, he depleted it rapidly, and then began haranguing his mother again for more money for his books and his living expenses. Ralph was never gainfully employed, but for two years in the early 1940s he managed the rental of the family home in Berkeley, California. Twelve tenants occupied two apartments in seven rooms of the house.

While I have said that the fifty-seven file boxes form one of the most comprehensive accounts of a man's life, and that his letters, notes, and comments on his illness and his money anxieties consume many pages and files, the files also contain many other valuable records.

More than fifty per cent of the files are devoted to the purchases of books, prints, skins, eggs and journals from book dealers, collectors, auctioneers, and scientific societies. Ralph's negotiations in the 1930s and 1940s with book dealers and auctioneers like the London firms of Wheldon and Wesley, Henry Sotheran, Bernard Quaritch, and the American firms of Dawsons, American Art Association, John Howell, H. T. Kraus, and Jake Zeitlin, could form the nucleus of a book in itself. His correspondence with book collectors, authors, ornithologists, and the directors of institutions and scientific societies obviously includes all the important, and some not so important, personalities of that time. While the value of Ellis's rare book and manuscript collection is obvious, this correspondence and record of purchases and sales is an additional very valuable and unique part of this fabulous collection.

Two copies exist of my typescript record of this cataloguing.

SAUER, GORDON C. 1976. 'Gouldiana in the British Museum Library'. *Books and Libraries at the University of Kansas* 13 (part 2): 12.

F. C. Sawyer (1971) listed the Gould manuscript material in the British Museum (Natural History) Library. Omitted was a set of six volumes noted by me in 1975. These volumes are entitled 'John Gould's manuscript for "The Birds of Europe" incorporating various letters and notes from other naturalists'. Volume 5 is missing from this set. [This missing volume is in the manuscript collection of the Academy of Natural Sciences of Philadelphia (Sauer 1978).]

Since these volumes contain handwritten notes by Gould and his secretaries, plus notes from many other naturalists, with clippings from newspapers and magazines on specific birds, these books could be properly called the 'filing cabinets' for his *Birds of Europe* manuscript.

SAUER, GORDON C. 1976. 'John Gould, artist? Testimony of the Yellow-billed Cuckoo'. *Books and Libraries at the University of Kansas* 13 (part 2): 8-12.

A pencil sketch of a Yellow-billed Cuckoo is in manuscript collection 71 of the Academy of Natural Sciences of Philadelphia. This sketch was drawn by Gould when he visited the Academy in May 1857. None of his artists accompanied him on this U.S. trip, so this sketch is the only known example of a drawing of a bird that definitely can be attributed to John Gould.

Further comments are made on Gould's artistic ability and on the evidence that supports the facts that he was a capable ornithological artist.

SAUER, GORDON C. 1977. *John Gould bird print reproductions.* Kansas City, Mo., privately printed.

The purpose of this booklet of seventy-six pages was 'to assist others in correctly identifying the origin of a particular Gould bird print'. This is a compilation of the known reproductions of Gould's prints by sixteen different companies with descriptions and measurements of these reproductions.

The short introduction contains a history of Gould's works with special reference to his hand-coloured lithographs. Mention is also made of some uncoloured prints purchased by Ralph Ellis of California in 1938 from Gould's estate and then coloured by artists employed by Ellis.

There are twenty-six full-page illustrations.

SAUER, GORDON C. 1978 'The missing volume of John Gould's working notebook for his "Birds of Europe." ' *J. Soc. Biblphy. Nat. Hist.* 9(1): 15-16.

After publishing an earlier article (Sauer 1976) on manuscript volumes in the library of the British Museum (Natural History), it was noted that a missing 'Volume V' was in the library of the Academy of Natural Sciences of Philadelphia. This article is concerned with this volume and its contents. The label on the spine of this volume is 'Europe/Insessores/Vol. V/Frigillidae/Pars'.

A single page is devoted to each bird species, with the scientific name, or infrequently the common name, as a heading. Notations, or even sketches (seven in this volume) are by various hands.

Five letters are set-in the book.

SAUER, GORDON C. 1980. *John Gould's prospectuses and lists of subscribers to his works on natural history: with an 1866 facsimile.* Kansas City, Missouri, privately printed.

PART FOUR BIBLIOGRAPHY OF JOHN GOULD, HIS FAMILY AND ASSOCIATES

This small forty-six page booklet contains a short introduction to the life of John Gould, plus information on the very rare five editions of Gould's lists of his subscribers. The 1866 edition, with a listing of 1010 possessors and subscribers for ten of Gould's works, is reprinted in facsimile.

SAUNDERS, JOHN. [1875.] *List of the books, memoirs, and miscellaneous papers of Dr. John Edward Gray* ... London, Taylor and Francis.

A compilation of Gray's papers and books covering 1162 items.

Item 240 directly, and other papers indirectly, refer to specimens of reptiles given to Gray by Gould, mostly collected by Gilbert.

SAVAGE, CHRISTOPHER. 1952. *The Mandarin Duck*. London, Adam and Charles Black.

Plate X, *Aix galericulata*, facing page 27 is from Gould's *Birds of Asia*, not coloured.

SAWYER, FREDERICK C. 1971. 'A short history of the libraries and list of manuscripts and original drawings in the British Museum (Natural History)'. London, Bulletin of the British Museum (N.H.), Historical Series, vol. 4, no. 2.

This important collection of Gould material is listed as follows:

Autograph letters, manuscript notes and drawings relating to Australian Mammals and Birds also autograph letters to Sir W. Jardine, c. 1848-1869. Two framed water-colour paintings of Trout, dated 1862. These were presented by Mrs. Edelsten in 1967 and are attributed to J. Gould (Z).

A pastel vignette of Gould dated 1875, signed by Marian Walker a miniaturist who exhibited at the Royal Academy between 1854 and 1877 was purchased in 1938 (Z).

Z refers to these items residing in the Zoological section library.

Gould is also referred to on pages 82 and 203.

[Very unfortunately the 'autograph letters, manuscript notes, etc.' has not been catalogued. *See also* [Gould, John] Box of manuscripts, and also Slater 1973.]

SCHERREN, HENRY. 1905. *The Zoological Society of London*. London, Cassell.

A handsome book with twelve lovely reproductions of watercolours by Charles F. Flower. They portray the Zoological Gardens in Regents Park.

Of necessity this book by Scherren must be compared with one on the same subject published 24 years later by P. Chalmers Mitchell.

Scherren, as a result of his research into the early files of the Linnean Society and the Zoological Society, concluded that 'the foundation of the Zoological Society of London was a natural development from the Zoological Club of the Linnean Society'. Mitchell in his book goes to considerable length to refute this. He supports the idea that the Zoological Society evolved *sui generis*. From the evidence presented by both authors, I would favour Mitchell's belief.

Gould references appear frequently, far exceeding the three listed in the index. On page 30 is a note on the giraffe given to George IV and later stuffed by Gould; page 33 has the first specific reference to Gould as follows: 'John Gould's connection with the Society began this year [1828] by his appointment as Curator and Preserver to the Museum'; page 46 refers to Gould's stuffing the giraffe in 1829; pages 53-4:

In 1833 Gould was appointed Superintendent of the ornithological department of the Museum, over which he presided four years, when he resigned in order to go to Australia in search of material for his great work on the birds of the island continent. He did not, however, leave England till the following year; and before embarking wrote thus to the Council:

JOHN GOULD — THE BIRD MAN

With regard to the Society's ornithological collection, as I have at all times taken a great interest in it, and have ever done my utmost to increase its value, I hope that on my return to England, I may be allowed to resume the care of it, should I be desirous of so doing.

To this application a favourable reply was sent, and Gould was elected a Corresponding Member of the Society. On that occasion the Council recorded their sense of the great scientific value of his work, and expressed the earnest hope that his present undertaking might be crowned with that success which had hitherto accompanied his efforts. (Gould took up the Fellowship in 1840, and was afterwards a Member of Council and Vice-President.)

On his return, however, he did not take up his old duties, but devoted his energies to the production of his famous books. Mr. G. R. Waterhouse was appointed Curator of the Museum in 1836, and fulfilled the duties of that post till 1843, when he obtained an assistantship in the British Museum. Mr. Louis Fraser succeeded him, and after his resignation in 1845 the Museum was in charge of subordinate officers.

Further references are as follows: page 78: Gould's communications to the scientific meetings of the Society over a decade were only exceeded by Professor Richard Owen; pages 89, 94, 98, 100, 105-6 (on page 106 is a map of the 1851 plan of the gardens showing the building for Gould's humming-bird exhibit (Figure 37.)), 119 (Gould and Darwin), 120, 124, 125, 146, 200, and 226.

SCLATER, PHILIP LUTLEY. 1857. 'Notes on the birds in the Museum of the Academy of Natural Sciences of Philadelphia, and other collections in the United States of America'. *Proc. Zool. Soc. London,* pp. 1-8.

Sclater states that he 'recently returned from a few months' excursion to the United States of America' so he undoubtedly was there in 1856. This paper was read on 13 January 1857. One wonders if this visit by Sclater influenced Gould to also go to the United States, which he did in 1857.

Sclater stated that 'The collection of the Academy of Natural Sciences of Philadelphia is certainly the best zoological collection in the New World, and in the particular department of Ornithology, . . . is probably superior to every museum in Europe, and therefore the most perfect in existence'. He attributed much of the value of the collection of birds to the munificence of Dr Thomas B. Wilson who gave the Academy the Gould type of Australian birds.

SCLATER, PHILIP LUTLEY. 1857. 'Description of a new tanager of the genus Euphonia (*Euphonia Gouldi*)'. *Proc. Zool. Soc. London,* p. 66.

Sclater begins with this paragraph:

Mr. Gould having placed in my hands for examination some specimens of Euphonia, which he has lately received from Guatemala, I am enabled to exhibit to the Society examples of both sexes of what I believe to be a hitherto uncharacterized species of that genus. I am no friend to the too frequent practice of calling animals after individuals, but I feel that I shall meet with approbation in this instance if I confer on the present bird the name of one of the most eminent naturalists of the day, to whom moreover I am indebted for numerous acts of kindness from the period when I first had the pleasure of his acquaintance. I therefore propose to call this species *Euphonia Gouldi.*

SCLATER, P. L. 1881. 'Letter to the editor of the Times.' *Nature* 23: 491.

A note on the letter by Sclater follows:

Writing to the Times on Friday last, Mr. Sclater calls attention to the fact that the collection of birds of the late John Gould, the ornithologist, have been offered to the trustees of the British Museum for 3000 pounds, and expressed the hope that there will be no difficulty on the part of the Treasury in sanctioning about 1500 mounted and 3800 unmounted specimens of humming-birds, being the types from which the descriptions and figures in the celebrated "Monograph of the Trochilidae" were taken. There are besides, 7000 other skins of various groups, amongst which are splendid series of the families of Toucans, Trogons, Birds of Paradise, and Pittas.

SCLATER, P. L. 1896. 'On drawings in the Knowsley Library'. *Proc. Zool. Soc. London,* pp. 981-9.

PART FOUR BIBLIOGRAPHY OF JOHN GOULD, HIS FAMILY AND ASSOCIATES

Mr. Sclater exhibited two bound volumes of original watercolour drawings by Wolf and Waterhouse Hawkins, belonging to the Knowsley Library, which had been kindly lent to him for examination by the Earl of Derby. These drawings were of very great interest to zoologists, as containing many of the originals from which the figures in the two volumes of the 'Gleanings from the Knowsley Menagerie' and Wolf's 'Zoological Sketches' had been taken.

The first and larger-sized volume (29 in. by 22 in.), lettered on the back 'Wolf's Original Drawings,' contained twenty-two watercolour drawings by Wolf, of which a manuscript list in the volume, written by Mr. T. J. Moore in 1871, gave the following particulars.

Then follows a list of twenty-two mammals and birds which except for perhaps three 'were living at Knowsley at the breaking up of the Collection in 1851, and these sketches were doubtless made *from those* specimens either before or after their removal. — T. J. Moore, April 28, 1871.'

The second volume, which was lettered on the back 'Knowsley Menagerie. Original Drawings by W. Hawkins and Wolf' (size 25 in. by 20 in.), contained sixty-nine original drawings by those artists. There was no manuscript list attached to this volume, but Mr. Sclater had prepared the subjoined account of its contents.

Then follows an annotated listing of sixty-nine drawings mainly of animals and a few birds.

[The two volumes of original drawings and eight other volumes were examined by me in 1974. The Collection included one volume of 'Drawings in Natural History/Birds by Mr. & Mrs. Gould' with seventy-three sheets; four volumes of 'Original Drawings/Birds./Gould, Richter, Hawkins, Wolf & Newton' on 191 sheets; two volumes of 'Gould's Original Drawings', volume one containing those for *Birds of Australia* on seventy-eight sheets, and in volume two the drawings for some of Gould's other works on fifty-three sheets; and one volume of 'Australian Animals/Original Drawings/by Gould & Richter' on seventy sheets. some being lithographs. I made notations and listings of all of the published plates represented by these original drawings.]

SCLATER, P. L. & EVANS, A. H. 1909. *The Ibis jubilee supplement.* London, Porter.

A supplement of *The Ibis* issued as volume II, 1908, ninth series. There are four parts to the work. Part 1 is on the proceedings of the special Jubilee Meeting of the British Ornithologists' Union held in December 1908; part 2 is 'A Short History of the British Ornithologists' Union' by P. L. Sclater beginning with its founding by Professor Alfred Newton in 1858; part 3 consists of biographical notices of the original members and principal contributors to the First Series of *The Ibis*, and of officers, with portraits; and part 4 is a list of members of the B.O.U. from 1858 to 1908.

It is interesting and surprising but Gould was never a member. One wonders why, and also what thoughts may have transpired over this obvious rebuff to him, unless he had declined to join.

On page 3 of this volume there is a reference to publications by various British ornithologists and Gould's monograph *Trogonidae* is mentioned with works by others. Also on page 129 is a note about the first meeting of Sclater and Gould at Hugh Strickland's home in Oxford.

Several references are made to Joseph Wolf, especially on the fact that Wolf was responsible for the well-known wood-block of the Ibis, the sacred bird of Egypt, that has ever since ornamented the cover of *The Ibis* magazine.

A biography of Sharpe appears on page 199.

SERVENTY, D. L. & WHITTELL, H. M. 1948. *A handbook of the birds of Western Australia.* Perth, Paterson Press.

There is a thirty-two-page section on 'The history of Western Australian ornithology'. This is a shorter version of Whittell's history of Australian ornithology than that which appears in his 1954 book, *The literature of Australian birds*, covering the dates from 1618 to 1850. In this present book, however, Whittell does elaborate some on the later collectors and naturalists who worked in Western Australia after the 1850s.

SERVENTY, D. L. & WHITTELL, H. M. 1962. *Birds of Western Australia.* 3rd edition. Perth, Paterson Brohenska.

In this 1962 edition there are only two minor changes in the section on history as it relates to Gould and Gilbert.

SERVENTY, VINCENT. 1966. *A continent in danger.* London, Deutsch.

The prophetic words of Gould are frequently quoted with regard to the probable extermination by man of many of the Australian mammals and birds.

Reference is made to Gould's remarks on the Tasmanian Wolf or Thylacine; Gilbert's comments on the Grey Kangaroo; Gould's interest in the Lyrebird, the Budgerigar which he brought back to England in 1840, the Emu, and many others. The index to Gould is adequate.

Finally, several pages are devoted to the Gould League of Birdlovers. The aim of this society is 'Conservation'.

SHARPE, RICHARD BOWDLER. 1891-98. *Monograph of the Paradiseidae, or Birds of Paradise, and Ptilonorhynchidae, or Bower-birds.* 2 vols. London, Sotheran.

In many respects this monograph is an extension of Gould's works, since many of the plates are printed from the same stones used in Gould's *Birds of New Guinea*.

The work appeared in eight parts with seventy-nine hand-coloured lithographic plates. The contents and date of each part are listed by Zimmer (1926).

There were ten plates in parts I to part VI. In part VII there were also ten plates but two species were illustrated with two plates each, and two additional species were discussed but not illustrated. In part VIII there were nine plates but two species of birds were figured on each of two of the plates, and sixteen additional species were presented but not illustrated.

Thus there were sixty species illustrated in the first six parts, eight in part VII, and eleven illustrated in part VIII. Eighteen additional species were not illustrated but were discussed by Sharpe. A total of ninety-seven species were reported on in the two volumes.

SHARPE, R. BOWDLER. 1893. *An analytical index to the works of the late John Gould, F.R.S.* London, Henry Sotheran.

Contains the frequently quoted and very valuable, though short, personal biographical memoir of Gould. Sharpe became acquainted with Gould in 1862 when Gould was 58 and Sharpe was 15. As time passed, Sharpe became the very well-qualified ornithologist enlisted to carry on to completion two of Gould's major works, *Birds of Asia* and *Birds of New Guinea*, plus the one-volume *Supplement to Trochilidae*. He played only a minor role in the publication of *Pittidae*.

Following the biographical memoir, Sharpe lists Gould's major works, then his 'Opuscula' or published papers. This latter list of 298 papers, published by Gould from 1829 to 1880, plus four minor works, is identical with the list published by Count Salvadori. The majority of articles were published in the *Proceedings of the Zoological Society of London* and in the *Annals and Magazine of Natural History*. Twenty-nine articles were published in both of these journals.

PART FOUR BIBLIOGRAPHY OF JOHN GOULD, HIS FAMILY AND ASSOCIATES

The major part of Sharpe's volume is devoted to an 'analytical index to the complete series of Gould's works'. Sharpe lists alphabetically every bird or mammal figured by Gould under its scientific name, both under genera and under species, and also with its popular name, indexed both with first and last names. Following this animal listing is the name of the volume in which the illustration appeared, with volume number and plate number.

After such a monumental index, I in turn felt that the persons mentioned by Sharpe in the memoir and the listing of papers, should be indexed. This follows.

A
Abbot, C. C., xxix, xlii
Aiton, J. T., x
Allen, Lieut, xxix

B
Baker, x
Belcher, Sir Edward, xxxvi
Blanchard, Laman, xiii
Briggs, Mr, xxii, xliii
Broderip, x
Buckley, Mr, xlvii
Burrows, James, xxiii

C
Cabot, Dr Samuel, xli
Charlotte, Queen, x
Chubb, Charles, vi, viii
Cleave, John, ix
Coxen, Elizabeth, xi
Coxen, Nicholas, xi

D
Darwin, Charles, xxxi, xxxiv
Denham, Capt., xli
Derby, Earl of, xxxii
DeVitre, x, xxii
Douglas, David, xxxi

F
Fitzroy, Capt., xxxiv
Folliot, xxix
Franklin, Lady, xvii
Franklin, Sir John, xvii

G
Gerrard, Edward, x, xii
Gilbert, John, x, xvi, xxxvi
Gould, Charles, xvii
Gould, Elizabeth Coxen, xii, xv, xvii
Gould, Franklin, xvii
Gould, Henry, xvii
Gould, John, senior, x
Gould, Louisa, ix
Gould, Sai, ix
Gray, George Robert, xiv
Gunther, Dr, xix

H
Hart, W., xii, xxi, xxii
Hauxwell, xl
Hodgson, B. H., xi, xxix

I
Ingleby, Sir William, x

J
Jerrold, Blanchard, xii

L
Lear, Edward, xii, xv
Legge, Mr, x
Leichhardt's expedition, xvi
Loddiges, George, xix

M
Mantell, Walter, xli
M'Gillivray [sic], John, xxxix, xli
Moon, Mrs, ix, xxiv

P
Prince, Edwin C., xvi, xxiii, xxiv

R
Reade, Sir Thomas, xxx
Richter, H. C., xii, xvii
Robertson, Mr, xxiv

S
Salvadori, Count, ix, xxv
Schomburgk, Sir R. H., xliii
Sharpe, R. B., xxi, xxii, xxiii, xxvii, xxviii
Shore, Hon. C. J., xiii
Sotheran, Messrs., v
Swainson, Wm., xi
Swinhoe, Robert, xliv
Sykes, xi

T

Temminck, C. J., xv, xxx, xxxi

V

Venables, Mr, xxii
Vigors, N. A., x, xi, xiii, xiv

W

Wallace, A. R., xlii

Walsingham, Lord, v
Waterhouse, F. H., xxv
Wharncliffe, Earl of, v
Wilder, Harry, xxii
Wilson, Dr Thomas, xix
Wolf, Joseph, xxi, xxv

A large paper copy of Sharpe's *Analytical Index* was also published. This edition was limited to 100 copies and some apparently were numbered. Mine is not. Appended to this volume are twelve pages of advertisement by Henry Sotheran of a complete set of Gould's works in an 'elegantly carved case', plus descriptions of separates of Gould's sets that were available for sale (Figure 75).

Figure 75 Three consecutive pages from Advertisement section of R. B. Sharpe's 1893 *Analytical Index*, large copy edition.

A COMPLETE SET OF MR. GOULD'S MAGNIFICENT SERIES OF ORNITHOLOGICAL & OTHER WORKS,

Uniformly Printed in Imperial Folio size,

AND COMPRISING—

THE BIRDS OF EUROPE, with 449 Coloured Plates	5 Volumes.
THE BIRDS OF AUSTRALIA, with the Supplement, 683 Coloured Plates	8 Volumes.
THE MAMMALS OF AUSTRALIA, with 182 Coloured Plates	3 Volumes.
A CENTURY OF BIRDS FROM THE HIMALAYAN MOUNTAINS, with 80 Coloured Plates	1 Volume.
THE BIRDS OF GREAT BRITAIN, with 367 Coloured Plates	5 Volumes.
THE TROCHILIDÆ, OR HUMMING BIRDS, with the Supplement, 418 Coloured Plates	6 Volumes.
THE RHAMPHASTIDÆ, OR FAMILY OF TOUCANS, with 51 Coloured Plates	1 Volume.
THE TROGONIDÆ, OR FAMILY OF TROGONS, with 47 Coloured Plates	1 Volume.
THE ODONTOPHORINÆ, OR PARTRIDGES OF AMERICA, with 32 Coloured Plates	1 Volume.
THE BIRDS OF ASIA, with 530 Coloured Plates	7 Volumes.
THE BIRDS OF NEW GUINEA AND THE PAPUAN ISLANDS, with 320 Coloured Plates	5 Volumes.

FORMING TOGETHER FORTY-THREE VOLUMES IMPERIAL FOLIO.

UNIFORMLY BOUND IN THE BEST STYLE IN GREEN MOROCCO SUPER-EXTRA, HANDSOMELY GOLD-TOOLED, GILT EDGES.

The whole enclosed in Two Elegantly Carved Mahogany Cabinets, with Plate-Glass Doors.

** *An Engraving of one of the Cabinets, showing the appearance of the Volumes when arranged therein, appears on the next page.*

PART FOUR BIBLIOGRAPHY OF JOHN GOULD, HIS FAMILY AND ASSOCIATES

SHARPE, R. B. 1896. *Lloyd's natural history.* 4 vols. London, Lloyd.

There is one paragraph on Gould on page iv.

SHARPE, R. BOWDLER. 1898. *Wonders of the Bird World.* New York, Stokes.

This is a compilation of lectures given by Sharpe throughout the United Kingdom in the previous ten years. It is a popular accounting, with a goodly addition of scientific information on the birds of the world, their migration, colouring, nesting, courtship, sounds, and distribution.

There are of course many Gould references. Sharpe indexes only eight but there are eighteen, namely pages 25, 33-6, 39, 77, 92, 99, 126, 129, 130, 132, 135, 199, 206, 213 (illustration), 216 (illustration), 223, 309 and 310.

SHERBORN, CHARLES DAVIES. 1897. 'Notes on the dates of "The Zoology of the 'Beagle'".' *Ann. Mag. Nat. Hist.* 6th ser. 20: 483.

The numbers of the nineteen separate parts, the contents (mammals, aves, etc.), number of plates, pages, and date of issue in months and years for each part are given

> Introduction to the Birds of Australia, 8vo. cloth (SCARCE), 1848   10s. 6d.
> Introduction to the Mammals of Australia, 8vo. cloth (SCARCE), 1863   10s. 6d.
> Introduction to the Birds of Great Britain, 8vo. cloth (SCARCE), 1873   10s. 6d.
> Introduction to the Trochilidæ, or Family of Humming Birds, 8vo. cloth (SCARCE), 1861   10s. 6d.
> Only a very limited number of the above four useful works were printed for private distribution among the author's scientific friends, and very few copies are now left
>
> Pittidæ, or Short-Tailed Thrushes, with 10 coloured plates. Part 1 (all published)   £3. 3s.
>
> *The Edition limited to 250 Small and 100 Large Paper copies.*
>
> ## ANALYTICAL INDEX
> TO THE
> ## COMPLETE SERIES
> OF
> # GOULD'S ORNITHOLOGICAL WORKS.
> BY
> ### R. BOWDLER SHARPE, LL.D., F.Z.S., etc., etc.,
> DEPARTMENT OF ZOOLOGY, BRITISH MUSEUM
> (By whom Mr. Gould's Works were completed after his death.)
> CONTAINING
>
> **UPWARDS OF 13,230 CROSS REFERENCES TO ALL THE SPECIES FIGURED IN MR. GOULD'S GRAND WORKS—A BIOGRAPHICAL MEMOIR AND PORTRAIT OF THE EMINENT ORNITHOLOGIST—AND A BIBLIOGRAPHY.**
>
> One Volume, impl. 8vo. cloth, top edges gilt   price £1. 16s Net.
> Large Paper copies, impl. 4to., half morocco, top edges gilt   price £4. 4s Net.
>
> *The Birds of Europe.* With 449 Coloured Plates. 5 vols. Imperial Folio.
> *The Birds of Australia.* With the Additional five supplementary Parts. 683 Coloured Plates. 8 Vols. Imperial Folio.
> *A Century of Birds from the Himalaya Mountains.* With 80 Coloured Plates, exhibiting 100 figures of Birds. Imperial Folio.
> ∗⁎∗ Of the three works named above, only single second-hand copies can be supplied as occasion offers; they having long been out of print, and very scarce Messrs H S & Co will have much pleasure at any time to report copies they may have in stock.

in columns by Sherborn, from a copy provided through the courtesy of Messrs. Smith, Elder, & Co., the publisher.

SHERBORN, CHARLES DAVIES. 1922. *Index animalium*. London, British Museum.

This is the second part of Sherborn's monumental index of animal names. On page lxi of the bibliography, appearing in volume one of this second part, is a list of Gould's works up to 1850.

SHUFELDT, R. W. 1902. 'Professor Alfred Newton, F.R.S.' *Osprey* 1: 30-2.

This short account of Newton, with a portrait, was written when Newton was 73 years old, as a tribute to him. Newton was a younger contemporary of Gould.

SINGLETON, M. E. 1938. 'Gouldians'. *The Foreigner* 5: 12-16.

Directions for successfully maintaining and rearing Gouldian Finches, or 'as they are called in America, "Lady Goulds" after the wife of their discoverer', is the subject of this article.

PART FOUR BIBLIOGRAPHY OF JOHN GOULD, HIS FAMILY AND ASSOCIATES

Gould was the first to bring these birds from Australia to England to be raised in aviaries.

SITWELL, SACHEVERELL. 1947. *The hunters and the hunted.* London, Macmillan.

In the section entitled Book III, The Kingdom of the Birds, there are numerous references to Gould and his works, especially in footnotes. Also there are six black-and-white illustrations from Sharpe's *Monograph of the Paradiseidae.*

The Sitwell style is flowery. There is no index.

SITWELL, SACHEVERELL. 1948. *Tropical birds from plates by John Gould.* London, Batsford.

With typical long-winded Sitwellian prose he waxes poetic for nine introductory pages, on 'these wonderful and obvious beauties'. There is barely a mention of Gould. Then there is a section on 'Notes to Plates', of which there are sixteen coloured. The text concludes with a single page comment on Gould and his works.

SITWELL, SACHEVERELL. 1949. *Audubon's American birds.* London, Batsford.

This is a companion volume to Sitwell's *Tropical Birds* containing plates by Gould. In this Audubon edition sixteen of his plates are finely reproduced in colour. There are also several references to Gould and comparisons made between him and Audubon.

Sitwell states that Gould's bird plates are not works of art, as are Audubon's, but 'they are perfect ornithological illustrations'.

Lear's contribution to the Gould publications is also mentioned.

SITWELL, SACHEVERELL AND BLUNT, WILFRID. 1956. *Great Flower Books, 1700-1900.* London, Collins.

They make the statement that the botanical draughtsman, Walter Hood Fitch, was somewhat of a parallel to Gould, in that Fitch had his hand in almost every botanical project during the half-a-century of Queen Victoria's reign, as Gould did in ornithology.

SITWELL, SACHEVERELL; BUCHANAN, HANDASYDE AND FISHER, JAMES. 1953. *Fine bird books 1700-1900.* London and New York, Collins and Van Nostrand.

This magnificent folio volume is not only beautifully and colourfully composed but also contains worthwhile historical information with astute evaluations of fine bird books.

The contents are in three sections. The first on 'Fine Bird Books' is by Sitwell. It is historical, with more pages devoted to Gould's volumes than any others, but Gould published more volumes than anyone else.

The second section is on 'The Bibliography' by Buchanan and Fisher. Interestingly they gave stars to the finer books, 'finely produced, well printed on hand-made paper, preferably finely bound', pictures well drawn, perfectly reproduced, and as near to life size as possible. Naturally, Gould's works received more stars than any others, but he, in fairness again, produced more than anyone else, and they were fine.

The third part of this volume is 'A List of Species Illustrated in the Plates' by Fisher.

A complete index concludes this beautiful-to-behold volume. The colour reproduction is the finest I have seen.

SKIPWITH, PEYTON. 1979. *The great bird illustrators and their art 1730-1930.* New York, A & W.

A fine book with excellent colour reproductions and an adequate text. A short historical sketch of each of the fifteen artists precedes a selection of one, or, in the case of John Gould, twelve, representative colour plates. Four of Lear's plates are also from Gould's works.

Skipwith writes that Gould was 'second only in importance to Audubon as a bird illustrator'. However, he is careful to point out that Gould only drew rough sketches as models for his artists.

Hart is called William, but there is no evidence that W. stood for William.

SLATER, EDERIC. 1973. 'John Gilbert – rediscovery of his descriptive bird and mammal notes'. *Royal Australasian Ornithologists' Newsletter* #17, p. 1.

This short note concerns the discovery in the British Museum (N.H.) of a large box of notes, manuscripts and letters written to John Gould primarily by John Gilbert, MacGillivray, Bynoe, Bennett, Gray, Krefft and others.

This valuable manuscript collection must be indexed and published.

[*See* [Gould, John] Box of manuscripts. [Several dates], and also Sawyer 1971.]

SMITH, ROGER C. AND PAINTER, R. H. 1966. *Guide to the literature of the zoological sciences.* 7th ed. Minneapolis, Burgess.

'This book is not primarily a bibliography, but a listing and brief discussion of the kinds of help, the time-saving sources of information, and the functions and uses of these aids in keeping informed.'

Under the section on 'Bibliography of the Zoological Sciences', beginning on page 52, there are short notes on Engelmann's 1846 volume, the Carus and Engelmann 1861 two-volume bibliographical reference, Taschenberg's seven volumes, and the Zoological Record. Many other references are of value for the amateur bibliographer.

SNOW, JOHN. 1936. *Snow on cholera, being a reprint of two papers by John Snow, M.D. together with a biographical memoir by B. W. Richardson, M.D. and an introduction by Wade Hampton Frost, M.D.* New York, Commonwealth Fund.

Dr Geoffrey Edelsten, the great-grandson of John Gould had told me that the Goulds moved from 20 Broad Street, Golden Square because of the cholera epidemic of 1854 traced by Dr Snow to the Broad Street pump. The family moved in 1859. However, the cholera epidemic may have been a contributing factor to this decision.

Aside from the move question, I began to investigate Dr Snow's papers because of interest in the Broad Street, Golden Square of 1854. The first article received on this Broad Street pump epidemic was by Barton (1959) to which I have alluded earlier.

I wanted to read the original Snow articles, both from my Gould interest and my medical interest. This volume of the reprints of two of his articles came to hand. The first article is in reality the 1855 second edition of Snow's booklet on cholera. To my surprise and thrill, as I was reading along about the Broad Street epidemic, here was a long reference to John Gould, Mr Prince and Gould's servant. From pages 52 to 53:

> I inquired of many persons whether they had observed any change in the character of the water, about the time of the outbreak of cholera, and was answered in the negative. I afterwards, however, met with the following important information on this point. Mr. Gould, the eminent ornithologist, lives near the pump in Broad Street, and was in the habit of drinking the water. He was out of town at the commencement

of the outbreak of cholera, but came home on Saturday morning, 2nd September, and sent for some of the water almost immediately, when he was much surprised to find that it had an offensive smell, although perfectly transparent and fresh from the pump. He did not drink any of it. Mr. Gould's assistant, Mr. Prince, had his attention drawn to the water, and perceived its offensive smell. A servant of Mr. Gould who drank the pump water daily, and drank a good deal of it on August 31st, was seized with cholera at an early hour on September 1st. She ultimately recovered.

Gould's discernment undoubtedly prevented him from getting cholera. I am also sure he had been exposed through current reading, or possibly personal contact, to Dr Snow's theory that cholera came from contaminated drinking water.

In this reprint, the map was also reproduced that showed the Broad Street pump area and the deaths in each home or shop (Figures 56 & 57).

For some additional insight into the severity of the epidemic, and the Broad Street of Gould's day, here are some additional quotes from Snow's article.

From page 38:

The most terrible outbreak of cholera which ever occurred in this kingdom, is probably that which took place in Broad Street, Golden Square, and the adjoining streets, a few weeks ago. Within two hundred and fifty yards of the spot where Cambridge Street joins Broad Street, there were upwards of five hundred fatal attacks of cholera in ten days. The mortality in this limited area probably equals any that was ever caused in this country, even by the plague; and it was much more sudden, as the greater number of cases terminated in a few hours. The mortality would undoubtedly have been much greater had it not been for the flight of the population. Persons in furnished lodgings left first, then other lodgers went away, leaving their furniture to be sent for when they could meet with a place to put it in. Many houses were closed altogether, owing to the death of the proprietors; and, in a great number of instances, the tradesmen who remained had sent away their families: so that in less than six days from the commencement of the outbreak, the most afflicted streets were deserted by more than three-quarters of their inhabitants.

There were a few cases of cholera in the neighbourhood of Broad Street, Golden Square, in the latter part of August; and the so-called outbreak, which commenced in the night between the 31st August and the 1st September, was, as in all similar instances, only a violent increase of the malady. As soon as I became acquainted with the situation and extent of this irruption of cholera, I suspected some contamination of the water of the much frequented street-pump in Broad Street, near the end of Cambridge Street.

From page 40:

I had an interview with the Board of Guardians of St. James's parish, on the evening of Thursday, 7th September, and represented the above circumstances to them. In consequence of what I said, the handle of the pump was removed on the following day.

The workhouse on Poland Street was nearby, and a Brewery on Broad Street. On page 48 is further information on the type of neighbourhood involved in the epidemic:

The limited district in which this outbreak of cholera occurred, contains a great variety in the quality of the streets and houses; Poland Street and Great Pulteney Street consisting in a great measure of private houses occupied by one family, whilst Husband Street and Peter Street are occupied chiefly by the poor Irish. The remaining streets are intermediate in point of respectability. The mortality appears to have fallen pretty equally amongst all the classes, in proportion to their numbers. Masters are not distinguished from journeymen in the registration returns of this district, but, judging from my own observation, I consider that out of rather more than six hundred deaths, there were about one hundred in the families of tradesmen and other resident householders.

The entire paper would be interesting reading for a student of the times.

The second paper that is reprinted in this volume is more scientific. Dr Snow struggled to convince other physicians and public officials of the truth of his theories on contagion of cholera from fecal contaminated water. One of the most popular theories at this time was that cholera was transferred through the air or by 'Humours'.

SOMERVILLE, J. D. 1939. 'Gould queries'. *Emu* 39: 149-52.

Somerville ends the confusion between Dr Thomas Braidwood Wilson of New South Wales and Dr Thomas Bellerby Wilson of Philadelphia (Cleland *Emu* 38: 536).

Somerville then attacks the problem of 'James' in an admirable detective-like manner (Hindwood *Emu* 38: 237). James Benstead was Gould's servant who went back to England with Gould, came back to Australia with George Grey (Chisholm *Emu* 38: 218) and married Mary Watson in 1841. These facts would also corroborate the statement by Cleland (*Emu* 38: 536) that Grey wrote Item 6 (Hindwood *Emu* 38: 237).

The problem of 'who was Mr. Dark' (Whitley *Emu* 38: 159) is approached next. Somerville says he was John Charles Darke, a surveyor, who was speared by Aborigines in 1844.

SOMERVILLE, J. D. 1941. 'Gould in South Australia'. *Emu* 40: 425.

In this Letter to the Editor, Somerville mentions three places in South Australia named after Gould: Gould's River, Mount Gould, and Gould's Range. These are situated between twenty-five and thirty kilometres north-east of Adelaide.

SOMERVILLE, J. D. 1942. 'Mount Gould'. *Emu* 42: 59.

Concerning information on Mount Gould in Western Australia. It was discovered and named for John Gould by F. T. Gregory in 1858, according to the evidence presented in this article by Somerville.

SOTHERAN, HENRY, LTD. 1862. A catalogue of upwards of fifty thousand volumes of ancient and modern books... by Willis and Sotheran, 136, Strand, London... 1862.

This catalogue was purchased by Ellis on 26 September 1936 for '0/2/0 cash'. This was the firm of G. Willis and H. Sotheran.

Three Gould sets were listed for sale, *Century of Birds* for £10.10s (published at £14.14s), the monograph *Trochilidae* with 300 plates, 20 parts (incomplete) at £46.10s (published at £63), and *Ramphastidae* (1833-35) for £4.10s (published at £6.5s).

This and all of the following Sotheran catalogues are in the University of Kansas Spencer Library–Ellis Collection.

SOTHERAN, HENRY, AND CO. 1884. A clearance catalogue of superior second-hand books... by Henry Sotheran and Co., 136 Strand and 2, Wellington Street... West-end establishment – 36, Piccadilly (opposite St. James's Church), London.

Ellis purchased this as a set of 3 parts, September, 1884 (A-De), October, 1884 (De-Mac), and December, 1884 (Mac-Z) with an additional small part on 'standard publications and remainders'.

A set of thirty-seven volumes of Gould's works in 'two elegantly carved cabinets with plate-glass doors' was listed for sale at £1000. This advertisement appears on the inside front-cover of the first part.

SOTHERAN, HENRY, AND CO. Correspondence. 1884. [Correspondence between H. Sotheran and Co., and the Australian Museum.] Collection: The Australian Museum, Sydney.

Here are four letters as follows: from H. Sotheran of 36 Piccadilly, London, 17 January 1884 offering some of Gould's works for sale to complete the museum's collection; from S. Sinclair, Secretary of the Museum to H. Sotheran, dated 7 March 1884 declining to purchase any volumes; and from E. P. Ramsay, Curator, to H. Sotheran, 7 August 1884, ordering, bound, Gould's *Birds of Asia, Odontophorinae, Trogonidae*

[no date], *Rhamphastidae* [sic] [no date], and *Handbook*. Audubon's *Birds of America*, original edition, was also ordered along with three other items. The final letter is from the Secretary, S. Sinclair, to H. Sotheran, dated 4 September 1884, and contained a bank draft for £47.4.6 for 'Books purchased by Mr. Ramsay when in London'. There was no listing of which books were purchased, but there was a check mark in front of all the Gould and other items listed in the letter of 7 August, but not in front of the Audubon item.

SOTHERAN, HENRY, AND CO. 1888. A catalogue of superior second-hand books ... Henry Sotheran and Co., 136, Strand & 2, Wellington Street, W.C. (next Waterloo Bridge), West-End House: 36, Piccadilly, W. ... London.

This volume contains two parts. The first, of 336 pages, has on pages 225 to 231 an elaborate section devoted to the sets and volumes of Gould's works for sale. Figures of carved cabinets housing his imperial-folio volumes are on pages 227 and 228.

A second small part of this volume, also dated 1888, where the 2 Wellington Street address has been removed, lists Gould's works for sale on the first twelve pages, also with illustrations of cabinets.

At the end of the volume is an illustration of the Sotheran firm at 36, Piccadilly.

SOTHERAN, HENRY, LTD. 1891. A catalogue of second-hand books ... 136, Strand, and 2, Wellington Street, W.C. ... West-end House: —36, Piccadilly, W. ... London ... originally established (in Tower Street, City) 1816.

Several catalogues are bound in this one volume in the Kansas-Ellis Collection. Gould material is listed in four of these, as follows:

22 October 1891 Catalogue, page 54: 'Wanted to purchase. Gould. Himalayan Birds.'

16 November 1891 Catalogue, page 24: 'The Birds of Great Britain ... and the Introduction to the Birds of Great Britain, 8vo, — 6 vol. folio and 8vo, £60'; Introduction to the Birds of Australia, £1.1s.; Introduction to the Birds of Great Britain 8vo, 15s.

11 December 1891 Catalogue, page 23: Gould Birds of Asia, 7 vol. £85. Gould and Sharpe, Birds of New Guinea, 25 parts [incomplete], 320 col. plates, 'cheap, £42'.

Christmas 1891 Catalogue, page 47: 'Asiatic birds: Fifty-four specimens (Hand-coloured after nature) selected from the Birds of Asia, ... impl. folio, half morocco ... £12.12s.'

SOTHERAN, HENRY, LTD. 1906. Catalogue No. 660. London, Sotheran.

With the publication of Sharpe's *Monograph of the Paradiseidae*, Sotherans say that Gould's 'magnificent series of works may now be considered complete, and justly described as the finest set of Natural History Books in the world'. They offer a set of forty-five volumes of Gould's works imperial folio with four other smaller volumes for £630.

On page 21 of this catalogue is the full-page announcement of Sharpe's *Monograph of the Paradiseidae*, limited to 350 copies, for £30 for the two bound volumes.

On the next page is a full-page ad for Sharpe's *Analytical Index* to Gould's works. 'The edition limited to 250 Small and 100 Large paper copies'. The small edition is priced at £1.16s, and the large edition at £4.4s.

SOTHERAN, HENRY, LTD. 1929. Catalogue No. 815. London, Sotheran.

This is the best illustrated of the Sotheran catalogues, except for 'Piccadilly Notes' Nos. 9 and 22. There are nine illustrations of Gould's plates on fine paper. The following sets are listed for sale in 1929:

Birds of Australia with Supplement, 8 volumes,	£235.
Synopsis of the Birds of Australia	£8.8s
A Century...	£21.
Birds of Europe, 5 volumes	£80
Monograph of the Trochilidae with Supplement, 6 volumes	£75
Trogons, second edition (1875)	£18.18s
Birds of Great Britain, 5 volumes	£73.10s

[SOTHERAN, HENRY, LTD] [c. 1930] 'Separate Plates/From Mr. John Gould's/Famous & Beautiful Work on/The/Birds/of/Great Britain/Size 21½ x 14½/Beautifully coloured by hand/offered for Sale at/4/- each/by/Henry Sotheran, Ltd./.../43 Piccadilly, London W. 1.'

On the cover is a stamp: 'Received by/Ralph Ellis/Oct. 27, 1930.' It is in the Ellis Collection at the University of Kansas.

This is an unbound pamphlet of sixteen pages. The size is 150 x 100 mm.

[SOTHERAN, HENRY, LTD] [c. 1930] 'Separate Plates/From Mr. John Gould's/Famous and Beautiful Work on/The Birds/of/Europe/Size 21½ x 14½ in./Beautifully coloured by hand./Offered for sale at/Prices ranging from 3/- each:/By/Henry Sotheran, Ltd. /Booksellers and Bookbinders to H.M. the King,/43 Piccadilly, London, W. 1.'

On the cover of this unbound pamphlet of sixteen pages is a stamp: 'Received by/Ralph Ellis/Feb. 1, 1931'. It is in the Ellis Collection at the University of Kansas. The size is 153 x 100 mm.

SOTHERAN, HENRY, LTD. 1933. 'Piccadilly Notes', No. 1. London, Sotheran.

This apparently is the first issue of 'Piccadilly Notes' and was published from 43 Piccadilly, London, W.1. It was edited by J. H. Stonehouse and contained his first essay on 'Adventures in Bookselling'.

There is nothing concerning Gould in this catalogue.

SOTHERAN, HENRY, LTD. 1933. 'Piccadilly Notes', No. 7. London, Sotheran.

This 1933 catalogue contains an index to the first seven issues of 'Piccadilly Notes'.

In a rather complete examination of these first seven 'Notes' nothing was noted on or about Gould.

SOTHERAN, HENRY, LTD. 1934. 'Piccadilly Notes', No. 8. London, Sotheran.

On page 13 is an announcement of:

Now on view/an exhibition/of several hundred magnificent/original water-colour drawings/for/Gould's Humming Birds,/Birds of New Guinea, etc./also for/Sharpe's Birds of Paradise./Prices from £1.1s to £4.4s each./

Also beautiful hand-coloured plates to Gould's/Birds of Great Britain, and the Humming Birds./Price 4s each./Plates to the Birds of Europe./Price 3s to 10s 6d each/ ...

Full-page illustrations of two of the original drawings accompany this announcement.

PART FOUR BIBLIOGRAPHY OF JOHN GOULD, HIS FAMILY AND ASSOCIATES

Figure 76 Cover for Sotheran's 1934 'Piccadilly Notes' No. 9.

SOTHERAN, HENRY, LTD. [1934]. Catalogue No. 9 'Piccadilly Notes'. London, Sotheran.

This is the first of two paper-bound catalogues issued by Henry Sotheran Ltd., London devoted almost entirely to Gould's works, separate plates, and original drawings (Figure 76). In 1881 Sotheran purchased from John Gould's executors the copyright and whole remaining stock of his works.

There is a portrait of Gould and a five-page text written by John Harrison Stonehouse, with illustrations, entitled 'Adventures in Bookselling – John Gould, "The Bird Man" '. On the last text page is a listing of seven stages in the sequence of development of the finished hand-coloured lithographs from Mr Gould's rough sketches.

Then follow forty-six pages, with illustrations, of a listing of the items for sale, including 'a collection of one thousand four hundred and seventy original water-colour drawings, pencil drawings and sketches by John Gould, H. C. Richter and William Hart...' for £350 [US$1400.00]. [The majority of this item is intact in the Ellis Collection at the Spencer Library at the University of Kansas.]

Also for sale were original bird drawings by Mrs Gould, seventy-four original botanical drawings by Mrs Gould, a MS. diary by Mrs Gould while in Australia, original drawings for a contemplated monograph on the Cassowary, more original water-colour drawings by W. Hart, Edward Lear, H. C. Richter and G. Krefft, etc. On page 118 is the comment 'Many of these fine drawings, [even] though somewhat defaced by Mr. Gould's corrections and remarks'. [! Oh, the envy of hindsight!]

Also included in the catalogue were hundreds of separate plates from Gould's works, and complete sets of essentially every one of his publications. For instance, the eight-volume *Birds of Australia* and *Supplement* was priced at £175 [around $700], the five-volume *Birds of New Guinea* at £105 [$420], the five-volume *Birds of Great Britain* at £45 [$180] and the six-volume set *Trochilidae* in the original parts [!] for £42 [$168].

The concluding pages of this booklet list other important natural-history books for sale, original drawings by Seebohm, Keulemans, E. Albin, autographed letters, etc.

The booklet has a total of ninety-four pages numbered 66 to 160.

Apparently several printings were made of this No. 9 catalogue, one copy in my collection having 'Price two shillings and sixpence' on the bottom of the front cover.

For completeness sake, an index according to the item number follows for John Gould and his associates. This will cover only the last pages of this 'Piccadilly Notes', from pages 124 to 160. The first part of the catalogue from pages 66 to 123 is entirely devoted to Gould and thus does not need to be indexed. This second part is entitled 'Other Important Natural History Books'.

John Gould items 2075, 2083, 2103, 2115, 2164 and an illustration on the back cover of an 'original drawing by John Gould'; W. Hart item 2055; Edward Lear item 2167; R. B. Sharpe items 2050, 2069, 2107, and 2110; and Joseph Wolf items 2050, 2075, 2077, 2116, 2161, 2163, 2164 and 2167.

SOTHERAN, HENRY, LTD. 1934. 'Piccadilly Notes', No. 10. London, Sotheran.

From page 200 to 207 is a section referring to No. 9 'Piccadilly Notes' and offering for sale for £63 a collection of 22 water-colour drawings, 13 coloured proofs, 6 uncoloured proofs, and 22 coloured plates for Gould's *Pittidae*.

Five Edward Lear letters (four to Gould) are also offered for sale from £1.1s to £3.3s.

SOTHERAN, HENRY, LTD. 1934. 'Piccadilly Notes', No. 12. London, Sotheran.

On page 359 is the announcement that 'owing to the continued success of the exhibition of the original drawings and beautiful hand-coloured plates from Gould's birds, etc. which we have been holding in our first floor gallery, we have decided to keep the Show open for the Christmas Season'.

Also here is a statement that Henry Sotheran & Co. in 1881 purchased not only all of Gould's copyrights and the complete stock of his publications, but also 'the house that he lived in'.

SOTHERAN, HENRY, LTD. 1934. 'Piccadilly Notes', No. 13. London, Sotheran.

This issue contains an index to the issues of 'Piccadilly Notes' eight to thirteen.

PART FOUR BIBLIOGRAPHY OF JOHN GOULD, HIS FAMILY AND ASSOCIATES

SOTHERAN, HENRY, LTD. [*c.* 1934] Leaflet. 'The Resplendent Trogon'. London, Sotheran.

This four-page leaflet describes the discovery of several unfolded double-sized plates of the Resplendent Trogon [Quetzal] and offers them for sale for £2.2s each. 'Mr. Gould's first rough Sketch for the Plate is also for sale, price £4.4s framed.'

This leaflet is in the Kansas–Ellis Collection.

SOTHERAN, HENRY, LTD. 1935. 'Piccadilly Notes'. London, Sotheran.

No. 14 has over three pages devoted to Gould's works. No. 15 on page 123 has an advertisement of one hundred framed Gould's birds on view in the Piccadilly Bird Room priced from three shillings upwards unframed. No. 16 has listed on page 165 the five volume *Birds of Great Britain* for £55, the five-volume *Trochilidae* for £30, and the *Supplement to Trochilidae* for £21, stating that the latter is scarce.

In this latter catalogue, the article by Stonehouse entitled 'Adventures in bookselling: 36 Piccadilly' is biographical.

SOTHERAN, HENRY, LTD. 1936. 'The Piccadilly Register'. London, Sotheran.

The first issue of this catalogue was published from 43 Piccadilly in 1936.

SOTHERAN, HENRY, LTD. 1936-37. 'Piccadilly Notes'. London, Sotheran.

Catalogue No. 19 records the change of address from 43 Piccadilly, London W.1, to Nos. 2, 3, 4, & 5 Sackville Street, W.1. Here also on pages 381 to 383 is a separate listing of some W. Hart original water-colour drawings for Gould's *Birds of New Guinea* and Sharpe's *Monograph of the Paradiseidae*. The prices range from £1.1.0 to £5.5.0.

No. 20 has an illustration of the new Sackville building.

No. 23 was apparently the last 'Piccadilly Notes'.

SOTHERAN, HENRY, LTD. [1937]. Catalogue No. 22 'Piccadilly Notes'. London, Sotheran.

No. 22 is the second sale catalogue issued by Henry Sotheran, Ltd devoted to the listing of material from John Gould's original stock purchased after his death in 1881 (Figure 77).

This publication consists of five pages of text on 'The migrations of Gould's birds and the dangers which beset them' written by J. H. Stonehouse, concluding with the statement that the majority of the 'drawings & the other original matter connected with the inception and progress of Mr. Gould's great series of works in natural history have been sold to another American Collector, and are on their way to America'. [This American Collector was Ralph Ellis, and this very valuable and irreplaceable collection is now in the Spencer Library at the University of Kansas.]

Then follows a list of separate hand-coloured lithographs available from Gould's works, varying in price from 3s [US 60 cents] to £1.10s [$6.00]; a few remaining original water-colour drawings by John Gould, Mrs Gould, and W. Hart; autograph letters by Lear and others, and other natural-history works.

This booklet has a total of fifty-four pages numbered 150 to 204. An index is not needed since this No. 22 'Piccadilly Notes' is 99 per cent Gouldiana.

SPILLETT, PETER G. 1972. *Forsaken settlement*. Melbourne, Lansdowne.

A most interesting account of the Royal Marine settlement at Victoria, Port Essington from 1838 to 1849. The weather was the main element that created unusual hardships

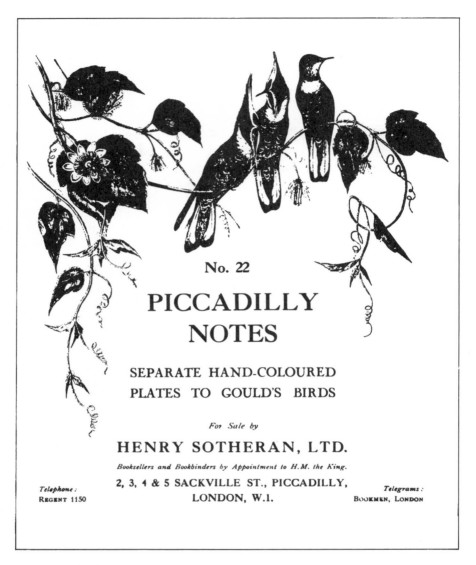

Figure 77 Cover for Sotheran's 1937 'Piccadilly Notes' No. 22.

for the military attachments and visitors, and eventually caused abandonment of the fort.

John Gilbert arrived at Port Essington on 12 July 1840 on the *Gilmore*. On 1 October, he and the botanist John Armstrong travelled on the *Lulworth* to Coepang for a month. On 15 March 1841 Gilbert left Port Essington on the *Pelorus* for Singapore and then took another boat to England.

During the eight months that Gilbert was in the Port Essington-Coburg Peninsula area he collected for John Gould 'over 200 specimens of birds, and numerous specimens of fish, reptiles and mammals'.

The exhausted and emaciated Leichhardt Expedition group arrived at Port Essington on 17 December 1845. John Gilbert had started out with the group, leaving Jimba on 1 October 1844 for their journey of exploration to Port Essington. On the night of 28 June 1845 Gilbert was speared during an attack on the camp by Aborigines.

PART FOUR BIBLIOGRAPHY OF JOHN GOULD, HIS FAMILY AND ASSOCIATES

SPINAGE, C. A. 1968. *The book of the giraffe.* London, Collins.

This book was examined at the Zoological Society of London when I was there in October 1974. On page 95 there is information about George IV's giraffe, stating that it arrived in London in August of 1827 and lived just over two years at Windsor. When the giraffe died, the skin and skeleton were given to the Zoological Society.

A portrait of the giraffe was painted by James Laurent Agasse and this oil painting was hanging at the Zoological Society in 1974.

On page 14 it states that Sir Everard Home dissected the giraffe. There is nothing in this book about Gould and his involvement in the stuffing of the giraffe.

STEVENS, LYLA. [No date] *Birds of Australia in colour.* Melbourne, Whitcombe and Tombs.

Gould is mentioned in reference to the author's comments about parrots. This is a quarto-sized book of sixty-one pages with mediocre drawings of birds in colour.

STONE, WITMER. 1899. 'Some Philadelphia ornithological collections and collectors, 1784-1850'. *Auk* 16: 166-77.

Several paragraphs relate the story of the purchases by Dr Thomas B. Wilson, through his brother, Edward Wilson residing in London, of the Prince Massena collection, the Gould collection of Australian birds and the Boys collection of Indian birds. [*See also* Meyer de Schauensee, Rodolphe. 1957.]

STONE, WITMER. 1899. 'A study of the type specimens of birds in the collection of the Academy of Natural Sciences of Philadelphia with a brief history of the collection'. *Proc. Acad. Nat. Sciences of Philadelphia* 51: 5-92.

Preceding the listing of the type specimens is an historical account of the ornithological acquisitions for the Academy. The story is related of the purchase of Gould's Australian birds by Dr Thomas B. Wilson. The number of birds from Gould is listed as 2000.

When the type specimens are listed in the article, however, Stone states that 'a complete manuscript catalogue of the Gould Australian specimens has been prepared, and may be published at some future time'. [*See* Stone with Mathews (1913) for this list.]

The only Gould type specimens listed as being in the Academy collection in the present article, are 'his types from localities other than Australia'. They number twelve skins for nine species.

Of related interest is the mention by Stone of the purchase by Dr Wilson of a collection of 1000 Indian birds from Captain Boys, of the British army. The Boys collection had been on loan to Gould for use in preparing his *Birds of Asia*. [*See* [Gould, John.] Manuscript. [No date]. Boy's [sic] Birds.]

STONE, WITMER. 1938. 'The Gould Collection in Philadelphia'. *Emu* 38: 231-3.

When, in 1847, Gould was unable to persuade the British Museum to purchase his Australian bird collection, he accepted an offer of £1000 from Dr Thomas B. Wilson of Philadelphia for them. Stone has written about and studied these skins carefully.

STONE, WITMER, WITH G. M. MATHEWS. 1913. 'A list of the species of Australian birds described by John Gould, with the location of the type-specimens'. *Austral Avian Record* 1: 129-80.

The great majority of Gould's type specimens of Australian birds are in the Academy of Natural Science of Philadelphia, having been purchased by Dr Thomas B. Wilson in 1847, and given to the Academy. Some of Gould's types are in the British Museum (Natural History), in Lord Derby's collection now in Merseyside County Museum in Liverpool, and a very few remaining are scattered in other collections.

These 'types' are listed by Stone and Mathews in this article as 'final'. Subsequently, more recent authorities have disagreed with some of these arbitrary selections of 'types' [Meyer de Schauensee 1957].

STRESEMANN, ERWIN. 1975. *Ornithology from Aristotle to the present.* Cambridge, Harvard University Press.

One of the most pleasant and rewarding aspects of my search for Gouldiana has been the unexpected 'discovery' of a book or letter or drawing that really added spice to my study. Such a find was this volume by Stresemann. It was edited by G. William Cottrell and expanded for American ornithology by Ernst Mayr. As one who has had almost no contact with professional ornithologists, I consumed this volume. And it provided real satiation of my appetite for a more scientific background of ornithological knowledge, especially related to the historical perspectives.

The ornithology and ornithologists of the 1800s are particularly well covered. Stresemann makes for pleasant reading, but this small volume also contains a staggering amount of information.

After reading this book, Gould, and also Audubon, seem somewhat insignificant when related to the overall field of ornithology. This is particularly true when one is made aware of the massive strides of scientific knowledge that developed in the field of ornithology in the 1900s. But in the 1800s the works of Gould and Audubon had an important, worthwhile and also long-term impact. Their main raison d'être, like that of Roger Tory Peterson of the 1900s, was, by their beautiful and dramatic depictions of birds, to provide a much needed stimulus in an awakening time for a general public who were beginning to thirst for scientific information.

References to Gould occur on pages 121, 144, 197, 210, 225, 243, 253, and 312.

The book is packed full with information and anyone interested in the development of ornithology in the 1800s will profit greatly by referring to this worthwhile source.

STRICKLAND, HUGH EDWIN. 1855. *Ornithological synonyms.* Edited by Mrs Hugh E. Strickland and Sir W. Jardine. London, Van Voorst.

This work is frequently mentioned in the Gould-Strickland correspondence collection in the University Museum of Zoology at Cambridge. Unfortunately Strickland was killed by a train before it was published.

Volume one is concerned with Accipitres and no further volumes were published.

STRONG, REUBEN MYRON. 1939-46. *A bibliography of birds.* 3 vols. Chicago, Field Museum of Natural History.

On pages 364-5 is a listing of twelve items by John Gould, two of which are only articles by him. Thus, this is a very incomplete bibliography.

STRZELECKI, SIR PAUL EDMUND DE. 1845. *Physical description of New South Wales and Van Diemen's Land.* London, Longman, Brown, Green, & Longmans.

The Libraries Board of South Australia, Adelaide, in 1967 produced a facsimile edition of this work as No. 19.

Strzelecki begins his book with a chapter on marine and land surveys, then covers the geology, climatology, botany, zoology, Aborigines and the agriculture of these two parts of Australia.

In the section on recent fauna he states that 'with the following list of mammals, inhabiting New South Wales, Van Diemen's Land, and Bass's Straits, as also with that of Birds, I was kindly favoured by John Gould, Esq. F.R.S.' The list by Gould of 'Birds of New South Wales and Van Diemen's Land' is on pages 317-29. Strzelecki refers the reader to Gould's 'splendid works', for further information.

A footnote on page 332 consists of an extensive quotation about Gould's works, and especially his 'princely work', the 'Birds of Australia'. This quote is from the 1841 'Tasmanian Journal of Natural Science, etc.' No. II.

STUART, JOHN MCDOUALL. 1865. *Journals of John McDouall Stuart during the years 1858-62.* London.

Not available to me but discussed in Cleland's 1937 article (*Emu* 36: 217-21). In the appendix to the volume is an excerpt from Gould's article of 9 June 1863 (*Proc. Zool. Soc.*) 'On a collection of birds from Central Australia'. Gould had been forwarded some of the birds collected by Frederick G. Waterhouse on Stuart's last journey and this article listed seventeen of them.

STURT, CHARLES. 1849. *Narrative of an expedition into Central Australia ... during the years 1844, 5 and 6 ... 2 vols.*

Not examined by me but referred to in Cleland's 1937 article (*Emu* 36: 206). Sturt with his companions, Browne and Stuart, sent Gould several birds from this expedition.

[STURT, CHARLES.] Review. 1849. 'Narrative of an expedition into Central Australia during the years 1844-5 and 6, etc. by Captain Charles Sturt, F.L.S.: with a botanical appendix by Robert Brown, D.C.L., F.R.S., F.L.S., and ornithological notices by John Gould, F.R.S.' *Ann. Mag. Nat. Hist.* 3 (second series): 56-61.

The majority of the review is concerned with comments on the botanical collections. The author is A.W.

Volume II contains several illustrations of birds by Gould and Richter. The reviewer wishes that Gould would publish his other works in octavo volumes, 'such a series of volumes would find an entrance where his larger works could never be seen'.

A final paragraph comments on the German publication of Gould's *Ramphastidae* on a reduced scale and adds that 'it is a pity that so spirited and talented a man should not have all the results of the profit of such books'.

STURT, MRS NAPIER GEORGE. 1899. *Life of Charles Sturt.*

This book has not been examined by me but is mentioned in Cleland's 1937 article (*Emu* 36: 205). Cleland quotes a paragraph in reference to Sturt's drawings Gould had admired, and which were later stolen.

SUTTON, J. 1929. 'Letters of John Gould to F. G. Waterhouse'. *S. Aust. Ornith.* 10: 104-15.

Sutton reproduces the complete contents of twelve letters written by John Gould to F. G. Waterhouse who was, at that time, Curator of the Museum of the South Australian Institute. These letters were found in the South Australian Museum, and are

mainly concerned with pleas for any new specimens, descriptions of some specimens that were sent to Gould, and mention of exchanges which Gould sent to Waterhouse. The dates are from 23 September 1863 to 30 September 1875.

Frederick George Waterhouse (1815-98) was the brother of G. R. Waterhouse who was Curator of the Zoological Society of London in 1836, and from 1843 to 1880 was in the Geological Department of the British Museum. He is referred to in letter No. 12, page 113.

SWAINSON, WILLIAM. 1840. *Taxidermy, with the biographies of zoologists and notices of their works.* London, Longman, Orme, Brown, Green and Longman's.

A reference to Gould appears on page 70 under a 'List of the most eminent zoological painters and engravers... [and] the names of the principal publications in which their designs appear... 3. Birds – England... Gould (various splendid publications)'.

Then follows a section entitled 'Part II A bibliograph of zoology with biographical sketches of the principal authors'.

Gould is not listed in this part but is listed in an appendix which begins on page 383:

The following names have been accidentally omitted in the regular alphabetical series...
Gould, John — Ornithology
A zealous and very able ornithologist, now travelling in Australia, who has published some valuable, although very expensive, works upon birds; the chief of these are —
1. Century of Birds from the Himalaya Mountains. 1 vol. folio. London, 1832-3.
2. Birds of Europe. 16 parts, at three guineas each.
3. Ramphastidae, Monograph of the 3 parts.
4. Trogonidae. Ditto.
We trust the author will hereafter reprint these expensive volumes in such a form as that they may be accessible to naturalists; and thereby diffuse science, instead of restricting it to those only who are wealthy.

[SWAINSON, WILLIAM.] Anonymous. 1900. Editorial: Swainson's correspondence. *Osprey* 5: 29-30.

This is an announcement of the paper by Dr Albert Günther which appeared in the *Proceedings of the Linnean Society of London* (1900) where he lists the correspondence of William Swainson from 1806 to 1840. [*See* Günther 1900.]

SWANN, C. KIRKE. 1956. 'Natural history books from a bookseller's point of view'. *J. Soc. Biblphy. Nat. Hist.* 3: 117-26.

A personal and informative dissertation on the specialty of natural-history bookselling, given by one of the Directors of Wheldon and Wesley, Ltd.

Mention is made of the sale of the Sturm brothers German edition of Gould's *Monographie der Ramphastiden.* Two letters from the Sturms to Gould were included. [Ralph Ellis bought this set, and it is now in the University of Kansas Spencer Library.]

SWANN, C. KIRKE. 1972. 'Natural history bookselling'. *J. Soc. Biblphy. Nat. Hist.* 6:118-26.

The sharp rise in prices for desirable books because of an increasing demand and inadequate supply is illustrated. Gould's *Birds of Great Britain* escalated from £350 in 1962 to £3500 in 1972. [In 1977 a set sold for £6000!]

TASCHENBERG, O. 1887-1923. *Bibliotheca zoologica II.* 8 vols. Leipzig, Engelmann.

Band 5, 1899 contains the monumental compilation of works relating to Aves and Mammalia. On pages 3797-8 (Aves) and page 4417 (Mammalia) are the main references to Gould's folio works and to several of his articles.

TATE, PETER. 1979. *A century of bird books.* London, Witherby.

A unique compilation, with notes, of books on birds published in Britain since 1875. Of especial value are the numerous photographs of authors, including John Gould, R. B. Sharpe, and Alfred Newton.

Gould's works are mainly included in the section on Fine Bird Books. Three folio works were published, in part, after 1875, namely *Birds of New Guinea, Supplement to Trochilidae* and *Pittidae.*

The book is not indexed! Gould references are on pages 18, 19, 20, 23, 32, 71, 194, and 234.

THAYER, EVELYN AND KEYES, VIRGINIA. 1913. *Catalogue of a collection of books on ornithology in the library of John E. Thayer.* Boston, private.

Eleven of Gould's folio works and four small works are listed.

THOMAS, G. ROSS. 1934. *Gould league songs and poems.* Sydney, New South Wales Gould League of Bird Lovers.

A diversified seventy-five page booklet containing numerous illustrations relating to birds (with two coloured plates by N. W. Cayley), also poems, songs, and several short articles by various authors.

A prefatory article on 'The Bird Man' is by Tom Iredale. Iredale writes some interesting vignettes about Gould, Mrs Gould and Vigors. The statements 'Mrs. Gould ... had painted some botanical pictures which Vigors had seen' and Vigors 'put all his power into the project of publishing paintings of these Indian [Himalayan] birds by Mrs. Gould' are intriguing, but I know of no source for them.

THORPE, ADRIAN. 1975. *The birds of Edward Lear.* London, Ariel Press.

A large 22-by-15 inch (559 x 381 mm) thin volume with twelve beautiful colour plates produced by Lohse of Frankfurt am Main, Germany. Thorpe selected two macaws, one toucan, three owls, and other fine assorted examples of Lear's work.

Lear's uncomplimentary comments about Gould are repeated. Thorpe also states that Lear did ten out of the thirty-three plates for Gould's monograph *Ramphastidae* and sixty-six of the 448 plates in Gould's *Birds of Europe.*

TROUGHTON, ELLIS. 1951. *Furred animals of Australia.* 4th ed. Sydney, Angus and Robertson.

An example of one of the several books on natural history of Australia which contain many references to Gould and Gilbert.

TURNER, W. J. 1946. *Nature in Britain.* London, Collins.

'The Birds of Britain' section was written by James Fisher and is identical with that in Fisher's 1942 'The Birds of Britain' book. The illustrations are decreased in number in this 1946 book.

TURNER, W. J. [No date] *A treasury of English wild life.* New York, Hastings House.
This volume is identical to Turner's 1946 'Nature in Britain', except for title.

V [VERREAUX, JULES.] 1847. 'Manuscript catalogue of the Gould collection'. Collection: Acad. Nat. Sciences of Philadelphia, #53.
Written in long hand in French, Verreaux, with others, lists 1847 specimens of birds on fifty pages. Included in columns is the generic name, species, sex, and locality of the specimen.
The list appears to be written by at least three different individuals.
Under number 1599 is the statement 'no Bird with this number', so the final total is 1847 specimens, not 1848 as listed by Meyer de Schauensee (1957). Even this figure of 1847 might be incorrect, since six sheets near the end of the list are damaged.

VIGORS, N. A. 1830. [Exhibition of several species of birds, apparently undescribed, from the Himalayan mountains.] Proc. comm. science and corresp. Zool. Soc. London. No. 1: 7-9.
'These [specimens] formed part of a collection which Mr. John Gould, A.L.S., had lately received from India, and of which he intended to publish coloured illustrations, to the number of one hundred figures. Several of the plates, representing some of the most interesting of the species, were laid upon the table.' Eleven species were named by Vigors. This presentation occurred at the meeting of 23 November 1830.
These earliest proceedings of the Zoological Society include only pages 1 to 44 and are bound in a volume of pamphlets in the University of Kansas Spencer Library-O'Hegarty Collection C595.

VIGORS, N. A. 1830. 'An address delivered at the sixth and last Anniversary Meeting of the Zoological Club of the Linnean Society of London, on the 29th of November, 1829'. *Mag. Nat. Hist.* 3: 201-26.
Valuable for the insight provided on the status of natural history, especially ornithology, prior to 1830.
There are many references to the *Zoological Journal* and the *Linnean Transactions.*
Gould's discovery of a new warbler (*Sylvia tithys*) for the British fauna is mentioned on page 213. Gould is referred to as a 'rising naturalist [with] critical knowledge'.

VIGORS, N. A. 1830. [Mr. Vigors exhibited several species of Humming-birds from the collection of Mr. John Gould.] *Proc. comm. science and corres. Zool. Soc. London.* No. 1: 12.
Presented at the meeting of 14 December 1830, Vigors reported that Gould named a new species for Mr George Loddiges as *Trochilus Loddigesii,* Gould.

VIGORS, N. A. 1831. [On birds from the Himalayan mountains.] *Proc. comm. science and corres. Zool. Soc. London.* No. 2: 22-3.
Thirteen new species were named at this meeting of 11 January 1831.

VIGORS, N. A. [On birds from the Himalayan mountains.] *Proc. comm. science and corres. Zool. Soc. London.* No. 3: 35.
At the meeting of 8 February 1831 Vigors named nine species.

PART FOUR BIBLIOGRAPHY OF JOHN GOULD, HIS FAMILY AND ASSOCIATES

VIGORS, N. A. 1831. [On birds from the Himalayan mountains.] *Proc. comm. science and corres. Zool. Soc. London.* No. 4: 41-4.

At this meeting of 22 February 1831 Vigors named three genera *Lanus, Collurio* and *Hypsipetes* and named nine species. One *Cinnyris Gouldiae* was named for 'Mrs. Gould, who executed the plates of these Himalayan birds'.

VON MÜLLER, DR BARON J. W. 1852. 'Balaeniceps rex, Gould'. Jardine's *Contributions to Ornithology*, 4: 91-92.

Dr Von Müller wrote in *Naumannia* of 1852 concerning this bird newly described by Gould. This is an abstract of that article.

VOSPER, ROBERT. 1953. 'A report on books and libraries in Kansas'. *California Librarian* 14: 1-3.

This is a short informative article on the wealth of several Kansas [and Missouri] libraries written by the then new University of Kansas Librarian. In essence he is trying to explain to his California friends why he chose to come to Kansas. [In a few years the California pull was too great and Vosper returned there.]

There is a short note on the Ellis collection with its 'vault-full of proof-sheets, original drawings, and extra runs' of Gould material.

VOSPER, ROBERT. 1961. 'A pair of bibliomanes for Kansas: Ralph Ellis and Thomas Jefferson Fitzpatrick'. *Papers of the Bibliographical Society of America* 55: 207-25.

Mr Vosper sifted through twelve file drawers of Ralph Ellis's voluminous and repetitive correspondence, notes, and business records and has written an interesting and awesome account of Ellis's bibliomania. [I know the extent of this Ellis collection because I did the same thing more recently to index at least some of it [Sauer 1976 'Summary and index...']]

The disease of bibliomania in Ralph Ellis began insidiously and fell upon a likely subject. The patient was sickly, spoiled, and eccentric. He was doted upon by wealthy parents. The disease did not run its full course however, because the patient succumbed at the early age of 37. [It has been thought however, that the disease has been passed on to others who have come into contact with the fomites.]

In the course of the disease, due to a virus according to Vosper, the Gould estate owned by Henry Sotheran of London was rediscovered. Ellis was there and, to his great credit, purchased almost all of the original sketches and drawings by Gould and his staff, thus keeping the Gould collection intact. The story of its arrival in Kansas is fascinating.

WADE, PETER, EDITOR. 1975. *Every Australian bird illustrated.* London, Hale. **W**

'The only book ever produced to illustrate all Australian birds in full colour action photographs and paintings.' A beautifully produced book with outstanding colour reproductions of excellent photographs, along with many paintings by John Gould and his artists (on 73 pages), Gregory Mathews' artists (on 26 pages), Peter Trusler (8 paintings), Margo Kröyer-Pedersen (7), Gracius Broinowski (4) and Neville H. P. Cayley (2 paintings).

WAGSTAFFE, R. 1978. *Type specimens of birds in the Merseyside County Museums.* Liverpool, Merseyside County Museum.

Two major collections form the nucleus of the bird skin collection in this Liverpool Museum, namely those of the 13th Earl of Derby, and Canon H. B. Tristram.

Lord Derby, through his friendship with John Gould, and thus John Gilbert, acquired by purchase, gift, or exchange many Gould specimens. Multiple references to Gould appear on pages 1, 6, 7, 8, 9, 10, 11, 14, 15, 16, 21, 22, 23 and 25. A syntype named by R. B. Sharpe is *Pelargopsis gouldi* on page 13.

Peter Morgan supplied a historical introduction to this short work.

WAGSTAFFE, R. & RUTHERFORD, G. 1954-55. *Letters from Knowsley Hall, Lancashire.* North Western Naturalist.

A small black-bound volume holds this series of articles. They were published as follows:
June 1954, part 1, pp. 173-83 in volume 25
September 1954, part 2, pp. 353-64
December 1954, part 3, pp. 487-503
March 1955, part 3 (cont.), pp. 9-22 in volume 26
June 1955, part 3 (cont.), pp. 169-72
September-December 1955, part 4, pp. 353-82.

Included is correspondence between Lord Derby and John Latham (13 letters), John Gould to Lord Derby (16 letters from 21 June 1830 to 18 January 1851, just before Lord Derby died), John Gilbert to Gould (3 letters, dated 13 September 1842, 27 March 1843 and 18 April 1842, which Gould loaned to Lord Derby), one letter from Captain Charles Sturt to Gould (also loaned by Gould), and the remaining 48 letters are from John Edward Gray to Lord Derby.

Chisholm (1964 *Emu* 63, parts 4&5) writes critically concerning the publication of these letters without consultation with Australian naturalists or historians. Chisholm's article must be read in conjunction with the Wagstaffe-Rutherford articles because important points are corrected or clarified by Chisholm.

WAINWRIGHT, M. D. AND MATTHEWS, NOEL. 1965. *A guide to western manuscripts and documents in the British Isles relating to South and South East Asia.* London, Oxford University Press.

The one listing for Gould is in the British Museum (N.H.) under '[1835-37]. Copies and originals of correspondence of Brian H. Hodgson (1800-94), Resident in Nepal 1820-44, with the Asiatic Society of Bengal, J. Gould, and others, concerning a projected work on the zoology of Nepal [1835-37] 46ff (Zoology)'.

[A typed copy has been made of the Hodgson-Gould material. *See also* [Hodgson, B. H.] Manuscript. 1837.]

WAKEMAN, G. 1973. *Victorian book illustration.* Newton Abbot, David and Charles.

On page 48 is the statement 'John Gould showed his method of reproducing the metallic colouring of humming birds' at the 1851 Exhibition.

WARREN, RACHEL L. M. 1966. *Type-specimens of birds in the British Museum (Natural History).* Vol. 1, Non-passerines. London, British Museum (N.H.).

References to the Gould types (1841) from Darwin's voyage in H.M.S. *Beagle* are made on page iv of the Introduction.

On page v are two paragraphs on John Gould stating:

PART FOUR BIBLIOGRAPHY OF JOHN GOULD, HIS FAMILY AND ASSOCIATES

Gould's famous collection of hummingbirds was acquired in 1881, many of them still mounted in the glass cases which had been displayed in the Crystal Palace* at the Great Exhibition of 1851. Gould rarely designated a holotype, and the syntypes of his species have mainly been selected from his study-skins, many of which are labelled in his own handwriting. References to particular specimens are often given in his *Monograph* (1849-1887), after the publication of the original descriptions, but there is still some doubt about the status of some of the supposed type-specimens.

The Museum acquired much of the rest of Gould's collection between 1837 and 1881, but his Australian Collection was sold to the Academy of Sciences, Philadelphia, in 1847 (Schauensee 1957), and there are syntypes of some species in both museums.

In 1971, vol. 2, Passerines was published by Warren and C. J. O. Harrison, and in 1973 vol. 3, a systematic index was published.

WATERHOUSE, CHARLES OWEN. 1902. *Index Zoologicus*. London, Zoological Society of London.

This is 'an alphabetical list of names of genera and sub-genera proposed for use in zoology as recorded in the "Zoological Record" 1880-1900'.

S. H. Scudder in 1882 issued the 'Nomenclator Zoologicus' containing the names in the *Zoological Record* volumes 1-16, from 1864 through 1879.

The years 1901-10 are covered in the 'Index Zoologicus No. II'.

See also Journal. *Zoological Record*.

WATERHOUSE, FREDERICK HERSCHEL. 1885. *The dates of publication of some of the zoological works of the late John Gould, F.R.S.* London, R. H. Porter.

The dates are taken from the original covers of the parts as published, but only the year.

The following works are included in this tabulation: *Mammals of Australia; Macropodidae; Birds of Great Britain; Birds of Asia; Birds of Australia* and *Supplement; Trochilidae* and *Supplement*, parts I-IV (not completed); *Odontophorinae;* and *Trogonidae* (revised edition). The species treated in each are listed alphabetically by genera. Following each specific name are given in tabular form the volume and plate-numbers, the number of the original part, and the year of issue.

Waterhouse was Librarian to the Zoological Society of London.

WATERHOUSE, F. H. 1889. *Index generum avium, a list of the genera and subgenera of birds.* London, Porter.

Lists 'about 7000 terms that have been employed or suggested by various authors, since the date of the twelfth edition of Linnaeus's "Systema Naturae," as generic and subgeneric names for birds, and of references to the places and dates of their publication'.

Many names were originated by Gould. Under the letter 'A' alone, twenty came from Gould's works.

WATERHOUSE, F. H. 1902. *Catalogue of the Library of the Zoological Society of London.* 5th ed. London, Taylor & Francis.

The following statement appears: 'The first Catalogue published in 1854 contained the titles of about 460 works. To this a Supplement, issued in 1864, added 1090 others, making a total of 1550 titles'. The second edition in 1872 contained about 2100 titles, and the third edition in 1880 had 3200 titles.

The fifth edition has about 11 000 titles, exclusive of periodicals.

All of Gould's works are listed.

---

*[The mounted hummingbirds were not displayed in the Crystal Palace, but in a specially built house on the grounds of the Garden of the Zoological Society in 1851.]

WELKER, ROBERT HENRY. 1955. *Birds and Men*. Cambridge, Harvard University Press.

An impressive group of essays on the bird art of Catesby, Wilson, and Audubon, among other philosophical considerations on natural history. The question 'Did Audubon paint the Audubon's' on page 85 is discussed.

Gould is mentioned on page 72, stating that Audubon's achievement of his *Birds of America* has not been matched since, 'although it has been approached in Britain by John Gould and his associates, and in the United States by Louis Agassiz Fuertes'.

WELSBY, THOMAS. 1935. 'The lost Leichhardt'. *Historical Society of Queensland Jour.* 2: 264-73.

This accounting of Leichhardt's several expeditions was read at a meeting of the Historical Society of Queensland in 1929. The story of Gilbert's spearing is repeated.

WENDT, HERBERT. 1959. *Out of Noah's Ark*. Boston, Houghton Mifflin.

An informative and pleasantly written book that provides a wealth of background information on many of Gould's contemporaries. The Gould references are indexed and five of his plates are used for illustrations.

WEST, GEOFFREY. 1938. *Charles Darwin, a portrait*. New Haven, Yale University Press.

Gould is referred to on pages 162 and 342 in reference to his volume on the birds of the *Zoology of the voyage of H.M.S. Beagle*.

WHITE, SAMUEL ALBERT. 1914-19. 'A sketch of the life of Samuel White — ornithologist, soldier, sailor, and explorer'. *South Australian Ornithologist*.

The son of Samuel White has written a series of twenty-one articles on his father's life in Australia, mainly centred around the Adelaide area. In 1920 this material was published in a book of 116 pages by Thomas of Adelaide [pers. corres. Mrs D. A. Lewis, Royal Society of Tasmania, 1 February 1978].

The first article appeared in July 1914 1: 6-7 and the last article in July 1919 4: 79-82.

There are a few references to Gould but the most extensive and interesting are in part V, July 1915, pages 73 to 76. The near-drowning of White is related when he discovered *Malurus callainus* (Turquoise Superb Warbler), which he then sent to Gould. Further on, the son relates that the elder White visited England in late 1869 and early 1870. 'A great deal of time was spent with John Gould, who gave a little dinner on one occasion in honour of the Australian ornithologist.'

WHITE, S. A. 1919. 'Gould's Types'. *Emu* 19: 72.

Ashby (*Emu* 18: 311) had stated in reference to a specimen that 'Gould would call anything within 100 miles Adelaide'. White states here that 'Gould was too careful to do this'. He mentions conversations Gould had with his father, Samuel White, about this bird, *Pachycephala rufogularis*.

WHITLEY, GILBERT P. 1938. 'John Gould's associates'. *Emu* 38: 141-67.

This is the longest article in the Gould Commemorative Issue of *Emu*. With Whitley's typical completeness essentially every Australian associate of Gould is researched,

PART FOUR BIBLIOGRAPHY OF JOHN GOULD, HIS FAMILY AND ASSOCIATES

from Gould's brother-in-law Charles Coxen, to 'an old sportsman' who remains unknown.

[All who are mentioned by Whitley are indexed under Journal. *The Emu.* 1938.]

WHITTELL, HUBERT MASSEY. 1938. 'Gould's Western Australian birds: with notes on his collectors'. *Emu* 38: 175-86.

In a thorough manner Whittell discusses the species of birds from Western Australia that were described by Gould. He lists twenty-seven species, sixteen of which were apparently collected by John Gilbert.

WHITTELL, H. M. 1939. 'The Night Parrot'. *Emu* 38: 419-20.

This note concerns some confusion on the collection and sale of a Night Parrot (Whittell, *Emu* 38: 184).

WHITTELL, H. M. 1939. 'John Gilbert in Western Australia'. *Emu* 38: 419.

This is an erroneous article regarding Gilbert's arrival at Tasmania, and was later corrected by Whittell (*Emu* 38: 534). The passenger on the *Ellen* was another 'Mr. Gilbert', not John Gilbert.

The name of the ship *Ellen* is correct, and it is not 'Helen' as in Chisholm (*Emu* 38: 189).

WHITTELL, H. M. 1939. 'Gilbert in Western Australia — A correction'. *Emu* 38: 534-5.

In a letter to the Editor, Whittell points out an error he made (*Emu* 38: 419) in saying that John Gilbert arrived at Fremantle from Tasmania on 2 April 1838. The passenger was not John Gilbert, but another 'Mr. Gilbert'.

WHITTELL, H. M. 1941. 'A review of the work of John Gilbert in Western Australia'. *Emu* 41: 112-29, 216-42, 289-305.

This is a very comprehensive attempt by Whittell to trace Gilbert's movements in Western Australia, and 'also to indicate the specimens he collected and to determine the localities whence they were obtained'. These three articles are based on the letters from Gilbert to Gould found by Chisholm in England in 1938, accounts in early Perth newspapers, and records on Gilbert's activities as contained in Gould's *Handbook*. The annotations are very complete, typical of Whittell.

WHITTELL, H. M. 1946. 'The ornithology of Francis Thomas Gregory (1821-1888)'. *Emu* 45: 289-97.

F. T. Gregory and his elder brother, Augustus C., made several exploratory trips into north-west Australia. This article concerns notes by Francis Gregory made while on such an expedition in 1858. Gould is mentioned several times, since the Gregorys sent specimens to Gould.

WHITTELL, H. M. 1947. 'Frederick Strange'. *Australian Zoologist* 11: 96-114.

A valuable record with some letters. Strange collected birds and mammals in Australia for Gould and for others and knew Gilbert and MacGillivray also.

In 1852 Strange with his family returned to England for a short time. [While there Gould helped him dispose of some of his collections, as noted in Gould's letters to Sir William Jardine.]

WHITTELL, H. M. 1949. 'The visits of John Gilbert, naturalist, to Swan River Colony'. *J. & Proc. Historical Soc. Western Australia* 4: 23-53.

In this talk given before the historical society of Western Australia, Whittell states that this paper has to do 'with Gilbert's visits as they concern the history of the state' and not with the natural history specimens collected by Gilbert. The latter accomplishments were related and published as a series in the *Emu* (1938, 1941).

Whittell, in his usual thorough manner, reviews what is known about John Gilbert and his travels in Western Australia on his two visits there, with additional notes on John Gould, Governor John Hutt, Ludwig Preiss, Francis Fraser Armstrong, James Drummond, and Dr Walker.

WHITTELL, H. M. 1951. 'A review of the work of John Gilbert in Western Australia'. Part IV. *Emu* 51: 17-29.

Ten years had elapsed since the first three parts of these papers on Gilbert were published (1941), and the publication of this fourth part in 1951. The information in this paper is the result of finding three 'new' items in the Queensland Museum in Brisbane associated with Gilbert.

The first was an altered volume on Marsupialia in Jardine's 'Naturalist's Library'. It contained only the text that referred to the Australian species of marsupials. The pages on the kangaroos had been removed. This was Gilbert's notebook and he had entered notes on interleaved sheets in his own handwriting on the species of marsupials he had collected.

The second item in the Queensland Museum was a manuscript index apparently to a similar volume on the kangaroos. But the location of this second notebook is unknown.

The third item relates to Australian ornithology and is Gilbert's notebook on birds. Whittell states that Gilbert had this volume of 681 pages made up for notes on information acquired on his second visit to Australia. The basis for the volume was the information contained in the first five parts of Gould's *Birds of Australia*, which Gilbert copied into this volume.

At some later date, Whittell relates, these pages must have been bound into the two volumes as they now exist. On the backs of the volumes is lettered 'Ornithological Notes, Gilbert'. These notebooks had been in the possession of the Coxen family.

Whittell concludes this article with excerpts from Gilbert's notebook as they relate specifically to certain birds.

WHITTELL, HUBERT MASSEY. 1954. *The literature of Australian birds.* Perth, Paterson Brokensha.

An outstanding and monumental contribution to world ornithology up to 1950. The *history* of Australian ornithology from 1618 to 1850 is covered in 116 pages of part I of Whittell's book. The *bibliography* of Australian ornithology, and also of course the ornithology and ornithologists of the entire world, is covered in the period from 1618 to 1950 in 786 pages of part II. The references number in the thousands.

There are so many references to Gould and his associates in this volume that an index has been compiled. This index includes any and all references to John Gould, except where the word 'Gould' appears after a species of bird. The index also includes all references to Gould's wife, their children, her relatives (the Coxens), John Gilbert, Edward Lear, W. Hart, H. C. Richter, R. B. Sharpe, E. C. Prince, and over fifty others associated in any way with the Goulds and John Gilbert.

PART FOUR BIBLIOGRAPHY OF JOHN GOULD, HIS FAMILY AND ASSOCIATES

Index

The following is listed: name, section in Whittell's book (I or II); page number, and the individual in whose bibliography the name appears. If the page number refers to the personal biography, the page number is italicised (i.e. *274*).

A

Aiton, Wm. T., II-180, A. Cunningham
Albert, Prince, II-294, J. Gould; II-*635*
Alexander, W. B., II-*5*
Alexandra, Princess, II-*8*
Allan, Wm., II-*8*
Allis, Thomas, II-*9*
Allport, Morton, II-*9*
Angas, G. F., II-*15*
Argent, , II-*17*
Armstrong, F. F., I-94; II-*17*
Ashby, E., II-*20*
Austin, R., II-*27;* II-*28*
Australian bird bibliography, II-492, G. M. Mathews; II-664, R. B. Sharpe

B

Baker, Thomas, II-*31*
Bankier, R. A., I-95; II-*32*
Barrett, C. L., II-*39*
Beauffort, L., II-*44;* II-297, J. Gould
Beck, H. N., II-*44*
Becker, L., II-*45*
Bell, J., II-*47*
Bennett, George (Dr), I-92, I-98; II-*49*; II-730, E. A. Vidler
Blackwood, F. P., II-*57*
Blandowski, W., II-*58*
Bonaparte, Charles L., II-*63*
Bourjot Saint-Hilaire, A., II-*69*
Bourke, R., II-*70*
Brete, Capt., II-*75;* II-284, J. Gould
Breton, H. W. (Lieut.), II-*76*; II-284, J. Gould
Broadbent, K., II-*77*; II-502, G. M. Mathews
Bryant, C., II-*85*
Bullock, W., II-89 & 90
Burges, L., II-*91*
Burton, Charles, II-*95*
Burton, Edward, II-*95*
Bynoe, Benjamin (Dr), I-96, I-100-105; II-*97*

C

Caley, G., II-*100*
Campbell, A. G., II-*105*
Campbell, A. J., II-*108* & *114*
Campbell, John, I-99
Cayley, N., II-*125*; II-346, K. A. Hindwood
Chamberlain, W., II-*128*
Chambers, W., II-*128*
Chisholm, A. H., I-98-99; II-*133*, 139, 140, 141
Clark, Wm. N., I-94
Cleland, J. B., II-*151*

Cockerell, James F., I-116; II-*154*
Cockerell, John T., I-116; II-*157*
Conigrave, C., II-*166*
Cooper, D., II-*168*
Coppinger, R., II-*169*
Coxen, Charles, I-90, I-93, I-116; II-*174*; II-362, R. Illidge; II-376, John Jardine; II-739, Eli Waller
Coxen, Elizabeth Tomkins, II-283
Coxen, Henry, I-88, I-99
Coxen, Nicholas, II-283
Coxen, Stephen, I-90; II-174, C. Coxen; II-175

D

Darke, J., II-*186*
Darwin, Charles, I-100
Davies, R. H., II-*187*
Day, W. S., II-*186*
Denison, , II-*192*
Derby, Earl of, II-*193*
Dickison, D. J., II-*200*
Diggles, S., I-116; II-*203*; II-499, G. M. Mathews
Dring, J. E., I-100-101; II-*213*
Drummond, James, I-94, 97
Drummond, Johnston, I-94; II-214

E

Elford, F. G., II-*225*
Elliott, H., I-88
Elsey, J. L., I-116; II-*228*
Emery, J. B., I-100; II-*229*
Enright, W. J., II-*230*
Ewing, T. J., II-*232*
Eyre, E. J., II-*233*
Eyton, T. C., II-*233*

F

Franklin, Lady Jane, I-88-90; II-348, Dr E. C. Hobson
Franklin, Sir John, I-88; II-179, W. L. Crowther
Franklin, Sister of Lady Jane, I-88
Froggatt, W. W., II-*261*

G

Gawler, Col., I-91
Gibson, F., II-*270*
Gilbert, John, I-3, I-88, I-90, I-93-99, I-107-108, I-110, plate 20; II-91, L. Burges; II-114, A. J. Campbell; II-128, W. Chambers; II-133, 139, 140 & 141, A.

Chisholm; II-166, C. Conigrave; II-174, C. Coxen; II-213, J. Drummond; II-214, Johnston Drummond; II-*271*; II-291-292, J. Gould; II-306, G. Grey; II-407, E. W. Landor; II-423, F. W. Leichhardt; II-451, H. A. Longman; II-622, J. S. Roe; II-689, F. Strange; II-770-771, H. M. Whittell; II-786, E. A. Zuchold
Godman, F. D., II-*280*
Goodwin, A. P., II-*281*
Gould, Charles, II-*282*
Gould, Elizabeth, I-88, I-92-93, I-104; II-133 & 140, A. Chisholm; II-174, C. Coxen; II-175, S. Coxen; II-*283*; II-343, K. Hindwood; II-418, E. Lear; II-613, H. C. Richter
Gould, Franklin, II-283
Gould, John, I-3, I-12, I-15, I-20, I-44, I-45, I-73, I-84-85, I-87-93, I-94-105, I-108, I-110-116, plates 19 & 20; II-5, W. B. Alexander; II-8, Princess Alexandra; II-8, Wm. Allan; II-9, Thomas Allis; II-9, Morton Allport; II-15, G. F. Angas; II-17, F. F. Armstrong; II-20, E. Ashby; II-28, R. Austin; II-31, T. Baker; II-32, R. A. Bankier; II-39, C. L. Barrett; II-44, L. Beauffort; II-44, H. N. Beck; II-45, L. Becker; II-49, George Bennett; II-57, F. P. Blackwood; II-58, W. Blandowski; II-69, A. Bourjot; II-70, R. Bourke; II-75, Capt. Brete; II-76, H. Breton; II-85, C. Bryant; II-91, L. Burges; II-95, E. Burton; II-97, B. Bynoe; II-105, A. G. Campbell; II-108 & 114, A. J. Campbell; II-125, N. Cayley; II-133, 139, 140 & 141, A. Chisholm; II-151, J. B. Cleland; II-154, J. F. Cockerell; II-174, C. Coxen; II-175, S. Coxen; II-186, J. Darke; II-187, R. H. Davies; II-192,   Denison; II-200, D. J. Dickison; II-203, S. Diggles; II-213, J. E. Dring; II-214, Johnston Drummond; II-225, F. G. Elford; II-228, J. L. Elsey; II-229, J. B. Emery; II-230, W. J. Enright; II-232, T. J. Ewing; II-233, E. J. Eyre; II-233 T. C. Eyton; II-261, W. W. Froggatt; II-270, F. Gibson; II-271, John Gilbert; II-282, Charles Gould; II-*283-297*; II-297, G. W. Goyder; II-300, G. R. Gray; II-306, G. Grey; II-308, R. C. Gunn; II-324, William Hart; II-327, 328, James B. Harvey; II-331, J. C. Hawker; II-343-346, K. A. Hindwood; II-360, Dr. George Hurst; II-364, J. M. R. Ince; II-367, Tom Iredale; II-371, S. W. Jackson; II-376, John Jardine; II-377, Wm. Jardine; II-392, G. A. Keartland; II-403, J. L. G. Krefft; II-414, E. L. Layard; II-417, Benj. Leadbeater; II-*418*, Ed. Lear; II-442, J. W. Lewin; II-451, Heber Longman; II-466, J. MacGillivray; II-473, George Mack; II-489, 494, 500-502, G. M. Mathews; II-505, Mathews & T. Iredale; II-572, J. K. Palacky; II-*584*, E. E. Pescott; II-591, H. Pompper; II-*593*, E. C. Prince; II-597, John Rainbird; II-599-602, E. P. Ramsay; II-608, F. M. Rayner; II-*613*, H. C. Richter; II-622, J. S. Roe; II-*631*, T. Salvadori; II-635, Saxe-Coburg (Prince Albert); II-638, 639, P. L. Sclater; II-643, A. W. Scott; II-*663*, R. B. Sharpe; II-670, S. Sidney; II-*672*, M. E. Singleton; II-*672*, S. Sitwell; II-*676*, J. D. Somerville; II-681, Owen Stanley; II-684, Wm. Stephenson; II-686, E. C. Stirling; II-*688*, Witmer Stone; II-*688*, Stone & Mathews, G. M.; II-692, P. E. Strzelecki; II-*693-694*, Charles Sturt; II-*698*, J. Sutton; II-722, E. L. G. Troughton; II-730, J. P. Verreaux; II-730, Queen Victoria; II-730, E. A. Vidler; II-738, Charles Waller; II-739, Eli Waller; II-741, Fred G. Waterhouse; II-758, Samuel White; II-762, Samuel A. White; II-765, G. P. Whitley; II-*770*, H. M. Whittell; II-*778*, T. B. Wilson; II-786, E. A. Zuchold
Gould, John Henry, I-88, I-92; II-283
Gould League, other references plus II-44, H. N. Beck; II-416, J. A. Leach; II-507, A. H. E. Mattingley; II-561, A. J. North; II-569, W. J. O'Neil; II-630, J. V. Ryan
Goyder, G. W., II-*297*
Gray, G. R., II-*300*
Grey, Lt. George, I-94 (2), I-100, I-116
Grey, Sir George, II-*306*
Gunn, R. C., I-89, I-90, I-116; II-*308*
Gurney, J. H., II-*310*

H

Hall, R., II-*313*
Hart, William, II-*324*
Harvey, J. B., I-92; II-*327-328*
Hawker, J. C., I-91; II-*331*
Hawkins, W., II-193, Earl of Derby
Higgins, E. T., II-*338*
Hindwood, K. A., II-*343-346*
Hurst, George, II-*360*
Hutt, John, I-110

I

Illidge, R., II-*362*
Ince, J. M. R., II-*364*
Iredale, Tom, II-*367*

J

Jackson, S. W., II-*371*
Janson, E. W., II-376, John Jardine
Jardine, John, II-*376*
Jardine, Wm., II-*377*
Jukes, J. B., I-98

PART FOUR BIBLIOGRAPHY OF JOHN GOULD, HIS FAMILY AND ASSOCIATES

K

Keartland, G. A., II-*392*
Kemp, R., II-*394*
Krefft, J. L. G., II-*403*

L

Landor, E. W., II-*407*
Langier, M., II-*413*
Layard, E. L., II-*414*
Leach, J. A., II-*416*
Leadbeater, Benj., II-417
Lear, Edward, II-69, A. Bourjot; II-193, Earl of Derby; II-*418*
Leichhardt, F. W. L., I-98, I-99; II-*423*
Lewin, A. W., II-*442*
Longman, H. A., II-*451*
Lushington, I-100

M

MacGillivray, John, I-107-116, plate 26; II-293-295, J. Gould; II-*465-466*; II-598, E. P. Ramsay; II-608, F. M. Rayner; II-689, F. Strange; II-718, Tilston; II-774, J. F. Wilcox
Mack, George, II-*473*
Maclaren, John, I-91
Mahoney, Eliza, I-92
Mathews, G. M., II-*489, 494, 500-2, 505*
Mattingley, A. H. E., II-507
McArthur, Capt., I-95
McIntosh, D., II-*472*
Munt, H., II-*543*

N

Natty, I-93
North, A. J., II-*561*

O

Ogilvie-Grant, W. R., II-*567*
Oliver, Gerald, II-*568*
O'Neil, W. J., II-*569*
Owen, R., I-90

P

Palacky, J. F., II-*572*
Parkinson, S., II-*574*
Perks, R. H., II-*582*
Pescott, E. E., II-*584*
Pompper, H., II-*591*
Preiss, J. A. L., I-94
Prince, E. C., I-97; II-31, T. Baker; II-290, J. Gould; II-343, K. A. Hindwood; II-*593-594*

R

Rainbird, John, II-*597*; II-598, E. P. Ramsay
Ramsay, E. P., II-296, J. Gould; II-*597, 599-602*
Rayner, F. M., II-*608*
Reichenow, A., II-*610*
Richter, H. C., II-*613*; II-694, C. Sturt

S

Salvadori, T., II-*631*
Saunders, W. R., II-*635*
Saxe-Coburg, see Albert, Prince, II-*635*
Sclater, P. L., II-*638-639*
Scott, A. W., II-*643*
Scott, G. F., II-*644*
Sharpe, R. B., I-12, I-29, I-116; II-17, Argent (This reference and most of those that follow are references to Sharpe's 'The History of the Collections contained in the Natural History Departments of the British Museum', London, 1906); II-27, R. Austin; II-310, J. H. Gurney; II-313, R. Hall; II-338, E. T. Higgins; II-394, Robin Kemp; II-413, M. Langier; II-472, D. McIntosh; II-543, Henry Munt; II-567, W. R. Ogilvie-Grant; II-568, Gerald Oliver; II-574, Sydney Parkinson; II-582, R. H. Perks; II-610, A. Reichenow; II-635, W. R. Saunders; II-644, G. F. Scott; II-*661-664*; II-669, Dr Sibbold; II-674, Metcalfe Smith; II-674, P. W. B. Smith; II-685, Samuel Stevens; II-742, Thomas Watling; II-757, J. B. White; II-783, C. W. Wyatt
Sibbold, Dr, I-95; II-*669*
Sidney, S., II-*670*
Singleton, M. E., II-*672*
Sitwell, S., II-*672*
Smith, J. E. (Sir), II-*674*
Smith, Metcalfe, II-*674*
Smith, P. W. B., II-*674*
Somerville, J. D., II-*676*
Stanley, Owen, Capt., I-116; II-*681*
Stephenson, Wm., II-*684*
Stevens, S., II-*685*
Stirling, E. C., II-*686*
Stokes, Capt., I-100, I-101, I-103-6
Stone, Witmer, II-*688*
Strange, Frederick, I-91, I-99, I-116
Strzelecki, P. E., II-*692*
Sturt, Charles, Capt., I-90, I-91, I-116; II-*693-694*
Sturt, Evelyn, I-92
Sutton, J., II-*698*

T

Tilston, , I-114
Troughton, E. L. G., II-*722*
Turner, , II-294, J. Gould

V

Verreaux, J. P., II-*730*
Victoria, Queen, II-*730*
Vidler, E. A., II-*730*
Vigors, N. A., II-*733*

W

Waller, Charles, II-*738*
Waller, Eli, I-116; II-502, G. M. Mathews; II-*739*

Waterhouse, Fred G., II-295, 297, J. Gould; II-597, J. Rainbird; II-*741*
Waterhouse, F. H., II-283, J. Gould
Watling, Thomas, II-*742*
Watson, Mary, I-88
White, J. B., II-*757*
White, Samuel, I-116; II-499, 501, G. M. Mathews; II-*758*
White, Samuel A., II-*762*

Whitley, G. P., II-*765*
Whittell, H. M., II-*770-771*
Wilcox, J. F., I-116; II-294, J. Gould
Wilson, T. B., II-*778*
Wolf, Joseph, II-301, G. R. Gray; II-*780*
Wyatt, C. W., II-*783*

Z

Zuchold, E. A., II-*786*

WHITTELL, H. M. 1945. Gilbert, Grey and MacGillivray. *Emu* 45: 175-6.

Further information on Gilbert's movements and associations in Australia are related.

Gilbert apparently met with George Grey, and, separately, with John MacGillivray on at least two occasions in Australia. Whittell's note is not clear with regard to where he obtained this information.

WILLIS AND SOTHERAN. *See* SOTHERAN, HENRY, LTD. 1862. Catalogue.

WOLF, EDWIN, II. 1977. *A flock of beautiful birds: the ornithological collection of Louise Elkins Sinkler.* Philadelphia, The Library Company.

An interesting, but brief forty-three-page essay on fine bird books, and their progenitors, in the Sinkler Collection. Wolf begins, naturally, with Mark Catesby, and concludes with Rex Brasher. He touches on George Edwards, Eleazar Albin, Francois LeVaillant, Alexander Wilson, elaborates on Audubon, Gould, and Daniel Giraud Elliot, and continues with Henry Dresser, R. B. Sharpe, W. Vincent Legge and Frederick D. Godman.

There are two illustrations from the works of John Gould, this 'prodigiously prolific' London ornithologist.

WOLLASTON, A. F. R. 1921. *Life of Alfred Newton.* London, John Murray.

Newton had a strong influence on natural history in Gould's time. Gould was 25 years his senior.

Wollaston says Newton carried on a correspondence with Gould [but only a few letters are extant]. On page 115 Newton refers to 'the great Gould'. When Newton was up for election to the Professorship of Zoology and Comparative Anatomy in 1865 he 'was armed with a powerful array of testimonials from Owen, Gould, Gray, Murchison, Sclater and others'. Newton won.

A letter written by Newton to his brother Edward on 25 April 1864 should be quoted:

By the way Gould always sends me his "Birds of Great Britain" to look over for him, and the utter ignorance they sometimes betray is amazing. He has no personal knowledge of any English birds, except those found between Eton and Maidenhead, and about these species he fancies no one else knows anything. It is most amusing to see how anxious he is to avoid committing himself about Darwin's theory. Of course, he does not care a rap whether it is true or not—but he is dreadfully afraid that by prematurely espousing it he might lose some subscribers, though he acknowledged to me the other day he thought it would be generally adopted before long.

This statement about Gould's knowledge of British birds is typically Newtonian, and refers to Gould's frequent fishing on the Thames between Eton and Maidenhead. Newton's remark about Gould not committing himself to any new theory, like Darwin's, is undoubtedly correct. In fact, in all the Gould letters I have read, he rarely makes any comment on postulates, politics or personalities.

PART FOUR BIBLIOGRAPHY OF JOHN GOULD, HIS FAMILY AND ASSOCIATES

This volume is one of the most illuminating ones on the times and personages of this era of natural history.

WOOD, CASEY. 1931. *An introduction to the literature of vertebrate zoology* ... London, Oxford.

This outstanding compilation of items in the McGill University Libraries contains many references to John Gould, his artists, and his scientific associates, along with association items, such as the Zoological Society of London, British Museum, Australia, etc.

Figure 78 Pages concerned with Gould's works from Casey Wood 1931. *An introduction to the literature of vertebrate zoology.*

JOHN GOULD — THE BIRD MAN

All of Gould's works are listed with comments. Also, there are two manuscripts listed, a sixty-page manuscript on the family *Odontophorinae* or partridges of America, and a MS. on 'Boy's [sic] Birds' with an 'Ornithological List'.

A set of twenty-two original watercolour drawings by John Gould and his wife mainly of Australian birds, is listed.

The McGill collection also contains two sets of Edward Lear original drawings mainly of parrots and a portfolio of Joseph Wolf drawings of pheasants.

John Gould is referred to on pages 46, 48, 72, 75, 77, 86, 87, 92, 127, 129, 144, 156, 157, 159, 160, 165, 170, 226, 277, 296, 310, 311, 319, 364-365, 369, 372, 385, 386, 405, 452, 454, 502 (original drawings), 505, 509, 544, 564, 565, 613, 620 and 643.

CATALOGUE OF TITLES IN McGILL UNIVERSITY LIBRARIES 365

1840–8. (The) birds of Australia. *7 vols. folio.* Vol. I, *36 pl. (col.). 3 figs. T. of c. index.* Vol. II, pp. *(4)+(208). 104 pl. (col.). T. of c.* Vol. III, pp. *(4)+(194). 97 pl. (col.). T. of c.* Vol. IV, pp. *(4)+(208). 104 pl. (col.). T. of c.* Vol. V, pp. *(4)+(184). 92 pl. (col.). T. of c.* Vol. VI, pp. *(4)+(164). 82 pl. (col.).* Vol. VII, pp. *(4)+ (170). 85 pl. (col.). T. of c.* London.
A truly magnificent treatise, issued in 36 parts. This work forms a thorough treatise on the birds of the Australian region illustrated with 600 hand-colored plates from drawings of the author and Mrs. E. Gould. Three years later the author commenced a supplement which was completed in 1869.

1841–2. A monograph of the *Macropodidæ*, or family of kangaroos. Pts. 1–2. *2 vols. 30 col. pl. with letterpress.* London.
This monograph was never completed.

1844. [Birds collected during the voyage.] See HINDS, R. B. The zoology of the voyage of H.M.S. Sulphur . . . during . . . 1826–42. Vol. 1.

[1844]. On the sub-family *Odontophorinæ*, or partridges of America. *8vo. pp. 60.* Original manuscript.
John Gould's original manuscript in his own handwriting of a paper on the Odontophorinae which he read at the British Association Meeting in 1844. It was never printed or published, only an abstract appearing in the British Association Report. Page 19 is unfortunately missing.

[1844]–50. A monograph of the *Odontophorinæ*, or partridges of America. *32 pl. (col.). T. of c.* London.
Published in 3 parts, of which pt. I appeared in 1844, pt. II in 1846, and pt. III in 1850, the first two parts with 10 plates each and the last with 12 plates. In the introduction will be found descriptions of three species which are not figured, thus bringing up the total number of known species in the group to 35, all of which are treated in the present work.

[1845]–63. Mammals of Australia. *3 vols. folio. 182 col. pl.* Each plate with descriptive letterpress. London.
Although not as sumptuously illustrated as some of his avian treatises the author's monographs on mammalian life, like the above, are wonderfully well done throughout.

1849–61. A monograph of the *Trochilidæ*, or family of humming-birds. *5 vols. folio.* Vol. I, pp. *(127)+84. 41 pl. (col.).* Vol. II, pp. *(2)+152. 75 pl. (col.).* Vol. III, pp. *(2)+178. 87 pl. (col.).* Vol. IV, pp. *(2)+162. 80 pl. (col.).* Vol. V, pp. *(2)+156. 77 pl. (col.).* London.
The most attractive of all Gould's publications, issued in 25 parts. The last part contained only the introductory matter, title-pages, and lists of contents for each of the projected volumes. This introduction was published in 8vo form in the same year as the last part appeared, under the title *An introduction to the Trochilidæ*. A supplementary volume was issued during 1880–7, completed by R. Bowdler Sharpe after Gould's death in 1881, under the title *A Monograph . . . Supplement.*

[1849]–87. A monograph of the *Trochilidæ*, or family of humming-birds. Completed after the author's death by R. Bowdler Sharpe. Supplement. *folio. pp. (8)+200. front. (col.). 57 pl. (col.).* London.

1850–83. The birds of Asia. Completed after the author's death by R. Bowdler Sharpe. Each plate accompanied by leaf with descriptive letterpress. *7 vols. folio.* London.
Published in 35 parts, of which parts I and II appeared in 1850, and pts. XXXIV and XXXV in 1883. The last three numbers and the Introduction were issued after the death of Gould in 1881. The total number of colored plates is 530, hand-colored, and mostly drawn by Gould; a few the work of J. Wolf.

[1851]–69. The birds of Australia. Supplement. *folio. pp. iv+(158). 81 pl. (col., 1 fold.). index.* London.
Issued in five parts, of which pt. I appeared on March 15, 1851, and pt. V on August 1, 1869, according to G. Mathews, *Birds of Australia*, Suppt. 4, pp. 48, 1925.

1854 [1852–4]. A monograph of the *Ramphastidæ*, or family of toucans. *folio. pp. 26+(106). 52 pl. (51 col.).* London.
A revised edition of Gould's earlier work, 1834, with new plates and discussions of various species not formerly treated.

1858–75. A monograph of the *Trogonidæ*, or family of trogons. 2nd ed. *folio. pp. (4)+v–xx+(98). 47 pl. (col.).* London.
Published in four parts, the first of which appeared in 1858, the second in 1869, the third in June and the fourth in September 1875. The present edition is the second, the first appearing in 1838. The copy in hand is not bound but enclosed in a strong cardboard case.

1861. An introduction to the *Trochilidæ*, or family of humming-birds. *8vo. pp. iv+216. bibliogr. 2 indexes.* London.
This volume embraces the introductory matter of Gould's *Monograph of the Trochilidæ*, 1849–61. The present copy is a presentation from the author to the Rev. William Rogers.

1862–73. The birds of Great Britain. *5 vols. folio.* Vol. I, pp. *cxl+68. 37 pl. (col.). 1 fig.* Vol. II (wanting). Vol. III, pp. *(2)+142. 76 pl. (col.).* Vol. IV, pp. *(2)+172. 90 pl. (col.).* Vol. V, pp. *(2)+174. 86 pl. (col.).* London.
A magnificent work (in 25 parts) with life-like portraits of the birds inhabiting the British Isles. The 367 hand-colored plates are mostly by Gould, a few by J. Wolf.

1863. The mammals of Australia. *3 vols. folio. 130 col. pl.* Each plate accompanied by leaf with descriptive letterpress. London.
It appears that an incomplete edition of the monograph (1845–63) has been published with 130 col. pl. instead of 182, with the above date.

1863. Introduction to the Mammals of Australia. *8vo.* London.
A separately printed tractate apparently unknown to collectors, taken from the larger work.

1865. Handbook to The birds of Australia. *8vo.* Vol. I, pp. *viii+636.* Vol. II, pp. *(6)+629. 3 figs. index.*
This is the text of Gould's *Birds of Australia*, 1840–8, and of the three Supplements thereto, 1851–69, with emendations and additions. He refers to it as 'a kind of handbook' to his folio work. The present copy is from the library of Professor Blasius, who has made many marginal notes and drawn, in pencil on a fly-leaf, a map of Australasia.

1873. An introduction to the birds of Great Britain. *8vo. pp. iv+135. front. (portr. inserted). 1 fig.* London.
The introductory matter of *The birds of Great Britain*, folio, 1862–73, set up in 8vo form for convenience of correction.

1875–88. The birds of New Guinea and the adjacent Papuan Islands, including many new species recently discovered in Australia. *5 vols. folio.* Vol. I, pp. *iii+108. 56 pl. (col.). T. of c.* Vol. II, pp. *(4)+(116). 58 pl. (col.). T. of c.* Vol. III, pp. *(4)+(144). 72 pl. (col.). T. of c.* Vol. IV, pp. *(4)+118. 59 pl. (col.). T. of c.* Vol. V, pp. *(4)+(150). 75 pl. (col.) T. of c.* London.
Issued in 25 parts. The distribution of the 320 fine hand-colored plates by Gould and W. Hart among their respective parts is given in the indexes. This truly magnificent work was completed by R. Bowdler Sharpe after Gould's death.

1880–1? Monograph of the *Pittidae*. Pt. I (all pub.). *folio. pp. (20, pt. I)+(2)+(20, pt. II). 10 pl. (col.).* London.
Partly written by Bowdler Sharpe after the author's demise.

1893. An analytical index to the works of the late John Gould. See SHARPE, R. B.

PART FOUR BIBLIOGRAPHY OF JOHN GOULD, HIS FAMILY AND ASSOCIATES

The entire sections pertaining to Gould's published works in this bibliography are reprinted here (Figure 78). The courtesy of the Blacker-Wood Library of McGill University in allowing this reproduction is gratefully acknowledged.

WOOD, WILLIAM. 1832. *Catalogue of an extensive and valuable collection of the best works on natural history.* London, Taylor.

Gould's *Century of Birds* is listed. The date is London, 1831. 'This elegant work is publishing in Nos., each containing 4 coloured Plates, 12s.'

WOODWARD, ELLEN SOPHIA. c. 1820-1910. Ellen Sophia Woodward autograph collection. Blacker-Wood Library of Zoology and Ornithology, McGill University.

---

386 THE LITERATURE OF VERTEBRATE ZOOLOGY

[HEWITSON, W. C. (contd.)]
1853–6. Coloured illustrations of the eggs of British birds, with descriptions of their nests and nidification. 3rd ed. 2 vols. 8vo. Vol. I, pp. xvi+289. 77 pl. (col.). Vol. II, pp. (4)+290–532. 72 pl. (col.). index.   London.
This is said to be the best of the three editions. It appeared in 38 parts from May 1853 to June 1856, being partly rewritten, enlarged, and supplied with new plates.

HEWITT, CHARLES GORDON [1885–1920].
1921. The conservation of the wild life of Canada. 8vo. pp. xx+344. 23 pl. 4 figs. 15 maps and charts (3 fold.). index.   New York.
A posthumous volume by the Dominion Entomologist. It covers the subject in a most excellent manner, being especially interesting to Canadians.

HEYDEN, CARL HEINRICH GEORGES VON [1793–1866].
1827. Reptilien. folio. pp. 2+24. pl. (Senkenbergische naturforschende Gesellschaft. Atlas zu der Reise im nördlichen Afrika, vol. 3.)
Frankfurt-a.-M.
1828. See RÜPPELL, W. P. E. Atlas zu der Reise im nördlichen Afrika, etc.

HICKSON, SYDNEY JOHN [1859–].
1889. A naturalist in north Celebes; a narrative of travels in Minahassa, the Sangir and Talaut islands. 8vo. pp. xv+392. 6 pl. 29 figs. 2 maps (col. fold.). T. of c. 3 append. index.   London.
1894. Fauna of the Deep-Sea. 8vo. pp. 11+169. 1 pl. illust. in text.   London.
An interesting volume from Sir J. Lubbock's *Modern Science Series*.

HIERSEMANN, KARL W. (Publisher.)
1927. Bibliographie der Germanistischen Zeitschriften.
A valuable work of reference.
1928. Inkunabeln. 4to. pp. 4+103. illust.
Leipzig.
A dealer's catalogue of incunabula, with many titles of interest to zoologists.

HIESEMANN, MARTIN.
1908. How to attract and protect wild birds. 8vo. pp. 17+86. 49 figs. T. of c. append.   London.
Translated from the German first edition, 1907, by Emma S. Buchheim.
1911. How to attract and protect wild birds. 2nd ed. 8vo. pp. 13+101. 54 figs. T. of c. append.   London.
1912. How to attract and protect wild birds. 3rd ed.   London.
1915. Lösung der Vogelschutzfrage nach Freiherrn v. Berlepsch. 6te ergäntzte und verbesserte Aufl. 8vo. pp. xii+172. 2 col. pl. num. figs. in text. index.   Leipzig.

HIGGINS, ELMER [1892–].
1930. Wild life. See REDINGTON, PAUL G.

HILDEBRAND, B. F. and SCHROEDER, W. C.
1927. Fishes of Chesapeake Bay. (Bull. U.S. Bur. Fisheries, vol. xviii, 1927, pt. I.)   Washington.
This work includes general statistics and remarks on the Fisheries of Chesapeake Bay; also a systematic catalogue, with keys to families, descriptions, etc.

HILGENDORF, F. 1825–48. See EHRENBERG, C. G.

HILL, JOHN [1716–75].
1748–52. A general natural history . . . of the animals, vegetables, and minerals, of the different parts of the world. 3 vols. folio. illust.   London.
Vol. III contains an account of the birds, with plates Nos. 17–24, representing 111 species.
1752. An history of animals. folio. pp. (8)+584. 28 pl. index.   London.
This volume is one of three (perfect without the others) comprising the author's general natural history, 1748–52.
1752. Essays in natural history and philosophy; containing a series of discoveries, by the assistance of microscopes. 8vo. pp. 8+415.   London.

HILL, RICHARD [1795–1872].
1847. The birds of Jamaica. See GOSSE, P. H.
1851. A naturalist's sojourn in Jamaica. See GOSSE, P. H.

HINDS, RICHARD BRINSLEY, ed.
1844. The zoology of the voyage of H.M.S. Sulphur, under the command of Captain Sir Edward Belcher . . . during the years 1836–42. 2 vols. 4to. Vol. I, pp. (2)+150. 64 pl. (62 col.). Vol. II, pp. (2)+72. 21 pl. (col.). index. London.
The ornithology of this scientific voyage (by John Gould) will be found in vol. I, pp. 39–50, illustrated with 16 colored plates, Nos. 19–34. Many new species were obtained, selections of which went to the British Museum and the collection of the Zoological Society. The present copy is from the Godman Library.

HINGSTON, RICHARD WILLIAM GEORGE [1887–].
1920. A naturalist in Himalaya. 8vo. pp. xii+300. 24 pl. (1 map). T. of c. index.   London.
1923. A naturalist in Hindustan. 8vo. pp. 7+292. front. 9 pl. T. of c. index.   London.
1925. Nature at the desert's edge; studies and observations in the Bagdad oasis. 8vo. pp. 299. 11 illust.   London.
1928. Problems of instinct and intelligence. 8vo. pp. 8+296. illust.   London.

HINROTH, O. and HINROTH, M.
1924–9. Die Vögel Mitteleuropas. Lief. 1–60. 4to. many col. pl.   Berlin.
This treatise, so far as published, is a fine example of scientific text and excellent illustration.

HINTON, MARTIN ALISTER CAMPBELL [1883–].
1925. Reports on papers by G. E. H. Barrett-Hamilton on the whales of South Georgia. folio. pp. 159.   London.
1926. Monograph of the voles and lemmings (Microtinae) living and extinct. Vol. I. pp. xvi+487. 15 pl. 110 figs.   London.

HIRC, MIROSLAV.
1908. Horologische und gonimatische Beziehungen der Art *Accipiter nisus* (L.). 8vo. pp. [2]+19.
Zagreb.
A monograph on the hawk, *Accipiter nisus*, as seen in the Balkans.
1908. Die Jagdfauna der Domäne Martijanec. 8vo. pp. 85.   Zagreb.
A local list of animals interesting to the hunter. 161 species of birds found in the Croatian county Varazdin, are briefly described.

In 1978 I hurriedly examined this twelve-volume manuscript collection of autographs, letters, and photographs concerned with over four hundred naturalists, primarily in the field of geology.

Those naturalists included in the collection who were associated with John Gould include Charles Darwin, Edward Forbes, Jr., Frederick DuCane Godman, Robert Godwin-Austen, George Robert Gray, John Edward Gray, Albert Gunther, Sir Richard Owen (51 letters), Lovell Reeve, Philip L. Sclater, and Joseph Wolf.

A single John Gould item is in volume 3. This is an ink sketch of the bills of two humming-birds, male and female (Figure 79). A notation on the drawing is as follows: 'A new form/among the *Trochilidae*/24th July 63/Genus Androdon/J. Gould.'

PART FOUR BIBLIOGRAPHY OF JOHN GOULD, HIS FAMILY AND ASSOCIATES

[The original drawing of *Androdon aequatorialis*, attributed to W. Hart, for the *Supplement to Trochilidae*, plate 1, is in my collection, MS. 33GS.]

YARRELL, WILLIAM. 1830. [Concerning a recent addition to the British Fauna of a species of Warbler (the *Sylvia Tithys*, Scop.) ... recorded in the 5th volume of the "Zoological Journal," page 102, by Mr. John Gould.] *Proc. Comm. Science and Corres. Zool. Soc. London.* No. 2: 18.

A reference to Gould's note on this bird, and that two other specimens of the species were obtained 'last summer'. This meeting date was 28 December 1830.

YOUNGER, R. M. 1970. *Australia and the Australians.* Sydney, Rigby.

A short reference (page 155) is made to 'the naturalist John Gould and his artist

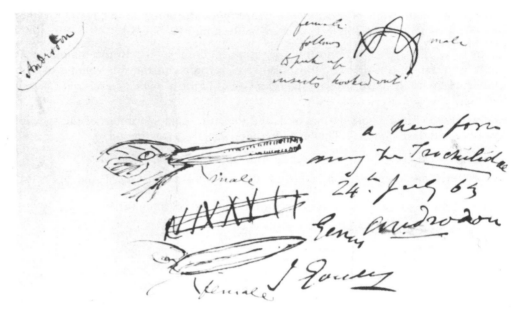

Figure 79 John Gould's pencil sketch of 'A new form/among the Trochilidae', from the Woodward Autograph Collection, Blacker-Wood Library, McGill University.

wife Elizabeth [being] enchanted with the scene' of the blue skies, and 'the weird and beautiful flowers, strange bright birds, and unbelieveable animals' of Australia.

Z ZEITLIN, JACOB. 1966. 'Natural history books from a rare bookseller's point of view'. From Buckman, T. R. *Bibliography and natural history.* Lawrence, Kansas, University of Kansas Libraries, pp. 140-8.

Comments on Zeitlin's personal contact with Ralph Ellis, the Gould collector, make this article most relevant.

In the 1930's I would often travel to Berkeley to be admitted by white-coated guards into the great house where Ralph Ellis ruled. From cellar to attic it was stuffed with books on only one subject: birds. Ellis hoarded Gould's folios by the dozens. If one was good, ten were better. He bought steel safes wholesale and pamphlet boxes by the hundred. When the collection came to Kansas, it occupied one fifty-foot freight car and one forty-foot freight car.

As Vosper [1961] says, its "impassioned, almost violent haste" marks the forming of his collection as it did all of Ellis' career. It was his ambition to acquire every book, journal or pamphlet of which birds were the major subject. He came very near to succeeding. In November, 1945 Ellis came to my shop in Los Angeles and asked if his mail could be sent there. When he died in December 17th of that year, the packages of books and letters addressed to him made a pile four feet high in my office.

Fate and the law have worked well together to bring and keep this great collection here and Robert Vosper and Thomas Buckman have worked diligently to make this place deserving of its custodianship. We are impatient to see the completion of the catalogue that Robert Mengel is preparing.

Zeitlin concludes with a listing of some bookseller prices compiled through the years on selected works. Two of Gould's works are included, 'A Century of Birds' ... listed at $110 to $160 in 1942, to £170 [$476] in 1964, and 'A Monograph of the Trochilidae', 5 vols. for $1,450.00 in 1942, to £2,250 [$5,300.00] in 1964. [Would that these could be bought now (1982) for those 1964 prices!]

ZIMMER, JOHN TODD. 1926. *Catalogue of the Edward E. Ayer ornithological library.* 2 parts. Chicago, Field Museum.

A classic catalogue of ornithological books, included in which is a good collation of all of Gould's works, except those on mammals. There was no Gould manuscript material in the Ayer library in 1926.

PART FOUR   BIBLIOGRAPHY OF JOHN GOULD, HIS FAMILY AND ASSOCIATES

References to Gould are on page 1 under A. L. Adams, pp. 157-159 under C. Darwin, 171-172 S. Diggles, 213-2 T. C. Eyton, *251-264*, 271 George Robert Gray, 283 Robert Hall (Gould illustrations), 304 Richard B. Hinds, 322, 327 William Jardine, 382 A. Lefevre, 394 John W. Lewin, 418 W. C. L. Martin, 420 G. M. Mathews, 479 Richard Owen, 577, 578, 581, 582 R. B. Sharpe, 608 Sturm, 657 N. A. Vigors, 660 Rudolf Wagner, and 664 F. H. Waterhouse.

The entire sections pertaining to Gould's published works in this bibliography are reprinted here (Figure 80). The courtesy of the Chicago Museum of Natural History in allowing this reproduction is gratefully acknowledged.

Figure 80 Pages concerned with Gould's works from John Todd Zimmer 1926 *Catalogue of the Edward E. Ayer ornithological library.*

1926.   CATALOGUE OF AYER LIBRARY—ZIMMER.   157

Cuvier, Georges Léopold Chrétien Frédéric Dagobert. (Blyth, Edward.)
 1840. Cuvier's | **Animal Kingdom,** | Arranged according to its Organization; | forming the basis for | a natural history of animals, | and | an introduction to comparative anatomy. |
   Mammalia, birds, and reptiles,   The Molluscous animals,
     by Edward Blyth.   by George Johnston, M.D.
   The fishes and Radiata,   The articulated animals,
     by Robert Mudie.   by J. O. Westwood, F.L.S.
   Illustrated by three hundred engravings on wood. | London: | Wm. S. Orr and Co., Amen Corner, Paternoster Row. | MDCCCXL.
  1 vol. 8vo, pp. I-VII+1, 1-670, text-figs. 1-209 (67-132 of birds), 1-142. London.
  An English translation of the second edition of Cuvier's "Règne Animal," 1829-30 (q.v.), with additions supplied by various editors entrusted with the different sections. Blyth is responsible for the ornithology which occupies pp. 154-267 and 67-132. Oberholser (Proc. Biol. Soc. Wash., **35**, p. 79, 1922) cites the genus *Habia* from this edition.

Cuvier, Georges L. C. F. D.
 1844-64? See Selby, Prideaux John, [**The Natural History of Parrots**], reissue.

Dam, D. C. van.
 1868. See Pollen and Dam, **Recherches sur le Faune de Madagascar.**

Danford, C. G.
 1889. See Rudolf Franz Carl Josef, Crown Prince of Austria, **Notes on Sport and Ornithology.**

Darwin, Charles. (Gould, John; Eyton, Thomas Campbell; Gray, George Robert.)
 1838-44. The | zoology | of | the voyage of H.M.S. Beagle, | under the command of Captain Fitzroy, R.N., | during the years | 1832 to 1836. | Published with the approval of | The Lords Commissioners of Her Majesty's Treasury. | Edited and Superintended by | Charles Darwin, Esq. M.A. F.R.S. Sec. G.S. | Naturalist to the expedition. | Part I [-V]. | Fossil Mammalia: [Mammalia (*Pt. II.*); Birds, (*Pt. III.*); Fish. (*Pt. IV.*); Reptiles, (*Pt. V.*)] | by | Richard Owen, Esq. F.R.S. [*etc.*, *2 lines*. (*Pt. I.*); George R. Waterhouse, Esq. *etc.*, *2 lines* (*Pt. II.*); John Gould, Esq. F.L.S. (*Pt. III.*); the Rev. Leonard Jenyns, M.A., F.L.S., &c. (*Pt. IV.*); Thomas Bell, Esq., F.R.S., F.L.S. *etc.*, *2 lines* (*Pt. V.*).] | Lon-

don: | published by Smith, Elder and Co. 65, Cornhill. | MDCCCXL [MDCCCXXXIX;MDCCCXLI; MDCCCXLII; MDCCCXLIII].

> [*Pt. III.*] Birds, | Described by | John Gould, Esq. F.L.S. | with | a notice of their habits and ranges, | by Charles Darwin, Esq. M.A. F.R.S. Sec. Geolog. Soc. | and with an anatomical appendix, | by T. C. Eyton, Esq, F.L.S. | Illustrated by numerous coloured engravings.

5 pts. in 3 vols, medium 4to. Pt. I, tit., pp. I-IV (pref.), I-IV (conts. and pll.), 1-111, pll. I-XXXII (3 fold.). Pt. II, 2 pr. ll., (subtit. and tit.), pp. I-IX+1, 1-97+1, 1 l. (index), pll. 1-35 (32 col.), 4 text-figs. Pt. III, 4 pr. ll. (subtit., tit., list of pll. and errata), pp. I-II, 1-156, 4 ll. [index], pll. 1-50 (col.; by J. Gould). Pt. IV, pp. I-XV+1, 1-172, pll. 1-29. Pt. V, pp. I-VI, 1 l. (list of spp. and pll.), 1-51, pll. 1-20. London.

The report on the zoological collections obtained by the members of the Beagle expedition. As indicated above, Pt. III is devoted to the birds. The "descriptions of the new species and names of those already known" were supplied by Gould, but owing to the incompleteness of his manuscript, Darwin was obliged to emend and enlarge certain portions of this part of the text. Darwin acknowledges the assistance of Gray in the matter of synonymy and general arrangement, and gives frequent references to him in places where his assistance has been used. He also states ("Advertisement," pp. I and II) that Gould is to be credited with all new descriptions of genera and species and he (Darwin) with all accounts of habits and ranges, although he has not indicated the division of text throughout the work. The case proves to be not so simple as it seems.

The generic names of some of the new species have been altered to accord with generic changes proposed herein by Gray, as, for example, *Myiobius magnirostris*, p. 48, pl. VIII. Gould's manuscript name, *Tyrannula magnirostris*, is cited in synonymy but appears on the plate, while the new combination is given without authority other than that it follows Gray's proposal of *Myiobius* for *Tyrannula* of Swainson. In this case, the authorship of the species may remain with Gould since the plate has priority over the text (see below), but Darwin appears to be properly the author of the new combination of names. In the case of *Myiobius parvirostris*, described on the same page but not figured, Darwin also becomes the author of the species.

In other cases, generic names have been altered in similar manner but the new names are used, likewise, on the plates whence they cannot be quoted as of Gould, since Gould never used the combination of terms, even in manuscript. Such cases, including *Pachyramphus albescens*, p. 50, pl. XIV, must bear Darwin's name as author. In still other cases, as that of *Opetiorhynchus nigrofumosus*, p. 68, Gould's manuscript specific name is subordinated owing to supposed synonymy, but it had already appeared on the plate (Pl. XX in the example given) whence it may still be quoted as of Gould. Most of the new species, fortunately, are described and accredited properly without discrepancy.

1926. CATALOGUE OF AYER LIBRARY—ZIMMER. 159

Among other material and matter ascribed to Gray is a page of "Corrigenda" (on a leaf following the list of plates). This is easily overlooked but important since it contains several new names which may have priority over Gray's "List of the Genera of Birds," second edition, 1841 (q.v.), from which they are usually quoted. Eyton's anatomical "Appendix" occupies pp. 147-156.

The entire report was issued in parts, of which the ornithology comprised Nos. 3, 6, 9, 11 and 15. The dates and contents of each were obtained by Sherborn from the publishers and given by him (Ann. Mag. Nat. Hist., (6) **20**, p. 483, 1897) as follows. Pt. 3, pp. 1-16, July 1838; Pt. 6, pp. 17-32, Jan. 1839; Pt. 9, pp. 33-56, July 1839; Pt. II, pp. 57-96, Nov. 1839; Pt. 15, pp. 97-164, March 1841. Each part contained 10 plates, presumably issued in numerical sequence; a review of the first three numbers on birds (given in the Rev. Zool. for Nov. 1839, pp. 338-339) cites pll. 1-30. Sherborn does not cite the allocation of title-pages and other preliminary matter (including the page with Gray's "Corrigenda"); it is presumed that they appeared with Pt. 15.

Daubenton, (le jeune), **Edme Louis.**

1765-80. [Planches enluminées d'histoire naturelle].
See Buffon, George L.L., Histoire Naturelle des Oiseaux, 1770-86.

Daubenton, Louis Jean Marie.

1824-31. See Buffon, George L. L.; and Daubenton, **Oeuvres complètes de Buffon.**

1828-33. Idem.

David, Armand; and **Oustalet, Emile.**

1877. Les | oiseaux de la Chine | par | M. L'Abbe Armand David, M.C. | ancien missionnaire en Chine, | Correspondant de l'Institut, du Muséum d'Histoire Naturelle, etc. | et | M. E. Oustalet | Docteur ès Sciencés, Aide-Naturaliste au Muséum, | Membre Correspondant de la Société Zoologique de Londres | Avec un Atlas de 124 Planches, dessinées et lithographiées | par M. Arnoul et coloriées au pinceau [Atlas] | Paris | G. Masson, éditeur | Libraire de l'Académie de Médecine | Boulevard Saint-Germain, en face de l'École de Médecine | M DCCC LXXVII.

2 vols. royal 8vo. Text, 2 pr. ll., pp. I-VII+1, 1-573. Atlas, pp. I-VI+1 l., pll. 1-124 (col,). Paris.

A catalogue of the birds of China so far as known to the date of publication, with synonymies, descriptions and notes on distribution, variation and other relative matter. The plates are clear and well drawn.

Davie, Oliver.

1885. An | egg check list | of | North American birds | giving accurate descriptions of the color and size of the eggs, | and locations of the nests of the land and water | birds of North America.

plan was to give a figure of each species described in the earlier work, but this was modified "by omitting such species as had been well figured before, in works easily available to the British public." Two species are figured (on pl. XXII and XLV) which are not mentioned in the text. The present volume has inscribed on the fly leaf "The late W. Yarrell's copy," and bears, also, the bookplate of Henry Wemyss Feilden.

Gosse, Philip.

1899. See Fitz Gerald, Edward A.; and Vines, Stuart, The Highest Andes.

Gould, John. (Vigors, N. A.)

1831-32. A | century of birds | from | the Himalaya Mountains. | By | John Gould, A.L.S. | London: | 1832.

1 vol. imperial folio, 6 pr. ll., 72 ll., 80 pll. (col.; by E. Gould). London.

This work appeared in two forms, one with the backgrounds of the plates uncolored, the other with them colored. The present copy belongs to the latter series. Although no actual dates of publication appear to be available, a remark by Vigors in an early number of the Proceedings of the Committee of Science and Correspondence of the Zoological Society of London, Vol. I. pp. 170 and 176, (meeting for Dec. 27, 1831) indicates that the work was being, or had been, issued in regular monthly parts. This would tend to show that publication was commenced in 1831 or earlier. Some of the plates were exhibited to the Society at a meeting held on Nov. 23, 1830; see Proceedings, Vol. I, p. 6. The plates were executed by Mrs. (E.) Gould from Gould's sketches and the scientific descriptions, with most, if not all, of the remaining letterpress, were written by Vigors to whom Gould makes acknowledgment in the "Advertisement" on a preliminary leaf. Whether or not the nomenclature of the various species should be quoted from this folio or from the Proc. Comm. Sci. Corr. Zool. Soc. Lond., Vol. I, in which the descriptions were published somewhat contemporaneously, depends on the actual dates of publication of the two works, but the authorship, in any case, is that of Vigors, not of Gould. The descriptions in the "Proceedings" were issued as follows, quoting the pagination of Vol. I. Pp. 7-9, Jan. 6, 1831; 22-23, Febr. 1, 1831; 35, March 2, 1831; 41-44, April 6, 1831; 54-55, May 6, 1831; 170-176, March 2, 1832. The 80 hand-colored plates contain 102 figures of birds, of which 2 represent species (and sexes) figured twice; the remaining 100 figures form the basis for the title of "Century." This was the first of Gould's famous folios.

Gould, John.

1832-37. The | birds of Europe. | By | John Gould, F.L.S., &c. | In five volumes. | Vol. I [-V]. | Raptores [Insessores; Insessores; Rasores.Grallatores; Natatores]. | London: | Printed by Richard and John E. Taylor, Red Lion Court, Fleet Street. | Published by the author, 20 Broad Street, Golden Square. | 1837.

5 vols. imperial folio. Vol. I, pp. I-XII, 1 l., pp. 1-4, 51 ll., 50 pll. (col.). Vol. II, 101 ll., 99 pll. (col.; nos. 51-149). Vol. III, 95 ll.,

93 pll. (col.; nos. 150-242). Vol. IV, 105 ll., 103 pll. (col.; nos. 243-345). Vol. V, 105 ll., 103 pll. (col.; nos. 346-449; 447 and 448 on one plate). London.

Issued in 22 parts, the first of which appeared June 1, 1832. The remainder were planned to appear on the first of each third month thereafter. The preface was written on August 1, 1837, which would have been in time for the issuance of Pt. XXII on September 1 in accordance with the program, but whether the program was followed throughout I am unable to determine. The plates in the bound volumes are in accordance with a "General List of Plates," widely differing from the order of publication. As to the plates in each number, I am able to allocate but a few. Pt. I apparently contained, among others, plates 23, 27, 61, 216, 233, 260 and 379; Pts. II and III included pll. 19, 53, 59, 97, 120, 148, 160, 204, 217, 246, 258 and 289; Pt. X contained pll. 28, 159, 282 and 314: Pt. XI included pl. 297. The plates, drawn by Gould and E. Lear, were hand-colored by Mrs. Gould.

Gould, John. (Owen, Richard.)

1833-35. A | monograph | of | the Ramphastidæ, | or | family of toucans. | By | John Gould, F.L.S. | London: | published by the author, 20, Broad Street, Golden Square. | 1834.

> {A | Monograph | of the | Family of Ramphastidae [Ramphastidae *(Pt. III.)*] | or Toucans; [or | Family of Toucans. *(Pt. III.)*] | By | J. Gould, | F.L.S. | Pt. 1. ["1" *crossed out and replaced by* "2" *(Pt. II.)*; *line omitted, but* "Pt. 3" *added by hand (Pt. III.).*] | To be completed in Two Parts. [Price £ 2-0 *(Pt. III.)*] | London. | Published by the Author, 20, Broad Street, Golden Square. | 1833 [1835 *(Pt. III.)*]. | Printed by C. Hullmandel.}

1 vol. imperial folio, 47 ll., 34 pll. (33 col.; by J. and E. Gould and E. Lear). London.

Published in three parts. Through the kindness of Dr. C. E. Hellmayr and of Dr. A. Laubmann of the Bavarian State Museum at Münich, I am enabled to transcribe the titles and give the contents of the several parts, from a copy in the library of the Münich Museum. They are as follows.

Pt. I contained text and plates of *Pteroglossus bitorquatus, Ramphastos carinatus, Pter. ulocomus, R.. culminatus, Pter. maculirostris, Pter. hypoglaucus, R. discolorus, Pter. prasinus, Pter. regalis, R. swainsonii, Pter. sulcatus,* and *Pter. bailloni.* Pt. II contained the title-page (dated 1834), the preface, introduction, etc., descriptions of the generic characters of the genera *Ramphastos* and *Pteroglossus* and the text and plates of *Pter. azarae, Pter. inscriptus, R. toco, Pter. castanotis, R. cuvieri, R. ariel (tucanus* Linne?), *Pter. culik, R. erythrorhynchus, Pter. aracari, R. vitellinus, Pter. viridis,* a new description of *R. culminatus,* and Owen's paper and plate of "Observations on the Anatomy of the Toucan." An insert-slip is attached to the cover, entitled "To the binder," giving the sequence to be followed in arranging the letterpress and plates, and closing with the statement, "The description of *Ramphastos culminatus,* given in Part I, to be cancelled, and the one in Part II inserted in its stead." Pt. III contained text and plates of *R. citreopygus, R. osculans, Pter. pluricinctus, Pter. humboldtii, Pter. langsdorffiii, Pter. nattererii, Pter. reinwardtii, Pter. derbianus, Pter. haematopygus* and *Pter. pavoninus*. There is another insert-slip in this part addressed "To the binder," giving a new order of arrangement for the contents of the work at the close of which is placed Owen's contribution; then follows the same note as in Pt. II regarding the cancellation of the original text of *R. culminatus* and the statement, "The Synopsis Specierum given in Part II to be cancelled, and the one in Part III inserted in its stead. N.B. The directions to the Binder in Part II to be disregarded." The present copy lacks the original text of *Ramphastos culminatus*.

In the Proc. Zool. Soc. London, 1834, p. 79, in a paper read July 8, 1834, (publ. Nov. 25, 1834) there is a statement to the effect that Gould's monograph of the Ramphastidae " is just completed"; this statement presumably refers to the completion of the two parts mentioned on the original wrappers of Pts. I and II and fixes the approximate date for Pt. II.

In 1852-54, Gould issued a new edition of the work (q.v.) under the same title as the present, and in 1855 published his "Supplement to the First Edition" (q.v.), giving the text and plates of the species included in the revised, but not the original, edition.

Gould, John.

1836-38. A | monograph | of | the Trogonidæ, | or | family of trogons. | By | John Gould, F.L.S., &c. | London: | Printed by Richard and John E. Taylor, Red Lion Court, Fleet Street. | Published by the author, 20 Broad Street, Golden Square. | 1838.

1 vol. imperial folio, 2 pr. ll., pp. I-VII+1, 39 ll., 36 pll. (col.). London.

Published in 3 parts. Part I is mentioned by Swainson in Vol. I of his "On the Natural History and Classification of Birds" in "The Cabinet Cyclopaedia" of Dionysius Lardner (q.v.), and therefore must date 1836 or earlier. I do not know what species were included in this part. In Vol. II (pp. 337 and 338), Swainson cites *Trogon elegans, T. ambiguus, T. melanocephala, Harpactes malabaricus, M. erythrocephalus,* and *Calurus resplendens* from Gould's work, so that these species were undoubtedly included in Part I or in Parts I and II if both were in print before July, 1837. A second edition, quite different from the first, was published in 1858-75 under the same title (q.v.).

Gould, John.

1837-38. The | birds of Australia, | and the adjacent islands. | By John Gould, F.L.S., &c. | Part I [II]. | [*List of contents.*] | London: | Printed by Richard and John E. Taylor, Red Lion Court, Fleet Street. | Published by the author, 20, Broad Street, Golden Square. | August, 1837 [February, 1838].

2 parts (all published) in 1 vol. imperial folio. Part I, cover-tit., 10 ll., 10 pll. (col.). Part II, cover-tit., 10 ll., insert-slip (notice to subscribers), 10 pll. (col.). London.

After the commencement of the present book, Gould found himself so handicapped by lack of specimens that he cancelled the work and set out for Australia to procure more and better material. After his return he began the publication of "The Birds of Australia" (q.v., 1840-48). Of the plates in the present work, Nos. 6, 7, 9 and 10 of Pt. I, and 3, 4, 8, 9 and 10 of Pt. II were reprinted in the later work with occasional minor alterations in colors; the remainder were redrawn. This unfinished folio is said to be the rarest of Gould's works.

Gould, John.

1837-38. A synopsis | of the | birds of Australia, | and the adjacent islands. | By | John Gould, F.L.S., &c. | Author of various works [*etc., 3 lines.*] | London: | Published by the author, 20 Broad Street, Golden Square. | 1837-38.

1 vol. imperial 8vo, 77 ll., pp. 1-8, 73 pll. (col.). London.

Issued in 4 parts: Pt. I (19 pll.), January, 1837; Pt. II (18 pll.), January, 1837; Pt. III (18 pll.), April, 1838; Pt. IV (18 pll.), April 1838. The work consists chiefly of diagnoses, descriptions, and synonymies of the species figured on the plates. Pages 1-8 at the conclusion of Pt. IV contain original descriptions of several new genera and 36 new species which were read before the London Zoological Society in December, 1837 but not published by that body until December 5, 1838 (P. Z. S. Lond. 1837, pp. 138-157), after the appearance of the present volume.

Gould, John.

1837-38. [Price £1 15s. (*Pt. II.*)] | **Icones avium,** | or | figures and descriptions | of | new and interesting species of birds | from various parts of the world. | By | John Gould, F.L.S., &c. | forming a supplement | to his previous works. | Part I. [Part II] | [*List of species, 5 lines in double column (Pt. I.*).; Monograph of the Caprimulgidæ, | part I. | *List of species, 4 lines, in double column (Pt. II.)* | London: | Printed by Richard and John E. Taylor, Red Lion Court, Fleet Street. | Published by the author, 20, Broad Street, Golden Square. | August, 1837 [August, 1838].

2 parts (all published), imperial folio. Part I, cover-tit., 10 ll., 10 pll. (col.). Part II, (cover-tit.), 8 ll., 8 pll. (col.).

This work was interrupted by Gould's departure for Australia in search of material for his work on the birds of that region, and was never resumed. Certain of the species were described in papers read in advance before the Zoological Society of London, but in most cases these descriptions were never published in the "Proceedings" of that body. Four species in Pt. I (Nos. 2, 3, 4 and 5 of the subjoined list), ascribed to "Proc. of Zool. Soc. Part V., 1837", without pagination, were not so published and must date from the present work; two species (Nos. 7 and 9) were published but are antedated by the folio. In Pt. II, five species (Nos. 2, 3, 4, 6 and 8), ascribed to "Proc. of Zool. Soc. Part VI 1838," without pagination, did not appear except in the present book.

1926. CATALOGUE OF AYER LIBRARY—ZIMMER. 255

The two parts contained the following species as listed on the wrappers. Pt. I:-*Eurylaimus (Crossodera) Dalhousieae, Todus multicolor, Ianthocincla phœnicea, Calliope pectoralis, Microura squamata, Paradoxornis flavirostris, Pteroglossus (Selenidera) Gouldii, Numida vulturina, Ortyx plumifera, Cursorius rufus.* Pt. II:-*Amblypterus anomalus, Nyctidromus Derbyanus, Semeïophorus (Macrodipteryx?) vexillarius, Lyncornis cerviniceps, Lyncornis macrotis, [Lyncornis] Temminckii, Batrachostomus auritus, Nyctibius pectoralis.*

Gould, John.

1838-41. See Darwin, Charles, **The Zoology of the Voyage of H. M. S. Beagle,** 1838-43.

Gould, John.

1840-48. The | birds of Australia. | By | John Gould, F.R.S., | F.Z.S., [etc., 5 lines.] | In seven volumes. | Vol. I [-VII]. | London: | Printed by Richard and John E. Taylor, Red Lion Court, Fleet Street. | Published by the author, 20, Broad Street, Golden Square. | 1848.

7 vols. imperial folio. Vol. I, 9 ll., pp. V-CII, 1-13+1, 37 ll., 36 pll. (col.), text-figs. 1-3. Vol. II, 106 ll., 104 pll. (col.). Vol. III, 99 ll., 97 pll. (col.). Vol. IV, 106 ll., 104 pll. (col.; 2 fold.). Vol. V, 94 ll., 92 pll. (col.). Vol. VI, 84 ll., 82 pll. (col.). Vol. VII, 87 ll., 85 pll. (col.). London. *600 pll.*

This sumptuous work was issued in 36 parts, of which Pt. I appeared in 1840, Pts. II-V in 1841, **VI-IX** in 1842, **X-XIII** in 1843, **XIV-XVII** in 1844, **XVIII-XXI** in 1845, **XXII-XXV** in 1846, **XXVI-XXIX** in 1847 and **XXX-XXXVI** in 1848. According to Mathews (Birds of Australia, Suppl. **4**, p. 48, 1925, the first part was dated December 1, 1840, and Pts. II-XXXII appeared regularly on March 1, June 1, September 1 and December 1 thereafter, except that Pts. XXXIII-XXXVI were all dated and (appeared on) December 1, 1848. The final arrangement in the completed volumes is irrespective of the sequence of issue. The actual year of publication of each plate may be ascertained from F. H. Waterhouse's "The Dates of Publication of Some of the Zoological Works of the late John Gould," 1885 (q.v.).

Owing to delay in the publication of the various sheets of the Proceedings of the Zoological Society of London, to which Gould submitted the descriptions of his new species for early publication, many of the new names must date from the present work.

This work forms a thorough treatise on the birds of the Australian region (including an occasional species from outlying territory) with 600 hand-colored plates executed from drawings by the author and Mrs. (E.) Gould. Three years after the completion of this work the author commenced a supplement (q.v.) which was completed in 1869. On August 1, 1848, the introductory matter of the present work (pp. V-CII and 1-13) was issued under the title of "Introduction to the Birds of Australia" (q.v.), preceding the folio containing the same material by four months. In 1865 the text of the entire work, including the supplement and even further enlarged to bring the subject matter to date, was issued without plates as a "Handbook to the Birds of Australia"

(q.v.). Prior to the present folio, the author issued his "A Synopsis of the Birds of Australia," 1837-38 (q.v.), and two parts of a folio entitled "The Birds of Australia and the Adjacent Islands," 1837-38 (q.v.). The latter work was cancelled, due to the insufficiency of specimens for study, and the author proceeded to Australia for the collection of more ample material, incorporating the results of his investigations in the present issue of the work.

Gould, John. (Sturm, Johann Heinrich Christian Friedrich; Sturm, Johann Wilhelm; Owen, Richard; Wagner, Rudolf.)

1841-47. J. Gould's | monographie | der | **Ramphastiden** | oder | Tukanartigen Voegel. | Aus dem Englischen Übersetzt, | mit Zusætzen und neuen Arten vermehrt | von | Johann Heinrich Christian Friedrich Sturm, | der kaiserl. naturforschenden Gesellschaft [*etc.*, *2 lines.*] | und | Johann Wilhelm Sturm, | der königl. bayer. botan. Gesellschaft [*etc.*, *3 lines.*] | Erstes [-Viertes] Heft | [*List of contents, 5 lines in double column, (Pts. I-III.); 3 lines double column and 4 lines full width, (Pt. IV.).*] | Nürnberg, 1841 [1841; 1842; 1847]. | Gedruckt auf Kosten der Herausgeber. | (Panierstrasse S. Nr. 709.).

1 vol. crown folio, 42 ll., 38 pll. (36 col., by J. Gould and Friedrich Sturm; 2 plain, by G. Scharf and A. Köppel). Nürnberg.

Issued in four parts, with dates as given above. Pts. I-III contained 10 plates and 10 leaves of text, each; Pt. IV contained 6 colored plates and the accompanying text, and a translation of Owen's article on the anatomy of the toucan with its plate, to which is added a second plate and further remarks by Rudolf Wagner. The principal part of the general text is based on Gould's Monograph of the Ramphastidae," 1833-35 (q.v.), of which it is, in places, a literal translation; but considerable of the matter is rewritten or revised and there are additional species described herein for the first time. The plates are sometimes redrawn and reduced from Gould, but frequently altered or designed afresh, while the illustrations of the new species are entirely new. There is no title-page, since the work was never carried to completion, and the title as quoted above, is taken from the wrapper of Pt. IV, bound with the present copy, and from facsimiles of the wrappers of Pts. I, II and III, obtained from copies in the Zoological Museum at Munich.

The contents of the four parts, as indicated on the wrappers, is as follows. I-*Ramphastos culminatus, R. Cuvieri, Pteroglossus melanorhynchus, P. Azarae, P. bitorquatus, P. prasinus, P. maculirostris, P. Gouldii, P. Nattereri* and *P. Reinwardtii.* II-*Pteroglossus castanotis, P. torquatus, P. pluricinctus, P. Humboldti, P. Langsdorfii, P. Wagleri, P. albivitta, P. atrogularis, P. Lichtensteinii* and *P. haematopygus.* III-*Ramphastos Toco, R. carinatus, R. vitellinus, R. Temminckii, R. dicolorus, Pteroglossus hypoglaucus, P. Sturmii, P. Humboldti* (foemina), *P. inscriptus* and *P. Derbianus.* IV-*Ramphastos Swainsonii, Pteroglossus Beuharnaisii, P. Azarae, P. Bailloni, P. piperivorus* and *P. sulcatus;* also the text to *Pteroglossus flavirostris* and *P. Wiedii* and the anatomical article by Owen and Wagner.

Gould, John.

1843-44. See Hinds, R. B., **The Zoology of the Voyage of H. M. S. Sulphur,** Birds.

Gould, John.

1844-50. A monograph | of | the Odontophorinæ, | or | partridges of America. | By | John Gould, F.R.S., | F.L.S., [*etc., 6 lines.*] | London: | Printed by Richard and John E. Taylor, Red Lion Court, Fleet Street. | Published by the author, 20, Broad Street, Golden Square. | 1850.

1 vol. imperial folio, 4 pr. ll., pp. 11-23+1, 33 ll., 32 pll. (col.). London.

Published in 3 parts, of which Pt. I appeared in 1844, II in 1846 and III in 1850, the first two parts with 10 plates each and the last with 12 plates. The dates and allocation of these plates is given by Waterhouse in his "The Dates of Publication of Some of the Zoological Works of the late John Gould," 1885 (q.v.), pp. 56 and 57. Pt. I was received by the Boston Society of Natural History on February 26, 1845, and Pt. II on June 20, 1846, according to information kindly furnished me by Dr. Glover M. Allen of that institution.

Gould, John.

1848. Introduction | to the | birds of Australia. | By | John Gould, F.R.S., | F.L.S., [*etc., 9 lines.*] | London: | Printed for the author, by Richard and John E. Taylor, Red Lion Court, Fleet Street. | 1848.

1 vol. post 8vo, pp. I-VIII, 1-134, text-figs. 1-3. London. August 1, 1848.

This little work consists of the preface and introduction to the author's folio work, "The Birds of Australia," 1840-48 (q.v.), Vol. I, pp. V-CII and 1-13, December 1, 1848. These according to the author, were set up in small type for facility of correction and issued in a limited edition in that form prior to the issue of the larger work. Sherborn (Index Animalium, Sect. 2, Pt. 1, p. LXI, 1922) appears to reject this volume and refers to the introductory "Notice" for a statement that "these are but proofsheets" and the folio the authoritative text. However, the "Notice" is worded somewhat differently and says that readers "must . . . still regard it more as a proof-sheet than otherwise, inasmuch as it contains many imperfections, most of which have been corrected in the folio edition." As issued in the folio, the matter is entirely reset and does not affect the validity of anything contained in the present issue which was "printed in an octavo form, for distribution among my scientific friends and others, to whom I trust it will be at once useful and acceptable." Various new names must be quoted from this 8vo edition and not from the folio. Mathews (Austral Av. Rec. 4, No. 1, p. 9, 1920) gives the date of publication as August 1, 1848; source of information not cited.

The present copy contains the bookplate of Frederick DuCane Godman, and was a presentation copy from the author.

Gould, John.

1849-61. A monograph | of | the Trochilidæ, | or | family of humming-birds. | By | John Gould, F.R.S., | F.L.S., [etc., 8 lines.] | In five volumes. | Vol. I [-V]. | London: | Printed by Taylor and Francis, Red Lion Court, Fleet Street. | Published by the author, 26 Charlotte Street, Bedford Square. | 1861. | [The author reserves to himself the right of translation.].

5 vols. imperial folio. Vol. I, 4 pr. ll., pp. V-CXXVII+1, 42 ll., 41 pll. (col.; nos. 1-41). Vol. II, 77 ll., 75 pll. (col.; nos. 42-116). Vol. III, 90 ll., 87 pll. (col.; nos. 117-203). Vol. IV, 82 ll., 80 pll. (col.; nos. 204-283). Vol. V, 79 ll., 77 pll. (col.; nos. 284-360). London. *360 pll.*

Published in 25 parts of which Pt. I was issued in 1849, II in 1851, III and IV in 1852, V and VI in 1853, VII and VIII in 1854, IX and X in 1855, XI and XII in 1856, XIII and XIV in 1857, XV and XVI in 1858 (May 1 and September 1), XVII and XVIII in 1859, XIX and XX in 1860 and XXI-XXV in 1861. The last part contained only the introductory matter, title-pages and lists of contents for each of the projected volumes. In the introduction, the nomenclature is revised and a brief review given of the various species of the group, including some not figured in the general work. The introduction was republished in 8vo form the year of appearance of the last part, 1861, under the title, "An Introduction to the Trochiliæ (q.v.). The allocation of the various plates to their respective parts is given by Waterhouse in his "The Dates of Publication of Some of the Zoological Works of the late John Gould," 1885 (q.v.). pp. 45-55. A supplementary volume was issued during the years 1880-87 (being completed by R. B. Sharpe after Gould's death) under the title, "A Monograph . . . Supplement" (q.v.).

Gould, John. (Sharpe, Richard Bowdler.)

1850-83. The | birds of Asia. | By | John Gould, F.R.S., | F.L.S., [etc., 8 lines.] | Dedicated to the Honourable East India Company. | In seven volumes. | Volume I [-VII]. | London: | Printed by Taylor and Francis, Red Lion Court, Fleet Street. | Published by the author, 26 Charlotte Street, Bedford Square. | 1850-1883.

7 vols. imperial folio. Vol. I, 4 pr. ll., pp. 1-9+1, 77 ll., 76 pll. (col.). Vol. II, 77 ll., 75 pll. (col.). Vol. III, 80 ll., 78 pll. (col.). Vol. IV, 74 ll., 72 pll. (col.). Vol. V, 85 ll., 83 pll. (col.). Vol. VI, 77 ll., 75 pll. (col.). Vol. VII, 73 ll., 71 pll. (col.). London. *530 pll.*

Published in 35 parts, of which Pts. I and II appeared in 1850, III-XXVIII at the rate of one each year from 1851 to 1866, Pts. XXIX and XXX in 1877, XXXI in 1879, XXXII in 1880, XXXIII in 1882 and XXXIV and XXXV in 1883. The last three numbers were issued after the death of Gould (1881) by R. B. Sharpe who wrote the introduction and whose initials appear after many of the discussions of the various species in the posthumous parts. The year of appearance of each plate is given in the index to the volume in which the plate was finally placed, and a general index in Vol. I assigns each species to its proper volume. The following further dates are available for certain parts, from current reviews in the Ibis: **XVII**, April, 1865; **XVIII**, April 1, 1866; **XIX**, May 1, 1867; **XX**, April 1, 1868; **XXI**, April 1, 1869; **XXII**, March 1, 1870; **XXVI**, August 1, 1874; **XXVII**, March 1875; **XXVIII**, July 1, 1876; **XXIX**, April 1, 1877; **XXX**, October 1, 1877; **XXI** July 1, 1879; **XXXII**, July 1, 1880. The 530 plates are hand-colored, most of them being drawn by Gould but a few being the work of J. Wolf.

Gould, John.

1851-69. The | birds of Australia. | By | John Gould, F.R.S., | F.L.S. [*etc.*, *13 lines.*] | **Supplement.** | London: | Printed by Taylor and Francis, Red Lion Court, Fleet Street. | Published by the author, 26, Charlotte Street, Bedford Square. | 1869.

1 vol. imperial folio, pp. I-IV, 79 ll., 81 pll. (col.; 1 fold.). London.

Issued in five parts, of which Pt. **I** appeared on March 15, 1851, **II** on September 1, 1855, **III** on September 1, 1859, **IV** on December 1, 1867 and **V** on August 1, 1869, according to Mathews, Birds of Australia, Suppl. 4, p. 48, 1925.

The final arrangement of plates in the volume is different from that of their issue. The latter may be ascertained from the list of plates in the volume and from F. H. Waterhouse's "The Dates of Publication of Some of the Zoological Works of the Late John Gould," 1885 (q.v.). The present work is intended to treat of the species discovered after the publication of the folio to which this volume forms a supplement, "The Birds of Australia," 1840-48 (q.v.).

Among the species included in Pt. I, several were also discussed in a paper by Gould on the "Researches in Natural History of John McGillivray, Esq." published in Jardine's Contributions to Ornithology for 1850, pp. 92-106+105*, which will hold priority for the new names unless a more exact and subsequent date can be proved for Jardine's magazine. The species in question are those numbered 5, 6, 12, 16 and 45 in the list of plates; Nos. 36 (Pt. III.) and 67 (Pt. II.) have undoubted priority in Jardine's "Contributions." Nos. 5 and 6 were also published in the Proc. Zool. Soc. Lond., 1850, p. 200 (publ. Febr. 28, 1851) which likewise antedates Pt. I of the folio.

Gould, John. (Owen, Richard.)

1852-54. **A monograph** | of | **the Ramphastidæ,** | or | family of toucans. | By | John Gould, F.R.S., | F.L.S. [*etc.*, *7 lines.*] | London: | Printed by Taylor and Francis, Red Lion Court, Fleet Street. | Published by the author, 20, Broad Street, Golden Square. | 1854.

1 vol. imperial folio, 4 ll., pp. 9-26, 2 ll., 52 pll. (51 col.). London.

A revised edition of Gould's earlier work of the same title, 1833-35 (q.v.), with new plates and with discussions of various species not formerly treated. The uncolored plate of anatomical details and its accompanying text, from the pen of Richard Owen, are unchanged. The text and plates of the species newly discussed in this edition were republished the following year, 1855 (q.v.), as a supplement to the first edition.

Carus and Engelmann cite this edition as having appeared in 3 parts. Pt. I is reviewed under date of 1852 in the Naumannia, 1853, pp. 237-240. It is said to contain text and plates of *Pteroglossus bitorquatus, Pter. flavirostris, Pter. Azarae, Pter. castanotis, Pter. viridis, Selenidera piperivora, Ramphastos Cuvieri, R. carinatus, R. erythrorhynchus, R. Ariel, Andigena hypoglaucus, A. nigrirostris, A. laminirostris* and *A. cuculeatus*. The preface, which probably appeared with the last part, is dated May 1, 1854.

Gould, John.

1855. Supplement | to | the first edition of | a | monograph | of | the Ramphastidæ, | or | family of toucans. | By | John Gould, F.R.S., | F.L.S. [*etc., 7 lines.*] | London: | Printed by Taylor and Francis, Red Lion Court, Fleet Street. | Published by the author, 20, Broad Street, Golden Square. | 1855.

> Supplement | To The | First Edition | of a | Monograph | of the | Ramphastidae, | or | Toucans. | By | John Gould, F.R.S., &c. [| Part II.] | Contents. | [*Pt. I, 7 lines., double column,-*] Ramphastos Inca. | " brevicarinatus. | " ambiguus. | " citreolaemus. | Pteroglossus erythropygius. | " Mariae. | Andigena nigrirostris. | Andigena cucullatus. | " laminirostris. | Selenidera Gouldi. | Aulacoramphus castaneorhynchus. | " albivitta. | " atrogularis. | " caeruleicinctus. [*Pt. II, 3 lines in double column and 1 line single column*;-Pteroglossus Azarae. | "flavirostris. | " Sturmi. | Pteroglossus poecilorostris. | " Wiedi. | Aulacoramphus caeruleogularis. | Title.-Introduction.-List of Plates.] | London: | Published by the Author, 20, Broad Street, Golden Square. | January 1st, 1855. | [Price Three Guineas (Price Two Guineas).].

1 vol. imperial folio, 2 pr. ll., pp. 9-26, 20 pll. (col.). London.

The present work consists of the plates and accompanying text relating to the species discussed by Gould in the second edition of his "Monograph of the Ramphastidae," 1852-54 (q.v.) but not included in the first edition, 1833-35 (q.v.). It seems to have been designed to add to the first edition to make it equivalent to the second edition, but as there are various alterations made in the species included in the first edition, of which changes no mention is made in the supplement, the equality is not maintained. The introduction, reprinted from the second edition, gives the classification as adopted in the complete work. The present copy is united with the volume to which it forms the supplement. There is, in addition, a copy of pl. 7 and its accompanying text, from Pt. I of Gould's "Icones Avium," 1837-38 (q.v.), inserted in the volume.

This supplement was issued in two parts, the cover-titles and contents of which are as given above, from a copy in the Bavarian State Museum, kindly examined by Dr. Q. Laubmann of that institution.

1926. CATALOGUE OF AYER LIBRARY—ZIMMER. 261

Gould, John.

1858-75. **A monograph | of | the Trogonidæ,** | or | family of trogons. | By | John Gould, F.R.S. &c. | London: | Printed by Taylor and Francis, Red Lion Court, Fleet Street. | Published by the author, 26 Charlotte Street, Bedford Square. | 1875.

1 vol. imperial folio, 2 pr. ll., pp. V-XX, 49 ll., 47 pll. (col.). London.

Published in 4 parts, of which Pt. I appeared in 1858, II in 1869, III in June 1875 and IV in September 1875. The allocation of plates is given by Waterhouse in his "The Dates of Publication of Some of the Zoological Works of the late John Gould," 1885 (q.v.), pp. 58 and 59. This is a second edition of the same author's work of 1838 (q.v.), but is entirely rewritten and is illustrated by new plates.

Gould, John.

1861. **An | introduction | to | the Trochilidæ,** | or | family of humming-birds. | By | John Gould, F.R.S., &c. &c. | London: | Printed for the author, by Taylor and Francis, Red Lion Court, Fleet Street. | 1861. | [The Author reserves to himself the right of Translation.].

1 vol. post 8vo, 4 pr. ll., pp. I-IV, 1-212. London.

This volume consists of the introductory matter of Gould's folio "Monograph of the Trochilidae," 1849-61 (q.v.), pp. V-CXXXVII, with an explanatory "Notice." The indices, at the close of the introduction, are improved by page-references, and the volume forms a handy reference book to the folio while giving a brief review of the classification of this group of birds. The present copy is one presented to P. L. Sclater by Gould, and contains numerous annotations by the former.

Gould, John.

1862-73. **The | birds | of | Great Britain.** | By | John Gould, F.R.S., &c. | In five volumes. | Volume I [-V]. | London: | Printed by Taylor and Francis, Red Lion Court, Fleet Street. | Published by the author, 26, Charlotte Street, Bedford Square. | 1873.

5 vols. imperial folio. Vol. I, 6 pr. ll., pp. I-CXL, 35 ll., 37 pll. (col.), 1 text-fig. Vol. II, 78 ll., 78 pll. (col.), text-figs. 1 and 2. Vol. III, 71 ll., 76 pll. (col.). Vol. IV, 87 ll., 90 pll. (col.). Vol. V, 88 ll., 86 pll. (col.). London. 367 pll.

Published in 25 parts. Pts. I and II appeared on October 1, 1862; thereafter two parts appeared annually (on August 1 and September 1 in 1866 to 1872 and probably also in the other years), and the last part, containing title-pages and indices for all the volumes, the preface and introduction, and certain plates, was issued in December, 1873. The allocation of the plates to their respective parts as issued has been made by Waterhouse in his "The Dates of Publication of Some of the Zoological Works of the Late John Gould," 1885 (q.v.), pp. 6-14. The 367 hand-colored plates are mostly by Gould; a few are by J. Wolf.

Gould, John.

1865. **Handbook** | to the | **birds of Australia.** | By | John Gould, F.R.S., etc. | Author of [*etc., 5 lines.*] | In two volumes. | Vol. I [II]. | London: | Published by the author, | 26 Charlotte Street Bedford Square. | 1865. | [The right of Translation is reserved.].

2 vols. 8vo. Vol. I, pp. I-VIII, 1-636, 1 insert-slip (advt.). Vol. II, 3 pr. ll., pp. 1-629, figs. 1-3. London. 1865 (September 1 and December 2?).

A reprint of the text (without the introductory matter) of the author's "The Birds of Australia," 1840-48 (q.v.), and the three published parts of the "Supplement," 1851-69 (q.v.), with such additions, corrections and alterations as were thought necessary. On pages 523-583 of Vol. II, the species treated by the author in his larger work which are not strictly Australian in distribution are collected and discussed separately. The dates of publication have been taken from an insert-slip by the publishers offering Vol. I as published "this day," September 1, and promising Vol. II for December 2. Both volumes are reviewed in the Ibis for Jan. 1866, pp. 111-113, as of completion in 1865.

Gould, John.

1873. **An** | **introduction** | **to the** | **birds of Great Britain.** | By | John Gould, F.R.S., &c., &c. | London: | Printed for the author, | by Taylor and Francis, Red Lion Court, Fleet Street. | 1873. | [Price Five Shillings and Sixpence.] | | [The Author reserves to himself the right of Translation.].

1 vol. post 8vo, 2 pr. ll., pp. I-IV, 1-135+1, 1-14 (list of subscribers), 1-4 (advt.), 1 text-fig. London.

This volume contains the lengthy introductory matter of the author's "The Birds of Great Britain," 1862-73 (q.v.), printed in 8vo form for ease of correction and issued as such. The date of issue may be later than that of Pt. XXV of the folio which contained the same matter (4 pr. ll. and pp. I-CXL), although the pages of advertising at the end of the present volume are dated August, 1873, and the preface (as in the folio) is dated November 1, the same year.

Gould, John; and Sharpe, Richard Bowdler.

1875-88. **The** | **birds of New Guinea** | and the | adjacent Papuan islands, | including many | new species recently discovered | in Australia. | By | John Gould, F.R.S. | Completed after the author's death | by | R. Bowdler Sharpe, F.L.S. &c., | Zoological Department, British Museum. | Volume I [-V]. | London: | Henry Sotheran & Co., 36 Piccadilly. | 1875-1888. | [All rights reserved.].

5 vols. imperial folio. Vol. I, 2 pr. ll., pp. I-III+1, 56 ll., 56 pll. (col.). Vol. II, 60 ll., 58 pll. (col.). Vol. III, 74 ll., 72 pll. (col.). Vol. IV, 61 ll., 59 pll. (col.). Vol. V, 77 ll., 75 pll. (col.). London.

Published in 25 parts, of which Pt. I was issued on Dec. 1, 1875; II, Jan. 1, 1876; III, May 1, 1876; IV, Jan. 1, 1877; V, June 1, 1877; VI, Febr. 1, 1878; VII, June 1, 1878; VIII, Oct. 1, 1878; IX, March 1, 1879; X, Sept. 1, 1879; XI, Febr. 1, 1880; XII in 1881; XIII in 1882; XIV and XV in 1883; XVI-XVIII in 1884; XIX and XX in 1885; XXI and XXII in 1886; XXIII in 1887; and XXIV and XXV in 1888. The distribution of the plates among their respective parts is given in the indices to the various volumes, which, with the introductory matter and title-pages, appeared with Pt. XXV. Gould's death in 1881 transferred the responsibility of the monograph to R. B. Sharpe who completed the work and to whom must be ascribed the subject-matter appearing in Pts. XIII-XXV. The 320 hand-colored plates are by Gould and W. Hart. The work was brought to a close long before the subject was exhausted.

Gould, John. (Sharpe, Richard Bowdler.)

1880-?. **Monograph | of the | Pittidæ.** | [*Vignette.*] | By | John Gould, F.R.S. &c. | Part I. | Contents. | Pitta bengalensis. Eucichla ellioti. | Eucichla cyanura. Leucopitta maxima. | - boschii. Cyanopitta steeri. | - schwaneri. Phœnicocichla arquata. | - gurneyi. Melampitta lugubris. | London: | published by the author, 26, Charlotte Street, Bedford Square, W.C. | October 1st, 1880. | [N. B. The Illustrations are principally taken from the Author's works on the 'Birds of Asia,' 'Australia,' and 'New Guinea.'] | [Price Three Guineas.].

1 vol. imperial folio, 10 ll. 10 pll. (col.), [10 ll.]. London. October 1, 1880 (-?).

All published of the author's projected work which was interrupted by his death. There appears to be some uncertainty as to the actual publication of the final 10 sheets without plates. These were written by Gould but edited by Sharpe, whose initials appear at the foot of each. Sharpe, himself, in "An Analytical Index to the works of the late John Gould," p. XXIII, 1893, says that a single part was published. The title, quoted above, is from the original wrapper (or board cover) where the first part alone is mentioned. The remaining pages may have been sold by the printer without authority.

Gould, John; and Sharpe, Richard Bowdler.

1880-87. A | **monograph** | **of** | **the Trochilidæ,** | or | family of humming-birds. | By | John Gould, F.R.S. | Completed after the author's death | by | R. Bowdler Sharpe, F.L.S. &c., | Zoological Department, British Museum. | Supplement. | London: | Henry Sotheran & Co., 36 Piccadilly. | 1887. | [The right of translation is reserved.].

1 vol. imperial folio, 104 ll., 58 pll. (col.; by Gould and Hart). London.

Published in 5 parts, of which Pt. I appeared in 1880, II in 1881, III in 1883, IV in 1885 and V in 1887. Gould died after the first part had been issued, but Pt. II appeared shortly thereafter in the form which he had planned. The com-

pletion of the work was then undertaken by the publishers who obtained the services of Sharpe to write the remaining text, Hart to prepare the plates, and Salvin to supervise the whole. Gould had planned to complete the work in four parts for which he had already prepared a large number of the plates. It was found that he had underestimated the probable extent of the project and that there were many species yet to be included if the work were to be a complete exposition of all the hummingbirds discovered since the publication of the original monograph (q.v.) in 1849-61. The resulting increase in the text and illustrations required an extra part.

The allocation of the plates of Pts. I-IV is given by Waterhouse in his "Dates of Publication of Some of the Zoological Works of the late John Gould," 1885 (q.v.), pp. 45-55; the remaining plates and the introductory matter appeared in Pt. V.

Gould, John.

1885. See Waterhouse, F. H., **The Dates of Publication of Some of the Zoological Works of the late John Gould**, F.R.S.

Gould, John.

1893. See Sharpe, R. B., **An Analytical Index to the Works of the late John Gould**, F.R.S.

Graham-Smith, G. S.

19__. See [Grouse], **The Grouse in Health and Disease.**

Grandidier, Alfred.

1874. See Milne Edwards, Alphonse, **Recherches sur le Faune Ornithologique Éteinte des Iles Mascareignes et de Madagascar.**

Grandidier, Alfred. (Milne-Edwards, Alphonse.)

1876-85. > Histoire | physique, naturelle et politique | de | Madagascar | publiée par Alfred Grandidier [*Period added (Vols. XIV and XV.).*] | Volume XII [-XV]. | **Histoire naturelle des oiseaux** | par | MM. Alph. Milne Edwards et Alf. Grandidier. | Tome I [-IV].-Texte [Atlas.-I; Atlas.-II; Atlas.-III]. | [*Blazon*.] | Paris. | Imprimé par autorisation de M. le Garde des Sceaux [*Comma added (Vols. XIV and XV.).*] à l'Imprimerie Nationale. | M DCCC LXXIX [M DCCC LXXV; M DCC LXXIX; M DCCC LXXXI].

4 vols. royal 4to. Vol. XII, 3 pr. ll. (half-tit., tit. and pref.), pp. 1-779. Vol. XIII, 2 pr. ll. pll. 1-104, 1A, 9A, 9A bis, 9B, 9C, 12A, 13A, 14A, 16A, 19A, 24A, 26A, 29A, 29B, 29C, 30A, 32A, 36A, 36B, 36C, 36D, 38A, 39A, 40A, 41A, 41B, 41C, 66A, 66B, 84A, 89A, 103A, 104A, 104B, 104C (61 col., by Keulemans, Huet and

Hilgert, Carl.
1908. Katalog | der | Collection von Erlanger | in | Nieder-Ingelheim a.Rh. | von | Carl Hilgert. | Berlin 1908. | Verlag von R. Friedländer & Sohn.

1 vol. 8vo, pp. XVII+1, 1-527, frontisp. Berlin.

A catalogue of 12,589 skins and 1,140 sets of birds' eggs in the collection of Carlo von Erlanger, the majority of which appear to be from portions of Africa.

Hill, J. W.
[1844.] [Illustrations for DeKay's "Zoology of New York."]
1 vol. (8½x11 in.). Not published.

A series of 94 original water-color drawings, 76 of birds and 18 of insects, the originals of the corresponding plates in DeKay's "Zoology of New York," 1844 (q.v.). The drawings are trimmed to approximately 4½x6½ in., and are mounted on cardboard.

Hill, Richard.
1847. See Gosse, Philip Henry, **The Birds of Jamaica.**

Hinds, Richard Brinsley. (Gould, John.)
1843-44. No. III (IV).] [Price 10s. | **The** | **zoology** | **of** | **the voyage of H. M. S. Sulphur,** | under the command of | Captain Sir Edward Belcher, R.N., C.B., F.R.G.S., etc. | during the years 1836-42. | Published under the authority of | the Lords Commissioners of the Admiralty. | Edited and Superintended by | Richard Brinsley Hinds, Esq., Surgeon, R.N. | attached to the expedition. | Birds, | by | John Gould, Esq., F.R.S., etc. | London: | published by Smith, Elder and Co. 65, Cornhill. | MDCCCXLIII [MDCCC-XLIV]. | **Birds.-Part I [II].** October, 1843 [January, 1844]. | Stewart and Murray, Old Bailey. [*Cover titles.*]

2 parts, royal 4to. No. III (Birds,-Pt. I.), pp. 37-44, pll. 19-26 (col.; by Gould and Waterhouse Hawkins). No. IV (Birds.-Pt. II.), pp. 45-50, pll. 27-34 (col.; by Gould). Copy in original wrappers. London.

The ornithology of the voyage (by Gould) forming Pts. III and IV of the complete zoological report, issued in (12) parts from 1843-45. A number of new species of birds described herein are cited as having been previously diagnosed by the author in the Proceedings of the Zoological Society of London for 1843. Those which occur in Pt. IV (Birds,-II) may be quoted from the Proceedings; those in Pt. III (Birds,-I) must be quoted from the present paper since the pages of the Proceedings on which they occur were not issued until December, 1843, two months after the date of this part of the present work (as shown by the covers).

PART FOUR BIBLIOGRAPHY OF JOHN GOULD, HIS FAMILY AND ASSOCIATES

ZOOLOGICAL SOCIETY OF LONDON. 1830. *Reports of the auditors of the accounts of the Zoological Society for the year 1829, and of the council read at the anniversary meeting, May 3, 1830.* London, Richard Taylor.

This pamphlet of sixteen pages is reported here for interest. It is in a bound volume of miscellaneous pamphlets in the University of Kansas Spencer Library-O'Hegarty Collection C595.

Also in this bound volume is a 'list of the members of the Zoological Society' as of May 1830, on twenty-four pages. Gould was not on this list.

The final item in this volume of miscellaneous pamphlets is pages 1 to 44 of the *Proceedings of the Committee of Science and Correspondence of the Zoological Society of London.* (*See also* the several articles by N. A. Vigors (1830-31) on the birds of the Himalayan mountains, and *see* Journal. 1830-32. *Proc. Comm. Science and Corres. Zool. Soc. Lond.*)

ZOOLOGICAL SOCIETY OF LONDON. 1830-44. [Reports, list of members, etc.] London, Zoological Society.

Here is a collection of papers beginning with an address by N. A. Vigors on the sixth and last meeting of the Zoological Club of the Linnean Society of London, dated 29 November 1829. On page 16 of this address is this statement: 'We are indebted to Mr. Gould of the Zoological Society, for the discovery of this addition [*Sylvia tithys,* a warbler] to our British Fauna. The bird had been sent to him as a common Redstart (*Sylvia phoenicura*), to which it bears a close affinity; and probably would have passed unnoticed as a specimen of that species, more particularly in consequence of its [female] sex . . . had not the critical knowledge of this rising naturalist detected the distinguishing characters'.

Then follow articles on the Charter of the Zoological Society, By-laws, annual reports of the auditors and the council, etc.

In the 1838 Report on page 9 is mention of a donation to the Society Library by Mr Gould, parts of his *Birds of Europe* and a work on the *Birds of Australia* just commenced.

'A List of the Fellows' as of May 1842 contains Gould's name: 'Gould, Mr. John, F.L.S., 20, Broad Street, Golden Square.' This is the only list of members in this volume.

This copy is in the University of Kansas Spencer Library-Ellis Collection G-6.

Postscript

For whom have I compiled and written this volume? First, for myself because I wanted to put down in print some previously very scattered data on John Gould. The evolution of the work has been gradual, easy, relaxing, and pleasant for me at most times, pressured only near the date of completion of this volume, and then the pressure was self-imposed.

Additionally, this compendium was written for ornithological historians, for general natural-history bibliographers and librarians, and for some students of the Victorian era. Finally, there will be a few persons, neither ornithologists nor bibliographers, who will find this compilation about John Gould — a most prolific naturalist — interesting and, I hope, enjoyable. This book should appeal to a British audience for obvious reasons; but Australians, if they remember the Gould Society of Bird Lovers of their youth, should be even more interested — here is a bibliography of an Australian pioneer naturalist.

As I look back over the years that have been devoted to this study of John Gould and his works, my intention from the start was to learn something about the man who had published those bird prints I first obtained from an Illinois antique dealer. What has evolved over the years, and is now committed to print, is a fairly broad study of the natural history of the nineteenth century, really quite encompassing, but, of course, always specifically related to one man, John Gould, the Bird Man.

Index

Any comprehensive reference work is only as good as its index. Some pertinent remarks concerning the coverage need to be made.

For Part One, Genealogy, the individual members of the Gould and Coxen families are not indexed unless they made a contribution to the genealogy.

For Part Two, John Gould's major published works, the works themselves are indexed under the abbreviated titles listed on page xi.

For the genera or species of any bird, mammal, fish, amphibian or reptile, they are listed under the appropriate group. Thus, the largest grouping is under 'Birds, genera and species'. Where there are multiple listings of birds, etc. only the page or the list is indexed.

In Part Three, Chronology, for Gould's works, only the *initial part* as issued is indexed.

The sources of the items mentioned or quoted in Part Two and Part Three are not indexed. A long correspondence series is indexed for the *initial* correspondence only.

In Part Four, the authors of the over 580 items are indexed, along with the general heading of the subject.

Abbot, C. C., 339
Abel, 305
Aberdeen, N.S.W., 273
Aberdeen on the Dartbrook, 117
Aboriginal native cranium, 238
   terms, 206
Aborigines, 124, 186, 202, 292, 352
Academy of Natural Sciences, Philadelphia, 25, 79, 126, 139, 160, 161, 183, 226, 230, 250, 251, 263, 283, 303, 306-11, 308, 320, 334, 336, 353, 358
  Library, 249
Actaeon Islands, 105
Adams, A. L., 160, 375
Adams, H. G., 160
Adelaide, 114, 215, 221, 233, 235, 362
Adelaide, Queen, 278
Agasse, James Laurent, 353
Agassiz, Louis, 160
Agnew, I., 215
Aiton, John Townsend, 241, 278, 325, 339
Aiton, William, 267, 365
Alabama coal shed, 299
Albertis, L. M. d', 161
Albin, Eleazar, 350, 368
Alexander, Messrs Wm. & Co., 109
Alexander, W. B. (two items), 161, 191, 365
Alexandra, Princess, 365
Allan, W., 278
Allan, Wm., 365
Allen, Glover M., 161

Allen, J. A., 84, 85, 152, 255, 275
Allen, Lieutenant, 339
Allen, W., 201
Allis, _____, 278
Allis, Thomas, 365
Allport, Morton, 278, 365
Alnwick Castle, 318
Alston, Edward R., 196
Altoona, Pennsylvania, 141
American Art Association, 184, 333
American collections, 302
  museums, 336
American Ornithologists Union, 152, 161, 275, 286
American ornithology, 354
American Philosophical Society Library, 212
Amhurst, Mrs, 82
ammunition, 232
amphibians, genera and species
  J. E. Gray list, 263, 264
  lizards, 294
Amsterdam, 57
Analytical Index (Sharpe), 183, 275, 277, 287, 292, 297, 338, 347
Anderson, C. J., 319
Anderson, Major, 278
Andrews, F. W., 278
Angas, George Fife, 221
Angas, George French, 278, 297
  correspondence with Gould, 221, 239, 365
Anker, Jean, 13, 159, 161

JOHN GOULD — THE BIRD MAN

Annals and Magazine of Natural History, 98, 275, 287, 338
Annals of Natural History, 275
anonymous writing on Gould, 162, 277
Ansdell, Richard, 319
anthology, 320, 321
antiquarian book trade, 180
antique dealers, 332
Archer, 219, 220
Argent (dealer in specimens), 179, 365
Argyll, Duke of, 278, 319
Armstrong, E. A., 162
Armstrong, Frances Fraser, 364, 365
Armstrong, J., 297
Armstrong, John, 352
Armstrong, Sir William, 318
arsenical soap, 276
Art of natural history, The (S. P. Dance), 197
Arthur (author), 290
Ashby, Edwin (two articles), 162, 310, 362, 365
Asia, Birds of, see *Birds of Asia*
Asia (Mail Steamer), 139, 142, 242
Asiatic Society of Bengal, 360
Aspinal, Mr, 232
Astrolabe (ship), 310
Athenaeum Club, 136, 262
Atlantic cable, 143
Atlantic Ocean, 276
auctions, natural history, 184
Audubon, John James, 21, 27, 28, 89, 90, 92, 94, 101, 162, 175, 176, 180, 184, 185, 191, 192, 197, 204, 205, 254, 264, 267-8, 272, 274, 277, 278, 295, 296, 308, 310, 317, 319, 321, 322, 343, 344, 347, 354, 362, 368
    death, 128
    folio sets, 205
    gift of dog from Gould, 193
Audubon, Len 27, 272
Audubon Magazine, The, 286
Audubon, Victor, 204
Auk, The (organ of American Ornithologists Union), 152
Auk, Great (egg), 146
Aurousseau, M., 167
Austin, Robert, 278, 365
Australia
    animals, 357
    bird bibliography, 365
    birds, 189, 206, 233, 302, 353, 359
    botany, 320
    colonies, 167
    drought, 111
    exploration, 204, 264, 315
    Gould's Australian bird types, 306-11
    Gould's Australian journeys, 34, 99, 118, 232
        I Tasmania-Hobart area, 103
        II Recherche Bay, 105
        III Northern Tasmania and Bass Strait Islands, 106
        IV Sydney-Maitland, Upper Hunter, Liverpool Ranges, 109
        V South Australia-Adelaide-Murray River, 114
        VI En Route from Hobart to Yarrundi, 115
        VII Liverpool Range-Mokai River, Namoi River, Brigalow, 117
        VIII Sydney-Wollongong-Illawarra District, 118
    return to England, 119
Australia and the Australians (Younger), 373

Australia, Birds of, see *Birds of Australia*
Australia, Handbook of Birds of, see *Handbook*
Australia, Mammals of, see *Mammals of Australia*
Australian Museum, Sydney, 346
Australian National Library, Canberra, see National Library, Canberra
Australian ornithology, 364
autograph collection (Woodward), 371
aviculture, 184, 331
Ayer, Edward E., Collection, 374
Azara, Don Felix de, 274

Bachman, John, 185, 193-4, 205, 279, 316
Backhouse, James, 167, 201, 297
Baillie, James, Collection, 275
Baily, William, 168
Baily, William Loyd, 139, 167-8, 249, 279, 303, 321
Bain, Mr, 68
Baird, Spencer Fullerton, 168, 197, 226, 279, 282, 310
    initial correspondence, 139
Baker, 339
Baker (gardener), 279
Baker, Joseph, 179, 194, 234, 279, 280
Baker, Mr (of island near Newcastle), 232
Baker, Thomas, 365
Baldamus, Eduard, 133
Balfour and Newton Library, see Newton Library
Baltimore, Maryland, 141
Bank, George, 99
Bankier, R. A. Dr, 279, 365
Barclay-Smith, P. (two items), 169
Barlow, Rev. John, 312
Barlow, Nora, 169
Barlow, T. Oldham, 312
Barrett, Charles (five items), 169, 170, 191, 292, 326
Barrett, C. L., 191, 365
Barry, Sir Redmond, 222, 223
Barton, M. T., 170, 344
Bartram's Gardens, Philadelphia, 139, 168, 303
Bartlet, Consul, 101
Bartlett, Abraham D., 319
Bass Strait, 106, 355
Bastin, John (two items), 173-4
Bateson, Charles, 173
Batrachian Collection, 180
Bauer, _____, 279
Bavarian State Museum, 72
Bay of Biscay, 100
Bayfield, _____, 279
Beagle, H.M.S., 32, 33, 40, 94, 97, 112, 169, 214, 264, 297, 315, 360
Beagle, Voyage of the, see *Zoology of the voyage of H.M.S. Beagle*
Beauffort, Leopold de, 326, 365
Beck, H. N., 365
Becker, Bernard H., 174
Becker, L., 279, 365
Bedford, 219, 220
Beechey, Captain, 169
Beinecke Rare Book and Manuscript Library, 64
    See also Yale University Library
Belcher, C. F., 191
Belcher, Captain, 169, 214
Belcher, Sir Edward, 339

INDEX

Belgrave, Lord, 209, 210
Bell, Alexander Graham, 149
Bell, J., 365
Bell, Thomas, 178, 274
Belltrees, N.S.W., 273
Bennett, Edward Turner, 174, 274, 312
Bennett, Dr George, 98, 112, 115, 116, 117, 118, 170 (three items), 174-5, 214, 218, 219, 221, 238, 279, 291, 297, 344, 365
Bennett, Mrs, 279
Benstead, James, 99, 106, 107, 219, 279, 280, 346
Bentham, G., 54
Bentinck, Lord Wm., 279
Berkeley, California, 332
Berlin, 57
Berrima, N.S.W., 118
Bettany, G. T., 212
Bewick, Thomas, 93, 273, 274
bibliographies reprinted, 160, 163-6, 327-9, 369-73, 375-92
bibliography, 181, 200, 201, 305, 326, 344, 354, 364-8, 374
    Australian, 202
    Australian entomology, 316
    British ornithology, 315
    (Mathews 1920-25), 300
    (Mathews 1925), 301
Bibliography, ornithological (Coues), 194
Bibliotheca zoologica (Carus & Engelmann), 183
Bicheno, J. E., 220, 274
Bickmore, A. S., 146, 226
Bidder, Mr, 137
bird eggs, 248
    See oology *and* eggs, birds
bird illustrators, 295, 344
bird specimen dealers, 179
    gifts by Gould, 255
    preservation, 147
bird synonyms, 354
birds, genera and species
    Acanthorhynchus, 256
    Accipitres, 354
    Adamastor cinerea (Great Grey Petrel), 102, 119
    Anthus spinoletta, 283
    Aegithaliscus anophrys, 284
    Aix galericulata, 335
    Albatross, 105
        Sooty, 119
    Alectura lathami (Brush Turkey), 118
    Amazilia, 282
    Anatidae (Duck Tribe), 202
    Androdon aequatorialis, 372, 374
    Anthochaera lunulata, 301
    Apteryx, 34
        A. australis, 290
        A. owenii, 290
    Artamus sordidus, 276
    Astur palumbarius, 282
    Auk, Great, 146, 320
    Australian, 189, 330
    Australian birds to Philadelphia, 126
    Australian list, 302
    Balaeniceps rex (Shoebill), 39, 128, 277, 290, 294, 295, 359
    Birds of Paradise, 143, 177, 202, 208, 272, 275
    Birds with *Gouldi*, 266
    Bittern, 115
    Boshas bimaculata, 276

Bonaparte article, 175
Bower-bird, 119, 208, 272, 298, 321
    Great Grey, 244
Bronze-wing Crested, 118
    Harlequin, 118
Budgerigars, 162, 331, 338
Budytes flava (Yellow Wagtail), 101
Bullfinch, 180
Bustard, Thick knee'd, 91
Cancroma, 128
Canthorhynchus, 256
Caprimulgidae (Goatsuckers), 38, 176, 177
    C. europaeus, 101
Cassowary, 233, 326, 350
Casuarius australis, 70
C. bennetti, 69, 175
C. uniappendiculatus, 70
Century of Birds list, 18-20
Channel-bill, 321
Cinnyris gouldiae, 359
Cockatiel, 187
Cockatoo, Rosebreasted, 182
Collurio, 359
Columba turta (Turtle Dove), 101
Cometes, 133, 177
Contemporary notes, 190
Coraciae, 178
Corvidae, 252, 290
Coturnix, 290
Curruca arundinacea (Reed-Warbler), 287
Coues 1878-80: 194
Cuckoo, Black-eared, 118
    European, 197
    Yellow-billed, 321, 334
Cypselus alpinus, 289
    c. australis (Australian Swift), 110
Daption capensis (Cape Petrel), 102, 119
Dendrortyx, 238
Dendrocitta leucogastra, 290
Diardigallus fasciolatus, 282
Diomedea chlororhynchos, (Yellow-nosed Albatross), 102
    D. exulans (Wandering Albatross), 102
Diurnal birds of prey, 265
Dodo, 265, 290
Eagle, 107
    Splendid, 232
    Whistling, 117
Egret, Australian, 118
Ellis Collection, 333
Emberiza (*Euspiza*) *melanocephala*, 284
Emu, 338
Eopsaltria griseogularis, 301
Euphonia gouldi, 336
Eurylaimus lunatus, 290
Falco tinnunculus (Kestrel), 101
Falcons, 180
Finch
    Gouldian, 342
    House, 185
    Spotted-sided, 117
Finches of Galapagos Islands, 291
Flamingo, 296
Fratercula placientius, 51
General characteristics, 233
Gerygone fusca (Brown Warbler), 162
Gilbertornis (*Pachycephala*) *rufogularis*, 162
Goose, Cape Barren, 108
    Pygmy, specimen, 303
Gould Australian types, 306-11
Gould Collection, 152, 336
Gould's Australian notebook, 236

Guinea Fowl, Verreaux's, 182
Hawks, 290
Heron, White-fronted, 104
Histriophaps histrionica (Flock Pigeon), 295
Honey-eater, Purple-gaped, 115
Hummingbird collection, 264
Hummingbirds, 85, 126, 128, 139, 150, 160, 167, 177, 180, 199, 240, 244, 270, 277, 286, 298, 321, 325
    Ruby-throated, 139
Hypsipetes, 359
Icones avium partial list, 36
in British Museum, 263
in Chip Boxes, 238
Lanus, 359
Leptotarsus eytoni, 292
Lesbia gouldi, 168
list by Captain Boys, 237
list by Gilbert, 188, 206, 207, 214
list by Gould, 214
list from G. R. Gray, 263
list from George Grey to British Museum, 235
list from Western Australia, 264
list of Australian birds (MacGillivray), 295
list of egg drawings, 258-62
list of families, 230
list of 33 original bird drawings, 244-8
list of skeletons, 238
lithographic stones of twelve species, 226
Longipennes, 232
Lovebirds, 242
Lyrebird, 338
Macroura, 56
Magpie, Red-billed Blue, 274
Mallee-Fowl, 200
Mallee Hen, 321
Malurus callainus (Turquoise Superb Warbler), 362
Menuras, 99, 117
    Menura alberti, 294
    M. superba, 111
Mississippi Kite, 310
Motacilla (Wagtail), 289
    M. boarula, 276
    M. flava, 289
    M. neglecta, 289
Muscicopa parva, 283
Musophaga rossae, 277
new Australian species, 300
new species, 156
Noisy Scrub-bird, 292
non-passerines, 361
Notornis, 127
    N. mantelli, 290, 318
number collected in Australia, 240
Odontophorinae (Partridges of America), 177, 370
Oreotroclulus, jamesonii, 276
original drawings, 191
Ortyx, 56
    O. plumifera, 253
Owl, 198, 290
    Spotted, 190
    Tengmalm's, 161, 317
Oystercatcher, Black, 106
Pachycephala rufogularis, 362
Paradiseidae (Birds of Paradise), 338
Paradoxornis, 39, 275
Paris meridionalis, 169
Parrots, 105, 187, 324
    Ground, 205
    Night, 205, 363

Partridges of America (Odontophorinae), 123
passerines, 361
Pelargopsis gouldi, 360
Penguins, 105
Petrel
    Black-bellied Storm, 102, 118
    Blue, 119
    Grey-backed Storm, 118
    Snow White, 106
    Stormy, 100, 101
    White-headed, 109
Phaebetria (*Diomedea*), *fulginosa* (Sooty Albatross), 102
Pharomacrus auriceps, 72
Pheasant, 143, 177
    Asian, 185
    St. Helena, 277
Phylloscopus rufus, 283
Picus imperialis, 289
Pigeon, Flock (Harlequin Bronze Wing), 117
Platycercus adelaidae, 162
Polytelis barrabandi, 34
Psaltria concinna
Psittacidae (Lear), 299
Ptilonorhynchidae (Bower-birds), 338
Ptilotis, 312
Puffinus cinereus
    P. obscurus, 101
raptorial, 265
Regent's Bird, 321
reviews, *see* [Gould, John] reviews, 251-5
Rhea darwinii, 169, 315
Robin, Buff-sided Scrub, 85
Sabre-wing, DeLattre's, 168
Scytalopus, 277
Skua, Great, 321
Sterna stoldida [stolida] (Noddy Tern), 101
Strepera arguta (Clinking Currawong), 104, 106
Strickland Collection, 332
Strix brachyotus (Short-eared Owl), 101
Sulphur, H.M.S., list, 51
Swan, Black, 106, 108, 295, 321
Swan River birds, 206
Swifts, Australian, 112
Sylphia, 168
Sylvia phoenicura, 393
    S. tithys, 93, 358, 373, 393
T (?). temminckii, 276
Thalassidroma pelagica (Stormy Petrel), 101
T. wilsoni, 101
Toucans, 212
Tri-coloured Ephthianura (Crimson Chat), 118
Trochilidae, 195, 200, 221, 231, 263, 282, 330, 336
Trochilus, 142, 275
    T. alexanderi, 140
    T. loddigesii, 358
    T. verticeps, 277
Trogon, Resplendent (Quetzal), 351
Trogonidae listed, 28
Trogans, 212, 325
tropical birds, 343
Tubinares (Petrels and albatrosses), 161
Tudus atrigularis, 284
type specimens, 353, 359, 360
Upupae, 178
Waterhen, Purple, 221
Western Australia, 338
Whimbrel, Little, 112
Wren, Emu, 105, 106

INDEX

Birds of Asia, 66, 200, 251, 277, 353
   initial part, 127
   Rutgers edition, 331
   temporary title pages, 66
Birds of Australia, 41, 184, 190, 194, 204, 221, 241, 244, 268, 292, 302, 310, 320, 321, 323, 337, 355, 364
   facsimile, 256
   initial part, 119
   manuscript, 237
   paper covers, 211
   pattern plates, 213
   prospectus, 187
   reviews (3), 254
   Rutgers edition, 330
   shipping and sale of, 216
   subscribers correspondence, 232
Birds of Australia and adjacent islands, 34, 253, 272, 393
   facsimile, 257
   initial part, 98
Birds of Australia (cancelled work), see *Birds of Australia and the adjacent islands*
Birds of Australia (Mathews), 299
   review, 301
Birds of Australia, Supplement (Mathews), 300
Birds of Australia, Supplement, 69, 277, 318
   facsimile, 256
   initial part, 128
   pattern plates, 69, 272, 302
Birds of Europe, 21, 211, 216, 241, 253, 263, 268, 317, 357, 393
   Audubon copy, 193
   initial part, 94
   manuscript, 229, 230
   notebook, 321, 334
   original drawings, 324
   prospectuses (2), 249
   reviews (3), 252
   Rutgers edition, 330
Birds of Great Britain, 74, 254, 265, 305, 368
   initial part, 145
   manuscript, 230
   prices, 356
   prospectus, 223
Birds of New Guinea, 82, 275, 338
   initial part, 149
   review, 255
   Rutgers edition, 331
Birds of the Colorado Valley, 195
Birds of South America (Rutgers), 331
Birds, types, 310
Blacker-Wood Library, McGill University, 23, 58, 63, 162, 182, 237, 238, 313, 324, 369, 371
Blackett, Mr, 111
Black Joke (ship), 114, 294
Blackwood, F. P., 365
Blanchard, Laman, 279, 339
Bland, Sarah, 239
Blandowski, W., 365
Blaxland exploration, 204
Blomefield, see Jenyns, Leonard
Blunt, Wilfrid, 343
Blyth, Edward, 198, 200, 279, 282
boats, listing, 294
boats used by Gould and party, see ships
Bolivia, 275, 330
Bombay, 137
Bonaparte, Prince Charles Lucien (or Lucian), 41, 44, 58, 61, 64, 67, 71, 168, 175, 185, 194, 202, 216, 264, 268, 279, 310, 322, 323, 365

Audubon comment, 194
   death, 142
   initial correspondence with Gould, 120
Bonaparte, Napoleon, 89
Bond, J., 310
Bong Bong, N.S.W., 118
Books and Libraries at the University of Kansas, 305
book auction, 321
   catalogue, 323
   collectors, 36
   dealers, 316, 333, 346-51, 356, 374
Booth, Lieutenant, 279
Borell, Adrey, 175
Borland, Hugh, 175
Borrer, Wm., 196
Bossep [?], Mr Lock, 220
Boston, 141, 142
botanical collectors, 297
   drawings, 350
Botanical Magazine, see Curtis's *Botanical Magazine*
Botany, 320
Boubier, Maurice, 175
Boucard, Adolphe, 272
Bourcier, M. J., 184, 279
Bourjot-St. Hilaire, A., 310, 365
Bourke, R., 365
Bovell, Dr James, 142, 275
Bowen, W. W., 310
'Bower-bird', 298
   See also birds, genera and species
Bowler, _____, 279
'Boy's [sic] Birds', 237, 370
Boys, Captain William J. E., 237, 353
Boys's Collection, 66
Brasher, Rex, 368
Brayley, E. W., Jr, 288
Brehm, Ludwig, 133
Bremer, Sir Gordon, 221, 279
Brennan, M. T., 176
Brete, Captain, 279, 365
Breton, H. W. (Lieutenant), 218, 219, 220, 279, 365
Brett, Stanley, 180
Brevoort Hotel, 140
Brewer, Dr, 139
Bridson, G. D., 176, 324
Briggs, Mr, 339
Briggs, William, 179
Brisbane, Public Library of, 162
Brisbane ornithologists, 148
Brisson, Mathruin-Jacques, 255
British Association for the Advancement of Science, 39, 58, 70, 123, 126, 127, 133, 143, 144, 145, 174, 176, 238, 254, 274, 276
   initial meeting, 94
   member list, 176, 178
British bird painters, 293
British birds, 203
British Isles areas visited by Gould, see Gould, John. Journeys
British Museum, see Museum, British
British natural history, 357
British Ornithologists Union, 143, 150, 282
British ornithology, 315
Broad Street, Golden Square (London), 344-5
   Map, 173
   Pump, 136, 170, 172, 314, 344
Broadbent, Kendall, 301, 365
Brockman, Mrs Robert, 183, 279
Broderip, Mr, 339
Brodie, Sir Benjamin, 178

Broecke, van den, 331
Broinowski, Gracius, 213, 359
Brooks, Allan, 161
Brown, ____, 254, 279, 292
Brown, George, 279
Brown, Richard, 180
Brown, Robert, 279, 355
Browne, 355
Brussels, 57
Bryant, C. E., 181, 277, 365
Bryant, H., 310
Buchanan, Handasyde, 181, 343
Buchanan, President, 141, 143
Buckland, Dr, 318
Buckley, Mr, 339
Buckman, Thomas R. (three items), 181, 305, 374
Buerschaper, Peter, 182
Buffalo Natural History Society, 182, 226
Buffalo, New York, 141
Buffon, G-L. L., 90, 321
Buller, W., 213
Bullock, W., 365
Bunsome, Robert, 279
  See Ransome, R.
Burgess, L., 279, 365
Burke and Wills expedition, 144, 315
Burmester, C. A., 182
  indexed, 159
Burrows, James, 339
Burton, Charles, 365
Burton, Edward, 301, 365
Butler, W., 182
Bynoe, Surgeon Benjamin, 112, 113, 239, 279, 344, 365

Cabot, S. Jr, 310
Cabot, Dr Samuel, 339
Cadwalader, C. M. B., 167
Calaby, J. H., 183
Caleb, 292
Caley, George, 279, 365
California box cars, 299
Calkin portrait of Wolf, 291
Callen, Terry, 174
Calvert, ____, 279, 292
Cambridge University, 206, 236, 237, 303, 332
  Library, 169
  Museum of Zoology, 354
Camden, N. S. W., 118
Cameron, Roderick, 183
Campbell, Archibald, 279
Campbell, Archibald George, 277, 365
Campbell, Archibald James (two items), 183, 191, 365
Campbell, John, 365
Canadian Journal of Industry, Science and Art, 142, 275
'Cancelled Parts', see *Birds of Australia and adjacent islands*
Caprimulgidae, see *Icones Avium* Part II
Carpentaria, Gulf of (Australia), 124
Carriker, M. A., Jr, 310
Carter, T. H., 209
Carus, Julius Victor, 183, 201, 344
Cassin, John, 140, 169, 183, 184, 197, 208, 279, 307, 310
Catamaran River, Tasmania, 105
Catesby, Mark, 162, 317, 323, 362, 368
Cayley, Neville H. P., 359
Cayley, Neville W. (three items), 184, 277, 357, 365

Celebes, Islands, 290
Century of Birds, 14, 212, 241, 243, 253, 269-70, 289, 346, 371
  Audubon copy, 193
  list of plates, 18
  Part I published, 93
  prices, 374
Chadwick, 172
Chalmers-Hunt, J. M., 184
Chamberlain, W., 365
Chambers, Captain William, 279, 307, 365
Chancellor, John, 185
Chapman, Abel, 298
Chapman, Frank, 161
Charlesworth, Edward, 101, 279, 287
'Charley' (Aboriginal with Leichhardt expedition), 292-3
Chatham collection, 269
  See also Fort Pitt
Chemies, England, 276
Chester, Mr, 233, 279
Children, J. G., 274
Childs, John L., 185
Chillingham cattle, 145, 318
Chisholm, Alexander Hugh, 162, 167, 181, 184
  (twenty-seven items), 185-91, 207, 215, 231, 235, 243, 258, 277, 302, 326, 360, 365
  obituary, 191
Chittenden, Carol, 21 (two items), 191, 305
cholera epidemic, 135, 136, 170, 344-5
Cleave, John, 196, 239, 339
Cleevely, R. J., 192
Cleland, J. Burton (two items), 192
Cleveland, Ohio, 141
Cleveland, William, 174
'Clonlea' homestead, South Australia, 296
Chubb, Charles, 339
Clark, William Eagle, 298
Clark, Wm. N., 365
Clark, Taylor, 248
Cleland, J. B., 216, 233, 365
Clerk, Sir George, 312
Clift, Mr, 238
Coburg Peninsula, northern Australia, 352
Cockerell, ____, 279
Cockerell, James F., 365
Cockerell, John Thomas, 148, 365
Coe, William Robertson, Collection, 326
Cohans, ____, 279
collectors, 273
  list of, 272
Collins, J., 168
colourists, 78
Colp, Ralph, Jr, 192
Colthard, ____, 279
Columbia River, United States, 310
Comet, 109, 214
Commissioners of Her Majesty's Woods and Forests, 222
Companion to the Botanical Magazine (journal), 275
Conigrave, C. Price, 192, 207, 277, 365
Contributions to Ornithology (Jardine), 70, 276
  indexed, 159
Contts, & Co., 225
conservation, 338
Cook, Captain, 304
Cookham, England, 244
Cooper, D., 365
Cooper's Creek, 315
Copenhagen, University Library in, 162
Coppinger, R., 365

INDEX

Cordeaux, John, 196
Cornelias, E. H., 174
Corning, Howard, 192
Corse, Dr B. M., 139, 160
Cotton, Catherine D., 194
Cottrell, G. William, 354
Coues, Elliott, 64, 65, 66, 152, 159, 194, 196, 263, 310
   indexed 1870-80 item, 159
   letter from British colleagues, 195
Coulthurst, Edmond, 287
Cowtan, Robert, 196
Coxen brothers, 34, 99, 117, 218, 233, 279
Coxen, Charles, 111, 117, 148, 187, 188, 190, 199, 220, 232, 271, 279, 298, 321, 363, 365
   portrait, 187
Coxen, Elizabeth (mother of Elizabeth Gould), 116, 186, 187, 189, 219, 234, 279, 339
Coxen, Elizabeth Tomkins, 365
Coxen family, 364
   genealogy, 184
Coxen, Henry, 365
Coxen, Henry Charles, 207
Coxen, Henry William, 4, 99, 111, 186
   portrait, 187
Coxen, Miss Marion, 4
Coxen, Nicholas, 96, 339, 365
Coxen, Stephen, 110, 117, 190, 219, 232, 268, 279, 365
   death, 124
   estate, 220
   suicide, 188
Coxen, Walter Adams, 4
Craufiurd, Mrs E., 279
Crawford, Mr and Mrs Harold, 4
Creeney, Wm. Ford, 210
Crimean War, 135
Crockett, Davy, 194
Crystal Palace, 133, 242, 360
Cuming, Hugh, 212
Cunningham, George, 196
Cunningham, Richard, 271
Currey, John, 256
Curtis's *Botanical Magazine*, 267, 322
Cutler, W., 179
Cuvier, Baron, 322

D'Albertis, _____, 279
Dale, Mr, 279
Dall, William H., 197
Dana, James D., 255
Dance, S. Peter, 197
Dark, _____, 279
   See Darke, J.
Darke, John Charles, 279, 346, 365
Darling Downs, 61, 123
'Dart Brook' at Yarrundi, 110, 117, 232, 273
Darwin, Charles, 39, 94, 97, 101, 115, 143, 144, 169, 175, 192, 196, 198, 200, 216, 279, 291, 305, 315, 319, 326, 336, 339, 360, 362, 365, 368, 372, 375
   death, 152
   Descent of Man, 198, 322
   library, 331
   Origin of Species, 143, 197
   works of, 205
   See also *Zoology of the Beagle*
Darwin, Mrs Emma, 120, 318
Darwin, Frances, 198
Davidson, Angus, 198
Davidson, Rev. E. J., 207

Davies, R. H., 365
Davis, Mr, 318
Davy, Sir Humphrey, 312
Day, W. S., 365
Dawsons (American firm), 333
de Azara, *see* Azara
de Beauffort, *see* Beauffort
Deignan, H. G., 311
Delacour, J., 311
Delafresnaye, Baron, 279
Denham, Captain, 339
Denison, 365
Denny, A., 162, 254
Denny Henry, 179
Denson, J., 252
D'Entrecasteaux's Channel, 105
Derby, Lord, 15, 36, 39, 44, 50, 58, 66, 67, 71, 176, 189, 257, 259, 279, 291, 307, 310, 312, 314, 319, 321, 337, 339, 354, 360, 365
   death, 130
   initial correspondence, 93
   manuscript, 297
   Museum, 205
   original drawing collection, 337
Derwent River, 104
Des Murs, M. A. F. O., 311
De Vitre, 339
Dickens, Charles, 198
Dickes, William, 273
Dickinson, _____, 279
Dickison, D. J., 199, 278, 365
Dictionary of birds (Newton), 316
Dieffenbach, Dr, 179, 279
Diggles, Sylvester, 148, 182, 187 (two items), 199, 283, 299, 326, 365, 375
Dixon, Joan M. (four works), 199-200, 257
dog for Bachman, 185
Donald, Wm., 232, 279
Donaldson, _____, 279
D'Orbigny, Alcide D., 40, 311
Douglas, David, 169, 253, 274, 279, 339
Draper, Lyman, 139, 160
Dresser, Henry E., 137, 196, 288, 298, 305, 319, 368
Dring, Purser John E., 112, 279, 365
Drummond, James, 200, 220, 267, 279, 297, 312, 364, 365
Drummond, Johnston, 279, 365
   death, 220
Drummond, Thomas, 267
Drummonds, Messrs, 222, 234
Dry, Mr, 219, 220
du Chaillu, Paul, 310
Duke of Corwall's Scholarship, 208
Dumont-Durville, 310
Dunn, J. H., 219, 230
D'Urban, William, 142
Dusseldorf clay, 226
Dwight, Jonathan, 305
Dyson, Mr, 273

East India Company, 135, 251, 262
Edelsten, _____, 279
Edelsten, Dr Alan, 4, 97, 149, 185, 186, 190, 191, 231, 279
Edelsten, Mrs (Alan) Grace, 4, 143, 145, 186, 190, 209, 257, 335
Edelsten, Doris (Twyman), 149
   See also Mrs Doris Twyman
Edelsten, Dr and Mrs Geoffrey, xxi

Edelsten, Dr Geoffrey, 4, 97, 143, 146, 147, 149, 170, 185, 188, 189, 190, 231, 248, 258, 262, 279, 344
Edelsten, Mrs Helen, 185, 187, 314
Edelsten, Ronald, 149
Edelston, Ernest Alfred, 149
Edinburgh, 276
  University of, 146
Edison, Thomas, 150
editions, second, 14
Edwards, 54
Edwards, Ernest, 324
Edwards, George, 322, 368
egg collection, 126
  drawings, 257
eggs, birds, 105, 106, 117, 179, 183, 188, 189, 232
  See also oology
Egham on the Thames, 120
Eiseley, Loren C., 200
Ekama, C., 200
Elford, F. G., 365
Elinor, 105
Ellen (ship), 363
Ellerston, 273
Elliot, Daniel Giraud, 65, 66 (two items), 200, 201, 208, 226, 319, 323, 368
Elliott, Henry, 279, 365
Elliott, Mr (fisherman), 279
Ellis Archives file boxes, 332
Ellis Collection, 14, 51, 66, 67, 175, 181, 191, 226, 241, 249-50, 255, 276, 277, 287, 288, 293, 299, 303, 305, 332, 333, 346-51, 359, 393
  catalogue, 305
Ellis, Irene S., 201
Ellis, Ralph, Jr, 175, 181, 191, 226, 248, 274, 276, 306, 332-4, 351, 356, 359, 374
  California coloured prints, 334
  collection, *see* Ellis Collection
  court records, 201
Elsey, J. L., 365
Elsey, Dr Joseph R., Jr, 182, 213, 279
Elwes, H. J., 196
Elwes, R., 279
Emery, J. B., 279, 365
Emery, Lieutenant James, 112
Emu, 1938 Commemorative Issue, 13, 181, 186, 190, 191, 231, 268-9, 277-81, 362
  indexed, 159
Encyclopedia Britannica, 316
Engelmann, Wilhelm, 183, 201, 274, 344
Enright, W. J., 201, 365
entomology, 316
epidemics, 90, 145
Erskine, Mr, 279
Eton, 368
Evans, A. H., 287, 337
Evans, David, 184
Everett, Michael, 202
Ewing, Rev. T. J., 216, 217, 219, 279, 365
Exhibition of 1851: 128, 180, 313, 326, 360
Eyre, Edward John, 192, 202, 279, 297, 326, 365
  Journals, 202
Eyton, Thomas Campbell, 78, 119, 128, 180, 202, 212, 273, 393, 311, 326, 365, 375
  death, 151
  ornithological library, 323

Fairfax, John, 187, 214, 217, 218, 219, 221, 279
Fairmount Monument Works, 228
Falmouth, Lord, 146
'Fauna Boreali-Americana', 169

Fawcett, Benjamin, 273, 315
Feilden, H. W., 196
Ferguson, John Alexander, 202
'Finnish wheelbarrow' (Mason), 299
Finsch, O., 311
fish genera and species
  flying fish, 100, 247
  Gilbert Collection, 325
  Machaerium subducens, 177
  Salmo faris, 244
  Serranus gilberti, 325
  Thames Trout, 145
  'Trout' paintings (Gould), 335
Fisher (of Hobart Town), 107
Fisher, James (two items), 203, 343, 357
Fishers, Mr, 232
Fitch, Walter Hood, 343
Fitter, R. S. R., 203
Fitzpatrick, Kathleen (two items), 203, 204
Fitzpatrick, Thomas Jefferson, 359
  collection, 299
FitzRoy, Captain Robert, 40, 94, 339
Flexner, Marion, 204
Flinders Island (Australia), 108, 232
Flower, Charles F., 335
Flower, Professor, 312
Flower, W. H., 196
Folliot, 339
Forbes, Edward, 133, 242, 372
  medal, 208
Ford, Alice, 204
Ford, G. H., 319
Forshaw, J. M., 205
Fortnum, Mr, 217
Fort Pitt Collection, 269, 300, 310
Fowler (acquaintance of Gould), 132
Frankfurt, 57
Franklin. Lady Jane, 105, 149, 182, 185, 204, 215, 218, 232, 279, 295, 339, 365
Franklin, Sir John, 41, 99, 110, 117, 203, 279, 339, 365
  death, 126
Franklin, Sir John and Lady, 109, 112, 214, 267
  correspondence, 296
[Franklin], Sister of Lady Jane, 279, 365
Fraser, Louis, 205, 313, 336
Freeman, R. B. (two items), 205
Fremantle, Western Australia, 122
French ornithological biography, 326
Friedmann, Dr Herbert, 298, 311
Friend, Lieutenant or Captain, 107, 109, 214, 218, 279
Fries, Waldemar, 205
Froggatt, Walter, 206, 365
Frost, Dr Wade Hampton, 344
Fuertes, Louis Agassiz, 161, 184, 279, 298, 317, 362
Fullagar, P. J., 205

G., A. T., 244
G., L., 244
G., L. P., 244
G., W. H. B., 244
Gaboon, 310
Galapagos Islands, 40, 192, 291
Galbraith, A. L., 279
Gallatin, Frederic, library, 206
Gambel, Wm., 184, 311
'Garden, The' (Robinson's weekly paper), 318
Gardner, J., 279
Garrod, A. H., 196

INDEX

Gatcombe, John, 230, 231
Gawler, George, 99, 115, 279, 296, 365
Gawler, Miss, 297
Genera of Birds (Gray, G. R.), 218
genera and species, *see* birds, mammals, amphibians, etc.
Gentry, A. F., 311
'genus-itch' (Coues), 65
Geoffroy-St. Hilaire, Isidore, 311
Geological Society (England), 208
geological survey of England, 208
George Town, Tasmania, 107, 232
German Athenaeum, 319
German collector, 279
German Ornithological Society, 133
German Ramphastiden, see *Ramphastiden, Monographie der*
German zoology, 324
Gerrard, Edward, 339
Gerrards, the Edward, 179
Gervais, F. L. R., 311
Gibson, F., 365
Gilbert, John, 51, 61, 99, 109, 120, 159, 167, 177, 182, 183, 185, 186, 187, 190, 200, 213, 214, 218, 219, 220, 221, 231, 237, 239, 263, 270, 276, 279, 294, 297, 298, 310, 312, 315, 316, 321, 326, 335, 338, 339, 352, 357, 360, 362, 363, 364, 365, 368
  centenary, 207
  correspondence, 189, 234
  death, 124, 292-3
  diary, 123, 191
  in New South Wales, 123
  in Perth area, 112, 192
  in Sydney, 118
  instructions from Gould, 233
  journal, 186, 206, 214
  manuscript, 206-7, 344
  mural tablet, 125, 207
  return to Australia, 122, 233
  return to London, 120
  River, named after, 175
  Swan River birds, 206
Gilbert, William, 235, 279
Giles, Ernest, 204
Gill, Mr, 296
Gill, Theodore (two items), 208
Gilliard, E. Thomas, 208
Gilmore (ship), 119, 352
Gipps, Governor Sir George, 99, 220, 232, 280
giraffe, 93, 288, 291, 315, 318, 319, 335, 353
Glasgow Art Gallery and Museum, 311
Glasgow Museum, 295
Gleeson's Station [South Australia?], 115
Gmelin, Johann Friedrich, 90, 300
Godman, Frederick DuCane, 196, 366, 368, 372
Godwin-Austen, H. H., 82, 196
Godwin-Austen, Robert, 372
gold discovery in Australia, 131
Goldie, 280
gold-leaf, 169
Goodwin, A. P., 366
Gordon, J., 174
Gosse, Philip Henry, 274
Government House, Hobart Town, 232, 296
Gould, Ann, 3
  death, 126
Gould, Charles (son), died in infancy, 96
Gould, Charles, 140, 141, 142, 147, 148, 178, 186 (two items), 208, 215, 233, 236, 240, 262, 280, 313, 330, 339, 366
  birth, 96
  death, 153
  geological paper, 145
  journey to Australia, 143
  journey to North America, 139
  obituary, 208
Gould, Eliza (daughter), 137, 141, 148, 215, 280, 314
  initial correspondence, 143
  married, 147
  See also Muskett, Eliza; Moon, Eliza
Gould, Elizabeth (mother), 3
Gould, Elizabeth Coxen (wife), 16, 34, 37, 40, 42, 93, 99, 106, 107, 109, 111, 116, 118, 148, 155, 182, 185, 186, 187, 188, 212, 215, 217, 231, 232, 244, 257, 274, 280, 290, 291, 292, 294, 302, 304, 339, 357, 359, 366, 374
  correspondence, 189
  death, 42
    certificate, 209
    puerperal fever, of, 120
  diary, 89, 115, 209, 268
  drawing birds and plants, 115
  marriage, 93
  nurse, 281
  original drawings, 190, 191, 245-7, 337, 350, 370
    botanical, 320
  portraits, (2) 186
Gould family 'news-clipping book', 185
Gould, Franklin Tasman, 112, 142, 143, 146, 215, 280, 339, 366
  born, 112
  death, 147
  in India, 209, 210
  manuscript, 209
'Gould, J. & E.' (lithographic stone reference), 14
Gould, John, 196, 280
  [John Gould appears on every page but selected references are indexed here.]
  agent for others, 193, 202, 216
  aircraft named after, 188
  artistic ability, xviii, 184, 219, 303, 304
    use of pencil and crayon, 219
  associates, *see* Gilbert, John; Drummond, Johnson
  Athenaeum Club, 136
  Audubon association, 185, 192-4, 205-6
    gift of dog, 193
  Australian bird collection, 184, 283, 306-11
  Australian journeys, *see* Australia
  Australian stamp, 255
  Australian venture 192, 212, 271
  baptism, 188
  bibliography, 157, 210
  biographical notes, 211, 212, 240, 317, 319, 324, 331, 366
  bird imitation, 318
  birth, 3, 91
  bronze plaque by Heyman, 127
  businessman, xviii
  character, xix-xx
  cholera epidemic, 344
  chronology, 87
  collection of specimens, 142, 180, 211, 336
  contribution to work of another, 293, 295, 355
  copyright, 211, 285
  correspondence, 189, 212-26, 215, 231, 234, 262, 264
  daily routine, 314
  Darwin association, 197-8, 331
    See also Darwin, Charles

dates of publications, 361
death, 85, 151
descendants, 185, 187
displays 'singing N.S.W. parrots', 120
elected Associate of Linnean Society, 94
   Fellow of Linnean Society, 96
   Fellow of Royal Society, 123, 214, 325
   Member, Torino Royal Academy, 331
facsimile, *Birds of Australia*; Birds of Australia, *Supplement* and *Handbook*, 256
   Macropodidae, 257
   Synopsis, 256
finances, of *Birds of Australia*, 218
   income, 204, 251, 284
   major works, 78, 80
fishing, 126, 368
genealogy, 5
health, 83, 85, 151, 285, 312
horse Georgie, 314
humming-bird exhibition, 128, 130, 199, 221, 240, 272, 277, 298, 313 323, 336
journeys
   Australia, 34, 99, 103-19 *passim*, 232
   British Isles, 94, 98, 120, 128
   Continent, 57, 64, 66, 94, 97, 120, 123
   Jardine Hall, 125
   Malta, 135
   North America, 139, 275, 303, 313, 321, 334
   Norway, 137
   Paris, 97
   Scotland, 96
   with son Henry, 125
library, 185, 285, 321, 323
lithographic stones, 226
mammals, authority, 199-200
manuscripts, 79, 182, 191, 230, 231, 234, 236, 237, 238, 297, 303, 334, 335, 358, 370
marriage, 93
membership, British Association, 176
   General Committee, 178
   German Ornithological Society, 133
   See also Gould, John. Elected
metallic colouring, 330
name on insects, 316
North American trip, *see* Gould, John, journeys
obituaries (six), 243-4
original drawings, 21, 132, 162, 182, 191, 241, 244, (33) 244-8, 292, 298, 301, 302, 303, 304, 306, 321, 326, 332, 334, 337, 359, 370, 372
   eggs, 248, 257
   for sale, 348-51
ornithologist, achievement as, xvii-xviii
pattern plates, 183, 272, 302
plates reproduced [many pages have multiple references], 160, 161, 169, 170, 181, 182, 183, 185, 190, 199-200, 202, 203, 204, 205, 206, 208, 256, 265, 275, 290, 291, 292, 295, 296, 297, 315, 317, 325, 330-1, 331, 332, 343, 344, 355, 357, 359, 362, 368
portrait, frontispiece, 127, 139, 144, 145, 148, 150, 162, 169, 170, 185, 202, 241, 249, 268, 303, 313, 324, 335, 357
power of attorney, 234
prices of works, 184, 185, 190, 199, 287, 346-51

print reproductions, 334
proof copy, 74
prospectuses, 249-50
   and list of subscribers, 250-1, 334
   See also prospectuses
publisher, 13, 192
residences, 197
   11 Broad Street, 92
   20 Broad Street, 93, 143
   26 Charlotte Street, 143
   96 Great Russell Street, 143
reviews (nineteen), 251-5, 282-7, 312
   See also reviews
sails, to Australia, 99
   from Australia, 118
scientific articles, 14, 34, 89, 93, 177-8, 183, 194, 205, 276, 287, 288, 289, 290, 294, 326, 330, 331-2, 338-9
   first article published, 93
second edition of a work, 74
specimens, 162, 176, 255
   desiderata, 294
   gifts of, 140, 233
   sale of Australian, 126
   stolen, 214, 215
   type, 311, 360, 362
subscription list of £143,000: 251
systematist, xviii, 303
taxidermist, 91, 214, 243, 295
will, 150, 225, 239
   codicil, 151
work(s), cancelled, see *Birds of Australia and adjacent islands*
work(s), major listed, 11-86 *passim*, 160, 179, 182, 185, 199, 200, 201, 202-3, 206, 222, 225, 226, 244, 266, 272, 294, 300, 312, 317, 330, 338-9, 340, 342, 343, 354, 356, 357, 361, 374-92
   financial aspects, 78, 80
   parts of, extant
      Birds of Australia and adjacent islands, 34
      Birds of Australia, supplement, 69
      Century of Birds, 17
      Icones Avium, 36
      Macropodidae, 49
      Mammals of Australia, 61
      Monographie der Ramphastiden, 47
      Ramphastidae, 26
      Synopsis, 30
      Trochilidae, 64
      Trogonidae, 26
      Voyage of the Sulphur, 51
      Zoology of the Beagle, 40
   second edition, 74
Zoological Society of London, 98, 312, 335
   'Curator and Preserver', 92
   in the chair, 275
   'Superintendent, Ornithological Department, Zoological Society Museum', 96
Gould, John, brother [?], 120
Gould, John (died in infancy), 93
Gould, John, (Senior), 3, 325, 339
Gould, John Henry, 99, 106, 111, 125, 135, 189, 232, 257, 258, 280, 330, 339, 366
   birth, 94
   death in Bombay, 137
   egg drawings, 188, 248, 262
Gould League of Bird Lovers, 181, 184, 190, 212, 280, 338, 366

INDEX

Melbourne, 153
New South Wales, 357
Wellington, 153
Gould, Lizzy, *see* Gould, Eliza
Gould, Louisa, 137, 141, 147, 153, 214, 239, 240, 257, 258, 280, 314, 339
  birth, 98
Gould, Mount, 346
Gould, Robert, 3
Gould, Sarah (Sai), 91, 141, 147, 153, 239, 240, 280, 314, 339
  birth, 120
Gould, Sarah (Wilson), 3
Gould, William, 234
Gouldiana, 191, 192, 269
Gouldian Finches, 342
Gould's Range, 346
Gould's River 346
Governor of Western Australia, 280
Gourlie, Wm., 280
Goyder, G. W., 366
Grafton, 271
Grant, Captain, 318
Grant, W. R. Ogilvie, 180
Granville, Lord and Lady, 209, 210
Gray (of Zoological Society), 265, 368
Gray (in index of *Emu* commemorative issue), 280
  See George Grey
Gray, Dr, 178
Gray, Mr (Sir George Grey's secretary), 280
Gray, George Robert, 40, 175, 198, 205, 218, 231 (four items), 263, 273, 280, 303, 311, 319, 339, 366, 372, 375
  death, 147
Gray, John Edward, 54, 90, 94, 263, 264, 274, 280, 288, 306, 319, 335, 360
  correspondence, 189
Gray, Mrs Maria Emma, 212
Great Auction Rooms, 320
Great Britain, The birds of (Lewin), 293
Great Exhibition, *see* Exhibition of 1851
Great Flower Books (Sitwell & Blunt), 343
Green Island, 104, 105
  in Bass Strait, 108
Greenewalt, C. H., 167
Gregory, Augustus C., 182, 213, 363
Gregory, Frances Thomas, 346, 363
Grey, Edward, Institute of Field Ornithology, 161
Grey, George, 99, 186, 192, 216, 233, 280, 297, 307, 326, 346, 368
  correspondence, 123, 235
  Journals, 264
Grey of Fallodon, 320
Griffith, Edward, 280
Gregory, F. T., 280
Grosvenor, Earl, 147, 209, 210, 319
  See also Westminster, Duke of
Grover, Mrs Rachel, 275
Guatemala, 336
Guildford, Surrey, 3, 91
Guillemard, Dr, 280
Gunderheim, 118
Gunn, Ronald, 105, 215, 219, 220, 232, 280, 366
guns, 90, 154
Gunther, Albert C. L. G., 196 (two items), 264, 309, 339, 356, 372
Gunther, Albert E., 264
Gurney, John Henry, 180, 196 (two items), 265, 311, 319, 366
Gurney, J. H. Junior, 265
Gyde, _____, 280

Hachisuka, Masauji (two items), 265
Hack, Mr, 216, 218, 280
Hack, Watson & Co., 217
Hagen, Mr, 218, 219
Hall, A. F. B., 272
Hall, Sir Benjamin, 173
Hall, Marshall, 274
Hall, Robert (two items), 265, 366, 375
Hallock, Mary, 266
Handbook (Gould), 80, 146, 215, 288, 363
  facsimile, 256
  prospectus, 224
Hanover, 57
Happoldt, Christopher, 316
Hardwicke, Thomas, 274
Harper, F., 311
Harris, E., 311
Harris expedition, 310
Harris, Harry, 266, 287, 325
Harris, J. I., 205
Harrison, C. J. O., 361
Harrison, H. F., 225
Hart, William [1982 note: Maureen Lambourne has received contact from the *William* Hart family. Thus, we now are sure that the 'W.' stands for William], 14, 85, 131, 145, 147, 159, 162, 202, 203, 208, 239, 240, 244, 280, 292, 294, 304, 314, 339, 344, 364, 366, 373
  original drawings, 246-8, 351
  for sale, 350
Hartert, Ernst, 178, 305
Harting, J. E., 196, 243
Harvard University, Houghton Library, *see* Houghton Library
Harvey, J. B., 215, 216, 217, 218, 280, 366
Hauxwell, 339
Havell, Robert Jr, 21, 94, 176, 193-4, 204, 268
Hawker, James C., 267, 280, 366
Hawkins, B. Waterhouse, 51, 98, 133, 242, 294, 319, 321, 366
  original drawings, 337
Hayden, Dr, 195
Hayman, Peter, 202
Heade, Martin Johnson, 317
Heermann, A. L., 311
Heine, F., 311
Helyar, L. E. James, 311
Hemsley, W. Botting, 267
Heney, Helen M. E., 267
Henry, Joseph, 140, 146, 226
Henry, T. C., 311
Henslow, J. S., 133, 212
Herrick, Edward C., 255
Herrick, Francis Hobart, 267
Herring, _____, 280
Heyman, J. A., 127
Heywood, Sir Benj., 214
Higgins, E. T., 366
Highley, Samuel (publisher), 192
Highley's (publisher), 28, 35, 41
Hills, _____, 280
Himalaya mountains, 289, 393
Himalayan birds, 176, 358
Hinds, Richard B., 54, 326, 375
Hinds, R. B. (*Voyage of the Sulphur*), 51
Hindwood, K. A., 192, 207, 213, 216, 231, 233 (five items), 268-9, 278, 366
Hirst, C., 99
History House, 190
Hobart, Tasmania [*also* 'Hobarton', 'Hobart Town'], 41, 99, 115, 174, 232, 296
  settled, 89

405

Hodges, Mr, 142
Hodgson, Brian Houghton, 269-70, 339, 360
Hodgson, Pemberton, 292
Hofer, Philip, 270
Holdsworth, E. W. H., 196
Hombron, J. B., 311
Home, Sir Everard, 318, 353
Honduras, Central America, 273
Hopson, John, 201
Hooker, 312
Hooker, Sir J. D., 198
Hooker, Sir W. J., 200, 275, 280
Hoopes, B. A., 311
Hope (entomologist), 316
Hope, Rev. F. W., 254, 270
Horsfield, Dr T., 39, 90, 274, 293
Hotel, Brevoort, 140
    Willard, 139
Houghton Le Pherne (ship), 235
Houghton Library, Harvard University, 147, 205, 317
Houghton, William, 326
Household Words (Dickens), 198
Howell, John, 333
Hoy, P. R., 311
Huber, W., 311
Hudleston, W. H., 196
Hudson, William Henry, 270, 320
Hughes, Ron, 206, 236
Hullmandel, C., 14, 51, 181, 280
Hullmandel and Walton, 14, 319
Humbolt, 64
Hume, Sir E., 238
Hume, Dr Portia Bell, 248
Humming-bird
    Brazilian, 317
    Collection, 331
    colouring, 330, 360
    exhibition, 128, 130, 199, 221, 240, 272, 277, 298, 313, 323, 336
    Gould's Monograph on, *see Trochilidae*
'Humming Bird, The' (article by Iredale), 272
Hunter River, 117, 201, 232, 271, 273
Hurst, George, 366
Hutt, Governor John, 99, 217, 280, 364, 366
Hutt, William, 280
Huxley, T. H., 144, 178
Hyman, Susan, 270

Ibix, The (journal), 143, 159, 255, 282, 287, 319, 322
    biographical index, 266
    indexed, 159
    Jubilee Supplement, 337
'Ichthylogy of Australia Contributions on' (articles), 325
Icones Avium, 36, 272
    initial part, 98
    Part 1, 253
    Part 2, *Caprimulgidae*, 38, 177, 194, 254
    'Part 3', 71, 219
Iguanodon (Hawkins's), 133, 242
Illawarra District (N.S.W.), 118
Illidge, R., 366
Ince, J. M. R., 280, 366
Ince, William, 214
Index Animalium (Sherborn), 342
Index generum avium, 361
Index Zoologicus, 288, 361
Indexed works, 159
India House, 39

Indian birds, 353
Indian Ocean, 276
Ingleby, Sir William, 91, 339
insects, genera and species
    Allecula gouldi, 316
    Cisseis gouldi, 316
    Coleoptera, 270
    Dineutes gouldi, 316
    Sericesthis gouldi, 316
    Stigmatium gilberti, 316
Institute of Painters in Water Colours, 319
'Introduction' books (Gould), 14
Introduction to Birds of Australia (Gould), 63, 126, 233
Introduction to Birds of Great Britain (Gould), 82, 148
Introduction to Gould's Birds of Asia (Gould), 85, 152
Introduction to Mammals (Gould), 79, 145
Introduction to Trochilidae (Gould), 73, 145, 255
Ipswich, Museum, 120, 127, 249, 313
Irby, Howard L.. 196
Iredale, Tom, 27, 28, 183, 213 (seven items), 271-2, 278, 280, 302, 303, 357, 366
Iredale, Tom and Hall, A. F. B., 272
Iridescent colouring, 330
Irwin, Raymond, 272
Isabella Island, 108

'J., H.' (initials in Jardine's 'Contributions'), 277
Jackson, Christine, 15, 28, 239, 240 (three items), 272-3
Jackson, Holbrook, 273
Jackson, Mr, 280
Jackson, Sidney Wm., 273, 280, 366
Jacquinot, H., 311
James, 280
    See also Benstead, James
Jameson, Robert, 194, 274, 276
Jamison, Sir John, 271
Janson, E. W., 366
Japan, 310
Jardine, John, 366
Jardine, Lady, 233, 280
Jardine, Sir William, 15, 17, 21, 24, 25, 26, 27, 28, 30, 34, 35, 36, 38, 39, 41, 42, 45, 49, 56, 57, 61, 64, 70, 71, 92, 94, 96, 98, 132, 146, 176, 180, 194, 202, 212, 216, 237, 252, 253, 254, 273, 274, 275, 276, 277, 280, 288, 299, 308, 311, 335, 354, 363, 375
    Audubon correspondence, 193
    Contributions to Ornithology, 126
    death, 148
    Gould correspondence, 93
    Marsupials, 207
    Naturalist's Library, 96, 207, 364
    ornithological library, 323
Jardine, Sir William and Selby, P. J., 274
Jefferson, Thomas, 89
Jemmy (companion on Liverpool Range exploration), 117, 280
Jenkins, Alan, 274
Jenyns, Leonard, 274
Jerdon, Dr, 82, 280
Jerrold, Blanchard, 280, 339
Johnston, Dr, 98
Jones, W. H., 311
Joop, Gerhard, 275

INDEX

Joseph (taxidermist), *see* Baker, Joseph
journals (in Index), *see* name of journal or magazine
Journals (in Bibliography), 275-90 *passim*
Jukes, J. Beete, 280, 291, 307, 366

Kangaroo Island, Australia, 115, 192
Kangaroos, Gould's Monograph on, *see Macropodidae*
Kansas-Ellis Collection, *see* Ellis Collection
Kansas, University of, 181, 191, 226, 305, 346
  Ellis estate, 201
  Libraries, 359
  See also Spencer Library
    Watson Library
    Zoology Library
Katherine Stewart Forbes, 115, 294
Kaup, Dr T. T., 276
Kay, Diana, 258
Keartland, G. A., 367
Kelso, E. H., 311
Kelso, L., 311
Kemp, R., 367
Kensal Green Cemetery, 149, 151
Kermode, _____, 280
Kerr, Robert, 107, 214
Keulemans, J. G., 202, 319, 350
Keyes, Virginia, 357
'Kickapoo Country' (Upper Missouri River), 310
King, Captain, 99, 105
King George III, 89, 91
King George IV, 91, 291, 315, 318, 335, 353
  death, 93
King of Portugal, 225
King, Philip Parker, 297
King William IV, 93, 97, 281
King's College, 143, 310
King Island, Australia, 232
Kinnear (ship), 118, 294
Kirby, Rev. William, 287
Kivell, Rex Nan, Collection, 182, 244
Kloot, Mrs Tess, 209
Knight, _____, 280
Knight, David, 291
Knip, P., 311
Knowsley Hall, 176, 360
  Gleanings . . ., 323
  library, 257, 321, 336
  See also Derby, Lord
Knox, A. E., 319
Kraus, H. T., 333
Krefft, J. L. G., 239, 280, 344, 350, 367
Krider, J., 311
Kröyer-Pedersen, Margo, 359
Kurrachee, India, 262

Laby, Sir Thomas, 255
Lack, David, 291, 326
Lafresnaye, F. de, 311
Lake Erie, 141
Lambert, A. B., 280
Lambert drawings, 303
Lambourne, Lionel S., 288, 291, 318
Lambourne, Maureen (three items), 291-2, 313
Landor, E. W., 367
Land's End, 100
Landsdowne, Marquis of, 288, 312
Landseer, Sir Edwin, 215, 319
Langier, M., 367

Lankester, Dr, 173
Latham, John, 90, 97, 162, 189, 280, 300, 303, 322, 360
La Trobe Collection, Victoria, 76, 222
Laudon (editor), 93
Launceston, Tasmania, 107, 123, 234
Laundry, Howard, 196
Laurence, W., 48
Lawrence, George N., 68, 74, 78, 140, 146, 169, 280, 282, 311
  Gould correspondence, 225-6
Lawson exploration, 204
Lawton, Mr, 232, 280
Layard, E. H., 311
Layard, E. L. C., 311, 367
Leach, J. A., 191, 292, 367
Leadbeater, Benj., 367
Leadbeater, Messrs, 176, 179
Lear, Edward, 14, 89, 90, 94, 97, 159, 161, 181, 192, 193, 198, 203, 244, 270, 273, 274, 280, 290, 291, 292, 294, 299, 304, 317, 319, 321, 323, 324, 339, 343, 344, 350, 364, 367
  birth, 91
  Book of Nonsense, 124
  death, 152
  Family of Psittacidae or Parrots, 93, 324
  obituary, 286
  original drawings, 247, 324, 370
  plates reproduced, 294, 295, 357
Lefevre, A., 375
Legge, Mr, 91, 339
Legge, W. Vincent, 280, 368
Leichhardt, F. W. Ludwig, 167, 175, 183, 186, 189, 190, 194, 221, 280, 292, 297, 362, 367
  expedition, 123, 124, 186, 188, 204, 219, 231, 310, 315, 339, 352
Leiden (Leyden), 41, 49, 57
Leidy, Dr Joseph, 139, 160
Leigh, _____ (lawyer), 280
Leipsig, 57
LeSouef, W. H. D., 191
Lesson, R. P., 65, 311
Levaillant, Francois, 90, 243, 274, 323, 368
Lewin, A. W., 367
Lewin, John William (two items), 293, 375
Lewin, William, 293
Lewis, Frank, 293
Lewis, Mr, 216, 232
Lewis, Mrs D. A., 362
Leycaster, A. A., 280
Lhotsky, Dr John, 271
L'Institut (journal), 326
Library of Congress, Washington, 140, 251, 256, 300
Lichtenstein, Martin, 133
Lilford, Lord, 305
Linda Hall Library, Kansas City, 58, 289, 326
Lindlow, G. L. E., 262
Link, Dr H. F., 324
Linnaeus, Carolus, 90, 123, 152, 233, 263, 274, 280, 300
  four letters, 276
Linnean Society of London, 91, 94, 96, 143, 212, 243, 287, 294, 335, 356, 393
  catalogue, 294
  index, 207
  Journal, 287, 294
  manuscript, 297
  Transactions, 294, 358
lists of names of persons (five lists), 250-1, 323
Lister, Joseph, 155
lithograph, 14, 176

407

lithographic stones, 226-9
lithography, 184, 272
Liverpool, England, 176, 315
  Museum, 295
Liverpool Plains, N.S.W., 190
Liverpool Range, 110, 117, 118, 232
Lizard Point, Cornwall, 100
Lloyd's Natural History, 341
Lloyd's Register, 294
Loddiges, George, 64, 168, 241, 280, 358
Lodge, George Edward, 273
Lolly, Edw's. [?], 214
London, 90, 153, 170-3, 174, 344-5
  balloon view of, 131
  bookshops, 180
  clubs, 262
  fire, 96
  influenza epidemic, 126
  museums, 295
  streets, 197
  Times newspaper, 336
  underground railway (tube), 145
  University of, 143
  See also cholera epidemic
Longman, Heber, 294, 367
Lonsdale, Mr, 280
Lord, Clive E., 295
Lorne, Lord, 319
Loudon, J. C., 252, 287
Lowe, Lieutenant, 280
Lulworth, 352
Lushington, Lieutenant, 280, 367
Lyme Regis, Dorsetshire, 3, 4, 91, 188
Lynd, Captain, 221
Lysaght, A. M., 295

McArthur, Captain, 232, 280, 367
McCall, G. A., 311
McClees, J. E., 139, 249
McClintock's expedition, 140
McCormick, Dr Robert, 179, 280
McCormick, Silas (reaper), 96
McCoy, Frederick, 62, 63, 65, 68, 69, 70, 73, 74, 77, 78, 80, 83, 142, 143, 146, 149, 222, 224, 280, 296
McDonald, N. J., 303
McEvey, Allan R., 213, 256, 257 (five items), 303-5
McGill, A. R., 191
McGill University (Montreal), 237, 303, 313, 324, 371
  See also Blacker-Wood Library
MacGillivray, John, 70, 180, 183, 212, 221, 239, 271, 276, 280, 295, 326, 339, 344, 363, 367, 368
MacGillivray, W. D. K., 295
MacGillivray, William, 194, 295
McIntosh, D., 367
McKellar, Captain J., 99, 174
McLean, Mr, 219, 220
MacLean, Eleanor, 23, 25, 63, 72, 237, 238, 313, 324
MacLeay, Mr, 118, 280
McLeay, 232
Macdonald, W. B., 324
Macintyre, Eliza, 119
Mack, George, 278, 296, 367
Mackaness, George, 159, 296
Mackworth-Young, Sir Robin, 225, 323
Maclaren, John, 367
Macquarie Plains, 104

Macropodidae, 48, 120, 217, 250
  facsimile, 257
Madeira, 100
Madsen, Cecil, 296
Magazine of Natural History, 93
Magazine of Natural History and Journal of Zoology, Botany, Mineralogy, Geology and Meteorology, 287
Magazine of Zoology and Botany, 98, 275
Maguire, T. H., 127, 249, 274
  portrait of Gould, 268
Mahoney, Eliza Sarah, 296, 367
Maiden, J. H. (three items), 297
Maidenhead, 368
Maitland, N.S.W., 110, 117, 201, 232, 268
Malaher, Dr Arthur, 187
Malherbe, Alf., 323
mammal collection, 180
mammals, genera and species
  Australian original drawings, 337
  Ellis Collection, 333
  Gilbert list, 207, 214
  giraffe, 93, 291, 318, 353
  Gray, J. E., list, 264
  hyaena, striped, or native tiger, 232
  Hypsiprymni, 49
  kangaroo, 41, 117, 183, 199
    grey, 338
    white, 297
  Lagorchestes conspicillata, 49
  Macropodidae, 120
  Marsupiala, 199-200, 364
  marsupials, 199-200
  monotremes, 199
  musk-ox, 140
  number collected in Australia, 240
  otter, 290
  Petaurista taguanoides, 59
  Petrogale xanthopus, 60
  Placental Mammals of Australia (Gould), 200
  reindeer, Barren Ground, 140
  rhinoceroses, 290
  skeletons, list of, 238
  wolf, Tasmanian (*Thylacine*), 186, 338
  wombat, hairy-nosed, 63
Mammals of Australia, 61, 124, 199-200, 219, 221, 241, 250, 325
  pattern plates, 213, 272, 302
Manchester Natural History Society, 193
Mander-Jones, Phyllis, 297
Manley, F. W., 196
Mannering, Eva, 297, 326
Manson-Bahr, Philip, 298
Mansuetti, Romeo, 298
Mantell, Walter, 339
Manuscripts, 320, 360
  on Australia, 297
  at Canberra, 302
  Gould's paper on *Odontophorinae*, 58
  Mitchell Library, 206
  on New Zealand, 297
  on the Pacific, 297
  Puttick and Simpson, 323
  Queen Victoria, 323
  Queensland Museum, 207
  University of Cambridge, 206
Maria Island, Tasmania, 109
Marion Watson (ship), 106
Mariners Museum, Virginia, 294
Marks, H. S., 196
Marriott, Rev., 280
Marsh, Peter, 256

INDEX

Marshall, A. J., 298
Martin, _____, 280
Martin, (lawyer), 280
Martin, Maria, 205
Martin, Mr, of the Zoological Society, 98
Martin, William Charles Linnaeus, 280, 298, 326, 375
Martin, William John, of Camden Town, 234
Mary Ann (ship), 116, 294
Maskelyne, Professor, 178
Mason, Alexandra, 181, 299
Masséna, Marshall André, 306
Masséna, Prince Victor, 308, 311
   See Rivoli Collection
Mathews, Gregory M., 44, 63, 64, 182, 278
   (eighteen items), 299-303, 306, 307, 353, 359, 367, 375
   autobiography, 302
   Collection, 182, 302
Matthews, Noel, 360
Mattingley, A. H. E., 191, 367
Maull & Fox, Messrs, 148, 249, 268
Maurot, Suzanne, 258
Maximilian, Prince, 280
Mayr, Ernst, 311, 354
Meade-Waldo, E. G. B., 298
Medical science in 1804, 90
   in 1881, 155
Medland, Lillian, 272
Meigs, Dr J. Aitkin, 139, 160
Meinertzhagen, R., 305
Melbourne, 143, 190
   Public Library, 222, 224
   University, 222, 224
      Library, 62
Mengel, Robert, 74, 181, 255 (four items), 305-6
Mercers Pool (trout pool), 244
Meryon, Dr Edward, 240
metallic colouring, 330, 360
Metropolis Local Management Act, 137
Meyer, 306
Meyer de Schauensee, Rodolphe, 230, 306, 310, 311, 321, 358
Millais, Sir John Everett, 311
Millarton, Sir and Lady, 215
Milligan, Alex W., 280, 312
Mintern Bros., 14
Mississippi River, 169
Missouri River, 310
Mitchell, David William, 21, 25, 26, 28, 29, 39, 222, 243, 254, 312
Mitchell Library, Sydney, 186, 190, 206, 209, 213, 221, 231, 234, 248, 268, 269, 303
   egg drawings, 258
   Gould correspondence, 216
Mitchell, Mrs Guy (Helen), 190
Mitchell, P. Chalmers, 187, 312, 335
Mitchell River, Queensland, 175
Mitchell, Mrs Robert, 116, 185, 187, 190, 213
Mitchell, Robert, 234, 280
Mitchell, Sir Thomas, 307
Mitchell, W., 54
Mitford, Captain R., 120, 274
Mivart, St. George, 196
Mokai River, N.S.W., 117, 118
mollusca, genera and species
   Cytherea, 292
   Loricates, 272
Molyneux, 239, 240
Montagu House, 119
Montevideo, Uruguay, 148

Montreal Natural History Society, 142, 313
Montreal, Quebec, 142
Moon, Eliza, 153, 280, 339
   'Reminiscences', 313
   See also Muskett, Eliza Gould
Moore, Frederic, 179
Moore, T. J., 314, 337
Moorehead, Alan (two items), 315
Moreton Bay, 292
Morgan, P. J., 176
Morgan, Peter, 360
Morris, Francis Orpen, 315
Morris, Dr J. C., 139, 160
Morse, Samuel (telegraph), 96
Morton, Mrs Nancé, 4
Mosquito Island, N.S.W., 117, 268
Mount Gould, Western Australia, 346
Moyal, Ann M., 315
Mueller, Ferdinand, 315
Muirhead, Dr, 280
Mullens, W. H., 315
Mumby, Frank A., 316
Munt, H., 367
Murchison, Sir Roderick, 178, 318, 368
Murie, James, 288
Murphy, John, 280, 292
Murray, Mate on *Parsee*, 99
Murray River, Australia, 192
Muscle Bay, Tasmania, 106
Museé Royal des Pays-bas, 96
   See also Museums, Rijksmuseum
Museum(s)
   American, 336
   American Museum of Natural History, 146, 225
   Bristol, 295
   British Museum, 98, 119, 124, 142, 147, 178, 179, 196, 235, 244, 263, 285, 307, 309, 310, 316, 325, 353, 356, 369
      first purchase of Gould's South Australian birds, 120
      Library, 185, 334
   British Museum (Natural History), 25, 79, 145, 174, 179, 180, 238, 244, 264, 276, 303, 313, 334, 344, 360
      indexed, 159
      Library, 229
      manuscripts, 297, 335
      purchased bird skins, 152
      purchased egg collection, 152
      type specimens, 360
      Zoology Library, 269
   Chicago Museum of Natural History, 375
   Dublin Museums, 295
   European, 273, 324
   Glasgow Museum, 295
   Liverpool Museum, 295
   Merseyside County, 176, 205, 354, 359
   National Museum of Melbourne, 222, 296
   National Museum of Victoria, 142
   Norfolk and Norwich Museum, 265
   Peabody Museum, 255
   Queensland Museum, Brisbane, 364
   Rijksmuseum, Leiden, 58, 96, 143
   Royal Ontario Museum, 182
   Royal Scottish Museum, 277, 299
   United Service Museum, 269, 310
   U.S. National Museum, 298
   Yale University, Peabody Museum, 255
   See also Zoological Society of London Museum
Musgrave, Anthony, 316
Musignano, Prince of, *see* Bonaparte, Charles L.

Muskett, Eliza Gould, 149, 225, 239, 240
  See also Moon, Eliza; Gould, Eliza
Muskett, Helen C. E., 97
  birth, 149
Muskett, John, 97, 147, 239
Mythical Monsters (Charles Gould), 153, 208

Namant [?], Edward, 223
Namoi River, N.S.W., 118, 271, 295
Napier, Lord, 141, 242
National Library of Australia, Canberra, 64, 182, 213, 244, 251, 302, 303
Natterer, Johann, 48, 280
Natty (companion on Liverpool Range exploration), 117, 281, 367
Natural history illustration, 317
Naturalist in La Plata, The (Hudson), 270
Naturalist's Library, see Jardine, Sir W.
Naumann, Dr J. F., 133
Naumannia, 71, 359
Nebraska farmhouse, 299
Neill, J. 'Patrick', 281
neossology, 78
Nepal, 360
  birds, 269
nests, birds, 183
Netherlands, 295
Neuffer, Claude Henry, 316
New Guinea, 276
New South Wales, 147, 174, 293, 294, 320, 354
New Zealand, 99, 291, 318
New York City, 139, 141
Newcastle, N.S.W., 39, 117, 232
Newman, Edward, 82
Newman, J., 216
Newspaper(s), 90
  Daily Mirror, Sydney, 242
  Evening Star, Washington, 139, 242
  Hobart Town Courier, 240
  Illustrated London News, 240, 241
  New York Evening Post, 139, 242
  Sydney Morning Herald, 188, 190
  Times, London, 162, 171, 185, 231
Newton, Alfred, 137, 196, 254, 281, 282, 283, 288, 298 (two items), 316, 319, 320, 337, 342, 357, 368
Newton Library, Cambridge, 185, 236, 237, 303
  manuscripts, 297, 323
Newton, original drawings, 337
Niagara Falls, New York, 141, 142
Nissen, Claus (three items), 317
Nissen IVB, 317
Nissen ZBI, 317
Noakes, Vivien, 317
Noble, J., 83
Nomenclator Zoologicus (Scudder), 361
nomenclature, 150, 152
  Linnaeus, 123
  Schlegel, 123
  Strickland, 123
Norelli, Martina R., 317
Norfolk Island, Pacific, 118
Norris, K. A., 331
North, A. J., 367
North American birds, 175
Northampton, Lord, 120, 318
Northern Territory, Australia, 297
Northumberland, Duke of, 145, 318
Norway trip, 318
Nundawar Range, 118, 295

Nuremburg, 57
nurse, Mrs Gould's, 281
Nuttall, Thomas, 311

Oberholser, H. C., 311
O'Brian, Dr, 232
Ocean, South Indian, 102
Odontophorinae (Gould), 55, 123, 177, 218, 277
  cost, 58
  manuscript, 238, 370
  prospectus, 249-50
Ogden, J. A., 311
Ogilby, William, 312
Ogilvie-Grant, W. R., 367
O'Hegarty Collection, University of Kansas, 289, 358, 393
Ohio collector, 184
Old Sportsman (in *Emu* Commemorative Issue index), 281, 363
Oliver, Gerald, 367
O'Neill, W. J., 367
Onkaparinga, South Australia, 192, 267
oological drawings, *see* egg drawings
oology, 152, 179, 183, 188, 248, 277
  See also eggs, birds
Oology of Australia, 308
Orand, Captain, 214
Orbigny, Alcide D. d', *see* D'Orbigny
Ord, George, 194
Origin of Species (Darwin), 143, 197
  See also Darwin, Charles
Orkneys, 289
  Ornithological Biographies (Audubon), 295
  'Ornithologist, The' (Millais painting), 311
ornithologists, North American, 194
ornithology, British, 195
  field, 156
  history, 354
Ornithology of Australia (Diggles), 299
Orton, Professor, 74
Ortyginae, 218
  See also Odontophorinae
Osteology, 202
Owen, Mrs Richard, 318
Owen, Richard, 25, 50, 71, 96, 98, 112, 115, 117, 118, 120, 127, 132, 145, 174, 198, 213, 215, 238, 242, 254, 273, 281, 288 (two items), 318, 319, 322, 336, 367, 368, 372, 375
Owen, Rev. Richard, 318
Oxford debate, 315

Pacific Railroad Survey, 310
Pagett, Agnes Olive, 3
Painter, R. H., 344
Palacky, J. F., 367
Palmer, A. H., 160, 181, 319
  The Life of Joseph Wolf, 319
Palmer, Everard, 182
Palmer, T. S., 161, 266, 281
Palmyra (ship), 218, 219
Papuan avifauna, 287
Paradiseidae, Monograph of (Sharpe), 202, 338, 343, 347
Paris, 49, 57
Parker, Shane, 320
Parker, W. K., 290
Parker, W. R., 196
Parkin, Thomas, 320
Parkins, Mansfield, 276

INDEX

Parkinson, S., 367
Parnell, Dr, 57
Parsee (ship), 98, 99, 174, 294
partridges of America, Gould's Monograph on, *see Odontophorinae*
Pasha of Egypt, 318
Pasteur, Louis, 156
Patterson, Mary, 233, 281
Peale, T. R., 311
Peattie, Donald Culross, 320
Peel River, N.S.W., 118
Pelorus, H.M.S., 235, 352
Pennant, Thomas, 281, 322
Perks, R. H., 367
Perry, Commander, 310
Persons listed, 176, 178
Perth, Tasmania, 107, 232
Perth, Western Australia, 112, 235
Pescott, Edward E., 320, 367
Peters, J. L., 161
Peterson, Roger Tory, 354
Philadelphia, 139, 141, 303
Phillips, M. E., 160, 263, 320
Phillips, Mr, 292
Phillips, V. T., 160, 263, 320
photography, 154
'Piccadilly Notes', *see* Sotheran, Henry
Piele, Jons, & Co., 214
Pitman, Mr, 70
Pittidae (Gould), 84, 151
  original drawings, 350
Pollard, Jack, 321
Pompper, H., 367
Port Essington, Australia, 118, 120, 177, 221, 235, 263, 270, 292, 301, 316, 325, 351-2
  expedition, 207
  See also Leichhardt expedition
Port Jackson-Botany Bay, N.S.W., 90
Port Phillip Bay, Victoria, 90
  District, 304
Portland, Maine, 142
Potentate (ship), 109, 294
Pouncey (paper co.), 281
Povey, Dorothy, 321
Preiss, Johann A. Ludwig, 214, 281, 297, 364, 367
Prestwich, 133, 242
Prevost, F., 311
Prince Albert, 80, 120, 130, 225, 278, 312, 318, 365, 367
  death, 145
Prince, Amelia A., 146, 149, 226
Prince, Edwin Charles, 33, 39, 57, 58, 61, 64, 74, 100, 139, 141, 194, 207, 213, 214, 216, 217, 219, 220, 221, 222, 224, 225, 233, 234, 237, 238, 239, 255, 263, 266, 281, 314, 339, 344, 364, 367
  as agent, 202
  correspondence with Gould, 234, 268
  death, 149
  onset as secretary, 94
  portrait, 186
Prince of Wales, 137
Princess of Wied, 65
printers, 14
  in England, 272
printing presses, 90
prospectus, 13, 63, 74, 81
  Birds of Australia, 187 (two items), 203
  Birds of Europe, 22
  Odontophorinae, 57
  Synopsis, 202
  Voyage of the Sulphur, 52

prospectus and list of subscribers, 67-73, 78, 334
public hanging ended, 146
publishers
  Highley, Samuel, 192
  Sowerby, 192
puerperal fever, 155
Purple Martin News (journal), 266
Puttick and Simpson, 185, 321, 323

Quaritch, Bernard, 54, 84, 85, 323, 333
Queen Charlotte, 89, 325, 339
Queen Elizabeth II, 225, 323
Queen of Prussia, 65
Queen of Saxony, 65
Queen Victoria, 41, 97, 129, 143, 145, 225, 281, 315, 323, 343, 367
Queensbury Hill Press, 256
Queensland Historical Society, 362

RAOU, *see* Royal Australasian Ornithologists Union
Rae, Dr, of Montreal, 140
Raffles, Sir Thomas Stamford, 173, 274, 281, 288, 312
Rafinesque, Constantine Samuel, 264
railroad, 90, 91
  report, 169
Rainbird, John, 281, 367
Ramphastidae (Gould), 25, 96, 241, 346, 357
  review, 252
Ramphastidae (Gould), 2nd edn., 70, 132
Ramphastidae, Supplement to (Gould), 72, 137
Ramphastiden, Monographie der, 46, 47, 120, 322, 355, 356
  See also Sturm Brothers
Ramsay, Edward Pierson, 146, 182, 213, 248, 269, 281, 301, 346, 367
  correspondence, 146, 269
Rand, B. Howard, 139, 160
Ransome, George, 120, 231, 249
Ransome, Robert, 233, 281
Rattlesnake, H.M.S., 183, 221, 276, 295
Rawnsley, Henry C., 281
Ray Society, 323
Rayner, F. M., 281, 367
Reade, Brian (two items), 324
Reade, Sir Thomas, 339
Recherche Bay, Tasmania, 204, 231, 295
Red Sea, 189, 262
Redoute, 321
Reeve, Lovell, 324, 372
Reeves, Thomas, 281
Reform Bill of 1867, 146
Regents Park, 92
Reichenbach, H. G. L., 311
Reichenow, A., 367
Reid, L., 304
Reids, *see* Mahoney, Eliza S.
Rendel (acquaintance of Gould), 132
reptiles, genera and species
  Boidae, 263
  collection, 180
  Gilbert collection, 325
  J. E. Gray collection, 263
  Lophognathus Gilberti, 264
  Nardoa Gilbertii, 263
reviews, 63, 325
  See also Gould, John. Reviews
Rhoads, S. N., 311
Richardson, Dr B. W., 344

Richardson, Dr John, 54, 177, 281, 325
Richter, Charlotte Sophia, 273
Richter, Henry Constantine, 14, 159, 162, 203, 218, 239, 240, 244, 273, 281, 290, 292, 294, 304, 314, 319, 332, 339, 355, 364, 367
    onset of Gould employment, 122
    original drawings, 245, 337, 350
Richter, Henry James, 273
Richmond, Charles Wallace, 300
Ride, W. D. L., 325
Ridgway, Robert, 311 (two items), 325
Ripley, S. Dillon, 13, 195, 311, 326
Ripley, S. D. and Scribner, L. L., 160
Ritger, Carolyn, 294
Rivoli, Duke of, 184
    collection, 306
Roberts, Dr, 230
Robertson, artist, 150
Robertson, Dr, 281
Robertson, Mr, 339
Robertson, Melbourne bookseller, 80, 281
Robinson, Coleman T., 182
Robinson, Dr D. J., 207
Robinson's *The Garden*, 318
Rodd, Edward H., 283
Roe, J. S., 281, 297
Rolfe, A. F., 274
Ronsil, René, 326
Roper, J., 281, 292
Ross, James, 281
Rossetti, W. M., 319
Rothschild, L. W., 311
Rothschilds, 305
Rowley, G. Dawson, 146, 320
Rowley, G. Fydell, 320
Royal Academy of Arts, 311, 318, 319
Royal Academy, Torino, Italy, Gould elected Corresponding Member, 122
Royal Archives, 225
Royal Australasian Ornithologists Union (RAOU), 191, 256, 277, 304
Royal College of Surgeons, 146, 174, 238
    manuscript, 297
Royal Commission, 326
Royal Gardens at Windsor, 241, 325
Royal Geographic Society, 174
Royal Library, Windsor Castle, 225, 323
Royal Society (Australia), 215
Royal Society of London, 174, 244, 330
    Fellow of, 119, 214
Royal Society of Tasmania, 362
Rowland, Viscount Hill, 82
Rüppell, Edward, 97, 216
Russell, William, 319
Rutherford, G., 360
Rutherford, H. W., 331
Rutgers, Abram (five items), 330-1

St. Boswell, 28
St. James Church, Sydney, 207
Sallé, A., 169
Salvadori, Conte Tommaso, 150, 275, 281, 306, 331, 338, 339, 367
Salvin, Mrs Osbert, 230
Salvin, Osbert, 66, 85, 178, 196, 231, 281, 284, 285, 288, 332
Salvin, Osgood, 226
Santa Cruz, Island of Teneriffe, 100
Sauer, Gordon Chenoweth, 230, 321 (eight items), 332-5, 373
    collection of original drawings, 244

Saunders, Howard, 286, 288
Saunders, John, 335
Saunders, W. R., 367
Savage, Christopher, 335
Sawyer, Frederick C., 239, 335
Saxe-Coburg, *see* Prince Albert
Saxe-Coburg, Duke & Duchess of, 130
scarlet fever epidemic, 145
Schafer, E., 311
Schafhirt, Frederick, 139, 160
Scherren, Henry, 222, 335
Schlegel, Hermann, 74, 96, 123
Schomburgk, Richard, 281, 297, 339
Science, Quarterly Journal of, 319
Sclater, Philip Lutley, 178, 196, 212, 226, 243, 281, 282, 285, 286, 287, 290, 308, 311, 312, 319 (five items), 336-7, 367, 368, 372
Scot, Alexander Walker, 232
Scott, A. W., 281, 367
Scott, G. F., 367
Scott Polar Research Institute, 203
    manuscript, 297
Scott, W. E. D., 311
Scribner, Lynette, 13, 326
Seebohm, Henry, 196, 305, 350
Segehoe, N.S.W., 273
Selby, Prideaux John, 15, 17, 28, 90, 92, 94, 98, 162, 193, 253, 274, 281
    death, 146
Selwyn, Mr, 143
Semmelweis, 155
Sergeant, J. Dickinson, 139, 160
Serventy, D. L. (two items), 337-8
    1948: 337
    1962: 338
Serventy, Vincent, 338
Severtzoff, Dr, 281, 319
Sharland, M. S. R., 281
Sharpe, Richard Bowdler, 66, 68, 86, 126, 145, 178, 180, 181, 183, 196, 202, 208, 214, 243, 255, 266, 268, 272, 275, 277, 281, 285, 286, 287, 288, 292, 297, 298, 302, 305, 308, 311, 319, 322, 326, 337 (four items), 338-41, 357, 360, 364, 367, 368, 375
Analytical Index, 160
initial correspondence with Gould, 150
plates reproduced, 343
works completed
    Birds of Asia, 66
    Introduction to, 85
    Birds of New Guinea, 82
    Pittidae, 84, 151
    Trochilidae, Supplement, 85, 151
Sheen Lodge, 318
Sheffield, Henry, 224, 225
Shell bird book, The (Fisher), 203
Shelley, G. E., 196
Shepherd, Chas. W., 196
Sherborn, Charles Davies, 64, 300 (two items), 341
Ships used by Gould and party, *see* under names listed here:
    Asia; Black Joke; Comet; Ellen; Gilmore; Helen; Katherine Stewart Forbes; Kinnear; Lulworth; Marion Walker; Marion Watson; Mary Ann; Palmyra; Parsee; Pelorus; Potentate; Susanna Ann; Starling; Van Sittart
shipwrecks, Australian, 174
Shore, J. C. [or C. J.?], 281, 339
Shufeldt, R. W., 342
Sibbald [Siebold, P. F. von?], Dr, 281, 367
Sibley, Fred, 255

INDEX

Sidney, S., 367
Simkinson, Mrs, 281
Simpkins, Diana M., 212
Simpson, Leslie G., 181
Simson, Clive, 184
Sinclair, S., 346
Singapore, 120, 235
Singleton, M. E., 342, 367
Sinkler, Louise Elkins, 368
Sitwell, Sacheverell, 326 (five items), 343, 367
size, imperial-folio, 14
skeletons, 238
Skinner, G. U., 281
Skipwith, Peyton, 344
Slater, Ederic, 344
Slijper, 331
Smit, Joseph, 208
Smith, A., 311
Smith, Andrew, 216
Smith, Aubrey H., 139, 160
Smith, Edward, 13th Earl of Derby, *see* Derby, Lord
Smith, Elder & Co., Messrs, 342
Smith, J. E., 367
Smith, Metcalfe, 367
Smith, P. W. B., 367
Smith, Roger C., 344
Smith, Thomas, 106
Smithsonian Institution, 126, 139, 226
Snooke, Mr, 281
Snow, Dr John, 135, 136, 170-3, 344-5
Soho area, 170
Somerleyton, 143, 282
Somerset House, London, 239
Somerville, J. D., 215 (three items), 345-6, 367
Sotheran, Henry, Ltd, 74, 84, 85, 180, 191, 209, 211, 213, 216, 226, 230, 237, 244, 255, 256, 268, 281, 285, 306, 316, 320, 326, 332, 333, 339, 340 (twenty-one items), 346-51, 359, 368
  original drawings purchased from, 245-8
  'Piccadilly Notes', 320, 348-51
  No. 9: 160, 209, 268
Souance, C., 311
South, Mr, 238
South American partridges, 56
South Australia, 192, 267, 296, 297, 320
  botanists, 297
  Institute (museum), 70, 145, 221, 310, 355
  State Library of, 215, 216
South Pacific Ocean, 118
Sowerby family, 192
Sowerby, G. B., 54
species, *see* Birds, mammals, etc. genera and species
Spencer, Sir Baldwin, 255
Spencer, Herbert, 305
Spencer, Kenneth, Research Library, Lawrence, Kansas, 14, 175, 191, 201, 226, 241, 244, 276, 289, 293, 303, 332, 351, 393
  Ellis Collection, *see* Ellis Collection
Spillett, Peter G., 351
Spinage, C. A., 353
Sprague, Henry S., 182
Spring Hill, Tasmania, 107
Stack, Lieutenant, F. R., 244, 247-8, 277
Stanley, Lord Edward Smith, *see* Derby, Lord
Stanley, Owen, 183, 213, 281, 295, 367
Starling (ship), 97
Steere, Professor, at Michigan, 226
Stephenson, Wm., 367
Stevens, Lyla, 353

Stevens, S., 367
Stevenson, Hy, 196
Stewart, James, 274
Stirling, E. C., 367
Stoke Hill near Guildford, 3, 91
Stokes, J. of Western Australia, 220
Stokes, John Lort, 99, 112, 179, 281, 367
Stokes, Mr, of Western Australia, 219
stolen birds, 187
Stone, Witmer, 161, 278, 301, 306, 307, 310, 311 (four items), 353-4, 367
Stonehouse, J. H., 348, 351
Strange, Frederick, 132, 180, 219, 221, 239, 281, 297, 363, 367
  death, 136
Stresemann, Erwin, 354
Strickland, Mrs Catherine D. M., 237, 277, 354
Strickland, Hugh Edwin, 42, 44, 63, 66, 123, 128, 160, 177, 212, 277, 281, 303, 311, 323, 332, 337, 354
  collection, 316
  death, 133
  initial correspondence, 119
  memoirs, 273
Strong, Reuben Myron, 354
Strzelecki, Paul Edmund de, 215, 267, 281, 295, 326, 354, 367
Stuart, Charles, 297
Stuart, John, 205
Stuart, John McDouall, 192, 281, 297, 355
  expedition, 145, 315
Stuart, R. L., 226
Sturm, ___, 281, 375
Sturm brothers, 46, 120, 356
Sturm, Frederic, 47
Sturm, John, 47
Sturt, Captain Charles, 114, 115, 180, 187, 189, 192, 205, 216, 217, 218, 281, 296, 297, 307, 355, 360, 367
  expedition, 315
Sturt, Evelyn, 296, 367
Sturt, Mrs Charles, 297
Sturt, Mrs Napier George, 355
Styche, J. W., 226
subscribers, *see* Prospectus and List of Subscribers
Suez Canal, 147
Sulphur, H.M.S., 97, 123, 326
Surgeon and wife on *Parsee*, 99
Susannah Anne (ship), 112, 294
Sutton, George M., 161
Sutton, J., 226, 355, 367
Swainson, William, 26, 30, 65, 185, 193, 194, 212, 273, 281, 293, 322, 339
  correspondence, 93, 264, 356
  death, 137
  works, 356
Swan River, Western Australia, 200, 235
  area, 218
  birds, 206
  Colony, 364
Swann, C. Kirke, 356
Swann, H. Kirke, 315
Swanston, Captain, 232
Swinhoe, Robert, 339
Sydney, 109, 232, 235
Sykes, 339
Sylvester, Dr John Henry, 210
Symers, Captain, 281
Synopsis (Gould), 30, 97, 207, 212, 249
  facsimile, 256
  reviews (four), 252

Systems avium Australasianarun (Mathews), 301
Systema Naturae (Linneaus), 361

Taglioni (ship), 216
Tankerville, Lord, 145, 318
Taschenberg, O., 201, 344, 357
Tasman, Abel, 295
Tasmania, 143, 186, 203, 295
  birds, 183
  geological survey, 145, 178, 208
  Royal Society of, 203
Tasmanian Natural History Society, 215
Tate, Peter, 357
taxidermy, 234, 242, 276
Taxidermy, with the biographies of zoologists . . . (Swainson), 356
Taylor, 254
Taylor and Francis, 66, 281
Taylor, Captain, 226
Taylor, Griffith, 255
Taylor, John, 169
Taylor, Richard (printer), 192
Tegetmeier, W. B., 196
telephone invented, 149
Temminck, Coenraad Jacob, 44, 58, 61, 65, 67, 71, 96, 97, 193, 254, 281, 300, 311, 340
  death, 143
Teyler Foundation, 200
Thames River, 120
Thayer, Abbott Handerson, 317
Thayer, Evelyn, 357
Thayer, John Eliot, 82
  collection, 357
Theakes, Captain, 281
Thomas, G. Ross, 357
Thomas, Oldfield, 180
Thompson, 274
Thompson, William, 274
Thorburn, Archibald, 184, 203, 273, 281, 305, 319
Thorpe, Adrian, 357
Throsby, C., 281
Tilston, 367
Todd, W. E. C., 311
Torino, Royal Academy, 331
Toronto, Ontario, 139, 142
  Thomas Fisher Rare Book Library, 275
Torrey, John 321
Townsend, J. K., 311
Toucans, Gould's Monograph on, *see* Ramphastidae
Tristram, Canon H. B., 196, 360
Trochilidae (Gould), 64, 126, 241, 277, 325-6, 346
  prices, 374
Trochilidae, Supplement (Gould), 84, 151, 255, 275, 373
Trogonidae (Gould), 26, 96, 241, 253, 272, 337
  Audubon copy, 193
  collation, 272
Trogonidae, 2nd edn, 72, 143
Trogons, Gould's Monograph on, *see* Trogonidae
Troughton, E. L. G., 367
Troughton, Ellis, 357
Trusler, Peter, 359
Tucker, Mrs Marcia Brady, 305
Tulk, Augustus, 222, 224
Turnbull, Mr, 226
Turner, 367

Turner, W. J. (two items), 357
Tweeddale, Marquess of, *see* Walden, Viscount
Tweedie, Alexander Forbes, 225, 239, 240
Twyman, Doris Edelsten, 4, 97
  See also Edelsten, Doris, 149
Tyndall, Professor, 318
type specimens, birds, 336, 353, 359, 362
  See also Birds, genera and species

United States, 153
  Civil War, 144
  See also Gould, John. Journey to North America
University Library at Copenhagen, 69
University of Kansas, *see* Kansas, University of
Upper Torrens River, 115

Van Diemen's Land, 102, 307, 354
  See also Tasmania
Van Muller, Dr Baron J. W., 277
Vansittart (cutter), 105
Van Voorst, John, 196
Veitch, John Gould, 267
Venables, Mr, 340
Verreaux, E., 58, 311
Verreaux Freres, 126, 161, 281, 307
Verreaux, Jules P., 271, 307, 311, 321, 358, 367
Verreaux, Maison, 179, 271
Victoria, birds, 265, 304
  book illustration, 360
  naturalists, 362
  State Library of, 222
Victorian era, 89
Vidler, E. A., 170, 281, 367
Vieillot, 300
Vigors, Nicholas A., 15, 17, 18, 90, 96, 241, 243, 274, 281, 289, 293, 312, 326, 340, 357
  (six items), 358-9, 367, 375, 393
  death, 119
Von Homeyer, 133
Von Müller, Dr Baron J. W., 359
Voorst Van, 319
Vosper, Robert, 181, 359, 374
Voyage of the Sulphur, 51, 123, 218

W., A., 48, 355
W., J. O., 252
Wade, Peter, 359
Wagler, 253
Wagner, Andr., 324
Wagner, Rudolf, 375
Wagstaffe, R., 189, 359
Wainright, M. D., 360
Wakeman, G., 360
Walden, Viscount, 180, 312
Walford, Edward, 324
Walker Art Gallery, 311
Walker, Dr, 364
Walker, Sir James, 147
Walker, G. W., 167
Walker, Marion, 230, 239, 292, 335
Wall, Thomas, 281
Wall, W. S., 281
Wallace, Alfred Russell, 143, 196, 198, 281, 340
Wallaston, A. F. R., 368
Waller, Charles, 367
Waller, Eli, 187, 281, 301, 367
Walsingham, Lord, 340
Walter, 14

INDEX

Walter and Cohn, 14
Ward, Henry, 204
Wards, Mr, 281
Warren, Rachel L. M., 360
Warszewiez, M., 281
Washington, D. C., 139, 141
Waterhouse, 77, 178, 286
Waterhouse, Alfred, 48
Waterhouse, Charles O., 288, 361
Waterhouse, Frederick George, 69, 80, 145, 226, 281, 297, 355, 368
  initial correspondence, 145
Waterhouse, Frederick Herschel, 22, 45, 313, 340 (three items), 361, 368, 375
  ... *dates of publication* ... (of) *works of the late John Gould*, 14
Waterhouse, G. R., 207, 313, 336, 356
Waterhouse Island, Bass Strait, 108
Waterman fountain pen, 152
Watermark, C. Wilmot, 231, 263
Watson, Dr George, 195
Waterman, 244
Waterton, Charles, 194, 274
Watling, Thomas, 302, 368
Watson Library, *see* Kansas, University of, libraries
Watson, Mary, 99, 281, 346, 368
wax flowers, 277
Webber, Dr, 223
Welker, Robert Henry, 362
Welsby, Thomas, 362
Wendt, Herbert, 362
Wentworth exploration, 204
West, Geoffrey, 362
West Indies, 56
Western Australia, 190, 192, 235, 264, 294, 337, 363
  birds, 363
  botanists, 297
Westminster, Duke of, 147, 209, 319
  See also Grosvenor, Earl
Westwood, J. O., 313
Wharncliffe, Earl of, 340
Wharton, C. Bygrave, 196
Wharton, Henry T., 196
Wheeler, I. R., 233, 281
Wheelwright, W., 231
Wheldon & Wesley, 15, 17, 333, 356
Whitaker, William, 208
White, J. B., 368
White Mountains, 142
White, Rev. Gilbert, 281
White, Samuel, 147, 281, 300, 320, 362, 368
White, Samuel Albert, 281, 299, 362, 368
Whitehead, Rev. H., 173
Whitley, Gilbert P., 278, 362, 368
Whittell, Hubert Massey, 160, 207, 235, 240, 278, 300, 308, 337-8 (eleven articles), 363-8
Whymper, Charles, 319
Whymper Engravers, 319
Wickham, Commander John C., 112, 281
Wilberforce, Bishop Samuel, 144
Wilcox, James F., 271, 281, 368
Wilder, Harry, 340
Wilkes, Captain, 310
Willard Hotel, Washington, D.C., 139, 242
Williams, George, 77
Willis, G., 346
Willis and Sotheran, 368
Wilmot, C, watermark, 231, 263
Wilmot, Henry, 281
Wilson, Alexander, 274, 311, 317, 362, 368

Wilson, Edward, 126, 180, 184, 208, 212, 281, 306-9, 311, 321
Wilson, H. M., 281
Wilson, James, 17, 193
Wilson, John, 3
Wilson, Sarah Gould, 3, 120
Wilson, Dr Thomas Bellerby, 126, 139, 160, 184, 208, 226, 271, 281, 283, 306, 321, 336, 340, 345, 353, 368
Wilson, Dr Thomas Braidwood, 345
Wilson, W. S., 139, 160
Windsor, Berks, 236
  Castle, 91, 225
  Royal Gardens, 3
Winterfield, Charles, 268
Wolf, Edwin II, 368
Wolf, Joseph, 14, 126, 159, 161, 175, 181, 182, 197, 202, 203, 208, 226, 266, 273, 281, 283, 290, 291, 292, 294, 304, 305, 317, 318, 321, 337, 340, 350, 368, 372
  birth, 91
  original drawings, 302, 337, 370
  Norway trip, 137, 215
  plates reproduced, 291, 295
  See also Palmer, A. H. for life of Wolf
Wolley, John, 137
Wongan Hills, 312
Wood, Casey A., 13, 160, 237, 238, 275, 369-73
wood engravings, 273
Wood, William, 15, 371
Woodhouse, Dr Samuel Washington, 139, 160
Woodward, Ellen Sophia, 371
Wollongong, N.S.W., 235
Woolmer, Thomas, 319
work(s), 'Cancelled', *see Birds of Australia and adjacent islands*
works, major, *see* Gould, John, works
World War II, 239
Wright, Charles A., 196
Würtemberg, Duke Ernest of, 130
Wyatt, C. W., 368

Xantus, Mr, 282

Yale University, Coe Collection, 54, 74
  Library, 250, 326
  Peabody Museum, 255
Yarrell, William, 33, 93, 101, 193, 204, 268, 273, 281, 290, 312, 319, 373
  memoir, 274
Yarrundi, 110, 117, 188, 190, 268, 273, 320
Yates, Miss, 139, 239
York, 58
Younger, R. M., 373

Zander, 133
Zeitlin, Jacob, 181, 333, 374
Zélée (ship), 310
Zimmer, John Todd, 13, 63, 64, 70, 72, 160, 178, 274, 289, 311, 374-92
Zoological Club, 91, 173, 287, 335, 358, 393
Zoological Journal, 91, 287, 358, 373
Zoological Magazine, 96, 288
Zoological Miscellany, 94, 263, 288
Zoological Record, 145, 288, 299, 344, 361
Zoological Society of London, 16, 113, 119, 173, 174, 192, 202, 205, 216, 231, 265, 266, 269, 275, 276, 288, 291, 309, 310, 312, 316, 319, 335, 353, 356, 358, 361, 369, 373, 393

centenary, 312
correspondence with Gould, 221
Council of, 142
Fellow, 119
first meeting, 92
Gardens, 92, 128, 198, 240, 271, 298, 312, 315, 316, 319, 323, 335
library, 130, 361
manuscript, 297
museum, 96, 295, 318
Proceedings, 142, 159, 275, 289-90, 310, 319, 330, 338

Proceedings of the committee of science and correspondence of, 288, 289
Transactions, 97, 159, 252, 275, 290, 319, 322
Zoologist, 123
Zoology Library, Kansas, University, Department of Ornithology, 266
Zoology of Voyage of the Beagle, 39, 101, 192, 198, 205, 291, 341, 362
Zuchold, E. A., 368